FUROPYRANS AND FUROPYRONES

This is the twenty-third volume in the series

THE CHEMISTRY OF HETEROCYCLIC COMPOUNDS

THE CHEMISTRY OF HETEROCYCLIC COMPOUNDS

A SERIES OF MONOGRAPHS

ARNOLD WEISSBERGER, *Editor*

FUROPYRANS and FUROPYRONES

Ahmed Mustafa

Chemistry Department
Cairo University

1967

INTERSCIENCE PUBLISHERS

a division of John Wiley & Sons

London - New York - Sydney

Library of Congress Catalog Card Number: 67–16130

The Chemistry of Heterocyclic Compounds

The chemistry of heterocyclic compounds is one of the most complex branches of organic chemistry. It is equally interesting for its theoretical implications, for the diversity of its synthetic procedures, and for the physiological and industrial significance of heterocyclic compounds.

A field of such importance and intrinsic difficulty should be made as readily accessible as possible, and the lack of a modern detailed and comprehensive presentation of heterocyclic chemistry is therefore keenly felt. It is the intention of the present series to fill this gap by expert presentations of the various branches of heterocyclic chemistry. The subdivisions have been designed to cover the field in its entirety by monographs which reflect the importance and the interrelations of the various compounds and accommodate the specific interests of the authors.

Research Laboratories
Eastman Kodak Company
Rochester, New York

ARNOLD WEISSBERGER

Preface

The furopyrans, and -pyrones occupy a prominent position among the plant phenols, and comprise a body of organic substances of extraordinary variety and interest. The use of certain plants as an aid in catching fish; the increasing use of derris products, including rotenone itself, as insecticides; the recognition of the active principles of Kellah plant (*Ammi visnaga* L.) by Egyptians centuries ago, as a home remedy against leucoderma, in relieving the pain of renal colic and uretral spasms, and in facilitating the passing of uretral stones; the production of antifungal compounds by host plants following the fungal infection, phytoalexins; and the isolation of metabolites from the mycelium of variant strains of mould, all present a rich field of scientific inquiry from which have come many interesting and important findings.

These compounds occupy a close structural and chemical interrelationship that appears to reflect a similarly close relationship in the processes by which they are formed in plants.

The varied nomenclature was most confusing. It is not unusual to find many names attached to each of the more widespread natural products, and two or three for each of the many plants. In general, the name given here is that apparently used most often; the use of trivial names has been adopted with the more complex ring systems in order to avoid unwieldy and cumbersome systematic names. In all instances, however, alternative nomenclature and numbering have been indicated so that no confusion should result. The synonyms can usually be found in the *Konstitution und Vorkommen den Organischen Pflanzenstoffe*, compiled by W. Karrer, which also gives the important references to historical matters and details of isolation. A good source of analytical data is *Moderne Methoden der Pflanzenanalyse*, edited by K. Peach and M. V. Tracey, and the *Merck Index* is as convenient a source as any for references to the medical and clinical aspects of many of the compounds to be dealt with. Finally, *The Chemistry of Flavonoid Compounds*, edited by T. A. Geissman, *Naturally Occurring Oxygen Ring Compounds*, compiled by F. M. Dean, the chapter on 'Naturlich vorkommende Chromone' by

H. Schmid in *Fortschritte der Chemie organischer Naturstoffe*, edited by L. Zechmeister, and on 'Compounds Containing Two Hetero–Oxygen Atoms in Different Rings' by W. B. Whalley in *Heterocyclic Compounds*, edited by R. C. Elderfield, discuss not only chemical matters, but also a variety of related subjects of economic importance.

Every effort has been made to include in the manuscript, papers indexed by *American Chemical Abstracts*, up to and including, 1964, and subsequent papers in the more important Journals up to December 1965. The *Chemical Abstracts* reference is listed in addition to the primary reference for any article not consulted in the original. Steroid sapogenins (spirostans), compounds, having a fused five-membered lactonic ring with a pyran nucleus, and those characterized by spiro rather than fused ring systems are outside the scope of this book.

It is an aim of this contribution to bring together the knowledge of these compounds that has so far been gained, and to present a systematic survey and bibliography of the present position from which further progress can be made. Present day studies on the synthesis, stereochemistry, physiological activity, and biosynthesis of furopyrones continue to add new information.

It is hoped that the arrangement and discussion of the closely related classes, included in the volume, will arouse greater interest and impart a new viewpoint to the chemistry of the individual substance.

Finally, I wish to acknowledge the understanding of my wife, Professor Dr. W. Asker, who not only suffered patiently all the problems of writing a book, but helped me to solve so many of them, by proofreading and indexing.

Department of Chemistry A. MUSTAFA
Faculty of Science
Cairo University

Contents

CHAPTER I

Furopyrans and -pyrones

The furan ring is commonly found in plant products as furans; a few are of fungal origin. Most of these compounds are terpenoids, whatever their source. The hydrofurans have rather complex structures; a few occur among the terpenes, and among the carotenoids. The lignans generally possess two fused tetrahydrofuran rings. Other hydrofurans are found in steroidal sapogenins, and yet others are found as complex alkaloids.

Whereas 2-methoxyfuran behaves as a true furan, 2-hydroxyfurans isomerize to butenolides (**1**) whose behavior is normally easily interpreted solely in terms of lactone structures.[1] 3-Hydroxyfurans, which are in fact vinylogous lactones, show that their ketonic forms (**3**) readily yield derivatives of the enolic forms, for example on acetylation, but the butenolides **1** and **2** resist even this type of conversion into true furans. Many examples among the terpenoids and the cardenolides carry the butenolide ring.

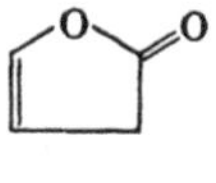

(**1**)

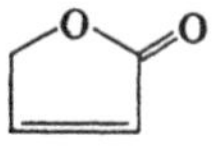

(**2**)

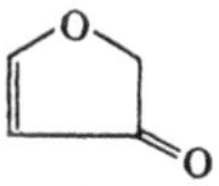

(**3**)

Although the parent pyran rings **4** and **5** are not known at all, they give rise to a considerable number of important natural products. A well-defined group of dihydropyrans forms a subdivision of the Rauwolfia alkaloids; the plant α-pyrones include the cardadienolides which have a steroid nucleus attached to the 5-position in **6**.

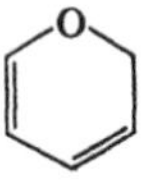

(**4**)

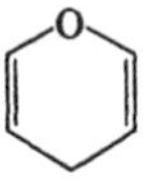

(**5**)

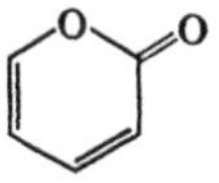

(**6**)

Fusion of the furan ring and/or the benzofurano nucleus with the pyran and/or the pyrono ring is a common one in natural products and appears in many forms. Among the types of such compounds, which serve as a guide to the chapters where they will be dealt with, are:

furopyrans and -pyrones
furocoumarins
furochromones
furoxanthones
furoflavones and -flavanones
the furoisoflavonoids
the rotenoids
less common furopyran ring systems

Compounds having a fused five-membered lactone ring with a pyran nucleus, e.g. patulin, rubropunctatin, monascorubin, rotiorin, monascin, and those characterized by spiro rather than fused ring systems, are outside the scope of this book.

Structural features of special interest are displayed by plumericin, which is the sole representative of the furopyran group known to occur naturally, and by anhydrotetrahydroaucubigenin which is not a natural product in the rigorous sense, and may be considered as a furopyran. Synthesis of a few examples of furopyrans and -pyrones has been achieved (Table 1).

I. Naturally Occurring Furopyrans

1. Plumericin

Plumericin (**7**), $C_{15}H_{14}O_6$, is a bactericide extracted from the roots of *Plumeria multiflora*[2] and from the roots of *Plumeria rubra* var. *alba*.[3,4] The latter source also affords two related compounds, isoplumericin (**8**) and β-dihydroplumericin (**9**, $R = CH_3$), together with β-dihydroplumericinic acid (**9**, R = H), and fulvoplumierin (**10**).[5]

Plumericin (**7**), on hydrogenation in ethanol in the presence of a $Pd/BaSO_4$ catalyst, gave α-dihydroplumericin (**11**), showing its sensitivity toward alkalis; chromic acid oxidation of **11** afforded acetic acid, propionic acid, and some butyric, but no succinic acid. Similar hydrogenation of isoplumericin (**8**) also gave **11**, which was converted with bromine in methanol into bromomethoxydihydroplumericin (**12**).[4] Hydrogenation of **7** in ethanol in the presence of Pd/C produced α-tetrahydroplumericin (**13**), again showing its sensitivity to alkali; appreciable

TABLE 1. Furopyrans and -pyrones

Compound (formula)	M.p. or B.p. (°C)	Solvent[a] for Crystallization	Remarks ((α)$_D$; U.v. spectrum max mμ (logϵ))	References
A. *Natural Furopyrans*				
Plumericin (**7**)	211.5–212.5 (dec.)	B, C/F, H	+197.5 ($CHCl_3$); 214–215(4.24)	4
Isoplumericin (**8**)	200.5–201.5(dec.)	F/H	+216.4 ($CHCl_3$); 214–215 (4.23)	4
β-Dihydroplumericin (**9**)	150–151	C/G, F/G	+257.5 ($CHCl_3$); 240 (3.976)	4
α-Dihydroplumericin (**11**)	191–192	C/G, F/G	+208.9 ($CHCl_3$); 239–240 (3.976)	4
α-Tetrahydroplumericin (**13**)	147.5–149	F/G	+92 ($CHCl_3$); 237 (4.02)	4
α-Hexahydroplumericin (**15**)	115–115.5	C/G	−29.1 ($CHCl_3$); 210 (1.90) end absorption	4
β-Tetrahydroplumericin (**16**)	86.5–88.5	C/G	+123.5 ($CHCl_3$); 235 (4.018)	4
β-Hexahydroplumericin (**17**)	88.5–89; 95	G	+19.6 ($CHCl_3$); 210 (1.96) end absorption	4
Bromomethoxy-α-dihydroplumericin (**12**)	148–150 (dec.)	F/G	Weak ultraviolet end absorption	4
Bromomethoxy-α-tetrahydroplumericin (**14**)	120–122.5	F/G		4
Anhydrotetrahydroaucubigenin (**29**)	90.5	C	Acetate (b.p. 90–100°); *p*-nitrobenzoate (m.p. 131–132°)	6
B. *Synthetic Furopyrans and -pyrones*				

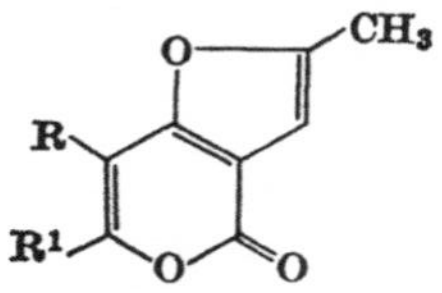

(*Table continued*)

TABLE 1 (*continued*)

Compound (formula)	M.p. or B.p. (°C)	Solvent for Crystallization	Remarks ((α)$_D$; U.v. spectrum max mμ (logϵ))	References
Substituents				
$R=COOEt, R^1=Me$	136	—		12
$R=COOEt, R^1=C_6H_5CH{=}CH$	138	A/I		12
$R=COOEt, R^1=p\text{-}MeOC_6H_4CH{=}CH$	178–180	E/I		12
$R=COOEt, R^1=p\text{-}Me_2NC_6H_4CH{=}CH$	180	E/I		12
$R=COOEt, R^1=3,4\text{-}({=}O_2CH_2)C_6H_3CH{=}CH$	184	E/I		12
$R=COOEt, R^1=C_4H_3OCH{=}CH$ (furylidene)	187	E/I		12
$R=COOEt, R^1=C_6H_5$	136	—		12
$R=R^1=CH_3$	125	—		12
$R=COOEt, R^1=C_6H_5CH{=}CHCH{=}CH$	150	A/I		12
$R=COOMe, R^1=C_6H_5$	132	—		12
$R=COOMe, R^1=OH$	168–170	—		12
$R=R^1=H$	112	—		12
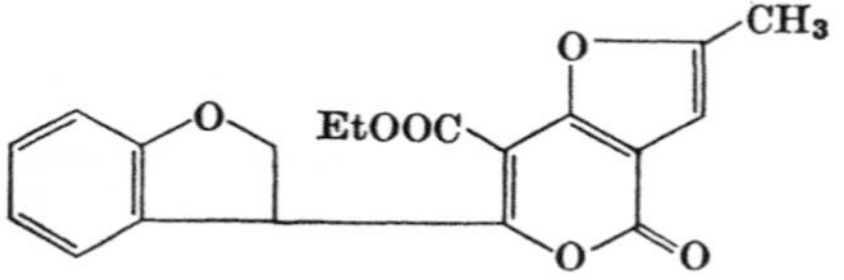	225	E/I		12

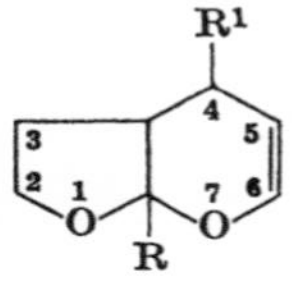

TABLE 1 (*continued*)

Compound (formula)	M.p. or B.p. (°C)	Solvent[a] for Crystallization	Remarks ($(\alpha)_D$; U.v. spectrum max mμ (log ε))	References
Substituents				
$R=R^1=H$	b.p.$_{20}$ 81–82		$n_D^{21.5}$ 1.4830; $d_4^{21.5}$ 1.113	14
$R=CH_3$, $R^1=H$	b.p.$_{20}$ 76–78			14
$R=H$, $R^1=CH_3$	b.p.$_{20}$ 88–90		n_D^{22} 1.4772, d_4^{22} 1.059	14
$R=R^1=CH_3$	b.p.$_{20}$ 86–88		n_D^{24} 1.4665; d_4^{24} 1.016	14
$R=R^1=H$, 5,6-dihydro	b.p.$_{20}$ 87–88		$n_D^{21.5}$ 1.4690; $d_4^{21.5}$ 1.068	14
$R=CH_3$, $R^1=H$, 5,6-dihydro	b.p.$_{20}$ 85–86		$n_D^{23.5}$ 1.4620; $d_4^{23.5}$ 1.036	14
$R=H$, $R^1=CH_3$, 5,6-dihydro	b.p.$_{20}$ 97–99		$n_D^{22.5}$ 1.4602; $d_4^{22.5}$ 1.015	14
$R=R^1=CH_3$, 5,6-dihydro	b.p.$_{20}$ 83–84		$n_D^{22.5}$ 1.4550; $d_4^{22.5}$ 0.990	14

[a] A, dioxan; B, ethyl alcohol; C, ether; D, ethyl acetate; E, methyl alcohol; F, methylene chloride; G, pentane; H, toluene; I, water.

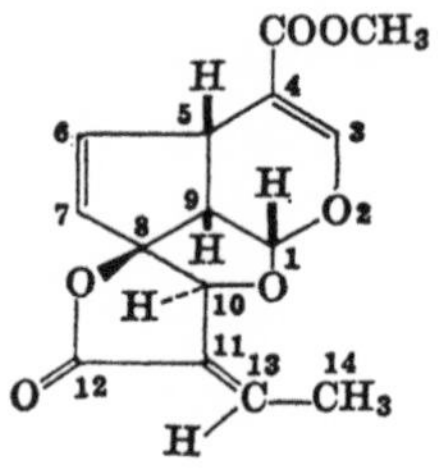

(7)

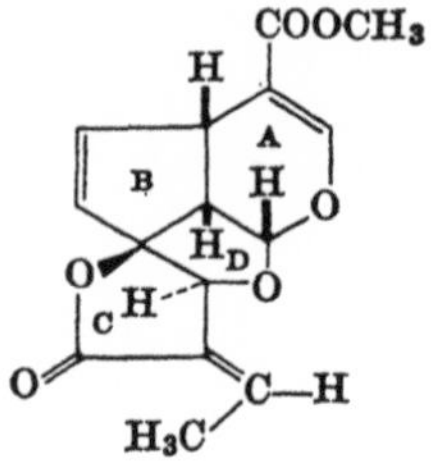

(8)

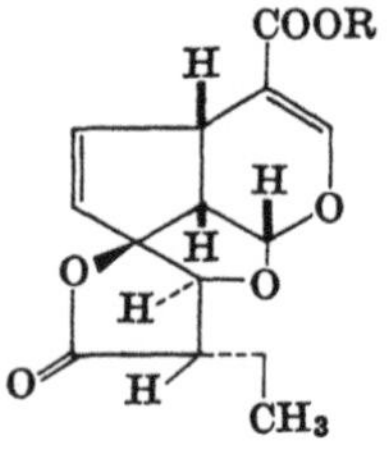

(9)

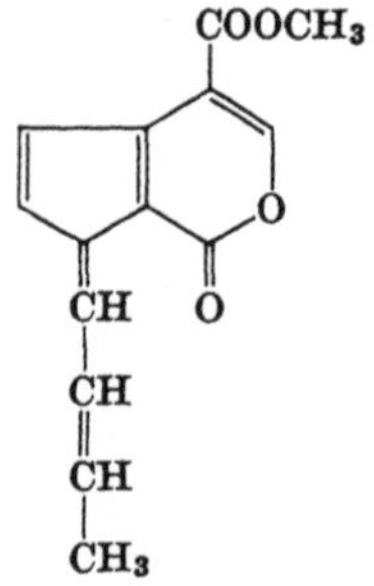

(10)

amounts of acetic acid and propionic acid resulted from chromic acid oxidation of **13** and succinic acid was obtained on boiling with chromic–sulfuric acids mixture. Similar hydrogenation of **8** gave **13**, which

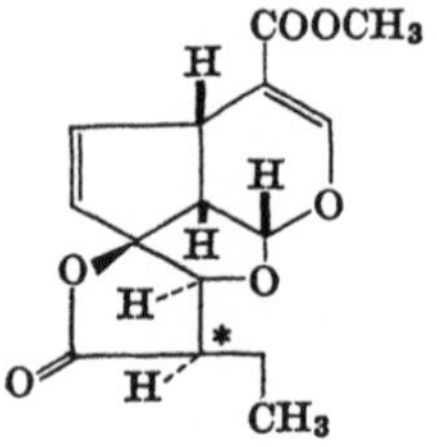

(11)

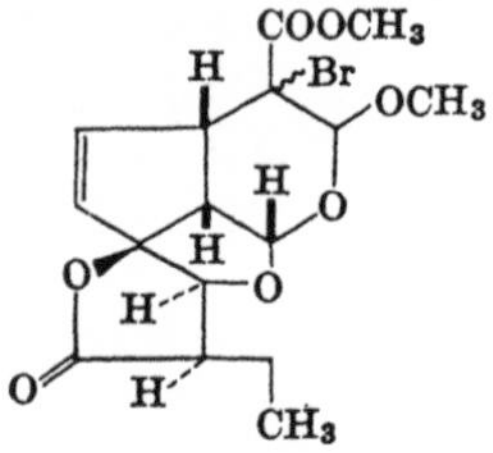

(12)

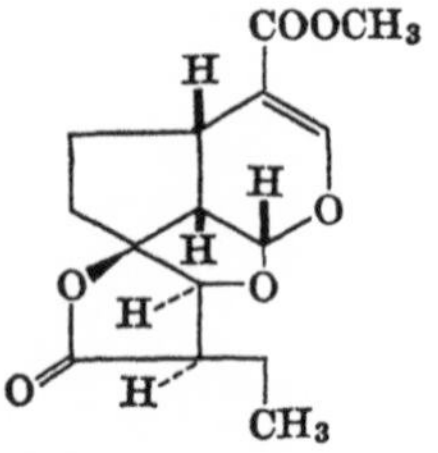

(13)

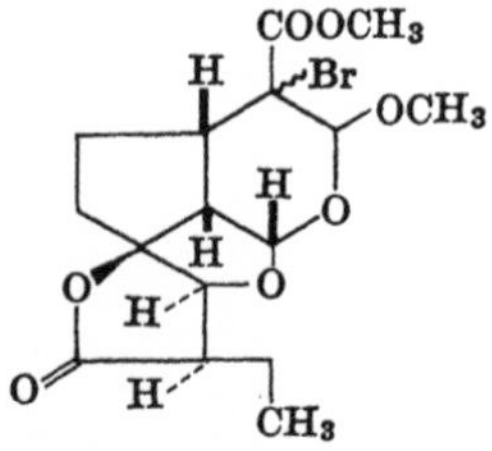

(14)

upon bromination in methanol yielded bromomethoxy-α-tetrahydro-plumericin (**14**).

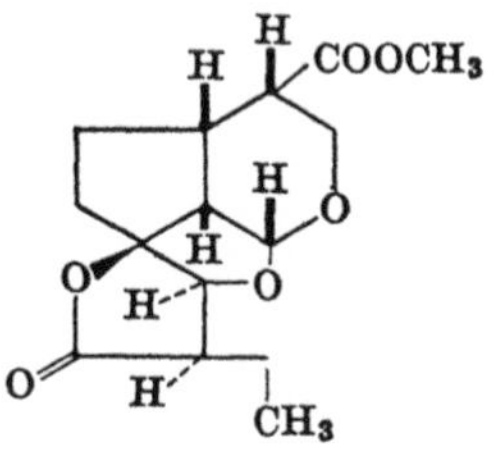

(15)

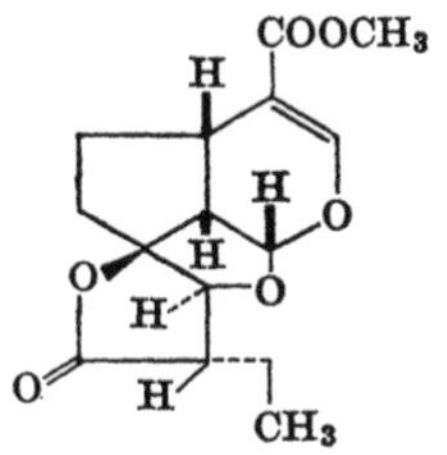

(16)

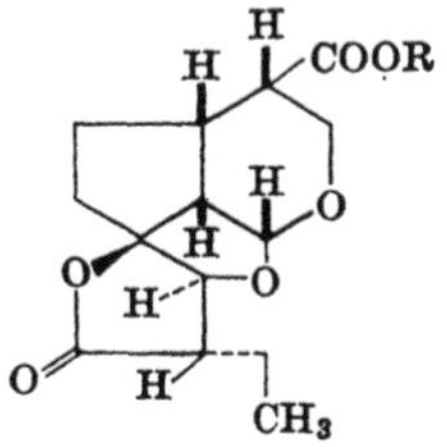

(17)

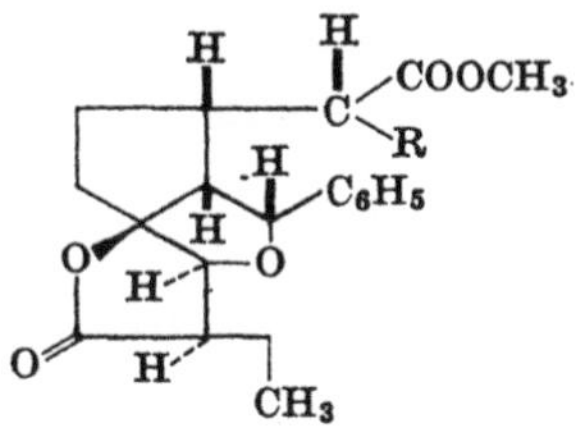

(18)

Hydrogenation of **7** with Rh/C in ethanol gave α-hexahydroplumericin (**15**); chromic acid oxidation of the latter gave acetic, propionic, and succinic acids. Under the same conditions, **15** was also obtained by hydrogenation of **11** and/or **13**.

Hydrogenation of β-dihydroplumericin (**9**, $R=CH_3$) yielded β-tetrahydroplumericin (**16**), which on further hydrogenation produced β-hexahydroplumericin (**17**, $R=CH_3$). The latter was hydrolyzed to β-hexahydroplumericinic acid (**17**, R = H); on the other hand, hydrolysis of **15** led to **17** (R = H).

Treatment of **17** ($R=CH_3$) with aluminum bromide in benzene effected ring A opening to yield **18** ($R=CH_2OH$). The latter gave an acetate (**18**, $R=CH_2OAc$); on oxidation it yielded the acid (**18**, R = COOH), which underwent thermal decarboxylation to yield **18** (R = H). Hydrogenation of the latter compound in presence of palladium black furnished **19** (R = H). Similar treatment of **17** with aluminium bromide in benzene gave **20** ($R=CH_2OH$) which underwent oxidation to **20** (R = COOH), and the latter decarboxylated to **20** (R = H).

Refluxing α-hexahydroplumericin (**15**) with methanolic hydrogen chloride gave **21** (R = H), which was converted by the usual method into **21** (R = Ac), and on hydrogenation in the presence of Rh/C gave **22** (R = H).

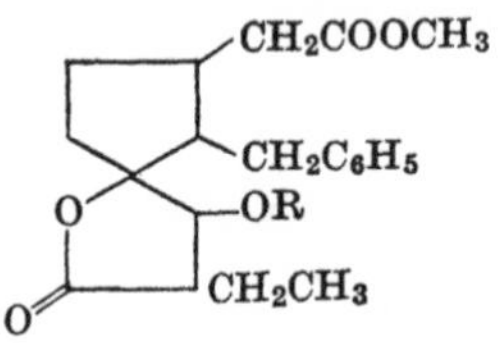

(19)

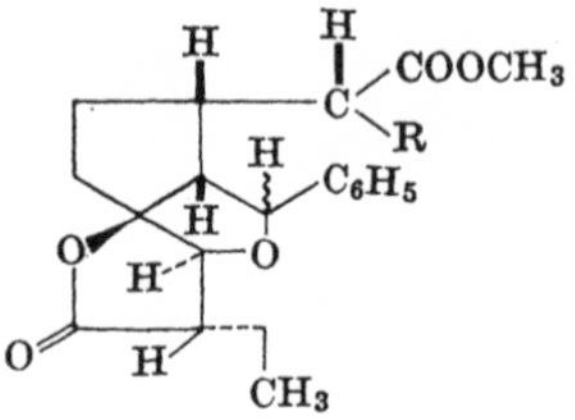

(20)

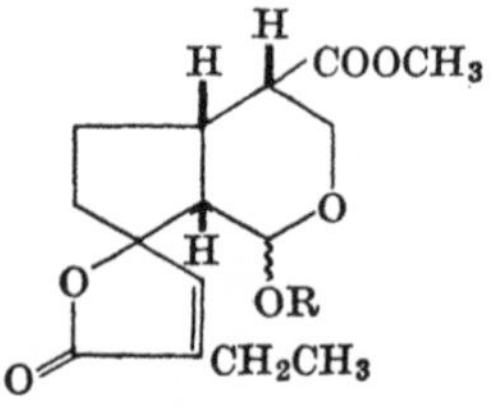

(21)

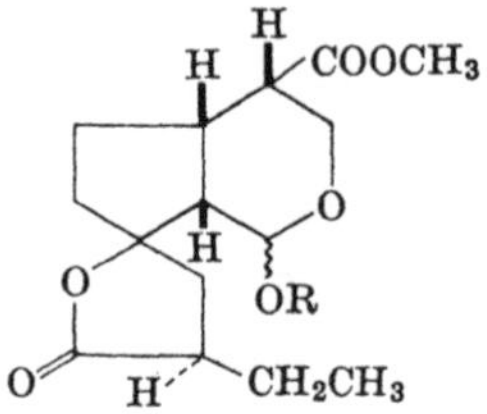

(22)

On oxidizing **21** (R = H) with chromic oxide in pyridine, **23** was obtained, which upon catalytic hydrogenation gave **24**. In this dilactone (**24**) sodium methoxide induced two β-eliminations which, after hydrolysis of

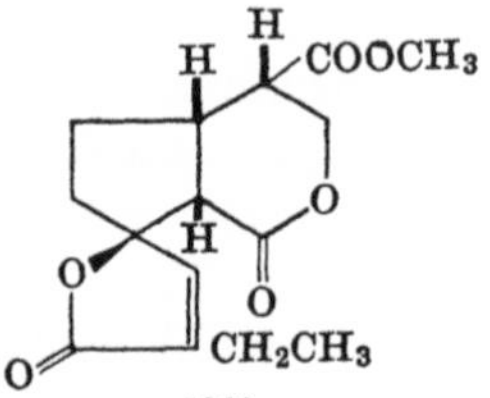

(23)

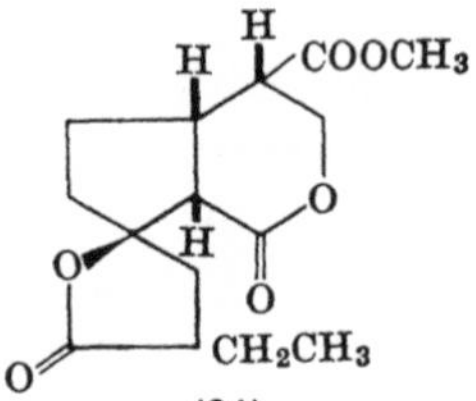

(24)

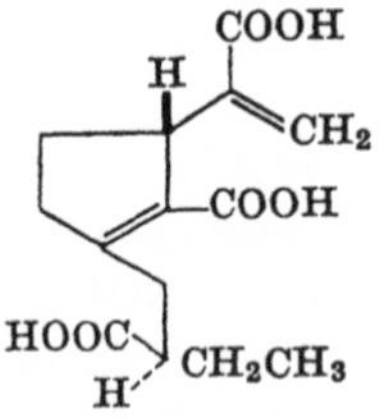

(25)

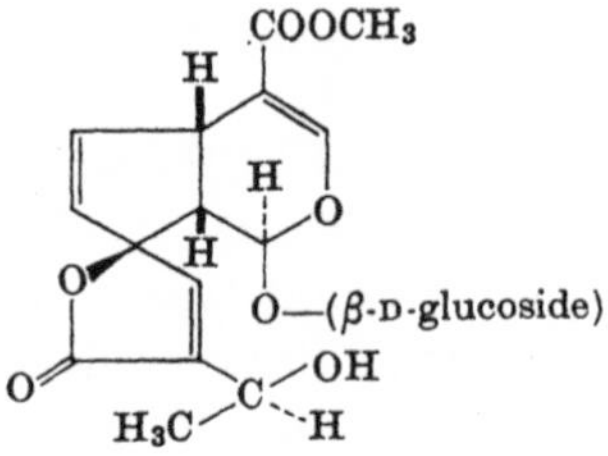

(26)

the ester functions, afforded the tribasic acid (**25**), identical with degradation product of plumieride (**26**). Taken with evidence derived from ultraviolet, infrared, and nuclear magnetic resonance spectra, these facts show that plumericin has structure **7**. The difference in reactivity between *exo*cyclic and *endo*cyclic unsaturation in a butenolide accounted for the fact that, whereas hydrogenation of plumieride (**26**) first affected the cyclopentene link, leaving an unsaturated lactone, hydrogenation of **7** afforded the saturated lactone (**11**) which retained the cyclopentene system. Lactone (**11**), α-dihydroplumericin, is epimeric with the naturally occurring β-compound (**9**); the difference involves only the stereochemistry at the starred position. This relation can be demonstrated by the fact (see above) that the β-dihydro-**9** affords the β-hexahydro derivative (**17**) which is isomeric with the α-derivative (**15**) but is transformed into it when warmed with alkali and then reesterified with diazomethane. The nuclear magnetic resonance spectrum is in accord with the suggested structure,[3, 4]

2. Anhydrotetrahydroaucubigenin

Anhydrotetrahydroaucubigenin, $C_9H_{14}O_3$, is the dehydrated product of tetrahydroaucubigenin. The latter is obtained by hydrogenation of aucubin hexaacetate in nearly neutral medium, followed by hydrolysis to the aglycone.[6] Fujise and coworkers,[7, 8] on the basis of their results, showed that aucubin, obtained from the seeds of *Aucuba japonica* L., and in at least seventy-five plants according to surveys,[9] could be represented by a structure of type **27**, tetrahydroaucubigenin can be given structure **28**, and anhydrotetrahydroaucubigenin **29**.[10]

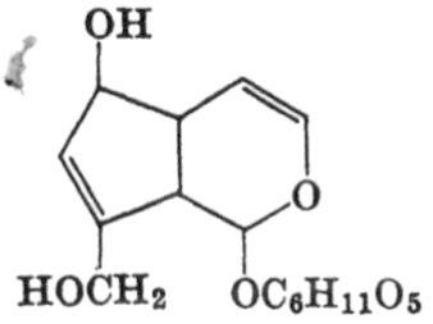

(**27**)

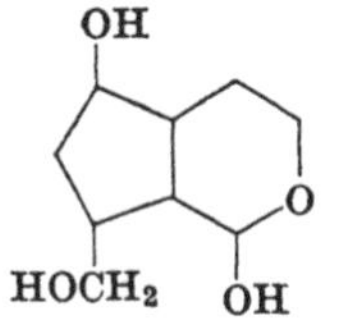

(**28**)

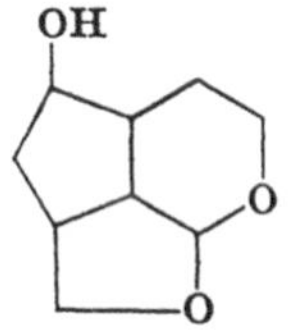

(**29**)

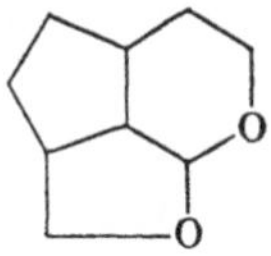

(**30**)

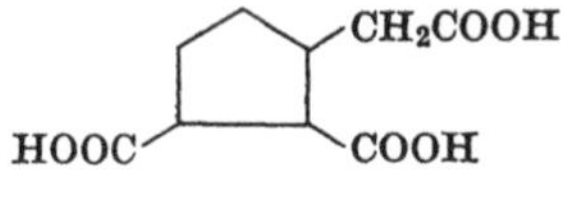

(**31**)

Structure **29** is consistent both with the inability of the anhydro compound to react with triphenylmethyl chloride, and with its oxidation to a ketone having ν_{max} 1742 cm^{-1} as do authentic cyclopentanones. Though no carbonyl bands appear in the spectrum of the anhydro

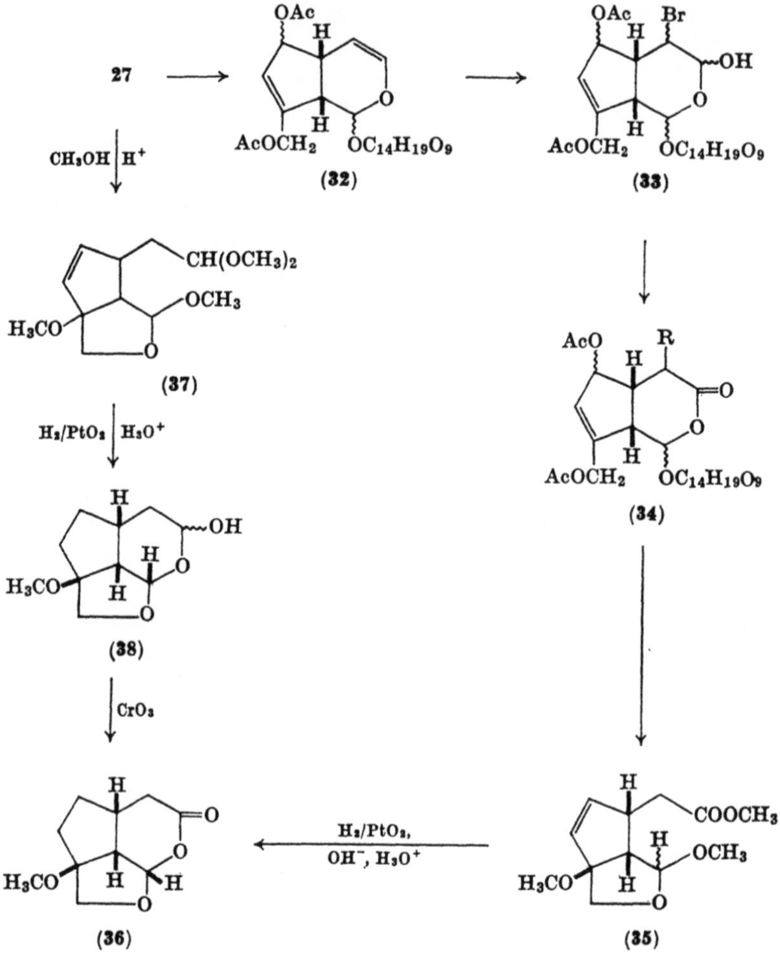

(1)

compound, it does react with 2,4-dinitrophenylhydrazine under forced conditions. The derivative is not a normal hydrazone, but resembles those obtained from hydroxy ketones that form ketals easily; thus anhydrotetrahydroaucubigenin may be regarded as having a ketal or an acetal grouping as in **29**. The anhydro compound has the appropriate characteristic bands of the OCHO system and thus the acetal grouping is preferred.

Pyrolysis of the methyl xanthate from **29**, followed by hydrogenation of the olefin produced by removal of the hydroxyl group, gives anhydrotetrahydrodeoxyaucubigenin (**30**) which still behaves as an acetal.

Oxidation of **30** with chromic acid gives a tricarboxylic acid, $C_6H_9(COOH)_3$, of type **31**.

Further degradation and nuclear magnetic resonance spectral determinations permitted a definite choice of the structure **27** for aucubin.[10] Treatment of aucubin hexaacetate (**32**) with bromine in aqueous tetrahydrofuran gave the bromohydroxy derivative (**33**), which when allowed to react with chromic oxide in acetic acid followed by reduction of the resultant α-bromolactone (**34**, R = Br), gave **34** (R = H).[11] Catalytic hydrogenation of the oily substance **35**, obtained via action of methanolic hydrogen chloride followed by alkaline hydrolysis, led to the formation of the lactone **36**. The latter was also obtained from aucubin (**27**) via **37** and **38** (Eq. 1).

II. Synthetic Furopyrans and -pyrones

Schulte and coworkers[12] have prepared a number of 3-propargyl-4-hydroxy-(5-R^1-6-R^2-substituted)-2-pyrone derivatives (**39**) according to the method of Boltze and Heidenbluth[13] by condensation of propargyl-malonic acid dichloride with β-dicarbonyl compounds, e.g. acetoacetic ester, benzoylacetic ester, acetylacetone, and benzoylacetone. On heating **39** with zinc carbonate at the melting point, ring closure took place to give the corresponding 3,4-(2-methylfuro)-2-pyrone derivatives (**40**) (Eq. 2). By reacting **39** ($R^1 = OC_2H_5$, $R^2 = CH_3$) with aromatic aldehydes in the presence of piperidine, the corresponding styryl derivatives (**41**) were obtained. With salicylaldehyde, 3,4-(2-methyl-furo)-5-ethoxycarbonyl-6-(2-coumaranyl)-2-pyrone (**42**) was produced.[12]

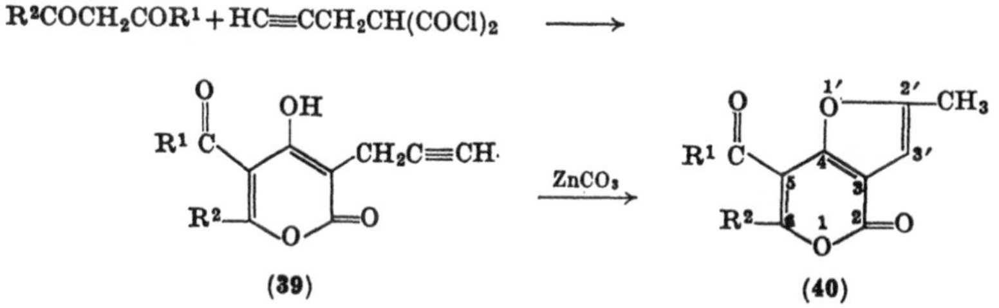

$$R^2COCH_2COR^1 + HC\equiv CCH_2CH(COCl)_2 \longrightarrow$$

$$\text{(39)} \xrightarrow{ZnCO_3} \text{(40)} \tag{2}$$

The condensation of a number of vinyl ether heterocycles with various dienes has been thought to yield most probably 1,7-dioxa-4,7,8,9-tetrahydroindanes (**43**).[14] 2,3-Dihydrofuran reacts with acrolein and with crotonaldehyde to give the adducts having probable structures **43** (R = R^1 = H) and **43**, (R = H, R^1 = CH_3), respectively. The same is true with 5-methyl-2,3-dihydrofuran, which yields the adducts **43**

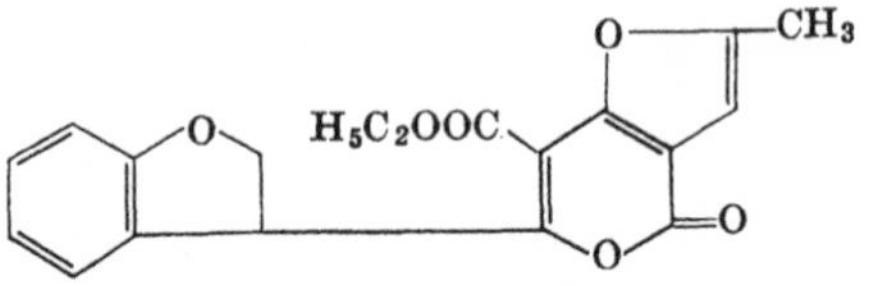

(42)

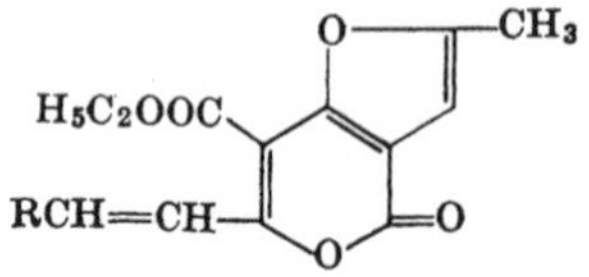

(41)

(R = CH_3, R^1 = H) and **43**, (R = R^1 = CH_3), respectively. Catalytic hydrogenation of these adducts in the presence of Raney nickel effects the formation of the corresponding 1,7-hexahydroindanes (**44**) (Eq. 3).

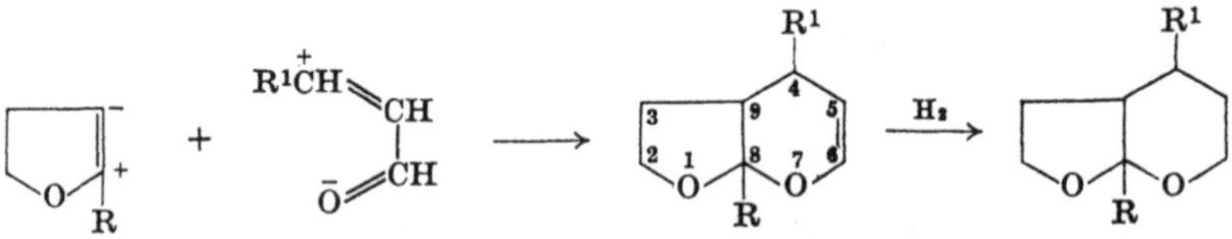

(43) (44) (3)

Worthy of mention is the fact that **45** has been isolated together with **43** (R = CH_3, R^1 = H); the formation of the former is attributed to the possible isomeriaztion of 5-methyl-2,3-dihydrofuran to methylene-2-tetrahydrofuran, which reacts with acrolein to give **45** (Eq. 4).

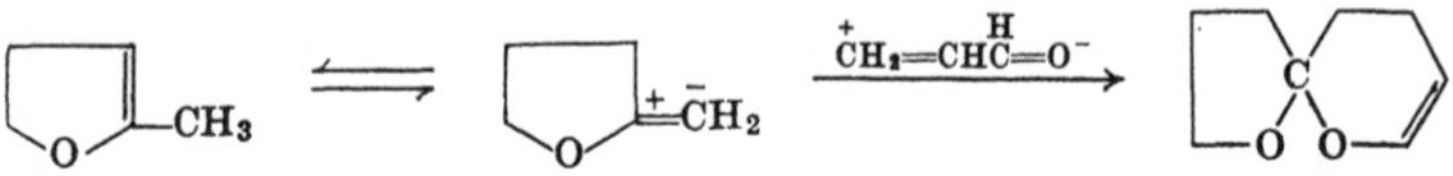

(45) (4)

1,7-Dioxahexahydroindanes (**44**, R = R^1 = H) and (**44**, R = R^1 = CH_3) are readily converted into the dialdehyde (**46**) and the monoaldehyde (**47**),[14] respectively, by the action of dilute hydrochloric acid.

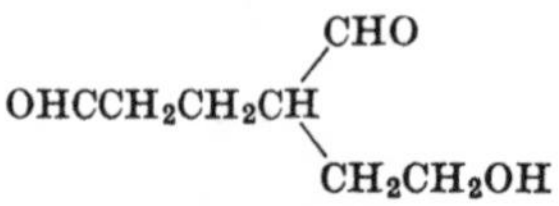

(46)

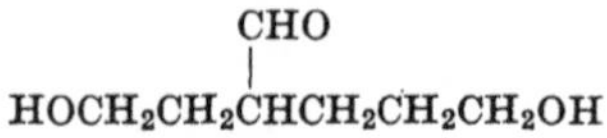

(47)

1,5-Dioxahexahydroindane (**50**) is readily obtained by catalytic hydrogenation of β-furylacrolein (**48**), together with 3-tetrahydrofuryl-propan-1-ol (**49**).[15] Treatment of **50** with hydrobromic acid effected pyran-ring opening to yield a dibromo compound, presumably having structure (**51**). The latter compound recyclizes to **50** via the action of a zinc–copper couple. Hydrogenation of **50** over nickel produced **49** (Eq. 5).

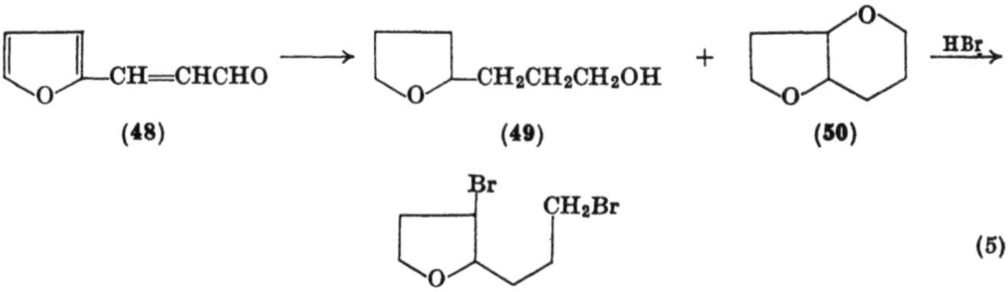

(48) (49) (50)

(51)

(5)

The fact that γ-furylpropionaldehyde (**52**) gave **50** under the same conditions which brought about transformation of **48** to **50** justifies the conclusion that **50** is formed primarily via **52**.

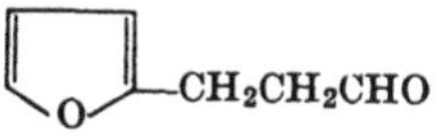

(52)

III. References

1. W. Hückel, *Theoretical Principles of Organic Chemistry*, Elsevier, Amsterdam, 1955.
2. J. E. Little and D. B. Johnstone, *Arch. Biochem. Biophys.*, **30**, 445 (1951); *Chem. Abstr.*, **45**, 5135 (1951).
3. G. Albers–Schönberg and H. Schmid, *Chimia (Aarau)*, **14**, 127, (1960).
4. G. Albers–Schönberg and H. Schmid, *Helv. Chim. Acta*, **44**, 1447 (1961).
5. H. Schmid and W. Bencze, *Helv. Chim. Acta*, **36**, 205, 1468 (1953).
6. P. Karrer and H. Schmid, *Helv. Chim. Acta*, **29**, 525 (1946).
7. S. Fujise, H. Uda, T. Ishikawa, H. Ohara and A. Fujino, *Chem. Ind. (London)*, **289**, 954 (1960); *Chem. Abstr.*, **54**, 2354 (1960).
8. S. Fujise, *J. Chem. Soc. Japan, Pure Chem. Sect.*, **74**, 725 (1953); Y. Iwanami, Y. Hotta, T. Kubota, S. Fujise, T. Ishikawa and H. Uda, *J. Chem. Soc. Japan, Pure Chem. Sect.*, **76**, 77 (1955).
9. R. Paris and M. Chaslot, *Ann. Pharm. Franc.*, **13**, 648 (1955).
10. W. Haegle, F. Kaplan and H. Schmid, *Tetrahedron Letters*, (3), 110 (1961).
11. M. W. Wendt, W. Haegle, E. Simonitsch and H. Schmid, *Helv. Chim. Acta*, **43**, 1440 (1960).
12. K. E. Schulte, J. Reisch and K. H. Kauder, *Arch. Pharm.*, **295**, 801 (1962); *Chem. Abstr.*, **58**, 11337 (1963).
13. K. H. Boltze and K. Heidenbluth, *Chem. Ber.*, **91**, 2489 (1958).
14. R. Paul and S. Tchelitcheff, *Bull. Soc. Chim. France*, 672 (1954); *Chem. Abstr.*, **49**, 9638 (1955).
15. H. E. Burdick and H. Adkins, *J. Am. Chem. Soc.*, **56**, 438 (1934).

CHAPTER II

Furocoumarins

Furocoumarins, discovered in plants, were for many years the only ones (Table 1) known, until synthetic methods were combined to produce virtually any furocoumarin desired. The natural product serves a powerful purpose in directing further synthetic effect and on occasion withstands the competition of its synthetic congeners, e.g. coumestrol and

TABLE 1a. Naturally Occurring Furocoumarins

No.	Name (Synonyms) (Formula)	$(\alpha)_D$ (Solvent,[a] °C)	References
1	Angelicin (Isopsoralen) (**1**)	—	51, 52, 54
2	Psoralen (Ficusin) (**16**)	—	64–66, 196
3	Bergapten (Heraclin, Majudin) (**25**)	—	77–78, 214
4	Bergaptol (**31**)	—	76–77, 85, 93, 215
5	Isobergapten (**44**)	—	85, 89, 139
6	Bergamottin (Bergaptin) (**46**)	—	94–95
7	Xanthotoxin (Ammoidin) (**47**)	—	97, 101, 216
8	Xanthotoxol (**54**)	—	115, 125–126
9	Isoimperatorin (**70**)	—	118, 217
10	Oxypeucedanin (**71**)	—	119–123a, 218
11	(+)-Oxypeucedanin (Prangolarin) (**71**)	+17(A,27)	123a–124
12	Ostruthol (**75**)	−18.3(E,15)	94
13	Imperatorin (Marmelosin, Ammidin) (**76**)	—	115, 127, 153, 219
14	Alloimperatorin (Prangenidin) (**77**)	—	128, 130, 220
15	Heraclenin (Prangenin) (**78**)	+22(E,32)	131–132
16	Isopimpinellin (**82**)	—	91
17	Phellopterin (**86**)	—	77, 133, 221
18	8-Hydroxy-5-methoxypsoralen (**87**)	—	222
19	(±)-Byakangelicol (**89**)	+34.77(E,—)	133, 135, 221
20	(±)-Byakangelicin (**91**)	+24.62(E,25)	6, 133, 138
21	Pimpinellin (**94**)	—	91
22	Sphondin (**95**)	—	90, 139
23	Halfordin (**103**)	—	141
24	Isohalfordin (**104a** or **104b**)	—	141
25	(−)-Nodakenetin (**110a**)	−22.6(A,18)	146–148, 143, 151
26	Nodakenin (**110b**)	+56.6(F,30)	146, 148
27	(+)-Marmesin (**110a**)	+24.5(A,18)	144, 151, 223

(Table continued)

TABLE 1a (*continued*)

No.	Name (Synonyms) (Formula)	α(D) Solvent,[a] °C)	References
28	Marmesinin (Amajin) (**110b**)	−60(C,25)	145, 152
29	Peucedanin (Oreoselone methyl ether) (**119**)	—	121, 153, 224
30	Athamantin (**127**)	+96(D,22)	156, 158, 187, 225
31	Oroselone (Kvanin) (**129**)	—	139, 156, 157
32	Discophoridin (**133**)	+20.4(C,20)	159
33	Edultin (**137**)	+41.5(E,10.7)	160
34	Columbianidin (**140**)	+26.5(B,27)	161
35	Columbianin (**145**)	+118(F,23)	161
36	Columbianetin (**141**)	+20(B,27)	161
37	Archangelicin (**146**)	+112.7(D,26)	157
38	Archangelin (**149**)	—	124, 166
39	8-Geranyloxypsoralen (**119b**)	—	21
40	Peulustrin (**145a**)	+278(D,25)	123b
41	Isopeulustrin (**145c**)	+273(D,25)	123c
42	Columbianidinoxide (**145b**)	+305(D,28)	123c

[a] A, chloroform; B, dioxan; C, ethyl alcohol; D, methyl alcohol; E, pyridine; F, water.

psoralen. Furocoumarins are found especially in the Umbelliferae, Rutaceae, Leguminosae families and also in other families of lesser importance.

Späth,[1] Dean,[2,3] and Reppel[4] have written comprehensive reviews about the chemistry of the naturally occurring coumarins. Mention should also be made of Karrer's[5] review in 1958 in which he lists the furocoumarins that had been isolated from natural sources up to the date of publication.

The selection of the proper plant parts is of great importance, for the furocoumarin may be present in one part and completely absent or present in minor percentage in another; for example phellopterin is present in the fruits and not the seeds of *Phellopterus lettoralis* Benth[6] (Table 2).

I. Isolation

For the most part, isolation procedures depend upon successive extraction with solvents of increasing polarity. Thus, petroleum ether is frequently used as the initial solvent and has the advantage that most of the oxygenated coumarinic materials are not particularly soluble in it. Petroleum ether may be omitted, and often is, and direct ether extraction employed. This, followed by the use of methanol or ethanol as the solvent often results in the fortuitous crystallization of coumarinic glycosides. Advantage may be taken of the alkali solubility and acid insolubility of

TABLE 1b. Natural and Synthetic Furocoumarins

A. Psoralen Type

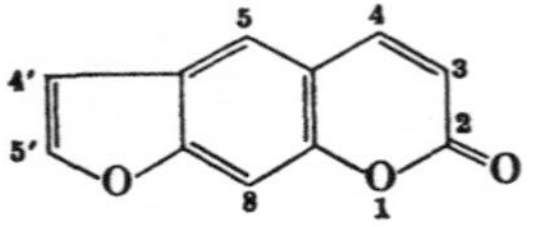

Substituents[a]	M.p. (°c)	Solvent for crystallization[b]	U.v. Spectrum (λ_{max} (mμ) (logϵ))	References
Unsubstituted (+ + + +)	162, 167, 171	J, K	246(4.37), 290(4.03), 328(3.08)	32, 66, 69, 140, 306, 317
4-Me	188.5	G	244(4.39), 289(3.99), 329(3.85)	32, 317
4'-Me(+ + +)	167	H		71
5'-CHMe$_2$ (anhydronodakenetin, deoxyoreoselone)	135–136, 138–139	C, J	211(3.89), 251(4.50), 294(4.05), 334(3.84)	144, 145, 147, 152
4-Ph	178–179	G	225(4.43), 248(4.41), 298(4.06), 331(3.83)	32, 317
3,4-Di-Me	235–236	H	245(4.38), 290(4.02), 329(3.95)	32, 70
4,4'-Di-Me (+ + +)	220, 222	H		71, 318
4,5'-Di-Me (+ + +)	162, 165, 171	H	245(4.28), 290(3.82), 340(3.68)	67, 73, 75
8,5'-Di-Me	176	J	250(4.39), 300(4.12), 335(3.83)	67
3-Et,4-Me	179–180	H	245(4.42), 290(4.06), 329(3.99)	32, 70
3-n-Pr,4-Me	100–101	H	244(4.30), 290(4.04), 328(3.83)	32
3-Iso-Pr, 4-Me	145–147	H	245(3.59), 290(3.96), 329(3.94)	32, 70
3-Me,4-Ph	160	I	225(4.356), 247(4.38), 296(4.05), 328(3.92)	32
3-Et,4-Ph	178	H	225(4.28), 247(4.32), 296(3.99), 328(3.84)	32
4-Me,4'-Et	177	—		319
4-Me,4'-Pr	175	—		319
4-Me,4'-Bu	158	—		320
4-Me,4'-Ph (−)	185	G	224(−), 250(−), 298(−), 325 inflex.(−)	72

Substituents[a]	M.p. (°c)	Solvent for crystallization[b]	U.v. Spectrum (λ_{max} (mμ) (logϵ))	References
3,4-Cyclohexyl	234	G	246(3.62), 298(4.01), 326(4.00)	32
3,4-Benzo	197.5–198.5	H		317
3,4,5-Tri-Me	203.5–204.5	C, H		73
4,8,5′-Tri-Me	234.5–235	E	250(4.35), 295(3.99), 325(3.80)	67
3-n.Butyl, 4,5′-Di-Me	119–120	H		73
3,4-Cyclohexyl,5′-Me	176.5–178	H		73
3,4-Benzo,8,5′-di-Me	232–233	C		73
5-CH_2CH═CMe_2	229	H		129
3-Me,8-OMe(+)	151	L	228(4.46), 252(4.52), 302(4.44)	104
4-Me,5-OMe(+)	175	J	223(4.35), 251(4.28), 278(3.78), 306(4.15)	200
4-Me,8-OMe(+++)	168	J	222(4.36), 248(4.49), 276(4.00), 300(4.24)	104
5-CH_2CH═CMe_2, 8-OMe	113	D		129, 321
5-CH_2CH═CMe_2, 8-Ac	137	H		129
4′-Methyl, 8-OMe(++)	155	H		71
5′-Iso-Pr, 4′-OMe (peucedanin)	82–87(dimor.)	—		153
3,4′-Di-Me, 8-OMe(+)	135	H	220(4.43), 248(4.43), 278(4.03), 302(4.27)	104
4,4′-Di-Me, 8-OMe(+)	182	C	218(4.37), 250(4.67), 268(3.83), 296(4.20)	104
4,5′-Di-Me, 8-OMe(+++)	159	J	214(4.29), 250(4.48), 278(3.78), 300(4.02)	104
4-Me, 5′-Ph, 8-OMe(−)	205	J	286(4.52)	104
8-OH (−)	238–240	A		100
4-OMe (xanthotoxol)	216–217	H		70
5-OMe (bergapten) (+++)	190	J		80
8-OMe (xanthotoxin) (+++)	146–147	J	219(4.32), 249(4.35), 276(3.81), 300(4.06)	15, 37, 101, 102, 140,
8-OEt	198	H		304
8-n-OPr	140	H		304
8-n-OBu	118	H		304
5-OCH_2CH═CMe_2(isoimperatorin)	108–109	L		217

(Table continued)

TABLE 1b (*continued*)

Substituents[a]	M.p. (°C)	Solvent for crystallization[b]	U.v. Spectrum (λ_{max} (mμ) (logϵ))	References
5-$OCH_2CH\overset{}{-}CMe_2$ (epoxide, O bridge)(heraclenin) (−)	111, 114.5	J		123, 124, 131, 132
5-$OCH_2CH(OH)C(OH)Me_2$(heraclenin hydrate, oxypeucedanin hydrate)	117–118, 131	E, G		123, 125, 132
5-$OCH_2COCHMe_2$ (isoheraclenin)	132–134	M		132
5-$OCH_2CHC(OH)Me_2$ (ostruthol) (−) with $OCOCMe{=}CHMe$ on CH	137.5	E/M		123
5-$OCH_2CH{=}C(Me)CH_2CH_2CH{=}CMe_2$	61	H	220(4.41), 250(4.28), 309(4.17)	95, 96
5-$OCH_2C_6H_5$ (−)	142	J	222(−), 250(−), 278(−), 310(−)	72
8-OCH_2COOH	248	H		304
8-OCH_2COOEt (−)	127	H		304
8-OAc	178, 185, 206	H		132, 304
5,8-Di-OH (−)	270, 275	H		100, 135
4-OH, 5-OMe	245	H		48
5-OH, 8-OMe	224–226	H		37, 322
8-OH, 5-OMe	221	—		135
4,5-Di-OMe	191–192	B		48
5,8-Di-OMe (isopimpinellin) (−)	149, 152–153	H		100, 135
5-OMe, 8-$OCH_2CH{=}CMe_2$ (phellopterin)	102, 105	H		135, 221
5-OMe, 8-$OCH_2CH\overset{}{-}CMe_2$ (epoxide, O bridge) ((±)-byakangelicol)	106	—		221
5-OMe, 8-$OCH(OH)C(OH)Me_2$ (byakangelicin)	116–117, 118, 120	G		137, 221
5-OMe, 8-OAc	180	—		135
8-OH, 5-NO_2	255	F	212(4.32), 254(4.34), 308(4.07), 345(3.87)	145, 304
8-OH, 5-NH_2	281	H		304
8-OMe, 5-NO_2 (−)	205–206, 231, 235–238	H		37, 322
8-OMe, 5-NH_2 (−)	234–235	H		37

Substituents[a]	M.p. (°C)	Solvent for crystallization[b]	U.v. Spectrum (λ_{max} (mμ) (logϵ))	References
8-OMe, 5-NHAc (−)	244.5	—		37, 100
8-OMe, 5-N=CHC$_6$H$_5$	189–200	H		37
8-OMe, 5-Cl (+)	187–188, 192–195	H, J	225(−), 253(−), 284(−), 306(−)	37, 72, 100
8-OMe, 5-Br	185–186	H	222(4.41), 246(4.01), 268(4.28), 308(4.14)	37
5-OMe, 8-COOH (−)	267(dec.)	J	227(4.28), 249(3.99), 273(4.21), 314(4.12)	81, 96
5-OMe, 8-COOMe (+)	213	J	235(4.33), 250(3.95), 274(4.27), 311(4.17)	81, 96
5′-Me, 4′-Ac	165.5–166.5	H		74
5′-Iso-Pr, 3-NO$_2$	209–211	H	210(4.38), 252(4.32), 330(3.92)	145
4,5′-Di-Me, 8-Cl	260.5–261	C		75
4,5′-Di-Me, 8-Br	261–261.5	A		75
4,5′-Di-Me, 8-NH$_2$	193–193.5	H		75
4,5′-Di-Me, 8-NHAc	304–305	A		75
4,5′-Di-Me, 8-NMe$_2$	142–142.5	L		75
4,5′-Di-Me, 8-CN	287–289	A		75
3-COOEt	148	G		317
8-Ac	206–207	D	230(4.3), 266(4.1), 341(3.8)	104
5-OEt (++)	142	H, J		86
5-OPr-n (−) (sunlight +)	115–116	H, J		86
5-OPr-iso (+)	122	H, J		86
5-OBu-n (−) (sunlight +)	117	H, J		86
5-OBu-iso	114–115	H, J		86
8-PO(OMe)$_2$	74–76	L		117
8-PO(OEt)$_2$	103–105	L		117
8-PO(OPr-iso)$_2$	102–104	L		117
8-PS(OEt)$_2$	131–133	L		117
5-PO(OMe)$_2$	102–104	L		117
5-PO(OEt)$_2$	86–88	L		117
5-PO(OPr-iso)$_2$	64–66	L		117
5-PS(OEt)$_2$	85–87	L		117
5-PO(OEt)$_2$, 8-OMe	70–72	L		117

(Table continued)

TABLE 1b (*continued*)

Substituents[a]	M.p. (°C)	Solvent for crystallization[b]	U.v. Spectrum (λ_{max} (mμ) (logϵ))	References
5-PO(OPr-iso)$_2$, 8-OMe	77–79	L		117
5-PS(OEt)$_2$, 8-OMe	86–88	L		117
5-PO(OMe)$_2$, 8-OMe	118–120	L		117
B. 4′,5′-Dihydropsoralen Type				
Unsubstituted	195.5, 201, 204	H, J		66, 68, 69, 317
4-Me	170–171	G	225(4.21), 254(3.57), 294(3.78), 322(4.26)	32, 317
4′-Me	137	J		69
5′-Iso-Pr (deoxydihydroöreoselone)	115–117	—	208(4.22), 348(4.31)	88, 145, 147, 151
5′-Me$_2$COH (nodakenetin, marmesin)	185, 189, 192	H	212(3.89), 248(3.50), 335(4.09)	144, 145, 147, 151, 152, 228, 323
5′-Me$_2$CO-Glucosyl(nodakenin, marmesinin)	218–219(dec.)	H		147, 152, 228, 323
4-Ph	202–203	A	235(4.13), 261(4.00), 292(3.61), 337(4.19)	32, 317
3,4-Di-Me	187.5	G, H	225(4.10), 296(3.75), 332(4.22)	32, 70
3-Et, 4-Me	143–144	H	226(4.11), 256(3.47), 295(3.76), 332(4.26)	32, 70
3-Iso-Pr, 4-Me	183–185	H	226(4.16), 255(3.57), 295(3.80), 332(4.28)	32, 70
3-n-Bu, 4-Me	89–90	H	226(4.08), 256(3.45), 296(3.73), 332 (4.23)	32
3-Myristyl, 4-Me	96–97	H		32
3-Me, 4-Ph	239–240	G	335(4.25)	32
3-Et, 4-Ph	206–207		335(4.16)	32
3,4-Cyclohexyl	185–186, 190.5–191.5	G	227(4.16), 255(3.60), 295(3.82), 332(4.30)	32, 317
4,8,5′-Tri-Me	230–231	H		324
4,8-Di-Me, 5′-CH$_2$I	144–145	H		324

2

Substituents[a]	M.p. (°C)	Solvent for crystallization[b]	U.v. Spectrum (λ_{max} (mμ) (logϵ))	References
4-OH	261.5–263.5 (dec.)	H		317
8-OH (dihydroxanthotoxol)	195	K		102
5-OMe (dihydrobergapten) (−)	158, 164	C, J	251(3.78), 334(4.21)	80, 81
8-OMe (dihydroxanthotoxin)	159, 161, 163	H, J	252(3.28), 262(3.44), 335(4.08)	37, 101, 102, 107, 157, 322, 323
8-OH, 4-Me	236–237	—		102
8-OH, 4-Et	210–212	—		102
8-OH, 4-$(CH_2)_5$Me	162–163.5	—		102
8-OH, 4-$(CH_2)_6$Me	145.7–147, 156–157	—		102
8-OH, 4-$(CH_2)_7$Me	139.6–140.6 148.6–149.6			102
8-OH, 4-$(CH_2)_8$Me	132.2–134.2	—		102
8-OH, 4-$(CH_2)_{10}$Me	138.2–138.8	—		102
8-OH, 4-$(CH_2)_{12}$Me	128.6–130	—		102
8-OH, 4-$(CH_2)_{14}$Me	127.2–128.4	—		102
8-OH, 4-$(CH_2)_{20}$Me	123.7–124.5	—		102
8-OH, 4-Ph	245–247	—		102
5-OMe, 4-Me (−)	154	J	213(4.47), 248(3.76), 258(3.69), 328(4.20)	200
8-OMe, 4-Me	149.5–151.2 156.2–157.1	—		102
8-OMe, 4-Et	119.6–121.4	—		102
8-OMe, 4-$(CH_2)_5$Me	90.2–91.6	—		102
8-OMe, 4-$(CH_2)_6$Me	90.3–91.9	—		102
8-OMe, 4-$(CH_2)_7$Me	91.2–92.6	—		102
8-OMe, 4-$(CH_2)_8$Me	75.6–78.1	—		102
8-OMe, 4-$(CH_2)_{10}$Me	74–75, 80.4–81.9	—		102

(Table continued)

TABLE 1b (*continued*)

Substituents[a]	M.p. (°c)	Solvent for crystallization[b]	U.v. Spectrum (λ_{max} (mμ) (logϵ))	References
8-OMe, 4-$(CH_2)_{12}Me$	80.6–81.5, 83.8–84.7	—		102
8-OMe, 4-$(CH_2)_{14}Me$	83.6–84.5, 87.4–88.8	—		102
8-OMe, 4-$(CH_2)_{16}Me$	91.2–92.3, 95.1–96.4	—		102
8-OMe, 4-$(CH_2)_{20}Me$	86.5–86.9, 89.6–89.9	—		102
8-OMe, 4-Ph	198–199	—		102
4′-OMe,5′-iso-Pr (dihydropeucedanin)	82–87(dimor.)	—		153
8-OMe, 5-NH_2	214–216, 243–245	H		37, 107
8-OMe, 5-NHCOOEt	216–216.5	H		107
8-OMe, 5-Cl	193.4	H		37, 100, 107
8-OMe, 5-Br	207.5	H	220(4.62), 254(4.42), 266(4.43), 330(4.52)	37, 100, 107
5-OMe, 8-COOH	251 (dec.)	J	217(4.41), 266(3.87), 284(3.44), 335(4.20)	81
5-OMe, 8-COOMe (−)	207	J	220(4.42), 264(3.98), 286(3.66), 332(4.24)	81
8-OMe, 3,5-di-Br	267.6–268.1	N		107
8-OMe,5,4′,5′-tri-Cl	202–203	B		100
8-OMe,5,4′,5′-tri-Br	166	J	220(4.62), 246(3.98), 268(4.27), 310(4.14)	37
5,8-Di-OMe (dihydropimpinellin)	156.6–157.4	H, K	253(3.82), 262(3.83), 326(4.17)	106
5,8-Di-OMe, 3-Br	231.2–233	H		106
5,8-Di-OMe, 3-COOH	292–293.5	H		106
4′-Me, 3-COOH	216	B/K		69
5′-Iso-Pr, 3-NO_2	209–211	H		145
5′-Iso-Pr, 3-NH_2	151	—		155
5′-Iso-Pr, 3-CN	207–208	—		151
5′-Iso-Pr, 3-COOH	211	—		151
5′-Iso-Pr, 3-COOEt	172	—		151
5′-Iso-Pr, 3-Br	159, 161 (dec.)	H		145

Substituents[a]	M.p. (°C)	Solvent for crystallization[b]	U.v. Spectrum (λ_{max} (mμ) (logϵ))	References
5′-C(OH)Me_2, 3-NH_2	248–250(dec.)	H		145
5′-C(OH)Me_2, 3-Br	231–233	E/H	208(4.49), 254(3.70), 325(4.09)	145, 200
5′-C(OH)Me_2, 3-COOH	223–224	—		151
3-COOH	241.5–243.5(dec.)	M		69, 317
3-COOEt	200.5–202.5	H		68
4′,5′-Dihydrothioxanthotoxin	165–167	H		37
3,4,4′,5′-Tetrahydro,4-Me psoralen	68.5–69.5	O		317
Deoxytetrahydroöreoselone	115–116	—		147, 151

C. *Angelicin* (*Isopsoralen Type*)

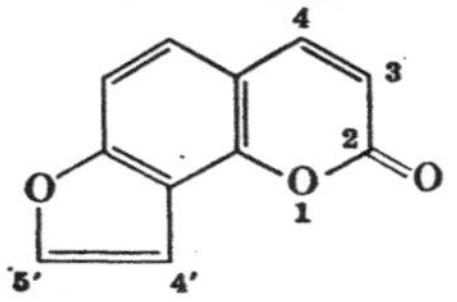

Substituents[a]	M.p. (°C)	Solvent for crystallization[b]	U.v. Spectrum (λ_{max} (mμ) (logϵ))	References
Unsubstituted (+)	137–138, 140	H	216(4.24), 248(4.05), 298(4.06)	60, 62, 95, 160, 306, 325
4-Me (+)	190–192, 194	J, L	218(4.27), 250(4.51), 296(4.21)	62, 325
5′-Me	153–154	J	250(4.36), 300(4.09)	67
5′-Iso-Pr (dihydroöroselone)	142–143	H	224(4.23), 252(4.45), 274(3.71), 302(4.06)	123, 158, 161
5′-C(Me)=CH_2 (oroselone)	178–180, 189–190	C		156–157, 160
5′-C(OH)Me_2 (oroselol)	149–151, 157–158	C	231(ϵ,9,450), 252(27,700), 273(4,000), 301(10,400)	158, 160
5′-C(OMe)Me_2	116, 118	L		156, 157, 160

(*Table continued*)

TABLE 1b (*continued*)

Substituents[a]	M.p. (°C)	Solvent for crystallization[b]	U.v. Spectrum (λ_{max} (mμ) (logϵ))	References
5′-C(OAc)Me_2	149–151	J	231(ϵ,7,500), 252.5(24,700), 301(8,600)	160
4′-Ph	121	H		326
4,5′-Di-Me	182–183	H	250(4.27), 297.5(3.93)	67, 73
4-Me, 4′-Et	137	—		319
4-Me, 4′-n-Pr	85	—		319
4-Me, 4′-n-Bu	89	—		320
4-Me, 4′-Ph	170	H		327
4-Me, 4′-$C_6H_4CH_3$-*o*	165	—		319
4-Me, 4′-$C_6H_4CH_3$-*m*	190	—		320
4-Me, 4′-$C_6H_4CH_3$-*p*	175	—		319
4-Ph, 4′-Me	153	—		319
4′-Ph, 6-Et	111	H		326
4,4′-Di-Ph	154	—		319
4,4′-Di-Me, 6-Et	157	H		301
6-OH	220–224	—	251(4.33), 295(3.84), 341(3.8)	328
6-OMe (sphondin)	189–191	—	249(4.3), 303(3.80), 341(3.7)	139, 328
5-OMe (isobergapten) (+)	223–224	G	223(4.20), 250(4.18), 308(3.90)	85, 92
5-OMe, 4-Me	288–289	H		60
4-Me, 5′-Ac	216–217	—		329
4-Me, 5′-COOH	295–299(dec.)	H		62
4-Me, 5′-COOEt	192, 194–195	H		62, 325
4′-Me, 6-Cl	219	H		326
5′-COOH	320–325	H		60,62
5′-COOEt	212–213	H		62
D. 4′,5′-*Dihydroangelicin Type*				
5′-C(OH)Me_2 (columbianetin)	163–164.5	J, K	210.5(4.28), 219(4.15)S, 252(3.33), 262(3.39), 329(4.13)	123a, 161
5′-C(O-glucosyl)Me_2 (columbianin)	275–276(H_2O)	J	216(3.97)S, 327(4.00)	161
5′-C(OAc)Me_2 (columbianetin acetate)	127.5, 125.5	H	218(4.11), 327(4.12)	123a, 161

Substituents [a]	M.p. (°C)	Solvent for crystallization[b]	U.v. Spectrum (λ_{max} (mμ) (log ϵ))	References
5′-C(Me_2)OCOCMe=CHMe (columbianidin)	118.5–119	M	219(4.35)S, 250(3.56), 261(3.59), 327(4.19)	123a, 161
5′-OCOC(OH)MeCH(OCOCMe=CHMe)Me (peulustrin)	129.5	J/M	$(\alpha)^{25}_{D}$ +278(MeOH)	123b
5′-OCOC(OH)MeCH(OCOCH=CMe_2)Me (isopeulustrin)	137.5–138	M	$(\alpha)^{25}_{D}$ +273(MeOH)	123c
5′-OCOCMe—O—CHMe (columbianidinoxide)	97	M/O	$(\alpha)^{28}_{D}$ +305(MeOH)	123c
5′-C(=CH_2)CH_2OH (discophoridin)	134–135	H	252(ϵ,2,557), 262(3,307), 328(14,440)	159
5′-C(=CH_2)CHOAc (discophoridin acetate)	71–72	H		159
4,5′-Di-Me	117.6–117.8	H		73
4′-OCOCH$_2$CHMe$_2$,5′-CMe$_2$OCOCH$_2$CHMe$_2$ (athamantin)	58–60	—		156
4′-OCOCMe=CHMe, 5′-C(OAc)Me_2 (edultin)	136, 142	C, E	219(ϵ,31,430),248(ϵ,6,770), 259(5,710), 299(9,350), 323(21,000)	160
4′,5′-Di-OCOCMe=CHMe (archangelicin)	100.5–102	O	258(3.55), 322(4.14)	157
4′-OH, 5′-di-COOEt	178–180	H		62
4′-OAc, 5′-di-COOEt	140–141	H		62
4-Me, 4′-OH, 5′-di-COOEt	175–176	H		62
4-Me, 4′-OMe, 5′-di-COOEt	189–190	H		62
3,4-Dihydro, 5′-C(OH)Me_2 (dihydrocolumbianidin)	111.5–112.5	H	281(3.29), 283(3.30), 288(3.32)	161
4′-Oxo	252.3	—		330
4-Me, 4′-oxo	153–154	H		331

E. *Furo*(3′,2′–6,7)*coumarin*

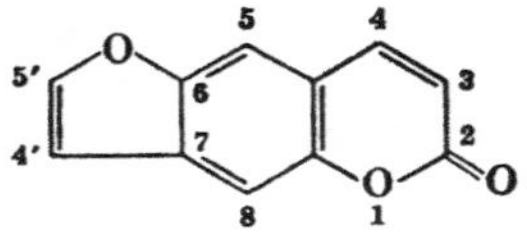

(*Table continued*)

TABLE 1b (*continued*)

Substituents[a]	M.p. (°c)	Solvent for crystallization[b]	U.v. Spectrum (λ_{max} (mμ) (log ϵ))	References
4-Me, 5-n-Pr	158–160	L		41
4′,5′-Di-Ph	286	A, H		332
4-OH, 4′,5′-di-Ph	288–289	H		45
4′,5′-Dihydro, 4-Me, 5-n-Pr, 4′-oxo	174–176	G		41
4′,5′-Dihydro, 4,5′-di-Me	131.5–132.5	L		41
F. *Furo*(3′,2′–5,6)*coumarin*				
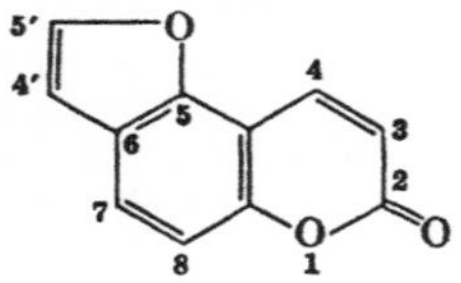				
4′-Me	138–140	K		333
4′-Et	150–152	K		333
4,4′-Di-Me	198	A, H		334
4-Me, 4′-n-Pr	129	A, H		334
4-Me, 4′-iso-Bu	101–102	A, H		334
4-Me, 4′-hendecyl	101–107	A, H		334
4-Me, 4′-pentadecyl	124	A. H		334
4-Me, 4′-heptadecyl	101–102	A, H		334
4-Me, 4′-Ph	197	A, H		334
4′-Me, 3-COOH	226–228(dec.)	H		333
4′-Et, 3-COOH	157–158	H/K		333
4,4′-Di-Me, 8-Et	158	A		335
4′,5′-Dihydro,4,7-di-Me, 8-CH_2I	166–167	—		318
4′,5′-Dihydro,4,7,8-tri-Me	205–206	—		318
4′,5′-Dihydro,4′-oxo-,4-Me	157–159	A		331

Substituents[a]	M.p. (°c)	Solvent for crystallization[b]	U.v. Spectrum (λ_{max} (mμ) (log ϵ))	References
G. *Furo*(2′,3′–5,6)*coumarin*				
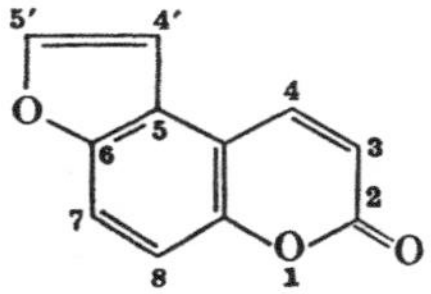				
4-Me	237–238	A		41
4,5′-Di-Me	207	C		41
7-OMe (allobergapten) (+)	207	J	220(4.29), 246(4.22), 255(4.24), 314(3.94), 344(4.04)	88
7-OMe, 4-Me (+)	236	J	219(4.39), 252(4.31), 308–309(3.95), 338(4.05)	200
7-OMe, 3-COOH	242	B		88
4-Me, 5′-COOH	330–333(dec.)	P		41
4′,5′-Dihydro,7-OMe	195	J	216(4.29), 236(4.07), 322(4.16)	200
4′,5′-Dihydro,4-Me, 5′-OH	265–266	H		41
4′,5′-Dihydro, 4,5′-di-Me	185–186	H		41
3,4,4′5′-Tetrahydro,7-OMe (−)	183–184	J	218(4.35), 277(3.15), 284(3.16), 318–320(2.97)	78, 200

(Table continued)

TABLE 1b (*continued*)

Substituents[a]	M.p. (°C)	Solvent for crystallization[b]	U.v. Spectrum (λ_{max} (mμ) (log ϵ))	References
H. *Furo*(2′,3′–7,8)*coumarin*				
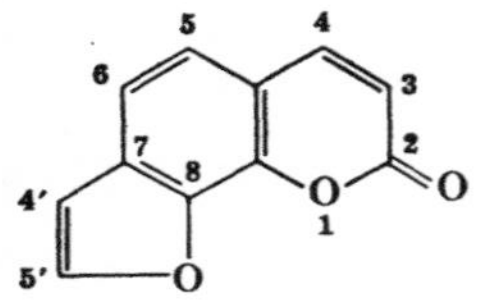				
5′-Me	158–159.5	H		336
5′-COOH	365–367	H		337
4′,5′-Dihydro,5′-CH_2Br	164–174	H		336

[a] Qualitative tests on human skin irradiated with long-wave ultraviolet light (3655Å) adapted from Musajo and Rodighiero,[138] (−) indicated inactivity; (+) indicated activity.

[b] A, acetic acid; B, acetone; C, benzene; D, benzene-petroleum ether mixture; E, chloroform; F, dioxan; G, ethyl acetate; H, ethyl alcohol; I, ethyl acetate- petroleum ether mixture; J, methyl alcohol; K, water; L, benzine; M, ether; N, butanol; O, cyclohexane; P, methylethyl ketone.

TABLE 2. Natural Sources and Furocoumarins Constituents

Natural Source	Distribution Part[b]	Constituents[a]	References
Aegle marmelos	F	13	128
Aegle marmelos	F	14	128, 130
Aegle marmelos, Bel	F	13	226
Aegle marmelos, Corea	B, F	13, 27	144, 223
Ammi majus	F	3	227, 228
	F	7	227–229, 152
	F	13	230
	F	16	152, 227, 228
	F	28	145, 152
Angelica anomale	R	3	231
Angelica archangelica var. *Litoralis*	S	1	22, 51
	S	7, 8, 13	14, 115
	S	3, 37, 38	22
Angelica archangelica var. *Norwegia*	S	1	22, 51
	S	37, 38	22
Angelica archangelica var. *decurrens*	F	9	232
	S	10	232, 233
	S	13	232–234
Angelica dahurica var. *dahurica*	R	17, 19, 20	235
Angelica dahurica var. *pai-chi*	R	17, 19	235
Angelica edulis Miyabe	R	33	160
Angelica formosana	R	3, 9, 10, 13, 17	231
Angelica gemelini	F	10	236
Angelica glabra	F	10	237
	F	13	136, 138
	F	17	237, 238
	R	19	238, 239
	R	20	138, 238, 239
Angelica glabra, Makino	F	13, 17	240
	F	19	239
	F	20	133, 239–240
Angelica japonica	R	2, 16, 20	241
	R	3	242
	R	9	118, 242
	R	18	222, 242
Angelica keiskei	R	2, 7	243
Angelica pubescens	R	3	244
Angelica Radix	R	1, 31, 37	245
Angelica saxicola	—	1	246
Angelica sylvestris	R	4, 9	247
	—	10	22, 247

(*Table continued*)

TABLE 2 (*continued*)

Natural Source	Distribution Part[b]	Constituents[a]	References
Athamanta oreoselium L.	R	30	158, 187
Apium graveolens	—	3	248–249
	—	7	178
Casimiroa edulis	B	3	250
	B	16	251
	S	18	251
Citrus acida	L	3, 16	252
Citrus aurantium natsudaidai	F	4	253
Citrus aurantifolia Swingle	—	2, 16	9, 212
Citrus bergamia (Bergamot oil)		3	93, 95, 214 254–255
		4	93
		6	94–95
Citrus limonum		10 (as hydrate)	256
		20	9, 257
		16	9, 257
		6, 39	21
Cornilla glauca	S	1	258
Fagara ailanthoides	S	16	259
Fagara flava (*Xanthoxylum flavum*)	—	1	150
Fagara schinifolia	S	3	259
Fagara xanthyloides	F	3	97
	F	7	216
Feronia elephantium	L	3	260
	R	27	260
	F	16	136
	S	19	136
Ficus carica L.	R	2	86, 261–266
	L	3	86, 261–263, 265, 266
Flindersia bennettiana	B	16	267
Halfordia kendack	B	23	141
Halfordia scleroxyla	B	23, 24	141
Heracleum candicans	R	15	132
Heracelum concanense	F	3, 16	268
Heracleum dissectum	R	1	269
	R	3, 21, 22	269, 270
	R	5, 16	270
Heracleum giganteum	F	3	271
Heracleum lanatum var. *asiaticum*	R[1]	3, 5, 13, 16, 21	272, 273

(*Table continued*)

TABLE 2 (*continued*)

Natural Source	Distribution Part[b]	Constituents[a]	References
Heracleum lanatum var. *nipponicum*	R	3, 5, 16, 21, 22	273, 274
Heracleum naplense	S	6, 20	220
Heracleum pances	R	3, 5, 16, 21	275, 276
	R	22	275
Heracleum sibricum	R	3, 5, 16, 21, 22	276–278
Heracleum sphondylium L.	S	3	279
	R	5, 16, 21, 22	59, 280, 281
Imperatoria dargest	—	8, 13	126
Imperatoria ostruthium	R	3	220
(*Peucedanum ostruthium*)	R	10	282
	R	12	125, 283
	R	13	118
Leptotaenia multifida Nuttall	R	35	284
Levesticum officinale Koch	R	3	285
Ligusticum acutilobum Sieb. et Zucc.	F	3	286, 287
Lonatium Columbianum	R	34, 35, 36	161
Luvanga scandens Ham	F, S	7, 16	288, 289
Parsnips	S	3, 7, 8, 13, 16	290
Pastinaca sativa	S	3, 13	291, 292
	S	7	292–294
	S	8, 16	291
Petroselium sativum (celery)	S	3	249
Peucedanum decursivum Maxim.	R	25, 26	295
Peucedanum japonicum	—	3	296
Peucedanum morisini	R	4, 9, 11	297
Peucedanum officinale	—	29	224, 283, 298
	—	10	217
Peucedanum palustre	R, F	9, 10 (hydrate), 12, 34, 40, 41, 42	123
Phebalium argensteum Smith	—	2	299
Phebalium nudum	B	18	134
Phellopterus lettoralis Benth	F	3, 13	300
	F	17	6
Pimpinella magna	R	3, 5, 16	7, 281
	R	22	281
Pimpinella major	R	5, 16, 21, 22	22
Pimpinella saxifraga	R	3	7
		5, 21	22, 281, 301
		16	22, 91, 281
		22	22, 281, 302
Prangos pabularia	R	13	131, 303
	R	10, 11	124, 304

(*Table continued*)

TABLE 2 (*continued*)

Natural Source	Distribution Part[b]	Constituents[a]	References
Psoralea corylifolia L.	S	1	52, 305, 306
	S	2	196, 305, 306
Ruta chalepensis	—	7	307
Ruta graveolens	F, L	3	307–310
Ruta montana L.	—	7	311, 312
Sarcobatus vermiculatus	—	16, 20	313
Seseli compestre L.	—	13	314
Seseli indicum	S	3, 16	315
Skimmia laureola	B	16	316
Thamnosa montana	—	16, 20	313
Velleia Discophora F. Muell.	L	32	159

[a] Numbers indicate name (synonym) of furocoumarins (cf. Table 1).
[b] F (Fruit), B (Bark), L (Leaves), R (Root), R[1] (Rhizome), S (Seeds).

most furocoumarins, in order to separate them from other alkali–insoluble constituents in the extract. Plant acids are easily separated from furocoumarins by sodium bicarbonate or sodium carbonate solutions. Furocoumarins present in the extracts should be identified before chemical treatment or sublimation[7] because of the possibility of structural modifications, a consequence of hydrolytic cleavages and thermal lability, resulting in artifacts.

Purification of crude furocoumarin fractions may be carried out by fractional crystallization or by separations employing column chromatography[8, 9, 10] or preparative thin–layer chromatographic procedures.[11, 12, 13] Examples of the use of chromatographic methods[7, 14–20] are the isolation of lemon oil constituents by Stanley and coworkers[21] and the identification of the components of Norwegian Umbellifers by Svendsen.[22] Useful devices in these cases have been the chromatostrip[23] for following the elution pattern from a column, and the paper–electrophoretic separation making use of the electrophoretic mobilities of the furocoumarin anions.[24] Gas–liquid chromatography has been successfully used in the separation of neutral furocoumarins from *Angelica archangelica* and *Heracleum sibricum*, on the other hand, isoprenoid ethers of furocoumarins decomposed at the column temperatures.[25]

II. Physical Properties

An interesting property of furocoumarins and coumarins, is their ability to fluoresce under ultraviolet light. Fluorescence of furocoumarins is not a unique feature of this group, but is, nevertheless, a useful diagnostic.[12, 26] The studies of Goodwin and Kavanagh[27, 28] on fluorescence as a function of pH are interesting in this respect, as are those of Feigl and coworkers[29] on the use of fluorescence as a sensitive test for coumarins. A greenish fluorescence is generally noted with many of the furocoumarins in a neutral state.

The absorption bands of furocoumarins in the ultraviolet region have been studied in the hope that these spectra might have structural significance. The ultraviolet absorption pattern of the coumarin nucleus can be attributed largely to the benz-α-pyrone structure, and Nakabayashi and coworkers[30] assume that the two absorption bands of the coumarin at 270, and 312 mμ are those of benzene at 200, and 240–260 mμ that have shifted to these regions. The α-pyrone moiety absorbs at about 300 mμ (log ϵ, ca. 4), and variations from this are attributed to the pronounced effect of substituents and their location. Furocoumarins showed a more intense band at the lower wavelength as compared to coumarins.[31] The study of the ultraviolet absorption spectra of psoralens substituted at various positions,[32] initiated by the demonstrated correlation between light absorption and photosensitizing activity of psoralen and derivatives,[33] has illustrated similar effect with the hydroxy and methoxy derivatives with a bathochromic effect, but at higher wavelength the curves resemble those of the corresponding coumarin derivatives.[34–36] Substitution of hydrogen by an alkyl group in the pyran ring (viz. at the 5-, and 6-positions) in psoralen, does not produce any significant change (cf. Table 1), on the other hand, a bathochromic shift as well as increase in log ϵ value is observed at a lower wavelength when the 5-position of psoralen is substituted by a phenyl group.

The furan ring has a hypsochromic effect on the umbelliferone system. The spectra of 2,3-dihydropsoralens have a band at 226 mμ (log ϵ, 4.08–4.20) due to the formation of a saturated ether (cf. Table 1) which causes a bathochromic effect.[36, 37]

The absorption curves of furocoumarins in acid solution resembled those in alcohol, indicating stability of the lactone ring. On the other hand, in alkaline solutions, the lactone ring was opened in several compounds and the original spectrum could not be restored. This change was hastened by light.[38]

Infrared analysis has been used in the structural characterization of furocoumarins. The particular value of such analysis, apart from detecting groups not fundamentally related to the coumarin nucleus, is its ability to assign a lactone function. Characteristic bands are found at 3175–3137 and 3137–3116 cm^{-1} (CH stretching vibrations of the furano nucleus), 1639–1616 cm^{-1} (C=C stretching mode of the furan ring), 1295–1266 cm^{-1} (C—O stretching), 880–864 cm^{-1} (CH out of plane bending), and 833–821 cm^{-1} (CH out of plane bending).[39] In carbon tetrachloride the carbonyl absorption is generally at 1742–1748 cm^{-1}, with some derivatives absorbing as high as 1744 cm^{-1}.[40]

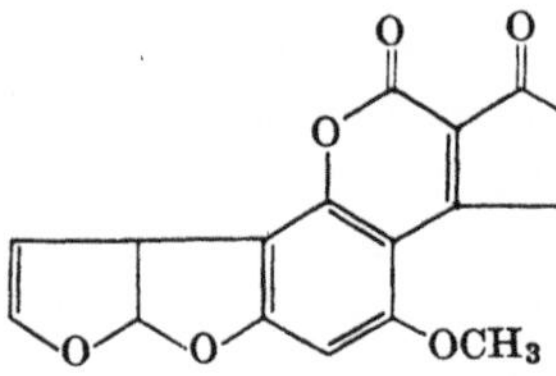

(i)

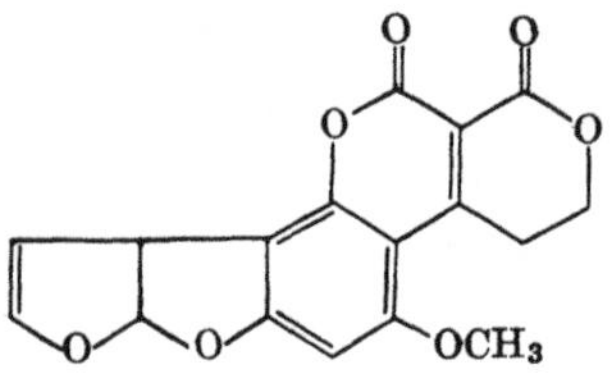

(ii)

High resolution nuclear magnetic resonance spectra have been of use in the structural definition of furocoumarins. In many cases, this analytical procedure has been useful in assigning a structure which it would have been difficult to determine chemically, for example in the case of aflatoxin B (**i**), and aflatoxin G (**ii**). N.m.r. spectra have been successfully used to assign the linear and angular structures of furocoumarins on the basis of their aromatic hydrogen absorptions. Kaufman and coworkers[41] have reported that in the angular isomer (**iii**) these two hydrogens are situated in positions 4 and 5 *ortho* to each other and hence should be split by spin coupling into an AB multiplet with a

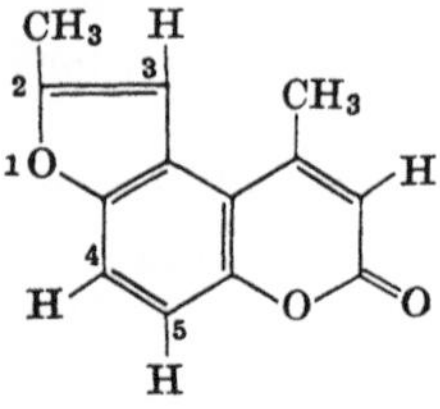

(iii)

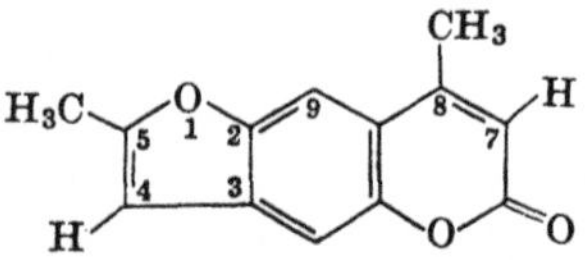

(iv)

coupling constant of 6–9 cps. In the linear isomer (**iv**) these two aromatic hydrogens are *para* to each other and should be split by less than 1 cps. if at all.[42] The 4-hydrogen of **iii** is favorably situated for maximum p–orbital overlap[43, 44] with the β-furan hydrogen and a coupling of 1 cps. across this 6-atom is found. Thus, the 4-hydrogen is actually a quadruplet. The 4-hydrogen of **iv** has a less favorable angle, but is one atom closer to the β-furan hydrogen; nevertheless, it was not clearly resolved, but was broadened.

The pyrone–ring hydrogen in **iii** and **iv** coupled with the pyrone–methyl hydrogens (J = 1.5 cps.) and the methyl absorptions were identified from their coupling constants. In the angular isomer, both methyl absorptions were about 8 cps. lower than those observed for the linear isomer.[42, 45]

Use of mass spectrometry in structural studies of furocoumarins has not been exploited to any degree. However, the technique has been used to determine the exact molecular weights of compounds which are scarce.[46]

Colorimetric determination of furocoumarins has been mainly based on their reaction with diazotized sulfanilic acid in alkaline medium,[47] and the deep-violet color developed with 8-amino-5-hydroxy-2-methyl-furo(4′,5′, 6,7)chromone in the presence of alkali.[48] Peucedanin has been determined recently by titrating its alcoholic solution in sodium hydroxide with sulfuric acid, using phenol red.[49]

III. Nomenclature

Several furocoumarin nomenclatures are currently in use; the literature abounds with trivial names used for brevity, but which can cause confusion.[50] Naturally occurring furocoumarins belong to two main categories: (1) psoralen type (linear type) (7*H*-furo(3,2-g)–1–benzopyran-7-one or δ-lactone of 6-hydroxy-5-benzofuranacrylic acid) (**A**); (2) isopsoralen type (angular type) (2*H*-furo(2,3-h)-1-benzopyran-2-one or δ-lactone of 4-hydroxy-5-benzofuranacrylic acid) (**B**). *The Ring Index* system (cf. **A** and **B**) is not in common use.

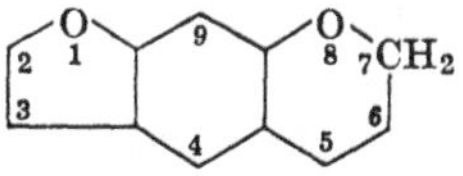

(**A**)

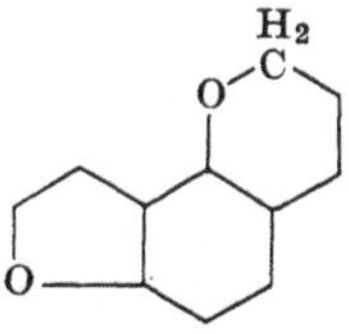

(**B**)

IV. Naturally Occurring Furocoumarins

1. Structure and Chemical Properties

A. Angelicin

Angelicin[51] (isopsoralen)[52](**1**), $C_{11}H_6O_3$, is one of the simplest naturally occurring furocoumarins, having no substituents. The presence of the furan ring was shown by oxidation with alkaline hydrogen peroxide to furan-2,3-dicarboxylic acid. Regulated oxidation with potassium permanganate led to umbelliferone-8-carboxylic acid, and then to 7-hydroxycoumarin (umbelliferone).[51] Therefore the rings are fused in an angular fashion. Hydrogenation and subsequent oxidation of 1 with nitric acid gave succinic acid from the α-pyrone ring, and alkaline dimethyl sulfate opened the coumarin ring, with formation of β-(4-methoxy-5-benzofurano)acrylic acid (2), which on oxidation with permanganate, gave the benzene-1,3-dicarboxylic acid (3). The latter was identified as the dianilide; further evidence for the angular structure of 1. The furan ring in 1 is readily opened by the action of benzene in presence of aluminium chloride to produce 8-(1,2-diphenylethyl)-umbelliferone (4).[53]

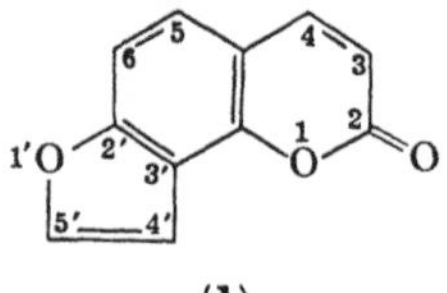

(1)

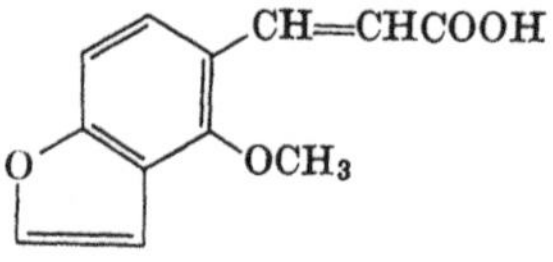

(2)

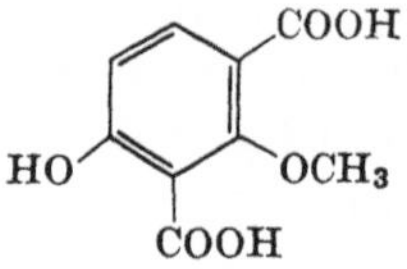

(3)

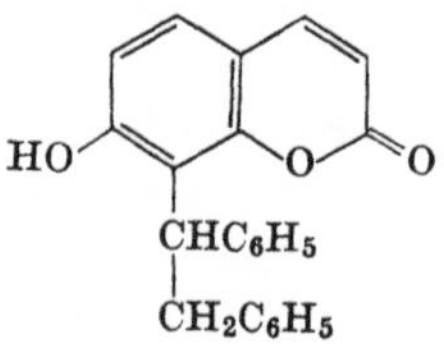

(4)

Angelicin has been synthesized by several methods. Treatment of the sodium salt of umbelliferone with the acetal of bromoacetaldehyde, followed by sublimation of the resultant ether (5) gave a low yield of 1.[54] This synthesis is ambiguous as ring closure may occur at position 6 or 8 of the coumarin residue; moreover, the acetal can act as an ethylating agent, effecting the formation of umbelliferone ethyl ether (Eq. 1).

$$+ \quad BrCH_2CH(OC_2H_5)_2 \quad \longrightarrow$$

(1)

$$\longrightarrow \quad 1$$

$(H_5C_2O)_2CHCH_2O$

(5)

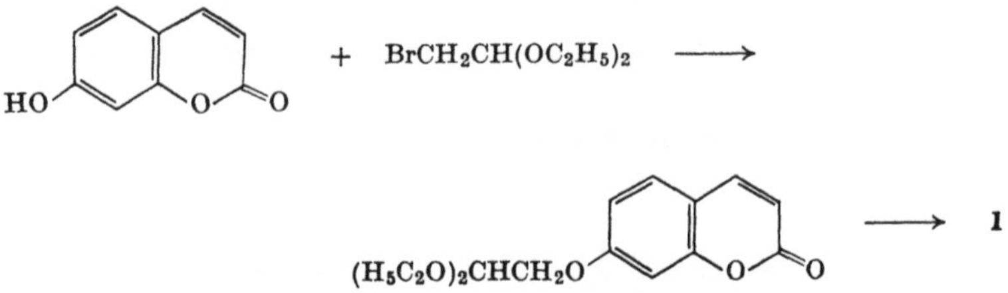

Seshadri and coworkers[55–57] used a method suggested by their hypothesis as to the biogenesis of benzofurans. 8–Allylumbelliferone (**6**) was converted by the action of ozone to 7-hydroxy-8-acetaldehyde (**7**), which when treated with phosphoric acid, cyclized to angelicin. A convenient modification is provided by the use of osmium trioxide and potassium periodate and ring closure of **7** by polyphosphoric acid, which also acts as a good solvent[58] (Eq. 2).

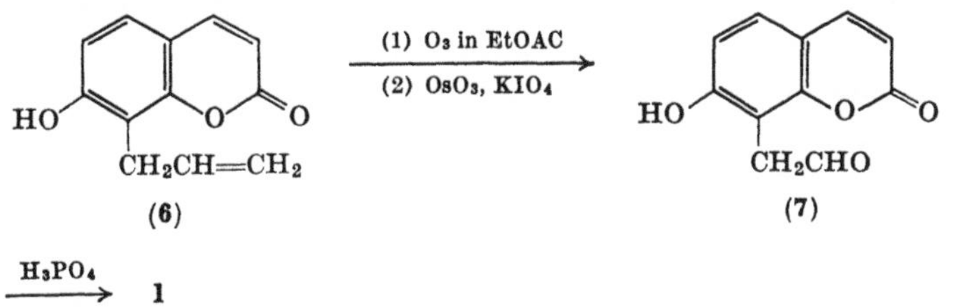

(2)

Treatment of 8-formylumbelliferone (**8**) with ethyl iodoacetate, followed by decarboxylative cyclization of the intermediate phenoxyacetic acid (**9**) yielded **1**.[59] Cyclization of **9** with a sodium acetate-acetic

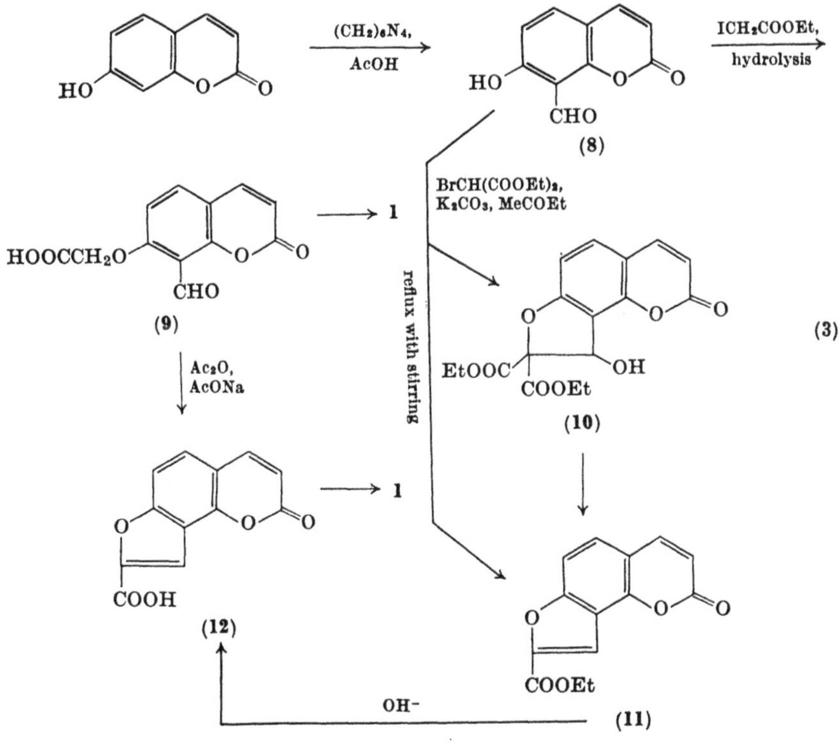

(3)

anhydride mixture effected the formation of 5′-carboxyangelicin (**12**), which underwent smooth decarboxylation to furnish **1**.[60] Using Tanaka's method[61] for benzofuran synthesis, Kawase and coworkers[62] have obtained angelicin. **8** was made to react with ethyl bromomalonate and potassium carbonate in ethylmethyl ketone and the hydroxy diester (**10**) so formed was converted by the action of phosphoric acid to the ester (**11**), which furnished the acid (**12**) (Eq. 3).

Transformation of the naturally occurring furoflavone, karanjin (**13**), to angelicin (**1**) has been achieved[63] by formylation of its degradation product, karanjol (**14**), to **15**, followed by cyclization by Perkin's procedure (Eq. 4).

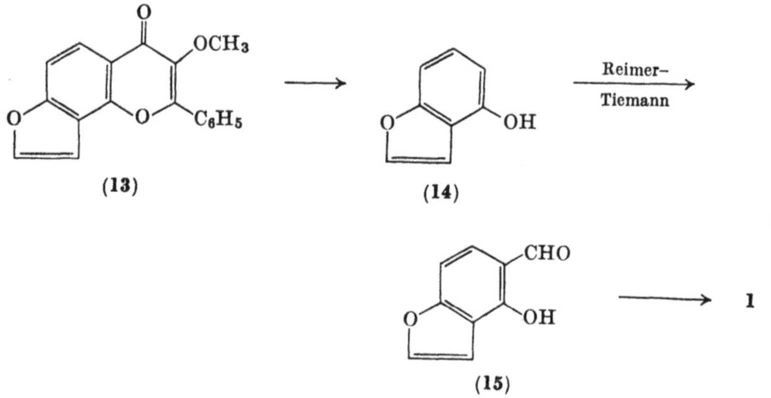

(**13**) (**14**) (**15**)

(4)

B. Psoralen

Psoralen (**16**), $C_{11}H_6O_3$, is identical with ficusin.[64, 65] It was allocated the linear type of structure[65, 66] on the grounds that it showed the lactone properties of a coumarin, could be oxidized to furan-2,3-dicarboxylic acid, and was degraded by permanganate to 4,6-dihydroxyisophthalic acid.[64] The furan ring is readily opened by the action of benzene in presence of aluminum chloride to produce 6-(1,2-diphenylethyl)-umbelliferone (**17**).[53]

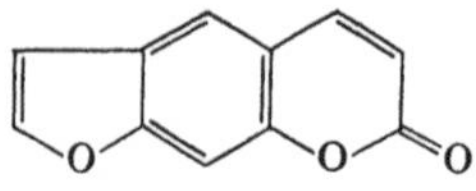

(**16**)

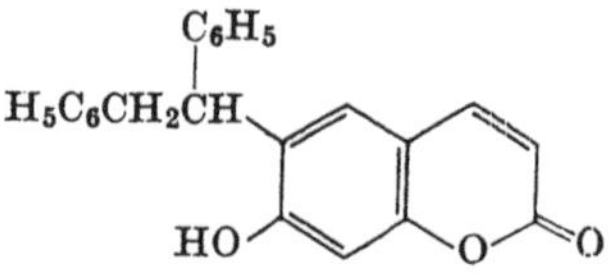

(**17**)

Structure **16** will be designated, and numbered throughout as shown, in agreement with the recommendation of the Food and Drug Administration.[50, 67]

The synthesis of the linear–fused system is not easy, as benzofurans are most readily substituted at the 2-position and umbelliferones at the 8-position, but if 6-hydroxycoumaran (**18**) is subjected to a Perkin condensation, followed by dehydrogenation over palladium, psoralen (**16**) is obtained[54, 66] (Eq. 5).

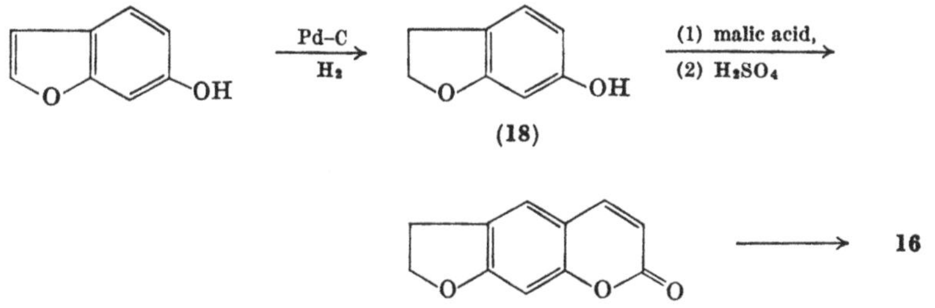

(5)

Horning and Reisner[68] made this approach more attractive by devising an improved synthesis of the key intermediate **18**. Furthermore, these workers found that 6-acetoxycoumaran, an intermediate for the preparation of **18**, could be used directly for the synthesis of 5-substituted 4′,5′-dihydropsoralens by condensation with a variety of β-keto esters.

Robertson and coworkers[69] devised an alternative route for the synthesis of psoralen. A series of 3-alkyl-4′,5′-dihydro-4-methylpsoralens were similarly prepared by condensing α-alkyl-β-keto esters with 6-acetoxycoumaran.[70] **18** was formylated in the 5-position, followed by condensation of the resultant compound **19** with cyanoacetic acid, to yield 3-carboxy-4′,5′-dihydropsoralen (**20**), which was then decarboxylated and dehydrogenated to **16** itself (Eq. 6).

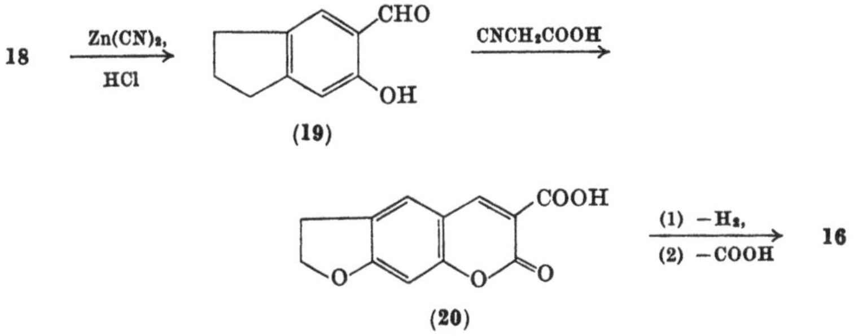

(6)

The problem of furocoumarin synthesis from the coumarin moiety of the psoralen molecule has been approached by Ray and coworkers,[71] who report the synthesis of 4′-methylpsoralen (**22a**). In this procedure, 7-acetonyloxycoumarin (**21**, $R = CH_3$) was prepared by treating umbelliferone with chloroacetone. Cyclization of **21** to **22** was accomplished by ethanolic sodium ethoxide. In this respect the ring closure was markedly different from the usual pattern. Treatment of 4-methylumbelliferone with ω-bromoacetophenone led to similar formation of 4-methyl-4′-phenylpsoralen (**22b**)[72] (Eq. 7).

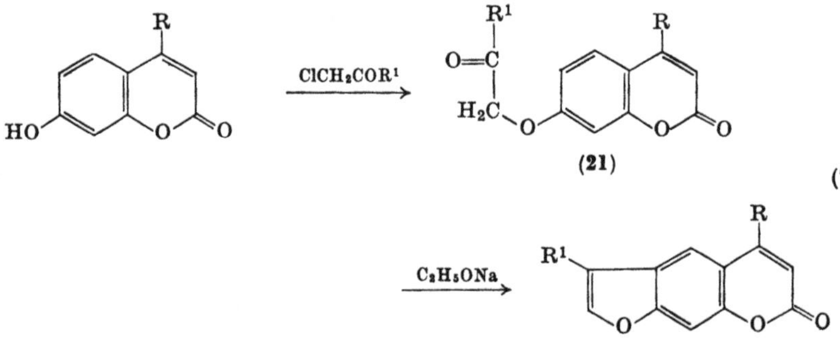

(**21**)

(7)

(**22a**) $R = H, R^1 = CH_3$;
(**22b**) $R = CH_3, R^1 = C_6H_5$

Seshadri and coworkers,[55] in addition to their synthesis of angelicin (**1**), have successfully obtained psoralen (**16**) by ozonolysis of 6-dimethylallyl-7-hydroxycoumarin (7-demethylsuberoin) (**23**), followed by cyclization of the aldehyde (**24**) formed with phosphoric acid (Eq. 8).

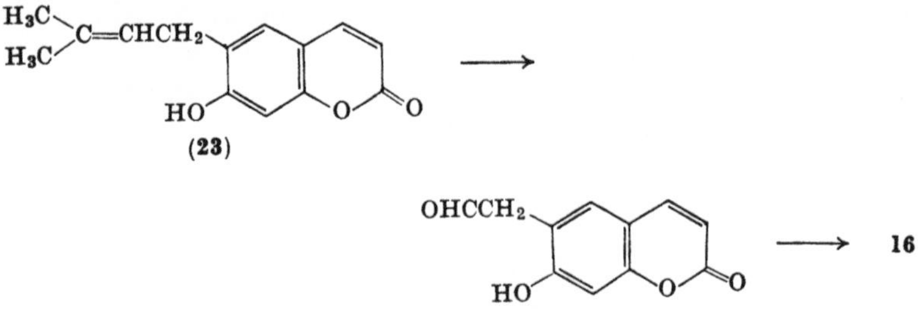

(**23**)

(8)

(**24**)

The value of *O*-allylhydroxycoumarins in the synthesis of furocoumarins has been stressed recently by Kaufman.[67] Thus, Claisen rearrangement of 7-allyloxycoumarins, having the reactive 8-position blocked by a methyl group or acetamido group, resulted in the expected 8-substituted 7-hydroxy-6-allylcoumarins, which were readily converted by the conventional procedure to 8-substituted 5′-methyl-, and 4,5′-

dimethylpsoralens.[73–75] 8-Amino-4,5′-dimethylpsoralen is a valuable starting material for the preparation of several derivatives of 8-substituted 4,5′-dimethylpsoralen, usually by diazotization; for example reduction of the diazonium salt with hypochlorous acid gave 4,5′-dimethylpsoralen. This is an example of the use of the amino group as a removable blocking group in the synthesis of psoralens from an umbelliferone.[75] It is worth noting that substitution of a hydroxyl group for the amino group of an aminomethoxypsoralen has been effected by diazotization in concentrated hydrochloric acid and ethanol.[37]

C. Bergapten

Bergapten (heraclin, majudin, bergamot camphor, 4-methoxy-7*H*-furo(3,2-g)-1-benzopyran-7-one or δ-lactone of 6-hydroxy-4-methoxy-5-benzofuranacrylic acid), (**25**), $C_{12}H_8O_4$, was orginially studied by Pomeranze,[76] whose work indicated that it was a furocoumarin derived from phloroglucinol. Thoms and Baetcke[77] showed it to have a linear type of molecule by nitration to nitrobergapten (**26**), followed by reduction to the amino derivative (**27**), and oxidation of the latter to 1,4-quinone, bergaptenquinone (**28**) (Eq. 9).

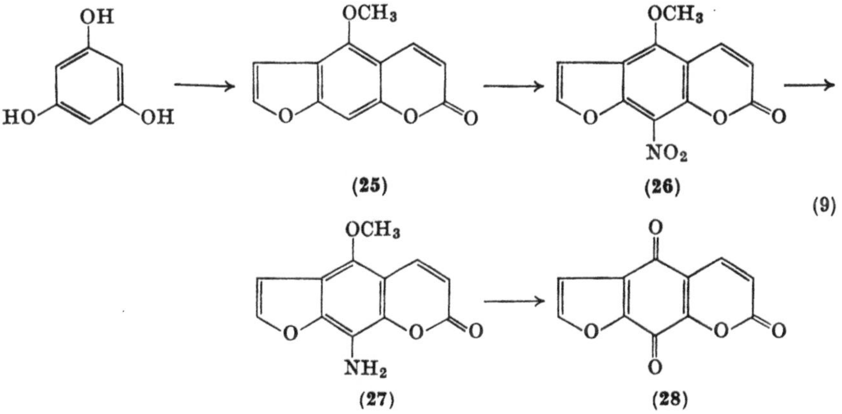

(25) (26) (27) (28) (9)

The synthesis of bergapten is beset by the same difficulties as that of psoralen. One method begins with the readily accessible 4,6-dihydroxycoumaran-3-one (**29**), which enolizes and so can be triacetylated. Hydrogenation of the triacetate can be regulated to give the corresponding courmaran (**30**). When this is heated with the sodio derivative of ethyl formylacetate in a sealed tube with the object of forming the pyrone ring, the coumaran ring splits off acetic acid at the same time, so that only a little bergaptol (**31**) results. The chief product is the angular isomer, allobergaptol (**32**), since the courmaran (**30**) forms the benzofuran so easily that it is this, which is mainly available for the condensation.[78] Furocoumarins of the allobergaptol type **32** are not known to occur in nature (Eq. 10).

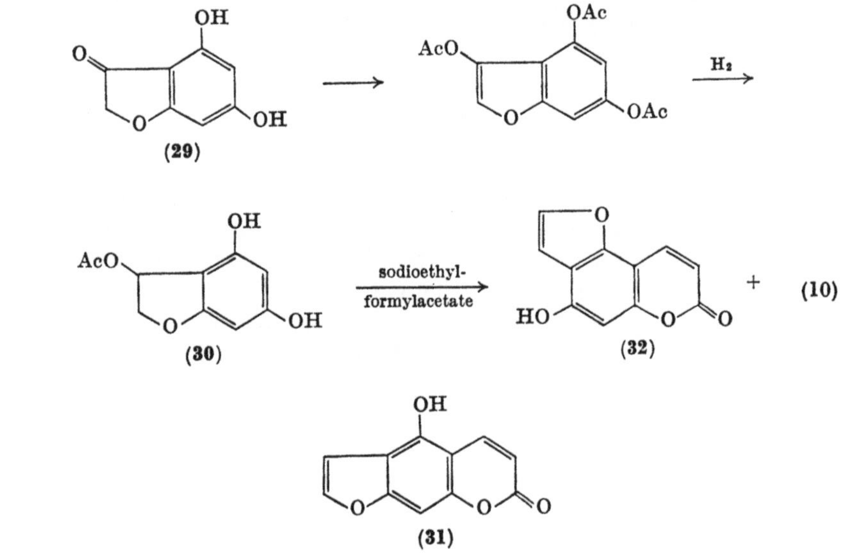

Howell and Robertson,[79] having obtained apoxanthoxyletin (6-formyl-7-hydroxy-5-methoxycoumarin) (**34**) by degradation of xanthoxyletin (**33**), built the furan ring as in the case of angelicin (**1**) (Eq. 11).

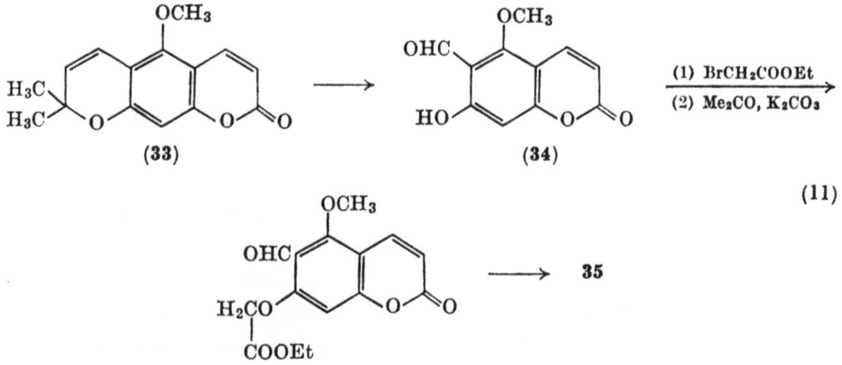

Caporale[80, 81] described a new synthesis of bergapten which solved the various problems (Eq. 12).

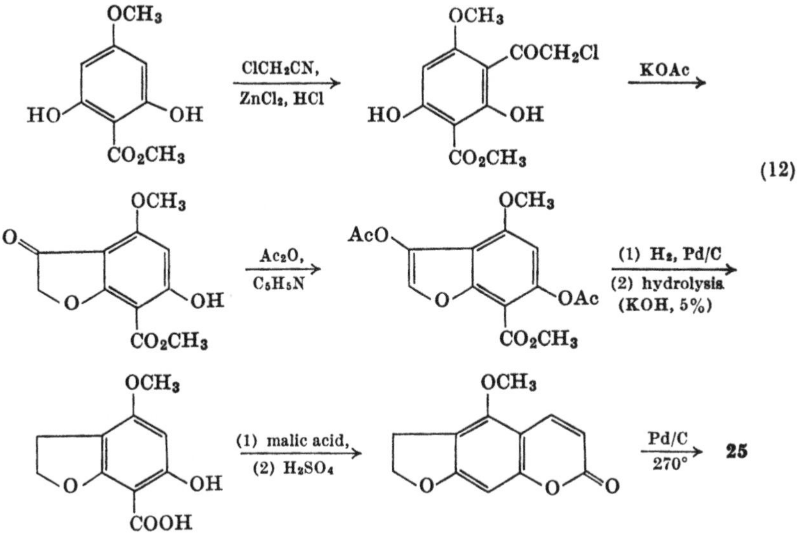

Bergapten is oxidized by chromic acid with the destruction of the furan ring to give the aldehydocoumarin, 6-formyl-7-hydroxy-5-methoxycoumarin, which is identical with **34**, the ozonolysis product of the natural coumarin, xanthoxyletin (**33**).[82] Replacement of the formyl group by a hydroxyl group by oxidation of **34** with hydrogen peroxide in sulfuric acid medium gave 6-hydroxy-5,7-dimethoxycoumarin (fraxinol) (**35**).[83]

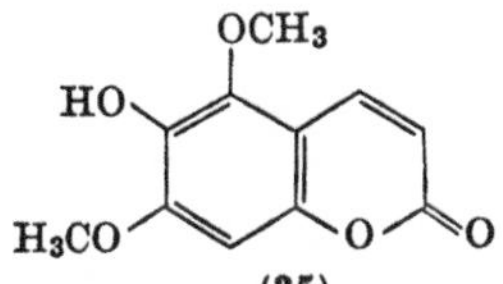

(**35**)

Bergapten readily underwent demethylation to bergaptol (**31**) by the action of magnesium iodide in absence of solvent, followed by decomposition with dilute sulfuric acid.[84]

Condensation of 4-hydroxybergapten (**37**), obtained by reaction of visnaginone (**36**) with ethyl carbonate, with formaldehyde gave the dicoumarol of bergapten (**38**)[48] (Eq. 13).

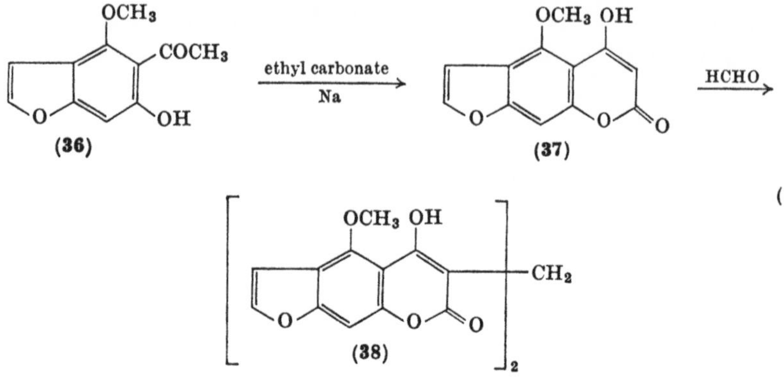

(13)

D. Bergaptol

Bergaptol (5-hydroxypsoralen, 4-hydroxy-7*H*-furo(3,2-g)-1-benzopyran-7-one) (**31**), $C_{11}H_6O_4$, belongs to the linear furocoumarin type. The determination of its structure is due to Pomeranze,[76] Thoms,[77] and Späth.[85] By methylation with diazomethane the methyl ether, bergapten, is formed. A number of other bergaptol ethers have been obtained by the action of diazoalkanes, and/or alkyl halides.[86] The presence of the furan ring in bergaptol (**31**) and also in allobergaptol (**32**), an alternative angular formulation, is proved by oxidation to furan-2,3-dicarboxylic acid.

The synthesis of **31** according to Späth's[87] method leads to a mixture of **31** and **32** and is ambiguous. Robertson and coworkers[88] have described an unequivocal synthesis of allobergapten (**39**), the methyl ether of **32**. The phenoxy ester (**40**) was cyclized with sodium ethoxide to yield ethyl 6-benzyloxy-4-methoxycoumarone-2-carboxylate (**41**, $R = CH_2C_6H_5$) which was debenzylated with hydrogen and palladium catalyst to ethyl 6-hydroxy-4-methoxycoumarone-2-carboxylate (**41**, R = H). Application of the Gattermann aldehyde synthesis of the latter substance gave the aldehyde **42**, which when decarboxylated gave 7-formyl-6-hydroxy-4-methoxycoumarin (**43**). Treatment of **43** with cyanoacetic acid gave allobergapten (**39**).

E. Isobergapten

Isobergapten (5-methoxy-2*H*-furo(2,3-h)-1-benzopyran-2-one or δ-lactone of 4-hydroxy-6-methoxy-5-benzofuranacrylic acid) (**44**), $C_{12}H_8O_4$, a natural furocoumarin,[89, 90] on reduction with sodium amal-

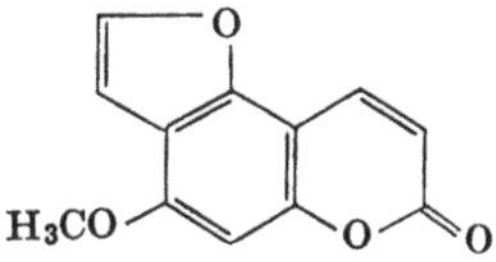

(39)

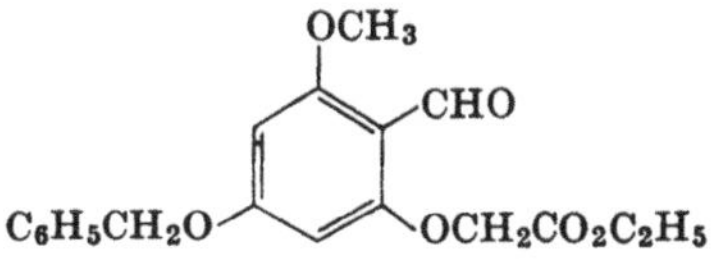

(40)

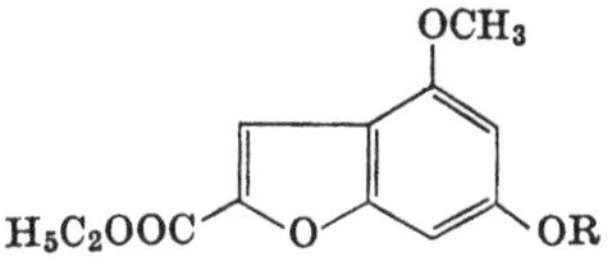

(41)

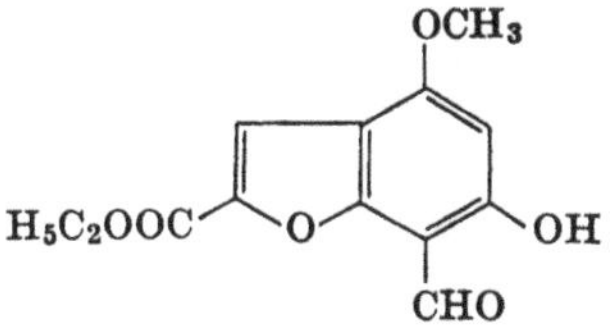

(42)

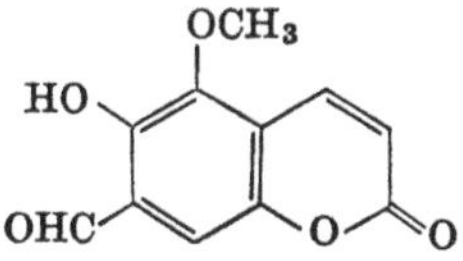

(43)

gam gave a dihydrocinnamic acid derivative, which after methylation, produced the same dimethoxy ester (**45**).[91] Therefore there is only one

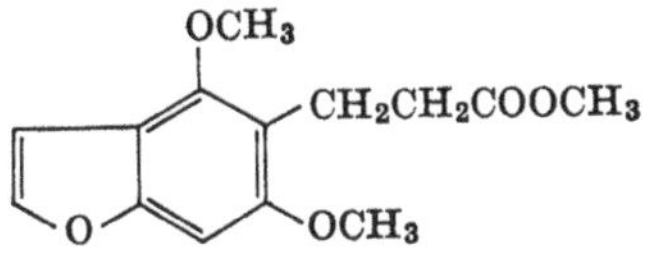

(45)

(**45**) structural possibility for isobergapten. The synthesis of **44** from phloroglucinol has been achieved (Eq. 14).[92]

Isobergapten was partially synthesized by methylation of bergapten (**25**) with methyl sulfate in alkali, owing to ring opening and relactonization at the alternative site.[85, 93]

F. Bergamottin

Bergamottin (bergaptin, 4-(geranyloxy)-7*H*-furo(3,2-g)-1-benzopyran-7-one or δ-lactone of 4-((3,7-dimethyl-2,6-octadienyl)-oxy)-6-hydroxy-5-benzofuranacrylic acid), $C_{21}H_{22}O_4$, was given the name bergamottin by Soden and Rojahn.[94] When warmed with acids, bergamottin (**46**) yields bergaptol (**31**) and geraniol.[95] It is not phenolic

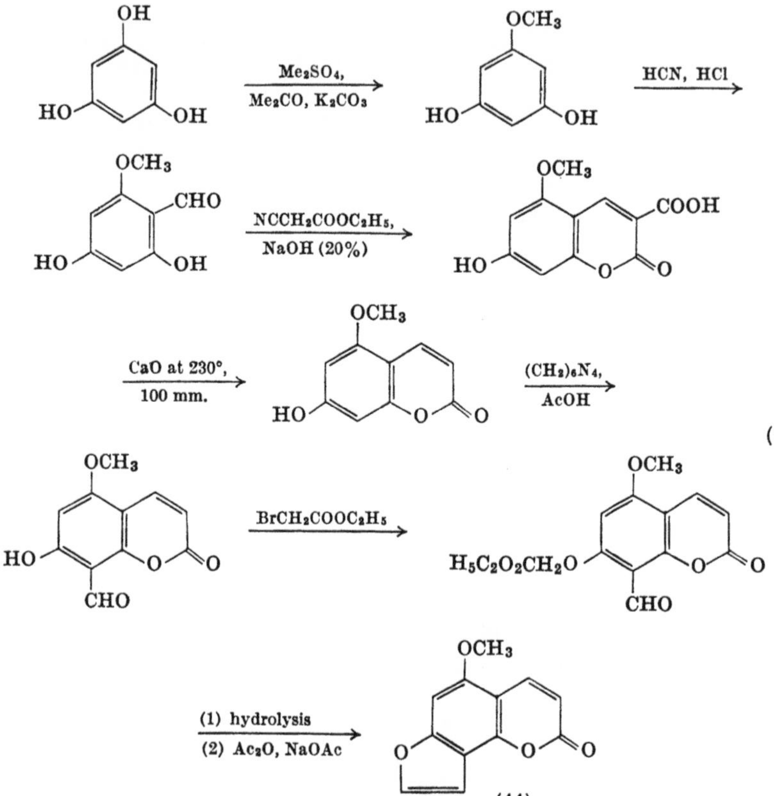

(14)

(44)

therefore it is the geranyl ether of bergaptol (**31**) as confirmed by its synthesis by the reaction of geranyl chloride with bergaptol.[96]

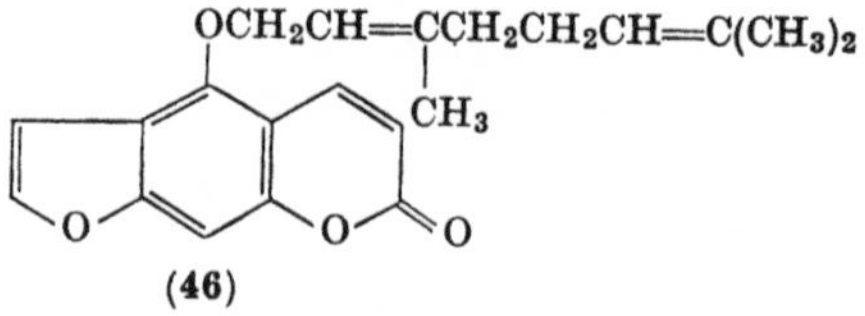

(**46**)

G. Xanthotoxin

Xanthotoxin (ammoidin, 8-methoxypsoralen, 9-methoxy-7*H*-furo(3,2-g)–1–benzopyran-7-one or δ-lactone of 6-hydroxy-7-methoxy 5-benzofuranacrylic acid) (**47**), $C_{12}H_8O_4$, had its structure determined mainly by Thoms.[97] The presence of the furan and the coumarin rings was recognized as for the isomeric bergapten (**25**). The α-pyrone ring is opened by prolonged boiling with sodium ethoxide, followed by acidification, to yield *trans* 6-hydroxy-7-methoxy-5-coumaranoyl-β-acrylic acid (**48**).[98] The latter compound was also obtained by treatment of **47** with sodium sulfite.[99]

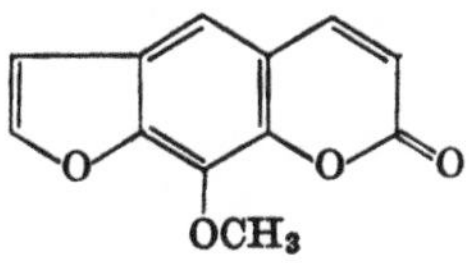

(**47**)

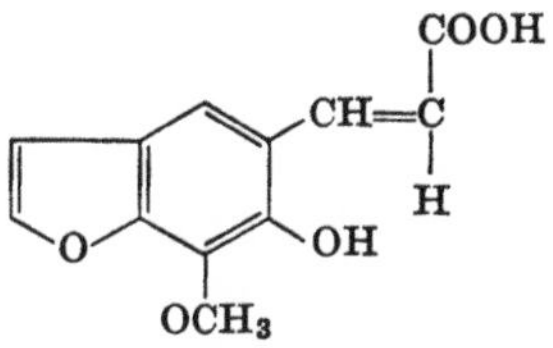

(**48**)

Pyrogallol-4-carboxylic acid and pyrogallol were obtained by caustic fusion of **47**. Nitration of the latter compound gave a mononitro derivative (**49**), isomeric with, but not identical with, the mononitro derivative of bergapten (**26**). The corresponding amino derivative (**50**) was readily converted by chromic oxide oxidation into bergaptenquinone (**28**). The same reuslt can be achieved by oxidizing xanthotoxin with chromic oxide in acetic acid. This is interesting, as the use of sodium

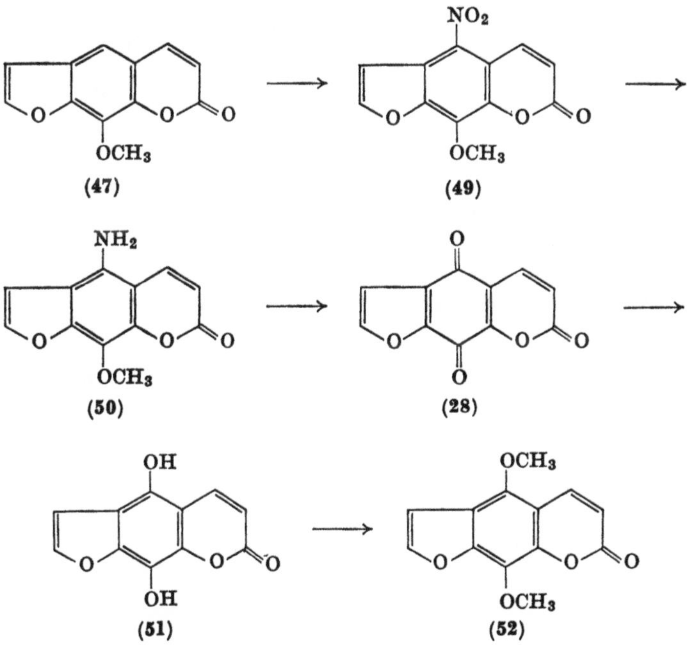

(**47**) (**49**) (**50**) (**28**) (**51**) (**52**)

(15)

chromate breaks the furan ring instead of attacking the benzene ring.[100] The structure of bergaptenquinone (psoralenquinone) was further confirmed by reduction to the hydroquinone (**51**), and subsequent methylation to isopimpinellin (**52**)[91, 100] (Eq. 15).

Späth and Pailer[101] accomplished the synthesis of xanthotoxin by reaction of 6,7-dihydroxycoumaran and malic acid, followed by methylation and dehydrogenation of the resultant dihydroxanthotoxol (**53**) on a palladium catalyst. The method is far from efficient,[102] but dehydrogenation of **53** gave xanthotoxol (**54**), a naturally occurring furocoumarin. Treatment of 6,7-dihydroxycoumaran-3-one with β-keto esters, similarly, led to the formation of a number of 4-substituted analogs of **52**, and subsequently to **55**[102] (Eq. 16).

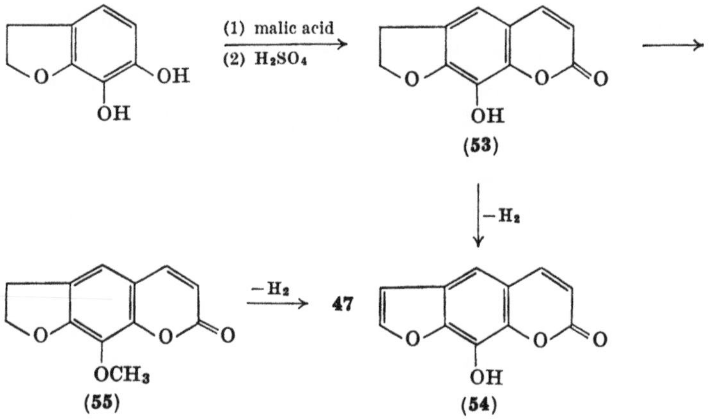

(16)

Another synthesis of xanthotoxin, using a coumarin as the benzofuran derivative, has been reported by Rodighiero and Antonello.[103] The synthesis has been successfully used for the preparation of a number of methyl–substituted xanthotoxins[104] (Eq. 17).

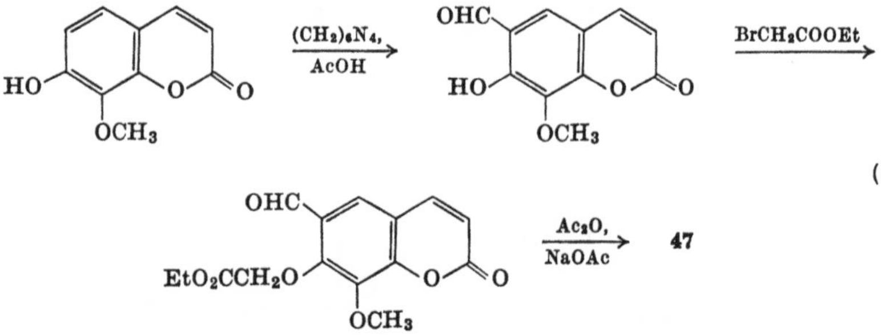

(17)

Xanthotoxin undergoes demethylation with magnesium iodide and sulfuric acid to the corresponding phenol, xanthotoxol (**54**).[98a] Aniline hydrochloride has been reported[98b] to be a superior demethylating agent but the correctness of this statement has been questioned by Brokke and Christensen.[37] Using Merchant and Shah's[105] procedure **54** is obtained in up to 40–45% yield.[100] Demethylation with aluminum chloride in benzene resulted in the opening of the furan ring in **47** and the formation of 6-(1,2-diphenylethyl)-7,8-dihydroxycoumarin (**56**).[37]

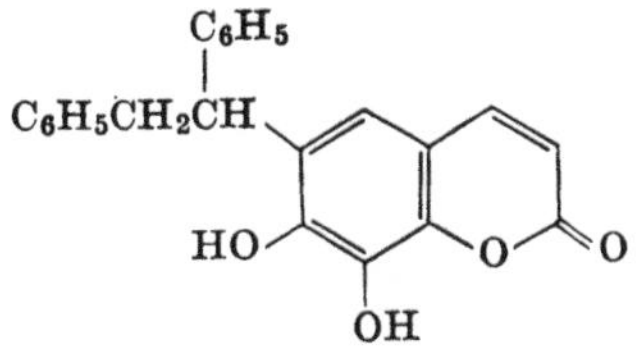
(**56**)

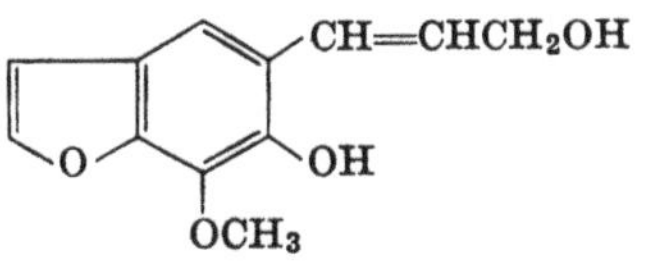
(**57**)

Palladium catalyzed hydrogenation of xanthotoxin affected the furan ring with the formation of **55**;[37] on the other hand, under normal conditions, lithium aluminum hydride cleaved the lactone ring to yield 6-hydroxy-7-methoxy-5-(3-hydroxy-1-propenyl)benzofuran **57**.[37] Nitration of **54** proceded similarly to **47**, yielding a mononitro derivative which resisted dehydrogenation.

Direct chlorination of xanthotoxin with chlorine gave only a trichloro derivative (**59a**), which can also be obtained by direct chlorination of 5-chloroxanthotoxin (**58a**). The latter compound has been produced either by chlorination of **47** with sodium hypochlorite, or by Sandmeyer reaction with **50**.[100]

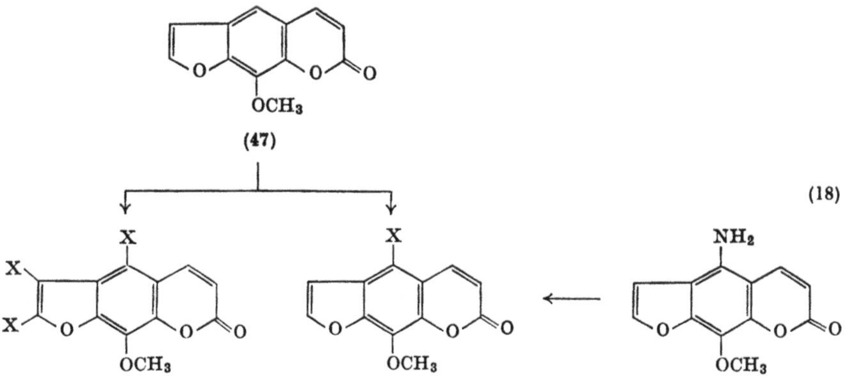
(**47**) (18)
(**59a**) X=Cl; (**59b**) X=Br (**58a**) X=Cl; (**58b**) X=Br (**50**)

Xanthotoxin, on direct bromination in chloroform, gave a mono-, and tribromo derivative (**59b**) which was readily converted to **58b** by conventional methods. Moreover, **58b** has been obtained by bromination of **47** with *N*-bromosuccinimide,[37] and by Sandmeyer reaction with **50**.[37] **58b** with hydrogen peroxide gave furan-2,3-dicarboxylic acid, and on ozonization, both hetero rings were cleaved producing isophthaldehyde (**60**), but, only one ring in xanthotoxin (**47**) was cleaved by direct ozonolysis, yielding 6-formyl-7-hydroxy-8-methoxycoumarin (**61**)[103] (Eq. 18).

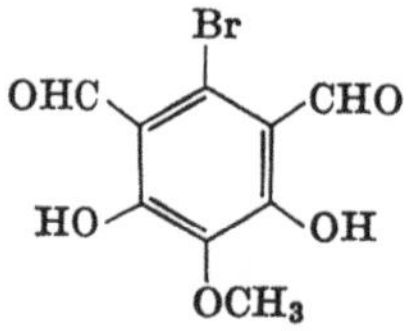

(60)

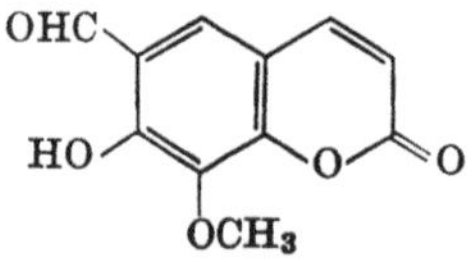

(61)

Dihydroxanthotoxin (**55**) was, likewise, observed to form a monobrominated product (**62b**), with bromine in chloroform solution, which is not identical with 5-bromo-4′,5′-dihydroxanthotoxin (**62a**).[37] Structure **62b** was confirmed as 3-bromo-4′,5′-dihydroxanthotoxin, which underwent ring contraction with sodium hydroxide to produce 6 - carboxy - 2,3 - dihydro - 8 - methoxybenzo(1,2 - b : 5,4 - b′)difuran (**63a**).[106, 107] Treatment of **54** with two equivalents of bromine gave the dibromo derivative (**62c**), which was also obtained by direct bromination of **62a**, and gave **63b** upon ring contraction with sodium hydroxide.

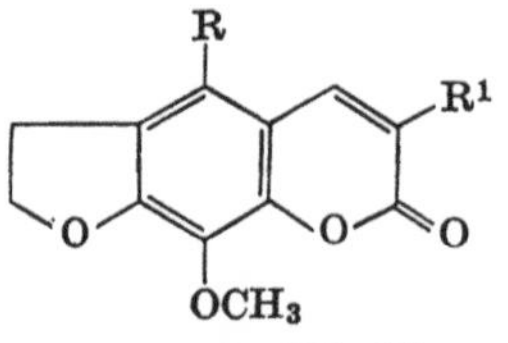

(62a) R = Br, R^1 = H;
(62b) R = H, R^1 = Br;
(62c) R = R^1 = Br;
(62d) R = SO_3H, R^1 = H;
(62e) R = SO_2Cl, R^1 = H

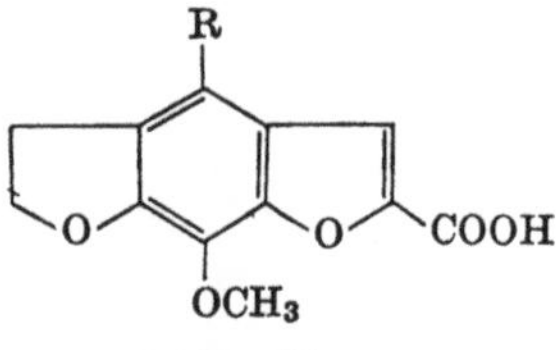

(63a) R = H;
(63b) R = Br

Sulfonation of xanthotoxin with chlorosulfonic acid produced either the free sulfonic acid (**62d**) or the acid chloride (**62e**). It was established that sulfonation occurred in the 5-position, by bromination, followed by nitration of **62d** to **62a** and **49**, respectively.[100]

Aromatic thiols add to the unsaturated system in xanthotoxin, to yield the adducts **64**.[108] The lactone ring is opened upon treatment of xanthotoxin with hydrazine[109] with the formation of the hydrazide (**65**).[37] Treatment of xanthotoxin, and its 5-substituted derivatives with phosphorus pentasulfide has been reported to give thioxanthotoxins (**66**).[37]

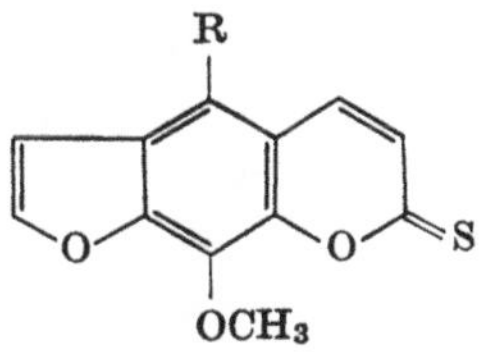

(66)

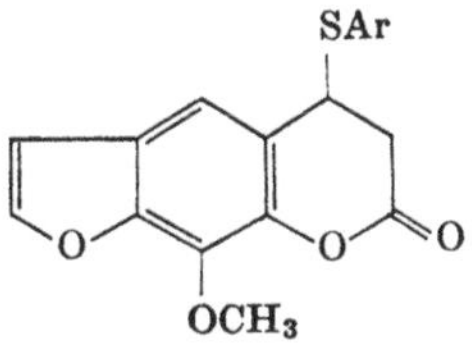

(64)

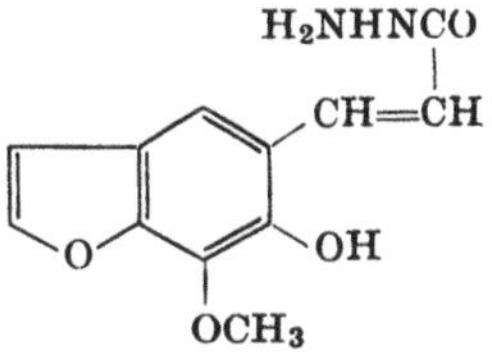

(65)

The study of the ultraviolet–induced photodimerization of the previously mentioned furocoumarins in relation to their photosensitizing properties on the skin has recently received much attention.[110] Photodimers of angelicin (**1**), psoralen (**16**), isobergapten (**44**),[18] and bergapten (**25**)[110] have the cyclobutane structure (cf. **67**)[111, 112] and depolymerize thermally to their monomers. The dimerization rate was psoralen > angelicin > bergapten. This order of reactivity differs from that of the photosensitizing properties since angelicin and psoralen dimers are ineffective on the skin. There seems to be no correlation between photochemical activity and dimerization.[110] In this connection, pimpinellin, a natural furocoumarin, structurally related to isobergapten, was

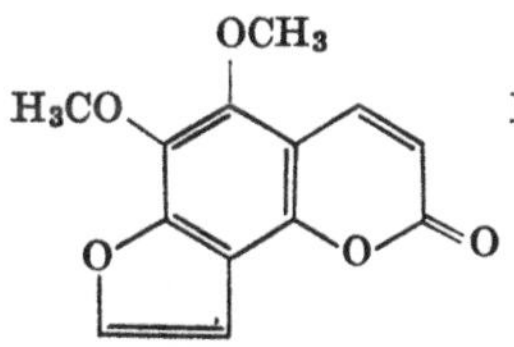

(Pimpinellin)

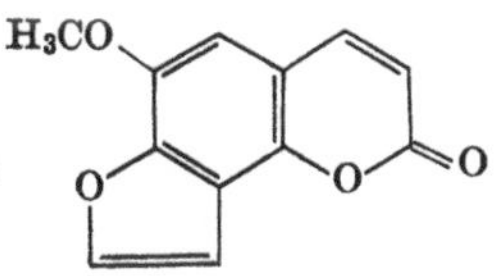

(Sphondin)

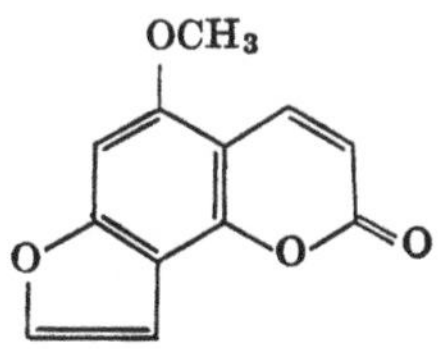

(Isobergapten)

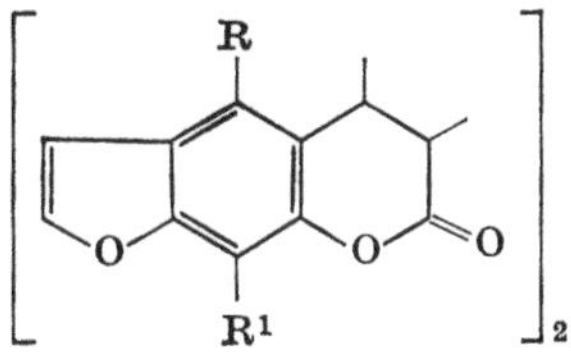

(67)

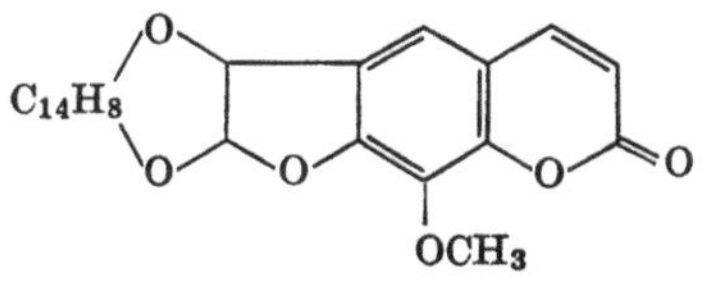

(68)

reported to yield a photodimer[113] in contrast to sphondin, which proved not to dimerize.[18] Xanthotoxin, which does not undergo a photodimerization reaction,[110] has been reported to react with phenanthraquinone, in sunlight, to yield the photoadduct (**68**).[114]

H. Xanthotoxol

Xanthotoxol (9-hydroxy-7*H*-furo(3,2-g)–1–benzopyran-7-one, or δ-lactone of 6,7-dihydroxy-5-benzofuranacrylic acid) (**54**), $C_{11}H_6O_4$, was discovered by Späth and Vierhapper.[115] On methylation it gave xanthotoxin (**47**). It was obtained by dehydrogenation of the synthetic 4′,5′-dihydroxanthotoxol (**53**).[101] Benzyl chloride reacts with **54** to give 8-benzyloxypsoralen[72] and similarly, 8–alkoxypsoralens are obtained by the action of diazoalkanes on **54**.[116] Phosphorylation of **54** with dialkyl phosphorochloridates gave the organophosphorus compound (**69a**),[117] which regenerates **54** on acid hydrolysis. The geranyl ether of **54**, 8-geranyloxypsoralen (**69b**), is a natural furocoumarin.[21]

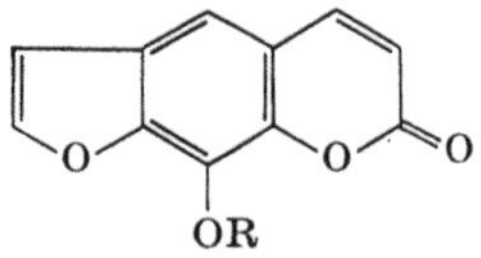

(**69a**) R = $PO(OR^1)_2$;
(**69b**) R = geranyl

I. Isoimperatorin

Isoimperatorin (4-(3-methyl-2-butenyloxy)-7*H*-furo(3,2-g)–1–benzopyran-7-one, or δ-lactone of 6-hydroxy-4-(3-methyl-2-butenyloxy)-5-benzofuranacrylic acid) (**70**), $C_{16}H_{14}O_4$, the phloroglucinol–based analog of the pyrogallol derivative imperatorin, on acid hydrolysis yielded bergaptol (**31**), and a substance, which after hydrogenation, was isolated as amyl alcohol.[118] **70** is oxidized by chromic acid to acetone, and therefore must be γ,γ-dimethylallyl ether of bergaptol. Späth and Dobrovolny[119] confirmed this fact by a partial synthesis from bergaptol (**31**) and γ,γ-dimethylallyl bromide, though they found that the hindered position of the hydroxyl group made etherification difficult (Eq. 19).

J. Oxypeucedanin

Oxypeucedanin (4-(2,3-epoxy-3-methylbutoxy)-7*H*-furo(3,2-g)–1–benzopyran-7-one, or δ-lactone of 4-(2,3-epoxy-3-methylbutoxy)-6-

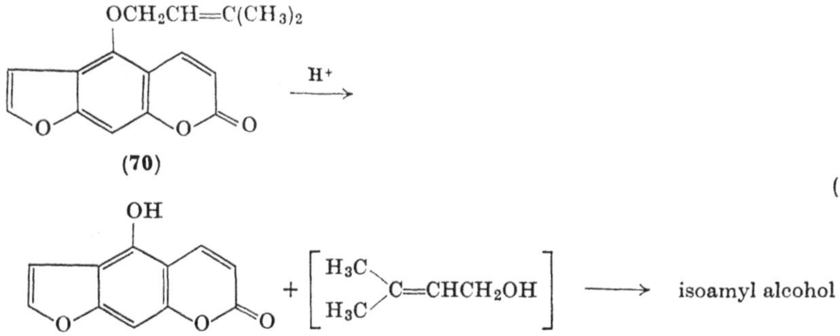

OCH$_2$CH=C(CH$_3$)$_2$

(70)

H$^+$ →

(19)

OH

(31)

+ [$\begin{array}{c} H_3C \\ H_3C \end{array}$ C=CHCH$_2$OH] ⟶ isoamyl alcohol

hydroxy-5-benzofuranacrylic acid) (71), C$_{16}$H$_{14}$O$_5$, is almost identical with the interaction product of isoimperatorin (70) with perbenzoic acid,[119] except that the naturally occurring 71 is slightly optically active. It was shown[120-122] that 71 gave phloroglucinol on alkali fusion, furan-2,3-dicarboxylic acid on hydrogen peroxide oxidation, succinic acid on hydrogenation, followed by nitric acid oxidation, and bergaptol (31) on acid hydrolysis. The presence of the oxide ring was shown both in its hydrolysis by aqueous oxalic acid to a diol, oxypeucedanin hydrate (72), and also in its isomerization by phosphorus pentoxide to a ketone, isooxypeucedanin (73), which gave isobutyric acid when oxidized. Chromic acid oxidation of 71 resulted in the formation of acetone, and oxypeucedaninic acid (74) (Eq. 20).

Nielsen and Lemmich[123] have isolated the *dextro* isomer of oxypeucedanin, which is similar to prangolarin, reported by Chatterjee and coworkers.[124] Hydrolysis of (+)-oxypeucedanin with boiling oxalic acid solution yielded (+)-oxypeucedanin hydrate. Acetic–sulfuric acid mixture produced bergaptol (31), and chromic acid oxidation gave acetone.[124] The structure of (+)-oxypeucedanin has been verified by its p.m.r. spectrum.

K. *Ostruthol*

Ostruthol (75), C$_{21}$H$_{22}$O$_7$, has the characteristics of a furocoumarin.[125] When heated with phosphorus pentoxide, ostruthol gave isooxypeucedanin (73). Hydrolysis of 75 with methyl alcohol/potassium hydroxide solution gave (±)-oxypeucedanin hydrate (72) together with angelic acid. Treatment with hydrogen peroxide showed the presence of an unsubstituted furan ring, and potassium permanganate removed the

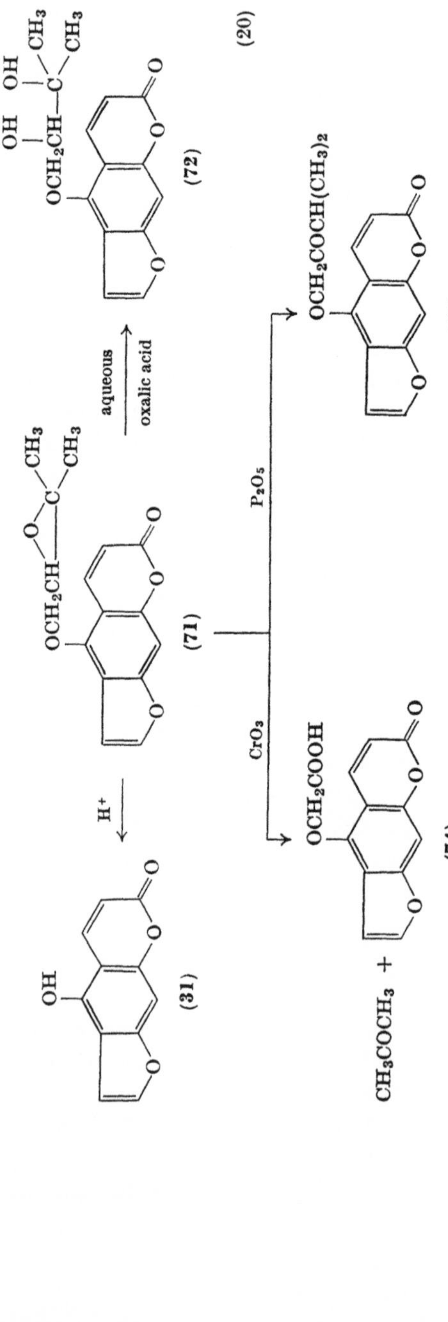

OCH2CH—C(CH3)2 (epoxide O)
(71)
aqueous
oxalic acid
OCH2CH—C(OH)(CH3)2 with OH
(72)
H+
OH
(31)
P2O5
OCH2COCH(CH3)2
(73)
CrO3
OCH2COOH
(74)
CH3COCH3 +

(20)

extensive side chain as α-hydroxyisobutyric acid. The presence of three reactive double bonds was shown by catalytic hydrogenation to hexahydroostruthol, which on nitric acid oxidation, gave succinic acid from the saturated α-pyrone ring. The formation of bergaptol (**31**) by glacial acetic–sulfuric acid fission of **75** indicated the structure of the nucleus and showed that it is a derivative of **31** (Eq. 21).

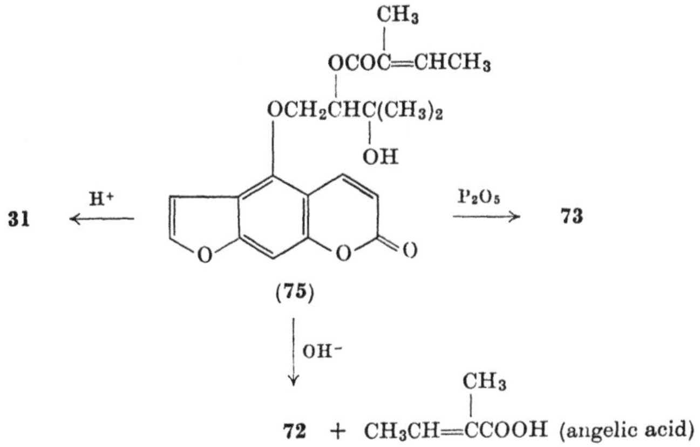

(21)

L. *Imperatorin*

Imperatorin (marmelosin, ammidin, 9-(3-methyl-2-butenyloxy)-7*H*-furo(3,2-g)–1–benzopyran-7-one, or δ-lactone of 6-hydroxy-7-(3-methyl-2-butenyloxy)-5-benzofuranacrylic acid), $C_{16}H_{14}O_4$, was identified as the γ,γ-dimethylallyl ether of xanthotoxol (**76**)[126] owing to its formation of xanthotoxol (**54**) upon either acid hydrolysis or hydrogenolysis, and acetone upon chromic oxidation. Having demonstrated the presence of fused furan and coumarin rings, Späth and Holzen[126] confirmed their views by synthesizing the compound from xanthotoxol and γ,γ-dimethylallyl bromide. Catalytic hydrogenation of **76** yielded a hexahydro derivative, which gave succinic acid and α-hydroxyisobutyric acid on oxidation. Oxidation of **76** with perbenzoic acid gave an epoxide,[127] which is isomeric with oxypeucedanin (**71**).

The characteristic melting point phenomenon of imperatorin (re-solidification and remelting as the temperature rises) is due to a Claisen ρ-migration of the allylic system forming alloimperatorin (**77**), which like **76** itself, is a naturally occurring furocoumarin (Eq. 22).

54 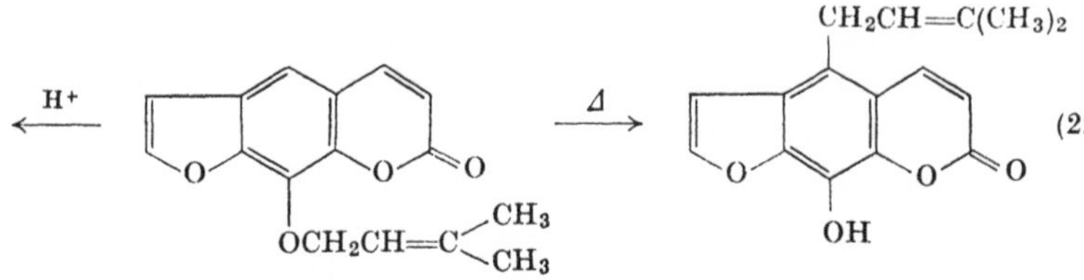 (22)

(76) (77)

M. Alloimperatorin

Alloimperatorin (prangenidin, 4-(3-methyl-2-butenyl)-9-hydroxy-7*H*-furo(3,2-g)–1–benzopyran-7-one, or δ-lactone of 6,7-dihydroxy-4-(3-methyl-2-butenyl)-5-benzofuranacrylic acid) (**77**), $C_{16}H_{14}O_4$, is a natural furocoumarin.[128–130] It forms a monomethyl ether, and a hexahydro derivative on catalytic hydrogenation. Oxidation of the hexahydro compound with nitric acid produced succinic acid and γ-methylvaleric acid from the side chain. With hydrogen peroxide **77** gave furan-2,3-dicarboxylic acid and acetone.

N. Heraclenin

Heraclenin (**78**), $C_{16}H_{14}O_5$, is identical with prangenin. It was isolated from the roots of *Prangos pabularia*[131] together with the product, described by Späth and Holzen[127] as imperatorin epoxide, which was also obtained by treatment of imperatorin (**76**) with perbenzoic acid. The structure assigned to heraclenin has been confirmed as 8-(β,γ-oxido-isoamyloxy)psoralen (**78**).[132]

Treatment of heraclenin with an acetic-sulfuric acid mixture gave xanthotoxol (**54**). Hydrolysis under mild conditions produced a diol, heraclenin hydrate (**79**). This, indicates that the C_5H_9O side chain moiety is similar to that in the isomeric oxypeucedanin (**71**). Oxidation with chromic acid gave acetone, and 8-ω-carbomethoxyfuro(3′,2′–6,7)-coumarin (**80**). On refluxing in toluene over phosphorus pentoxide, or on boiling with mineral acids, heraclenin was converted to a ketone, isoheraclenin (**81**), formed by opening and rearrangement of the epoxide ring. In contrast to oxypeucedanin (**71**), **78** does not form a diacetate with a sodium acetate–acetic anhydride mixture.

O. Isopimpinellin

Isopimpinellin (4,9-dimethoxy-7*H*-furo(3,2-g)–1–benzopyran-7-one, or δ-lactone of 6-hydroxy-4,7-dimethoxy-5-benzofuranacrylic acid)

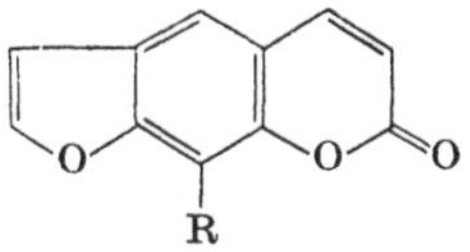

(78) R = $OCH_2CH\overset{O}{\frown}C(CH_3)_2$
(79) R = $OCH_2CH(OH)C(OH)(CH_3)_2$
(80) R = OCH_2COOH
(81) R = $OCH_2COCH(CH_3)_2$

(**82**), $C_{13}H_{10}O_5$, appears to be 5,8-dimethoxypsoralen.[91] The production of furan-2,3-dicarboxylic acid by hydrogen peroxide oxidation shows the presence of an unsubstituted furan ring. The presence of the α-pyrone ring and the relationship between pimpinellin (**84**) and isopimpinellin (**82**) is demonstrated by methylation of **82** with alkaline methyl sulfate to give β-(4,6,7-trimethoxy-5-benzofurano)acrylic acid (**83**), identical with the product of the same reaction of **84** (Eq. 23).

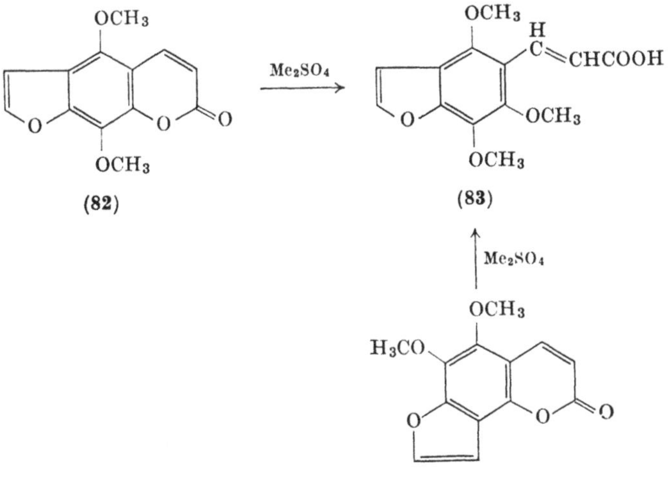

(82) (83) (84) (23)

Isopimpinellin was partially synthesized by methylation of the reduction product of bergaptenquinone (**51**), which can be obtained by chromic oxidation of bergaptol (**31**)[91] or by reaction of xanthotoxol (**54**)[100] with sulfur dioxide.

Horton and Paul[106] have reported the conversion of 1,2,3,5-tetramethoxybenzene, in a number of steps, two of which include selective cleavage of the methoxy group, to 4′,5′-dihydroisopimpinellin (**85**) (Eq. 24). Attempts to effect dehydrogenation were unsuccessful.

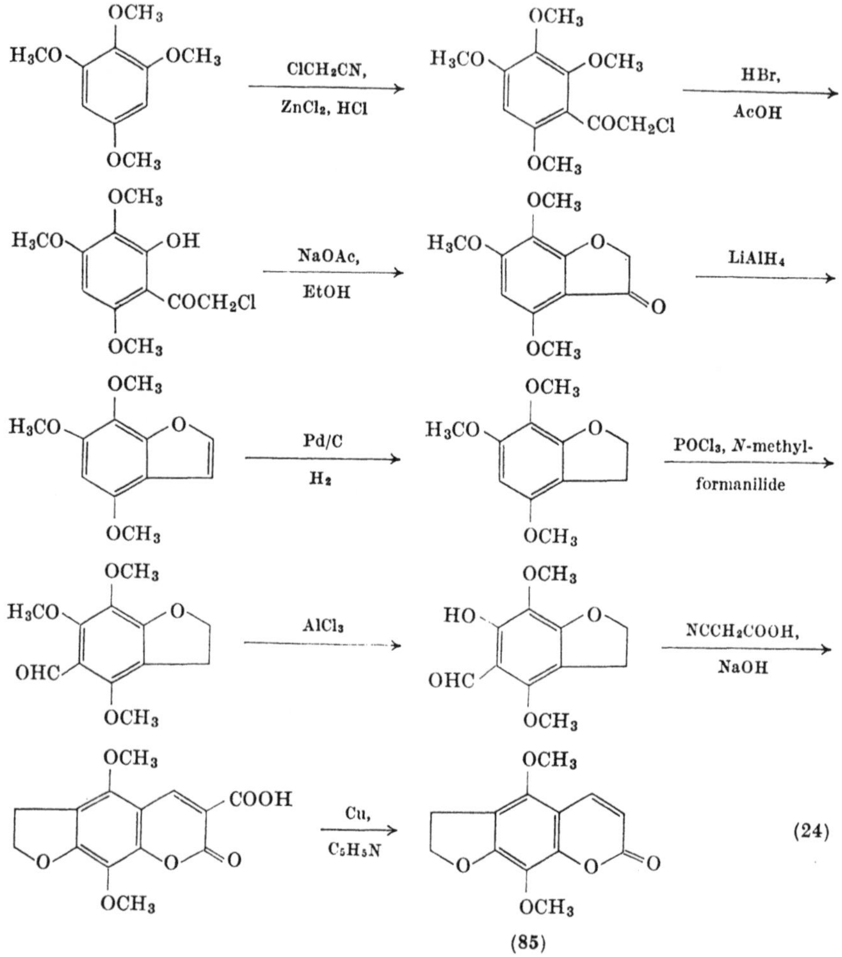

(24)

(85)

P. Phellopterin

Phellopterin (**86**), $C_{17}H_{16}O_5$, is shown to be 5-methoxy-8-(γ,γ-dimethylallyloxy)psoralen.[133] The structure is demonstrated by acid fission to 8-hydroxy-5-methoxypsoralen (**87**) together with γ,γ-dimethylallyl alcohol, and by chromic oxide oxidation to bergaptenquinone (**28**). Catalytic hydrogenation gave, after absorption of one molecular proportion of hydrogen, a mixture from which 8-isoamyl-5-methoxypsoralen (**88**) was obtained. Oxidation with benzoyl peroxide gave an epoxide, identical with *dl*-byakangelicol.

Phellopterin (**86**) has been partially synthesized from 8-hydroxy-5-methoxypsoralen (8-hydroxybergapten) (**87**), a naturally occurring furocoumarin,[134] which was readily obtained by treatment of 8-aminobergapten (**27**) with nitrous acid.[77] Etherification of **87** with γ,γ-dimethylallyl bromide yielded **86** (Eq. 25).

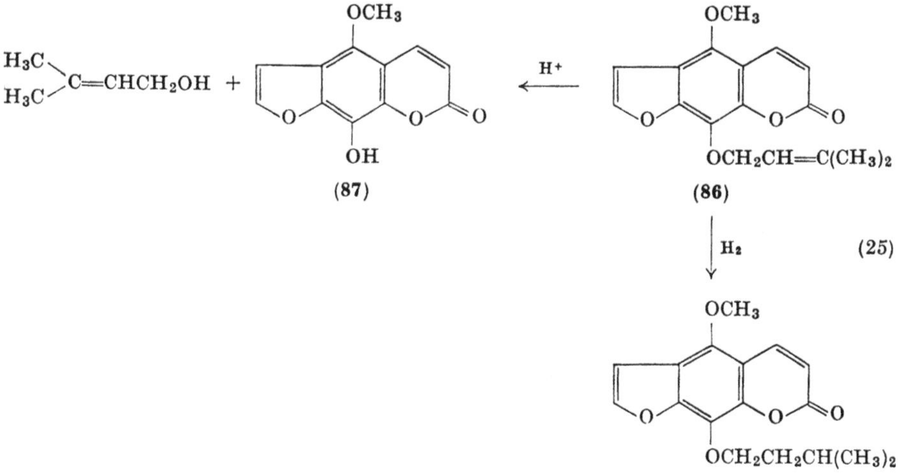

(87) (86) (88) (25)

Q. Byakangelicol

Byakangelicol (**89**), $C_{17}H_{16}O_6$, has been shown to be an epoxide[133, 135] since the racemate can be prepared by peracid oxidation of phellopterin (**86**), and because both natural and synthetic materials give isobyakangelicolic acid (**90**) with hot alkali. The latter acid, whose structure is not completely certain, is found with **89** in plants. Moreover, hydration of natural or synthetic **89** by treatment with aqueous oxalic acid gives the α-glycol, which is also a natural product, byakangelicin (**91**). Reduction of the α-pyrone ring with sodium amalgam produced the corresponding *o*-hydroxydihydrocinnamic acid, which gives succinic acid on oxidation. Oxidation of byakangelicol with chromic oxide gave bergaptenquinone (**28**) together with byakangelicinic acid (**92**); treatment with hydrogen peroxide gave furan-2,3-dicarboxylic acid, and potassium permanganate removed the side chain as α-hydroxyisobutyric acid. Fission of the ether side chain with glacial acetic-sulfuric acid produced 8-hydroxy-5-methoxypsoralen (**87**), a natural furocoumarin, which is converted to isopimpinellin (**82**) by methylation with diazomethane. The oxide ring

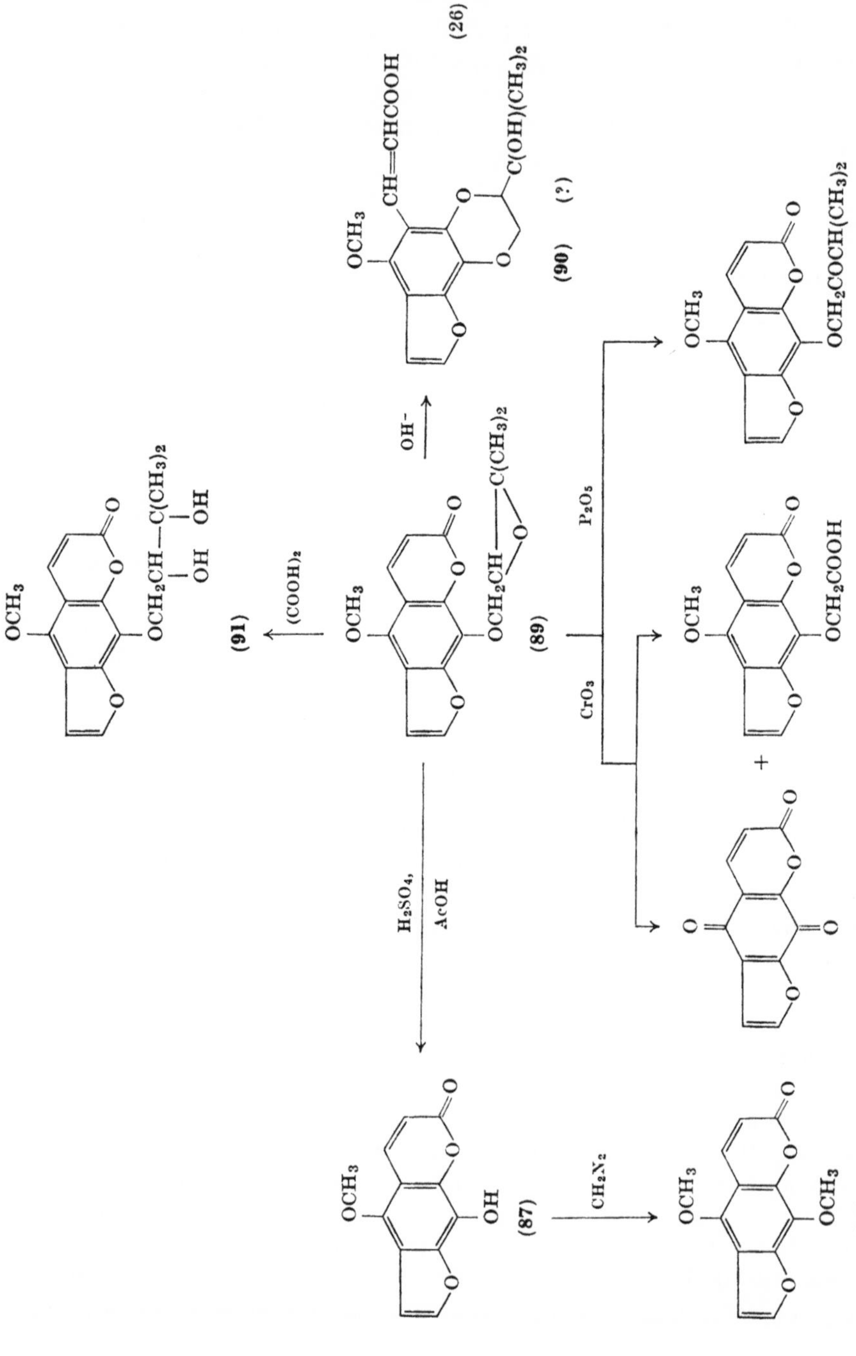

(91)
(COOH)₂
(89)
OH⁻
(90) (?)
(26)
H_2SO_4, AcOH
(87)
CH_2N_2
(82)
CrO_3
P_2O_5
(28)
(92)
(93)

in the side chain of **89** is opened readily by the action of a sodium acetate-acetic anhydride mixture, to form byakangelicin diacetate. When a toluene solution of **89**, containing phosphorus pentoxide is boiled, isomerization takes place to give anhydrobyakangelicin (**93**) (Eq. 26).

The claim that the formula previously assigned to byakangelicol (**89**) rightly belongs to ferulin,[136] isolated from the fruit of *Ferula alliacea* Boiss, has recently been verified by Chatterjee and coworkers.[137] The latter authors have shown that ferulin, on chromatographic elution with cyclohexane and benzene on formamide–impregnated cellulose powder, was resolved into two naturally occurring furocoumarins, phellopterin (**86**) and byakangelicol (**89**). Furthermore, they have synthesized **89** by treatment of 8-hydroxy-5-methoxypsoralen (**87**) with γ,γ-dimethylallyl bromide, followed by oxidation with monoperphthalic acid.

R. Byakangelicin

Byakangelicin (**91**), $C_{17}H_{18}O_7.H_2O$, was shown by Noguchi and Kawanami[6, 133, 138] to be very similar to byakangelicol (**89**). Acetone and α-hydroxyisobutyric acid were among its oxidation products. In addition, **91** formed a diacetate, and when heated with phosphorus pentoxide in toluene it lost the elements of water with the formation of anhydrobyakangelicin (**93**), which could be oxidized to isobutyric acid.

S. Pimpinellin

Pimpinellin (5,6-dimethoxy-2*H*-furo(2,3-h)-1-benzopyran-2-one, δ-lactone of 4-hydroxy-6,7-dimethoxy-5-benzofuranacrylic acid or 5,6-dimethoxyangelicin) (**94**), $C_{13}H_{10}O_5$, occurs naturally, together with isopimpinellin (**82**). Opening of the α-pyrone ring with methyl sulfate resulted in the formation of β-(4,6,7-trimethoxy-5-benzofurano)-acrylic acid (**83**). This together with other evidence, enabled Wessely and Kallab[91] to assign structure **94** to the compound, that is an angular isomer of **82**.

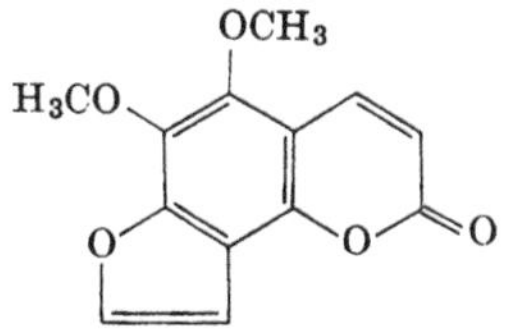

(**94**)

T. Sphondin

Sphondin (6-methoxy-2*H*-furo(2,3-h)-1-benzopyran-2-one, or δ-lactone of 4-hydroxy-7-methoxy-5-benzofuranacrylic acid) (**95**), $C_{12}H_8O_4$, is given a structure based on the following considerations of Späth and Schmid.[139] It contains one methoxyl group and on treatment with alkaline hydrogen peroxide, yields furan-2,3-dicarboxylic acid. Ozonolysis in chloroform produced a product, identified as 8-formylscopoletin (**96**), which gives a green color with ferric chloride, and fraxetin (**97**) upon treatment with alkaline hydrogen peroxide. Furthermore, **97** has been successfully obtained by formylation of scopoletin (**98**). Sphondin is thus assigned the formula 6-methoxyfuro(2′,3′–7,8)coumarin (Eq. 27).

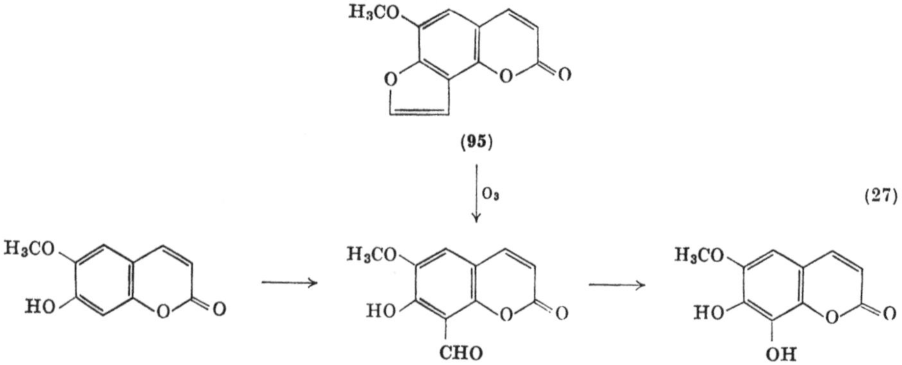

(95) (98) (96) (97) (27)

Seshadri and Sood[140] have synthesized sphondin, by the furan–ring closure method,[55] starting from 7-allyloxy-6-methoxycoumarin, cf. **99**, which was obtained from aesculetin (**100**), umbelliferone allyl ether (**101**), and/or scopoletin (**98**). The allyl ether underwent a Claisen migration to yield 8-allyl-7-hydroxy-6-methoxycoumarin (**102**), which when oxidised with osmium tetroxide–periodate, or allowed to undergo ozonolysis, gave the corresponding acetaldehyde compound. The latter was readily cyclized by polyphosphoric acid to sphondin (**95**) (Eq. 28).

Another method involved oxidation of angelicin (**1**), followed by methylation[140] (Eq. 29).

U. Halfordin and Isohalfordin

Halfordin (4,7,9-trimethoxy-6*H*-furo(2,3-g)–1–benzopyran-6-one, or δ-lactone of 5-hydroxy-α,4,7-trimethoxy-6-benzofuranacrylic acid) (**103**) and isohalfordin (3,5,6-trimethoxy-2*H*-furo(2,3-h)-1-benzopyran-2-one,

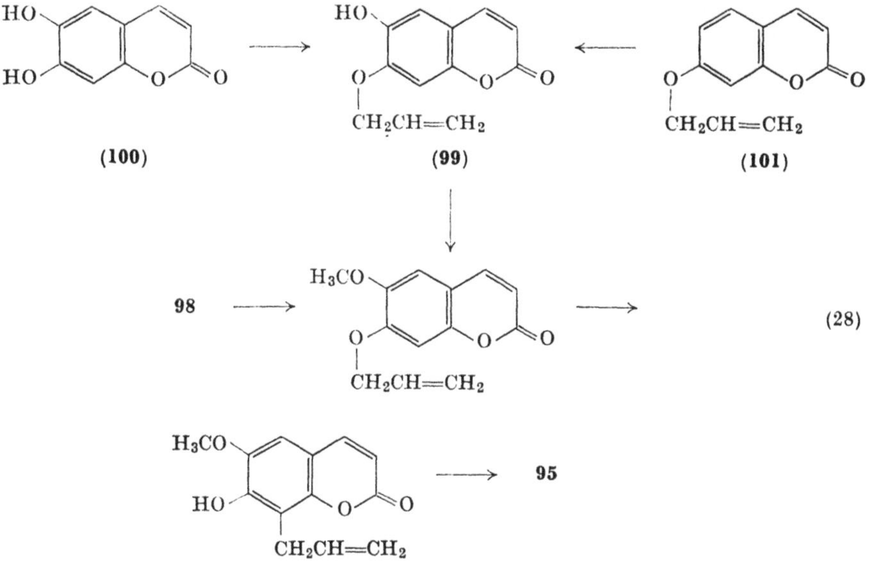

(100) (99) (101) 98 (28) 95

(102)

or δ-lactone of 4-hydroxy-α,6,7-trimethoxy-5-benzofuranacrylic acid) (**104a** or **104b**), isomers of the formula $C_{14}H_{12}O_6$, were isolated in 1956 by Hegarty and Lahey[141] from the bark of *Halfordia scleroxyla* F. Muell. Both exhibit blue fluorescence, and can be oxidized to furan-2,3-dicarboxylic acid. Each contain three methoxyl groups, and as there are only two free positions on the benzene ring, one methoxyl group must be attached to the pyrone ring. On ozonolysis the halfordins yield isomeric dialdehydes, therefore the methoxyl groups must be placed at the 3-position in each case. These compounds must be accounted the first naturally occurring 3-methoxyfurocoumarins to be discovered.

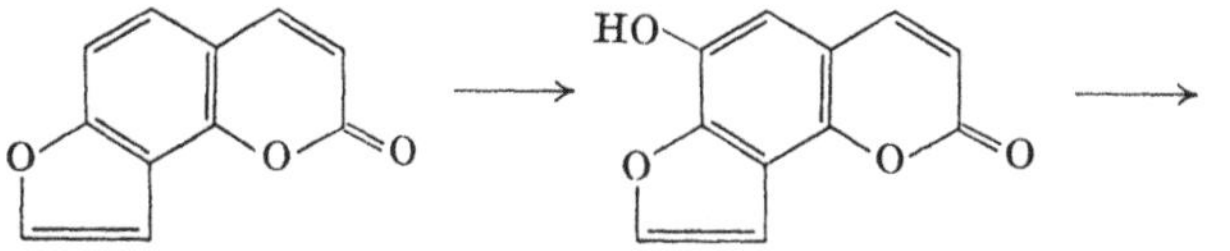

95 (29)

(1)

The dialdehyde (**105**) from **104** on methylation and oxidation yielded tetramethoxyisophthalic acid (**106**), identical with the authentic material. The dialdehyde from **103** does not respond to tests for an *o*-phthaldehyde, and after methylation, followed by oxidation, yields what must be terephthalic acid (**108**), since the compound is identical with neither the isophthalic acid (**106**) nor tetramethoxy-*o*-phthalic acid.

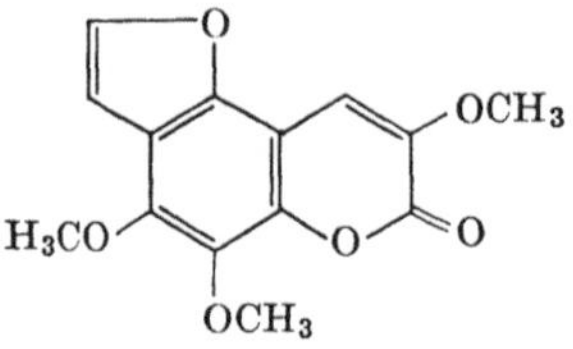

(**104a**)

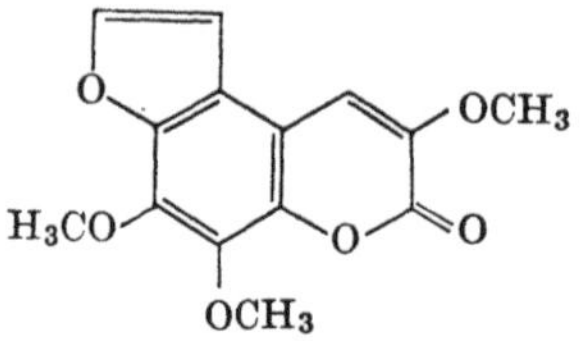

(**104b**)

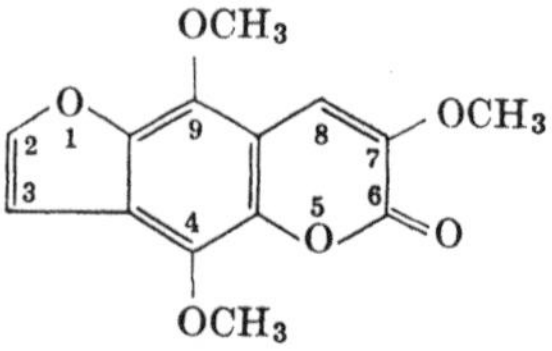

(**103**)

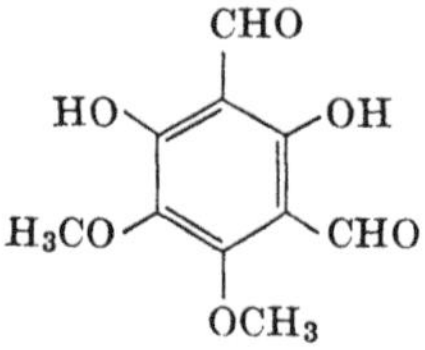

(**105**)

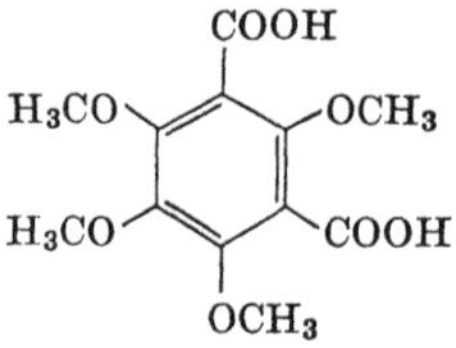

(**106**)

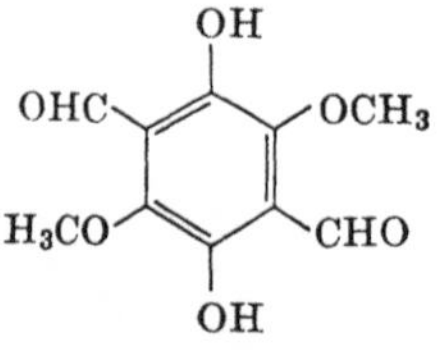

(**107**)

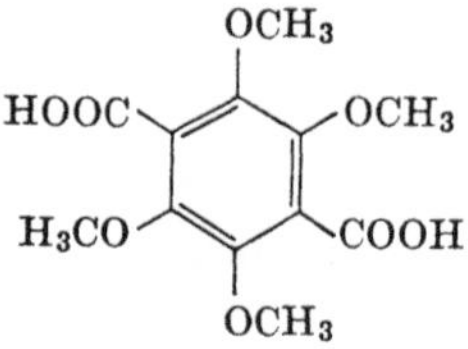

(**108**)

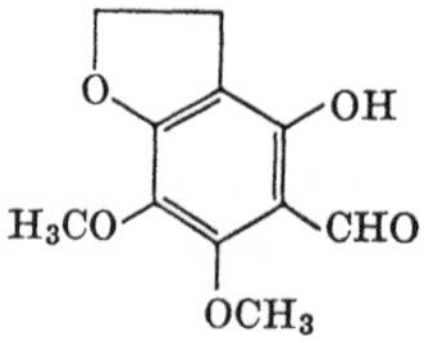

(**109**)

The dialdehyde from isohalfordin is readily volatile in steam, therefore must be fully chelated and have a structure (**105**), but that from halfordin is not, therefore its structure is probably **107**.[142] Thus, isohalfordin is **104a** or **104b**, and halfordin may be **103**. It should be noted that halfordin is the only natural furocoumarin not containing an umbelliferone nucleus.

Isohalfordin, like the authentic 3-methoxycoumarins, loses its methyl group when heated with acids, giving the corresponding 3-hydroxyfurocoumarin which has the typical green ferric reaction, and which with diazomethane, restores isohalfordin. It is also characteristic of 3-methoxycoumarins that they resist saturation of the double bond, as do 4-methoxycoumarins. Consequently hydrogenation of isohalfordin selectively attacks the furan ring, and the dihydro derivative produced can be oxidized to the aldehyde (**109**) or its isomer.

Isohalfordin, like other coumarins, gradually dissolves in warm alkali and is at once regenerated on acidification, whereas, hydrolysis of halfordin gives halfordic acid which does not cyclize when kept, or irradiated with ultraviolet light.

V. Nodakenetin

Nodakenetin (5′-(β-hydroxyisopropyl-4′,5′-dihydropsoralen, 2,3-dihydro-2-(1-hydroxy-1-methylethyl)-7*H*-furo(3,2-g)–1–benzopyran-7-one, or δ-lactone of 2,3-dihydro-6-hydroxy-2-(1-hydroxy-1-methylethyl)-5-benzofuranacrylic acid) (**110a**), $C_{14}H_{14}O_4$, is the aglycone of the glucoside nodakenin (**110b**).[143] Marmesin, the optical antipode of nodakenetin,[144] occurs in the free state and as the glucoside, ammajin. Ammajin is also known as marmesinin (**110b**).[145]

The most important reactions leading to the structure of nodakenetin,[146–148] were similar to those performed with marmesin[143] and include fusion with potassium hydroxide to give resorcinol. Oxidation with potassium dichromate gave umbelliferone-6-carboxylic acid (**111**), resulting from the degradation of the dihydrofuran ring, and thus revealed the coumarinic nature of nodakenetin. The nature of the side chain was shown by potassium permanganate oxidation of **110** to acetone, thus illustrating the presence of the isopropylidene group in the furan ring. The tertiary hydroxyl group is not so readily acylable but sodium acetate-acetic anhydride mixture gave the acetylated product.

Distillation of nodakenetin with phosphorus pentoxide in high vacuum resulted in the formation of anhydronodakenetin (5′-isopropylpsoralen) (**112**), which on hydrogenation is easily converted[147] into a coumaran, deoxydihydroöreoselone (**113**). Anhydronodakenetin has also been obtained by treatment of suberosin (**114**) with hydrobromic acid in the presence of red phosphorus[149] (Eq. 30).

Späth and Tyray[148] rejected the possibility that nodakenetin might be a 3-hydroxycoumarin on the gounds that such compounds are very susceptible to dehydration, and further, that **110** does not behave as a hemiketal, but yields acetone when oxidized. Nodakenetin behaves as expected towards alkalis to give a soluble *cis*-cinnamic acid salt, which lactonizes upon acidification and gives the stable *trans*-cinnamic acid (**115**).

Partial synthesis of nodakenin has been achieved by the action of β-pentaacetyl-D-glucose on nodakenetin (**110**) in the presence of a trace of *p*-toluenesulfonic acid,[148] and marmesinin is obtained by the action of α-acetobromoglucose on marmesin in the presence of silver carbonate.[145] King and coworkers[150] prepared racemic nodakenetin by epoxidation

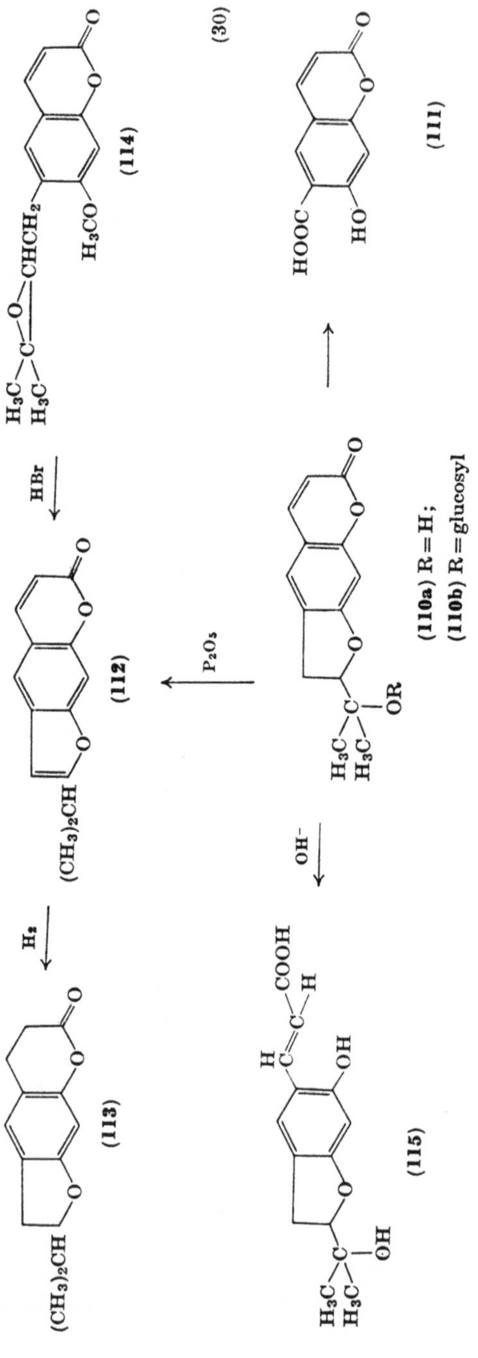
(30)
(114)
(111)
HBr
(112)
P2O5
(110a) R = H;
(110b) R = glucosyl
H2
OH⁻
(113)
(115)

of 7-demethylsuberosin (**116**) by the action of monoperphthalic acid, followed by cyclization (Eq. 31).

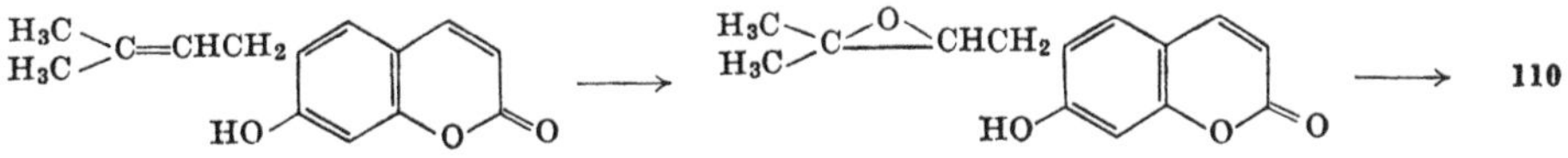

(116)

(31)

Nakajima and coworkers[151] have recently reported the synthesis of (−)-nodakenetin and the optical antipode, (+)-marmesin, from α-resorcaldehyde (Eq. 32).

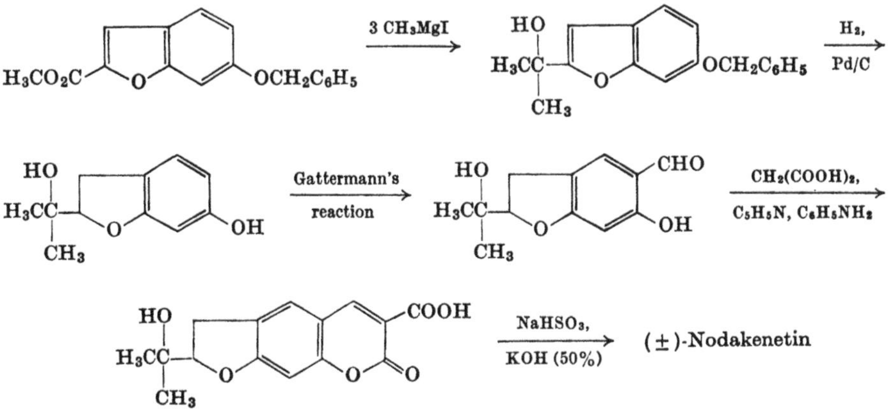

(32)

(117)

The brucine salt of racemic nodakenetin-3-carboxylic acid (**117**) gave an insoluble crystalline acid which on decarboxylation gave (+)-marmesin. The acid obtained from the filtrate gave (−)-nodakenetin upon decarboxylation.

Marmesin substitution reactions leading to 3-bromo,- and 3-nitro-marmesin ((**118a**) and (**118b**), respectively) have been described.[145, 146, 152]

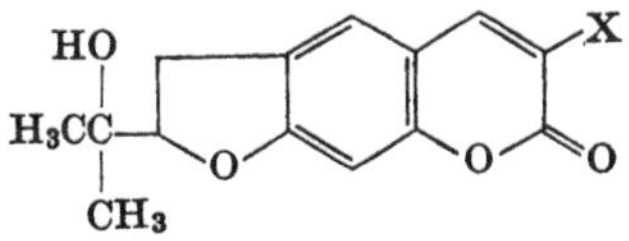

(**118a**) X = Br; (**118b**) X = NO_2

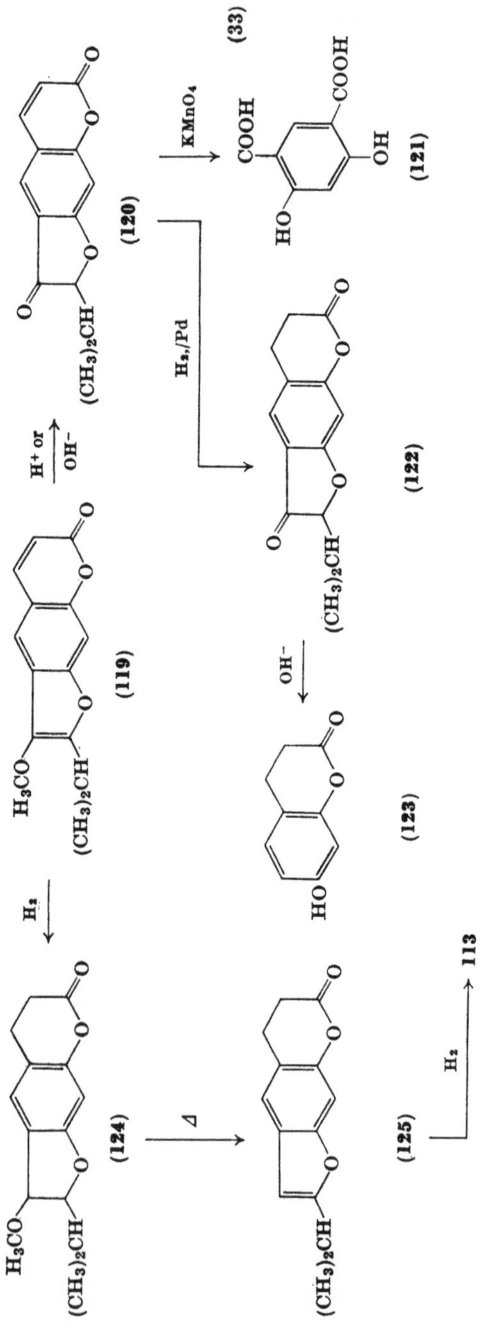
(33)
KMnO$_4$
COOH
COOH
HO
OH
(121)
(120)
(CH$_3$)$_2$CH
H$^+$ or OH$^-$
H$_2$,/Pd
(122)
H$_3$CO
(119)
OH$^-$
HO
(123)
H$_2$
(124)
Δ
(125)
113

W. Peucedanin

Peucedanin (4′-methoxy-5′-isopropylpsoralen, the methyl ether of the enolic form of oreoselone, 2-isopropyl-3-methoxy-7*H*-furo(3,2-g)–1–benzopyran-7-one, or δ-lactone of 6-hydroxy-2-isopropyl-3-methoxy-5-benzofuranacrylic acid) (**119**), $C_{15}H_{14}O_4$, undergoes acidic or basic hydrolysis to a methoxy-free compound, oreoselone (**120**), which is oxidized by permanganate to 4,6-dihydroxyisophthalic acid (**121**). In alkaline solution, oreoselone can be hydrogenated to a dihydro derivative that can be oxidized to succinic acid, thus indicating the presence of a coumarin ring. This result is confirmed by alkali fusion of dihydrooreoselone (**122**) to dihydroumbelliferone (**123**). Further oxidation of oreoselone, under controlled conditions, gives isobutyric acid, therefore it must contain an isopropyl substituent[121] (Eq. 33).

Oreoselone (2-isopropyl-7*H*-furo(3,2-g)–1–benzopyran-3(2*H*)-7-dione, or δ-lactone of 2,3-dihydro-6-hydroxy-2-isopropyl-3-oxo-5-benzofuranacrylic acid) (**120**), forms an enol acetate, and reduces Fehling's reagent. The enol ether of peucedanin loses its methyl group by hydrolysis in the production of oreoselone. The two double bonds of peucedanin can be saturated by hydrogenation, and the resultant tetrahydropeucedanin (**124**) on distillation eliminates methanol to produce 3,4-dihydro-5′-isopropylpsoralen (**125**), which on further hydrogenation yields deoxydihydroöreoselone (**113**).

In contrast to furocoumarins, oreoselone, which contains the coumarone system, gives no furan-2,3-dicarboxylic acid with hydrogen peroxide, but only isobutyric acid. Oxidation with alkaline potassium

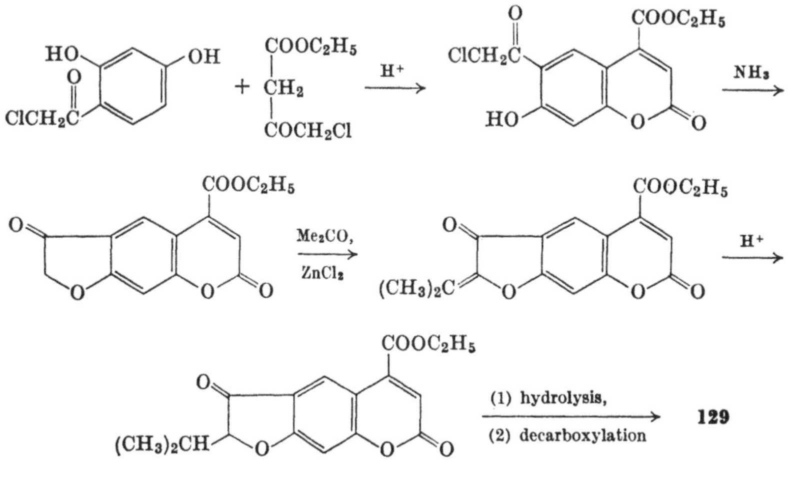

(34)

permanganate yields α-hydroxyisobutyric acid, together with resorcinol-4,6-dicarboxylic acid; thus establishing the linear structure of oreoselone.

Schmid and Ebnöther[153] have achieved the partial synthesis of peucedanin from oreoselone, using a solution of aluminum chloride in methanol. The synthesis of oreoselone has also been achieved[154] starting with ω-chlororesacetophenone (Eq. 34).

Recently, Nikonov[155] has reported the formation of 4′-amino-5′-isopropylpsoralen (**126**) by reduction of oreoselone oxime with zinc and hydrochloric acid to the corresponding amine. The latter, on nitration, gave the nitro derivative.

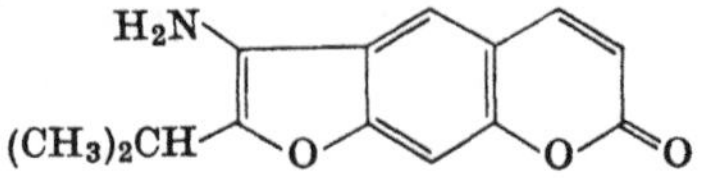

(**126**)

X. *Athamantin*

Athamantin (8,9-dihydro-9-hydroxy-8-(1-hydroxy-1-methylethyl)-2*H*-furo(2,3-h)-1-benzopyran-2-one diisovalerate, or δ-lactone of 2,3-dihydro-3,4-dihydroxy-2-(1-hydroxy-1-methylethyl)-5-benzofuranacrylic acid) (**127**), $C_{24}H_{30}O_7$, is easily split into two molecules of

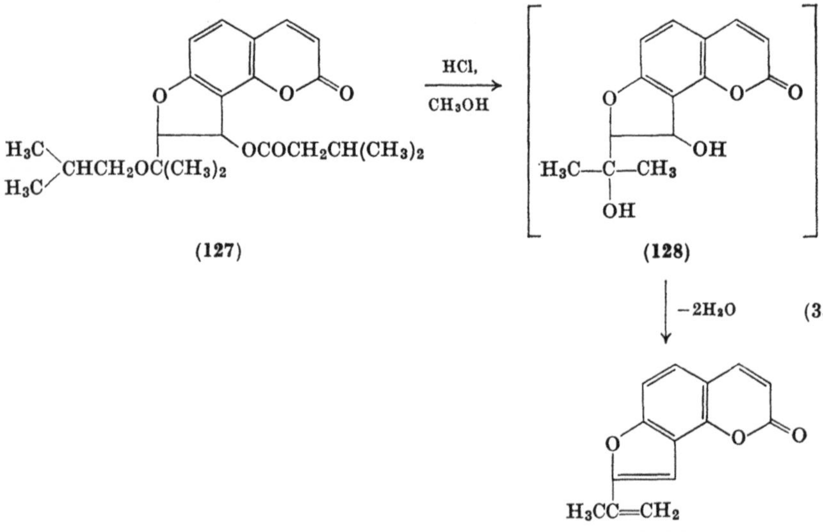

(**127**) (**128**) (**129**) (35)

isovaleric acid, leaving oroselone (**129**), which presumably is produced via the dihydroxy intermediate (**128**). The latter compound under the hydrolytic conditions loses two molecules of water to give **129**[156] (Eq. 35).

Alkaline hydrolysis of **127**, followed by acidification, and treatment with potassium permanganate, gave acetone from the isopropyl group on the furan ring of **128**; thus favoring the formation of **130** rather than **129** under these hydrolytic conditions.

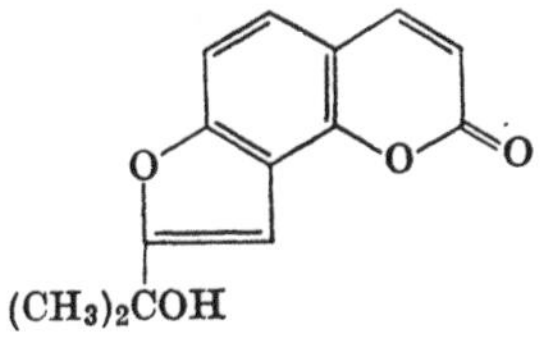

(130)

Oroselone (kvanin, 8-isopropenyl-2*H*-furo(2,3-h)-1-benzopyran-2-one, or δ-lactone of 4-hydroxy-2-isopropenyl-5-benzofuranacrylic acid) (**129**), $C_{14}H_{10}O_3$, and oroselol (**130**), $C_{14}H_{12}O_4$, are found with athamantin in plants.[157, 158] Oroselone is related to angelicin (**1**)[139] (5′-isopropylangelicin), since it is degraded by ozone to 8-formylumbelliferone. Hydrogenation of **129** readily gives successively, dihydroöroselone (by conversion of the isopropenyl to the isopropyl residue), tetrahydro-, and hexahydroöroselone. Ozonolysis of dihydroöroselone (**142**) gives isobutyric acid and umbelliferone-8-aldehyde. Oxidation of tetrahydrooroselone gives isobutyric acid (from the furan side chain) and succinic acid. Alkaline hydrogen peroxide oxidation of oroselone does not lead to furan-2,3-dicarboxylic acid, but ozonization of **129** produces resorcinol-2,4-dialdehyde and umbelliferone-8-aldehyde, thus establishing the angular formulation of oroselone (**129**).

Further evidence for the assigned structure of athamantin (**127**) has been given by Schmid and coworkers.[158] The action of sodium methoxide at 20° on **127** eliminates only one of the isovalerate groups completely, the other being converted into methoxyl. At the same time, methoxide adds to the pyrone double bond, and the ring opens so that **131** is obtained. At 70°, this methoxide group is reeliminated and a mixture of the coumarinic ester (**132**), and the coumarin (**133**) formed. 7-Methoxycoumarin behaves similarly toward methoxide therefore Schmid regards this as evidence for the mechanism of interconversion of coumarins and coumaric acids in basic media. Ozonolysis of the coumaric ester (**132**) to α-methylisobutyric acid is strong evidence for the location of the methoxyl group. and therefore for that of one isovalerate grouping in athamantin.

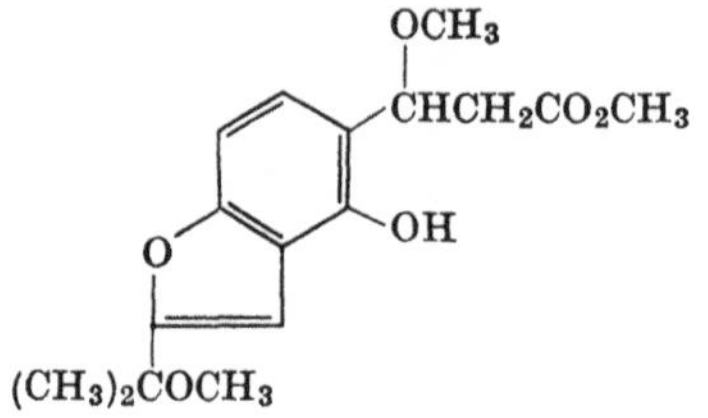
(131)

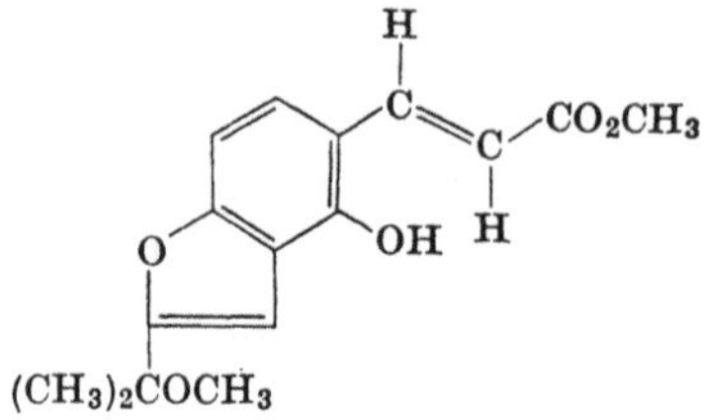
(132)

Y. *Discophoridin*

Discophoridin (**133**), $C_{14}H_{12}O_4$, displays the usual characteristics of a natural furocoumarin. On heating with potassium hydroxide,[159] it forms a yellow potassium salt of the coumarinic acid, which relactonizes on acidification. It readily forms a monoacetyl derivative from which **134** can be recovered by hydrolysis. The alcoholic group is not phenolic as is shown by the insolubility of the compound in cold alkali, and its failure to react with diazomethane or to give a ferric reaction. Its ultraviolet spectrum shows marked similarity to that of osthenol (**135**) and that of dihydrosesselin (**136**), indicating that there is no double bond conjugated with the aromatic ring. A study of the nuclear magnetic resonance spectrum has shown that one aromatic proton ($\tau = 3.3$) has an adjacent oxygen atom while the other aromatic proton ($\tau = 2.75$) has not. More-

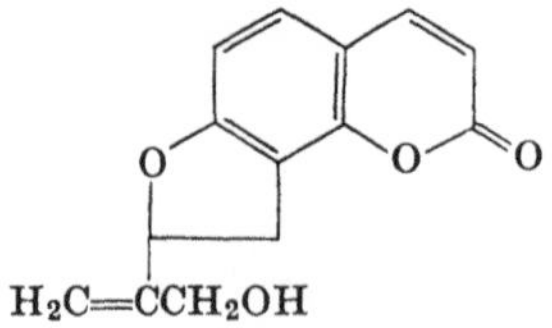
(133)

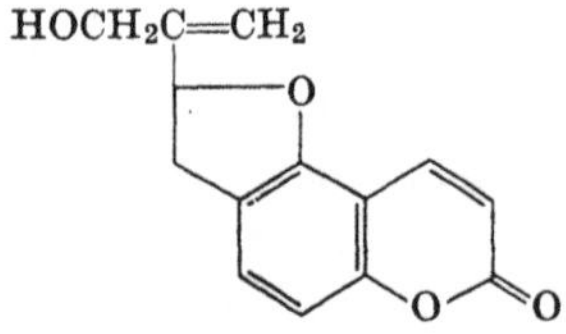
(134)

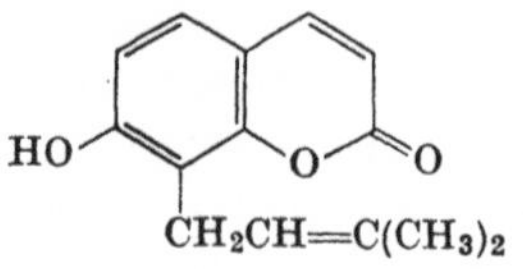
(135)

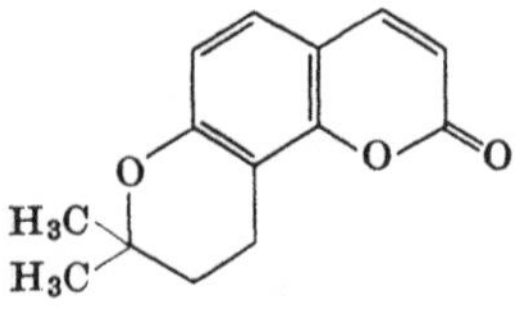
(136)

over, the structure **134** proposed by Bottomley[159] is favored on biogenetic grounds, which are supported by the isolation of **135**, found with **133** in the same plant and having a carbon skeleton identical to it. However the n.m.r. spectrum does not exclude the possible structure **134.**

Z. Edultin

Edultin (8-(1-acetoxyisopropyl)-9-angeloyloxy-2*H*-furo(2,3-h)-1-benzopyran-2-one, or δ-lactone of 3,4-dihydroxy-2-(1-hydroxy-1-methylmethyl)-5-benzofuranacrylic acid 2-acetate-3-angelate) has been identified as **137.**[160] The formula given to edultin is based on the following (Eq. 36).

137 —
- $\xrightarrow[\text{(7.5\%)}]{\text{NaOH-MeOH}}$ methyl ether of **130** + **129** + **1** + Angelic acid + AcOH
- $\xrightarrow{\text{N MeONa}}$ **133** + Angelic acid + AcOH
- $\xrightarrow[\text{(1:2)}]{\text{conc. HCl-MeOH}}$ **129** + Angelic acid + AcOH
- $\xrightarrow[\text{MeOH}]{\text{0.1 N NaOH-}}$ **130** + Acetate of **130** $\xrightarrow{\text{0.1 N NaOH-MeOH}}$ **130** + **129** + AcOH

(36)

The fact that angelic acid is obtained as the sole partial hydrolysis product of edultin, indicates that the C_8 or C_9 position of edultin is substituted by an angeloxy group. The C_8 of edultin, however, lacks semiketal behavior, and by comparison with the structure of athamantin (**127**),[156, 158] the formula 8-(1-acetyloxyisopropyl)-9-angeloyloxy-2*H*-furo(2,3-h)-1-benzopyran-2-one (**137**) is preferable to **138** or **139**.

AA. Peulustrin

Peulustrin (8(s)-(+)-8-(1-(2(R), 3(s)-3-angeloyloxy-2-hydroxy-2-methylbutyloxy)-1-methylethyl)-8,9-dihydro-2*H*-furo(2,3-h)-1-benzopyran-2-one) (**145a**), $C_{24}H_{28}O_8$, has been isolated from the ethereal extract of the *Peucedanum palustre* root.[123a, b] The coumarin character of **145a** was indicated by its blue fluorescence and by its ultraviolet, infrared, and p.m.r. spectra. Treatment of **145a** with methanolic sodium hydroxide gave 8(s)-(+)-dihydroöroselol (**141**), angelic acid, and 2(R), 3(s)-(+)-threo-2, 3-dihydroxy-2-methylbutyric acid.

Investigation of the coumarin content of the fruits of *Peucedanum*

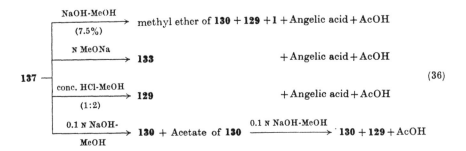

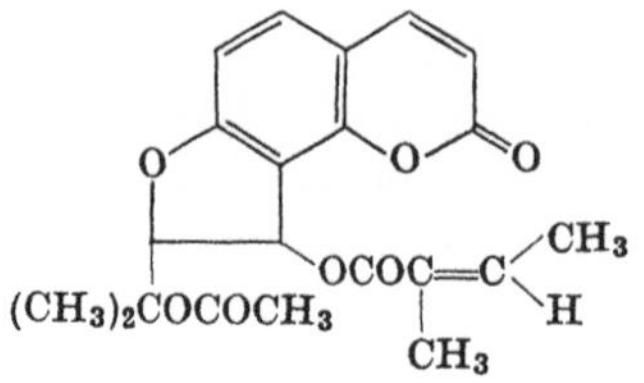

(137)

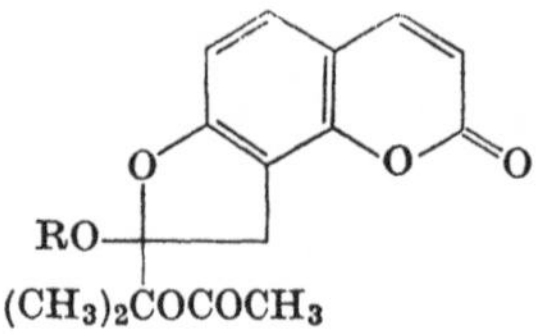

(138)

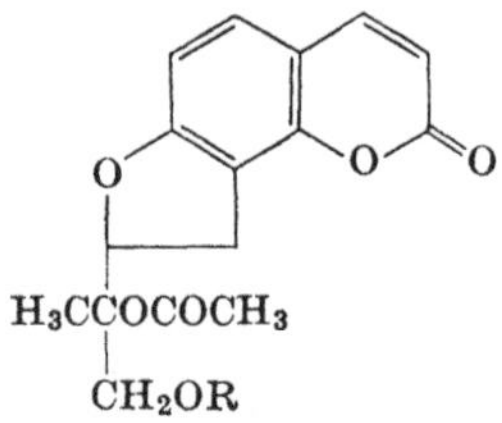

(139)

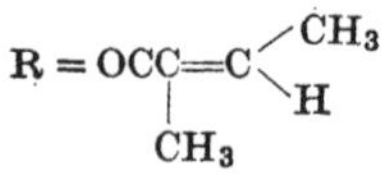

palustre has recently been reported.[123c] Two new blue fluorescent furocoumarins have been isolated in addition to isoimperatorin (**70**), columbianidin (**140**), (+)-oxypeucedanin (**71**), isoxypeucedanin (**73**), and peulustrin (**145a**).

Columbianidinoxide (8(S)-(+)-8-(1-(2(R)-2,3-epoxy-2-methylbutyryloxy)-1-methylethyl)-8,9-dihydro-2*H*-furo(2,3-h)-1-benzopyran-2-one) (**145b**), $C_{19}H_{20}O_6$, has an ultraviolet spectrum very similar to that of 8(S)-(+)-dihydroöroselol (**141**),[123d] and an infrared spectrum that indicates its coumarin character. Treatment of **145b** with methanolic sodium hydroxide yielded **141**, and 2(R), 3(S)-(+)-2,3-dihydroxy-2-methylbutyric acid. The nucleophilic attack of hydroxide ions on the epoxide ring in **145b** can take place only at C_3, the attack at C_2 being sterically hindered. Since it is known that this reaction at C_3 is accompanied by inversion, the formation of 2,3-dihydroxy-2-methylbutyric acid suggests that the glycidic acid moiety of **145b** is 2(R),3(R)-2,3-epoxy-2-methylbutyric acid. The methyl ester of the latter acid was synthesized. Further evidence for the structure **145b** has been found from the p.m.r. spectrum.

Isopeulustrin (8(S)-(+)-8-(1-(2(R), 3(S)-3-(3,3-dimethylacryloyloxy)-2-hydroxy-2-methylbutyryloxy)-1-methylethyl)-8,9-dihydro-2*H*-furo(2,3-h)-1-benzopyran-2-one) (**145c**), $C_{24}H_{28}O_8$, has very similar ultraviolet, infrared, and p.m.r. spectra to those of peulustrin (**145a**).[123b] Treatment of **145c** with methanolic sodium hydroxide yielded **141**, β,β-dimethylacrylic acid (senecioic acid), and 2(R), 3(S)-(−)-2,3-dihydroxy-2-methylbutyric acid.[123c]

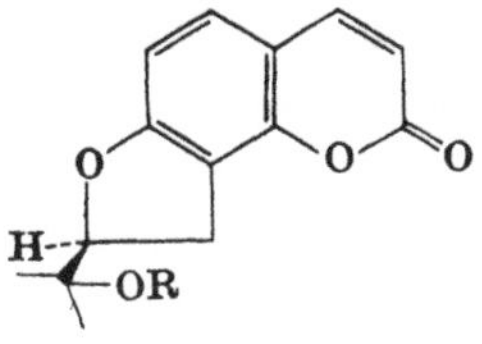

(**145a**)

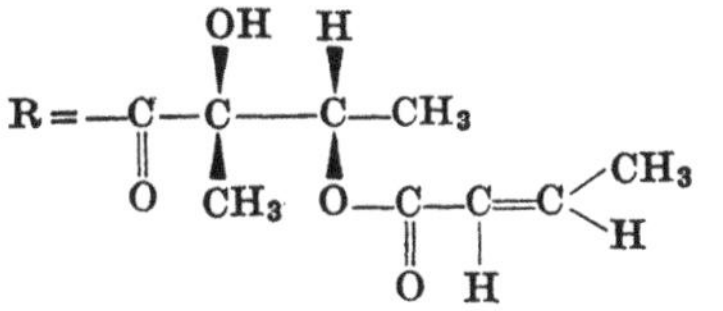

;

(**145b**)

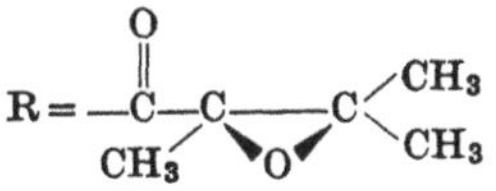

;

(**145c**)

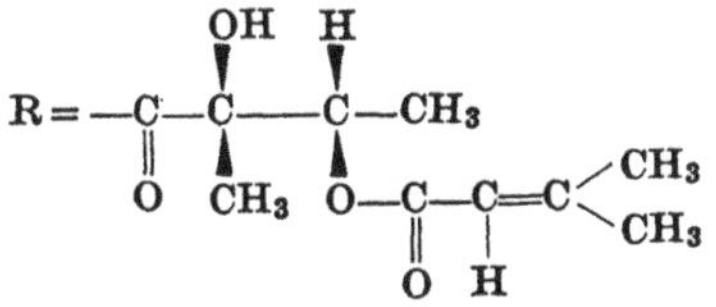

BB. Columbianadin and Columbianin

Columbianadin (**140**), $C_{19}H_{20}O_5$, isolated from the roots of *Lonatium columbianum* Mathias and Const., on alkaline hydrolysis gave tiglic acid, and an alcohol, columbianetin (**141**). Degradation of the latter gave dihydroöroselone (**142**), which could have been formed from the tertiary alcohol (**141**) or the secondary alcohol (**144**) by rearrangement. Hydrogenation of **141** yielded the dihydro alcohol (**143**), which is identical with a hydrogenation product of athamantin (**127**). Since a rearrangement is not possible during the hydrogenation, the location of the hydroxyl group in **141** is established.[161] This, in addition to ultraviolet, infrared, and nuclear magnetic resonance spectra, led to the establishment of **141** as (+)-8,9-dihydro-8-(1-hydroxy-1-methylethyl)-2*H*-furo(2,3-h)-1-benzopyran-2-one. Zosimol, recently described by the Russian authors[162] seems to be identical with columbianetin (**141**).

The tiglate of columbianetin has been obtained by treatment of **141** with tigloyl chloride, and the angelate by the action of 3-bromoangeloyl chloride, followed by hydrogenation.[163] Columbianetin angelate was identical to columbianadin (**140**). The tiglate gave dissimilar spectra and melting point depression. Comparison of n.m.r. spectra also confirmed the identity of the angelate to, and the dissimilarity of the tiglate from, columbianadin. Columbianadin is thus established as the angelate ester of columbianetin ((+)-8-(1-angeloyloxy-1-methylethyl)-8,9-dihydro-2*H*-furo(2,3-h)-1-benzopyran-2-one) (**140**), (angelic acid readily isomerizes to tiglic acid)[164, 165] (Eq. 37).

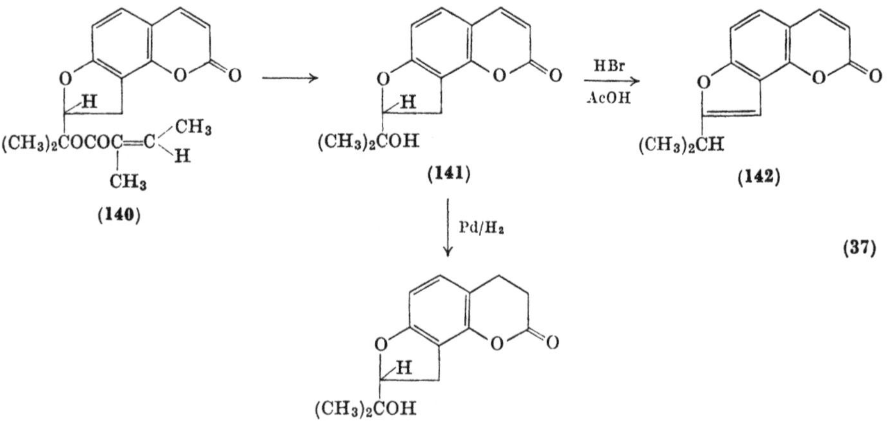

(140) (141) (142) (143) (37)

As evidence for the structure of columbianadin (**140**), Nielsen and Lemmich[123] have independently proposed a similar structure for a furocoumarin isolated from the roots of *Peucedanum palustre*, which has been supported by n.m.r. spectral studies.

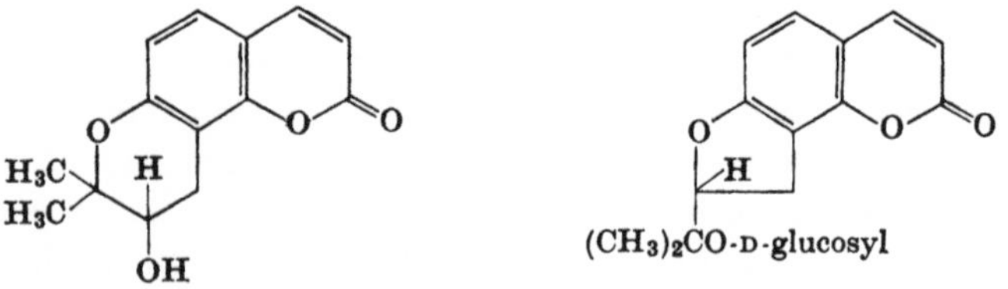

(144) (145)

Columbianin (**145**), $C_{20}H_{24}O_9 . 2H_2O$, is the glucoside of columbianetin (**141**), and its structure follows from its hydrolysis products which are mainly D-glucose and **141**.

CC. *Archangelicin*

Archangelicin (9-angeloyloxy-8-(1-angeloyloxy-1-methylethyl)-8,9-dihydro-2*H*-furo(2,3-h)-1-benzopyran-2-one) (**146**), has been isolated from the roots of *Angelica archangelica* subs *p. litoralis*.[157] The empirical formula was erroneously given as $C_{16}H_{20}O_4$. Treatment of archangelicin with sodium methoxide yielded oroselone (**129**), oroselol methyl ether (**130**), and the coumaric ester (**132**), (Eq. 38).

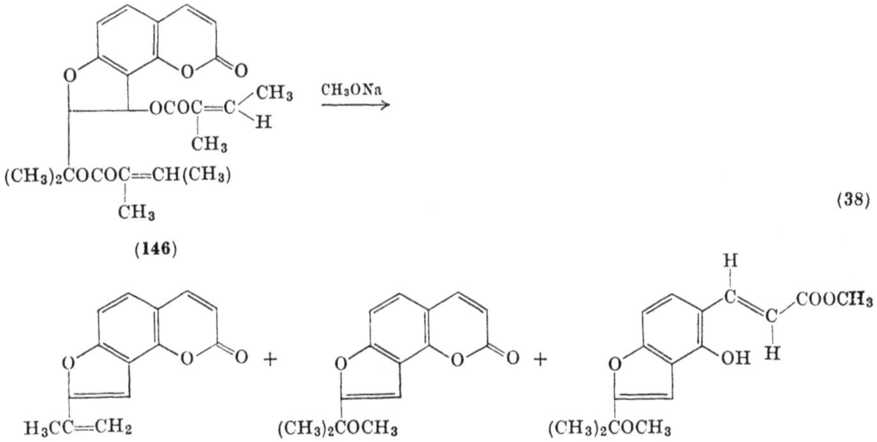

(146) (129) (130) (132) (38)

130 and **132** have been reported as the products formed by a similar treatment of athamantin (**127**),[158] and **129** and **130** as the products formed by saponification of edultin (**137**).[160] As for athamantin, treatment of archangelicin with sodium methoxide at 4–5° yielded the compound (**147**). The volatile acid formed by hydrolysis of **146** with phosphoric acid is angelic acid, and thus the presence of two moles of angelic

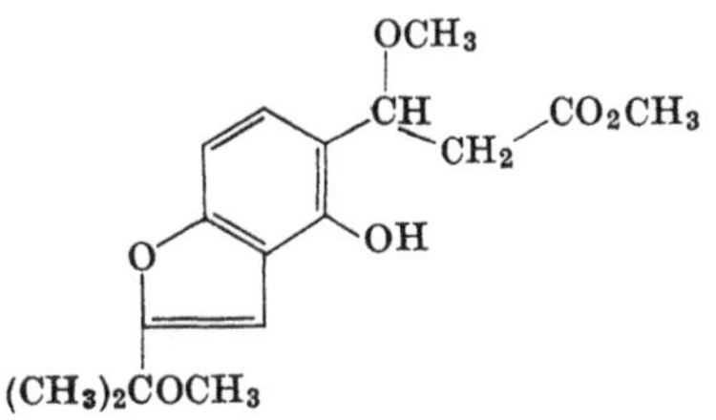

(147)

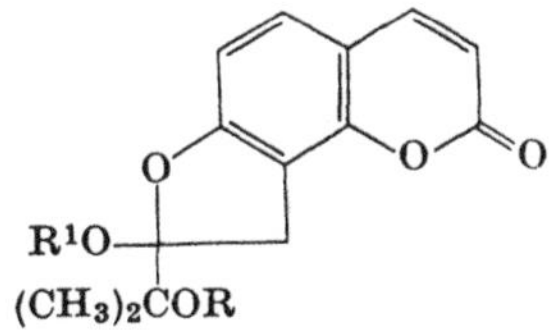

(148)

R = R[1] = angeloyl

acid per mole of **146** was confirmed. This evidence limited the structural possibilities for the coumarin to **146** or **148**, of which only **146** is reconcilable with the p.m.r. data, where signals at 316, 323, and 424, 431 cps. (J, 7 cps.) must arise from the two protons at C_8 and C_9 in structure **146**. The absence of CH_2 signals provides further support for structure **146**.

DD. Archangelin

Archangelin (**149**), $C_{21}H_{22}O_4$, has been isolated from the roots of *archangelica* Linn., together with angelicin (**1**) and prangolarin, the optical isomer of oxypeucedanin (**71**).[124, 166] It is a coumarin derivative as shown by its behavior towards aqueous alcoholic alkali. The n.m.r. spectrum shows a pair of doublets at 3.38 and 2.75 τ (position of α and β protons of furan). It contains an ethylenic double bond, which can be catalytically hydrogenated with the formation of isobergaptol. Furthermore, the latter compound is formed during pyrolysis and during cleavage. Oxidation of **149** with chromic acid yields acetone. Thus, taking into consideration the absence of an epoxide linkage as in prangolarin, and the absence of isopropylidene or $(CH_3)_2COH$ groupings in the compound, it is evident that archangelin must have a potential isopropyloxy group, involved in the ether linkage, which would give rise to acetone. These observations coupled with the mass fragmentation pattern and the n.m.r. spectra confirm the structure of archangelin as **149**.

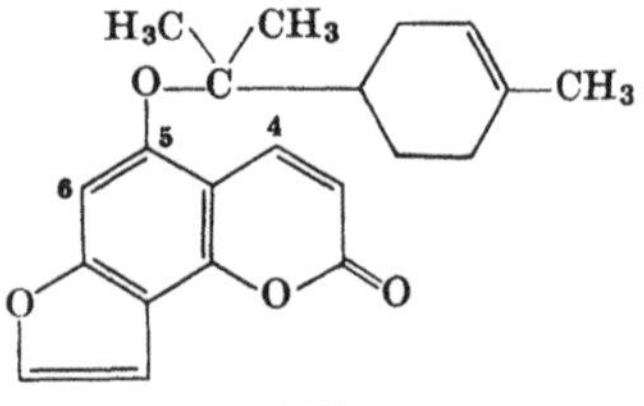

(149)

EE. Pyrocanescin

Pyrocanescin (**151**), $C_{13}H_{10}O_4$, is obtained by pyrolysis of the antibiotic canescin (**150**) from *Pencillium canescens* and *Aspergillus malignus*.[167] It gives a blue ferric test, and its monomethyl derivative gives a negative ferric test. The presence of a hydrogen–bonded enol δ-lactone in pyrocanescin, is supported by alkaline hydrolysis of methylpyrocanescin. After methylation the acid product gave a keto ester, $C_{15}H_{16}O_5$,

ν_{max} 1715, 1735 cm^{-1}; the former band is absent in 2,4-dinitrophenylhydrazone (λ_{max}^{EtOH} 360 mμ). Therefore hydrolysis has generated a carboxyl group and an unconjugated carbonyl; the latter according to n.m.r. spectra found in the CH_3CO group. The presence of an α-methylfuran ring is also suggested. A probable structure, for pyrocanescin, supported by n.m.r. evidence is **151** (Eq. 39).

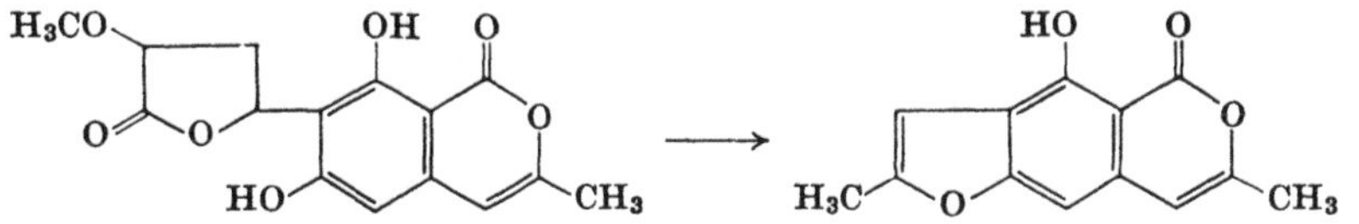

(150) **(151)**

(39)

FF. 4,5′,8-*Trimethylpsoralen*

Recently, 4,5′,8-trimethylpsoralen (**152**) and xanthotoxin (**47**) have been isolated from diseased cultivated celery (*Opium graveolens*), caused by the fungus *Sclertina sclerotiorum.*[168]

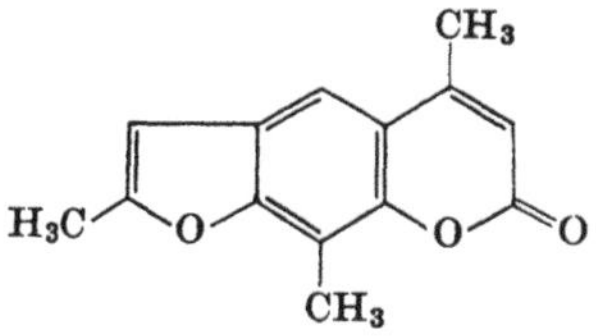

(152)

GG. Aflatoxins B and G

Aflatoxins B and G are two major metabolites of the fungus *Aspergillus flavus* Link ex Fries which causes unusually toxic outbreaks in several domestic animals.[169–173]

Catalytic reduction of aflatoxin B (**153**), $C_{17}H_{12}O_6$, over palladized charcoal in ethanol, was complete after three moles of hydrogen had been absorbed with the production of **154**, $C_{17}H_{16}O_5$. The nature of the sixth oxygen atom and of the remaining carbon atoms was revealed by the n.m.r. spectrum pattern of **153**, which can only arise from the four protons of a dihydrofuran ring, and an assignment is made to an aromatic and three methoxy protons. The spectrum of the reduction product, (**154**), lacked signals for vinylic protons, but the peaks due to the acetal, aromatic, and methoxyl protons were still present. Aflatoxin B, thus, has structure **153.**[46]

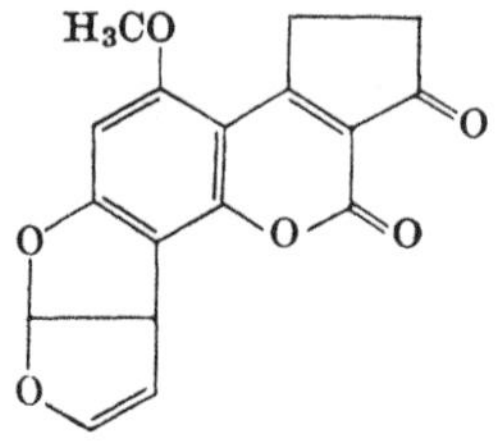

(153)

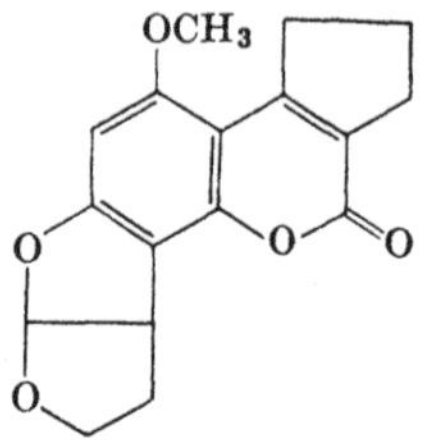

(154)

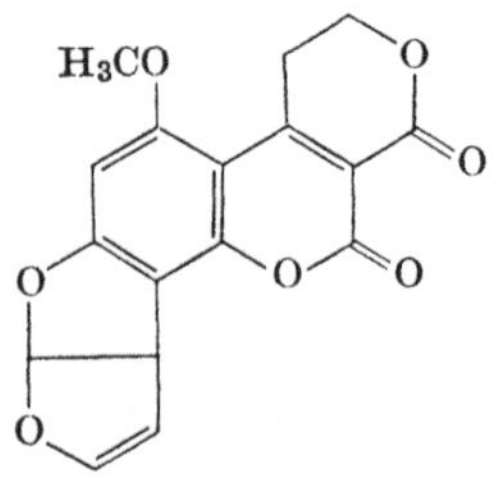

(155)

Aflatoxin G (**155**), $C_{17}H_{12}O_7$,[174] has an n.m.r. spectrum with an A_2X_2 pattern and the chemical shifts and multiplicities for all other protons are identical with those of aflatoxin B (**153**). It is concluded that aflatoxin G has structure **155.**

2. Configuration of Furocoumarins

Nielsen and Lemmich,[123] and Wilette and Soine[161] have shown that (+)-dihydroöroselol (**156**), obtained from columbianadin (**160**), and columbianin (**161**) has the same configuration at C_8 as archangelicin (**157**) and athamantin (**159**). These conclusions were based on the fact that the coumarins could be converted to (+)-tetrahydroöroselol (**165**). Further, the absolute configuration of **156** has been determined[175] by ozonolysis of **156** to the aldehyde (**166**), which upon oxidation yielded (+)-hydroxydihydrotubaic acid (**167**) (Eq. 39a).

(−)-Hydroxydihydrotubaic acid (**168**) was obtained from natural rotenone (**169**). Rotenone was cleaved by alkali to give (−)-tubaic acid **157**,[176] which upon dissolution in an acetic–sulfuric acid mixture at room temperature was converted into **168** (Eq. 39b).

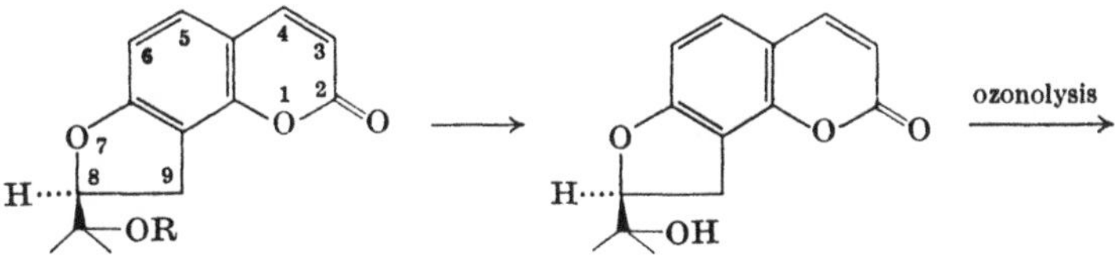

(**165**) H replaces OR
(**160**) R = angeloyl
(**161**) R = D-glucosyl

(**156**)

(39a)

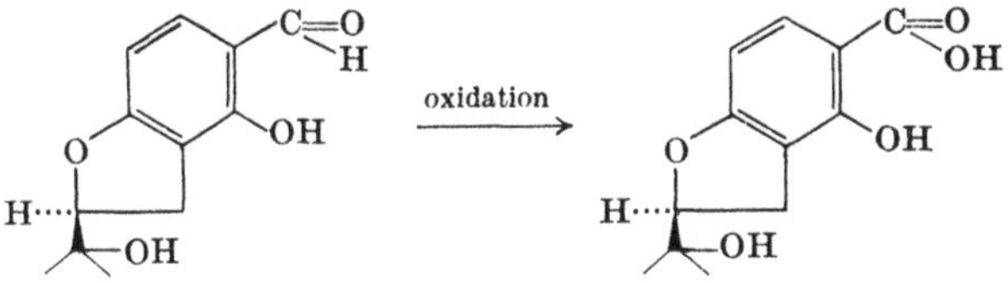

(**166**) (**167**)

As the configuration of (−)-tubaic acid (**157**) has been shown to be (R),[177] it follows that the configuration of (+)-dihydroöroselol (**156**) and hence of the coumarins **158**, **159** is 8(S).

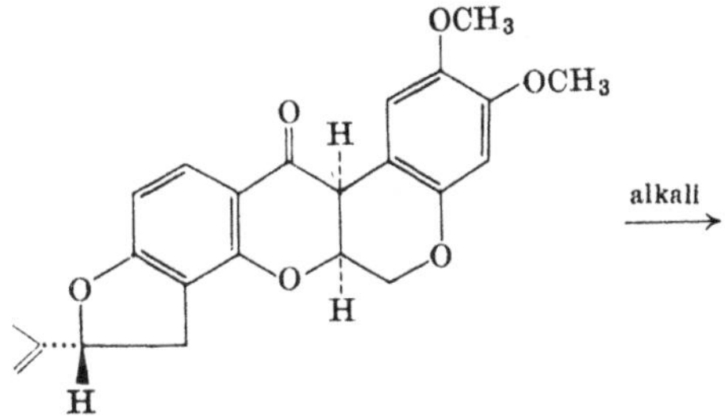

(**169**)

(39b)

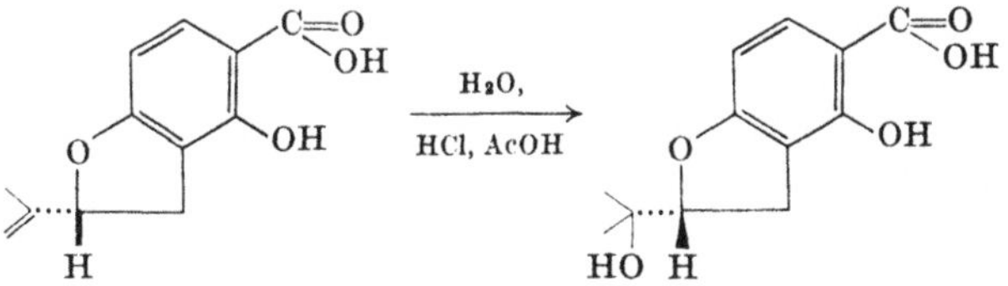

(**157**) (**168**)

A comparison of the rotation value of marmesin (**162**) and the natural chromone (**164**)[175, 178] allows the configuration 2(S) to be assigned to **162** and **164**. The coumarin, nodakenetin, (**163**) has been shown to be the optical antipode of marmesin (**162**).[144]

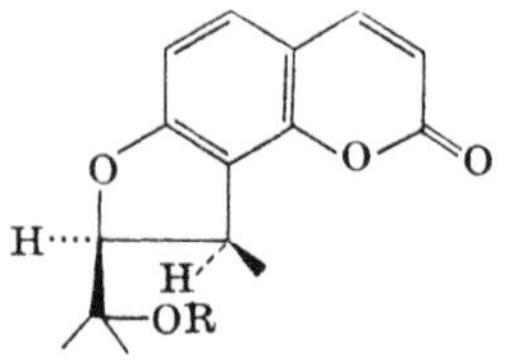

(**158**) R = angeloyl
(**159**) R = isovaleroyl

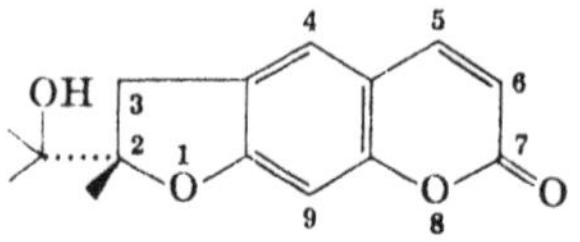

(**162**)

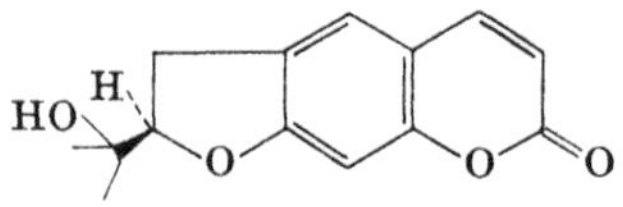

(**163**)

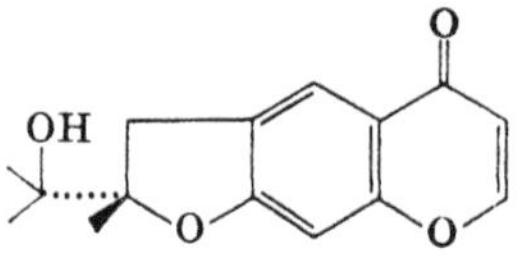

(**164**)

In accordance with the suggestions of Schmid and coworkers,[158] and owing to the molecular rotations,[175] the configuration 9(s) has been assigned to athamantin (**159**). By a comparison of the rotation values of archangelicin (**158**), athamantin (**159**), and columbianadin (**160**), it appears that archangelicin also possesses the configuration 9(s).

3. Biosynthesis of Furocoumarins

The majority of coumarins are formally derivatives of umbelliferone, though a considerable number contain a second hydroxyl group (alkoxyl group) at position 5, and a few at the 4–position.

Isoprenoid groups are present in many phenolic compounds; they may appear as *O*- or *C*–alkyl substituents or involved in ring formation with an adjacent hydroxyl group.[179] Multiples of the C_5 unit, such as the geranyl (C_{10}), are also found. There seems to be adequate justification for the suggestion that these C_5 units are introduced into the benzopyrone at almost the last stage. This is supported by the positions they occupy, and the fact that frequently the unsubstituted nuclei occur together with the C_5–substituted compounds. Methylation, including nuclear methylation,[180] of these compounds is analogous.

Little definite is known about the origin of these isoprenoid groups. Geissman and Hinreiner[181] have adopted the earlier ideas that the C_5 group is the result of condensation of C_2 and C_3. Robinson[182] has suggested senecioic acid (β,β-dimethylacrylic acid) (**170**) as the terpene precursor, and the carbonyl groups of **170** as the spearhead of the attack on the aromatic nuclei. More recent work has emphasized the importance of mevalonic acid (β-hydroxy-β-methyl-δ-valerolactone) (**171**).[183] The dihydroxy acid from **171** may be considered to undergo initial oxidation of the primary alcoholic group to the aldehyde and dehydration to produce a double bond, giving **172**.[184] Reaction of the aldehyde (**172**) with an activated nuclear position of a phenolic compound, e.g. phloroglucinol and decarboxylation, would lead to the formation of **173**, which could be regarded as the primary stage. Other methods of forming **173** are possible, for example the senecioic acid hypotheses[182] suggest the formation of a ketone which could be considered to undergo selective reduction. The aldehyde acid (**172**) suggestion appears more convenient, particularly for building up multiple C_5 units.

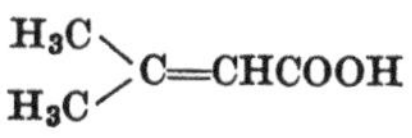

(**170**)

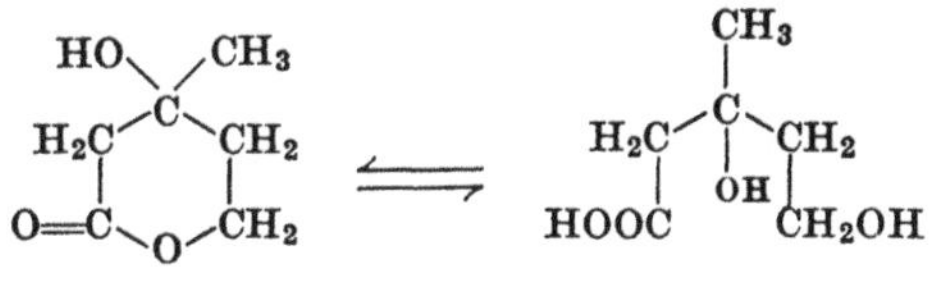

(**171**)

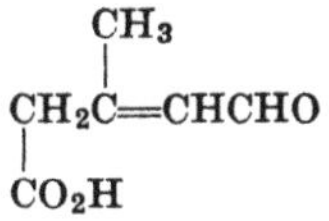

(**172**)

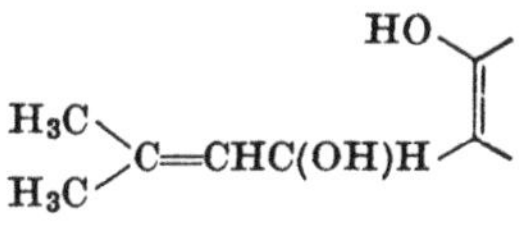

(**173**)

A. α-(β′-Hydroxypropyl)dihydrofurans and α-Isopropenyldihydrofurans

Cyclization of the epoxide (type **174**, R = H) and the glycol (type **175**, R = H) with an adjacent phenolic hydroxyl can give rise to the dihydrofuran structure found in visamminol (**177**)[170] and (**176**).[144] Nodakenin, a glucoside (**178**)[148] on hydrolysis gives nodakenetin, a stereoisomer of marmesin (**176**). Loss of water leads to the modified structure, isopropenyl dihydrofuran (**179**), which is characteristic of rotenone (**180**).[185] These changes can be effected in the laboratory, for example 7-demethylsuberosin (**181**) undergoes epoxidation and cyclization to yield (±)-marmesin (**176**).[150] The final dehydration, however, yields an iosmeric isopropylfuran (**182**) owing to the instability of the isopropenyl compounds in the presence of acids (cf. isomerization of rotenone into isorotenone).[185]

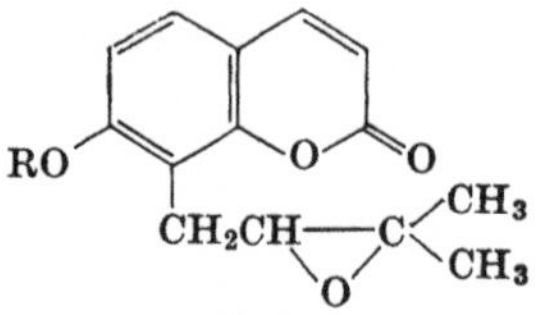

(174)

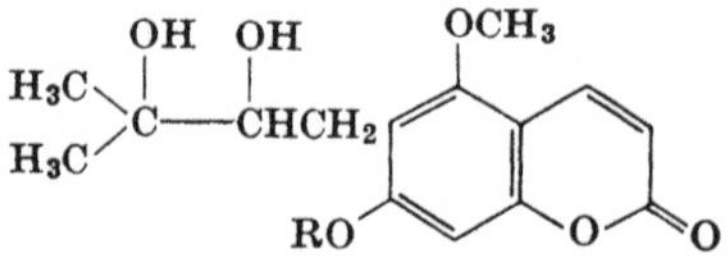

(175)

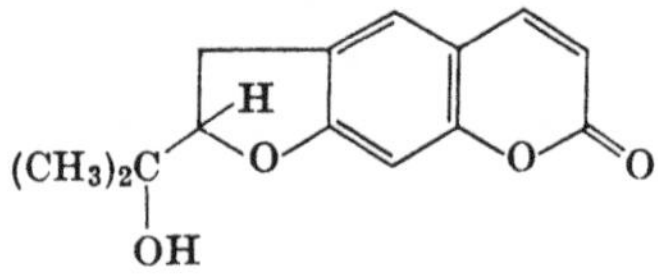

(176)

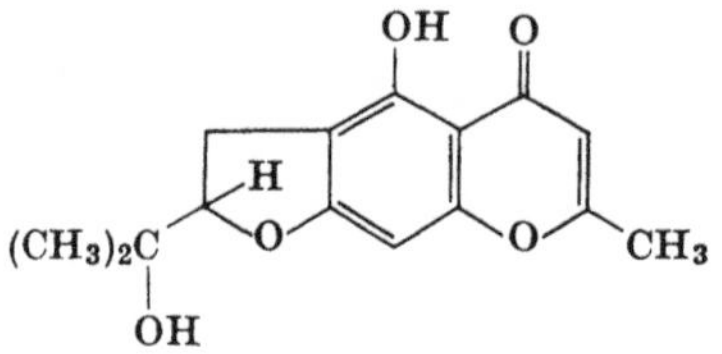

(177)

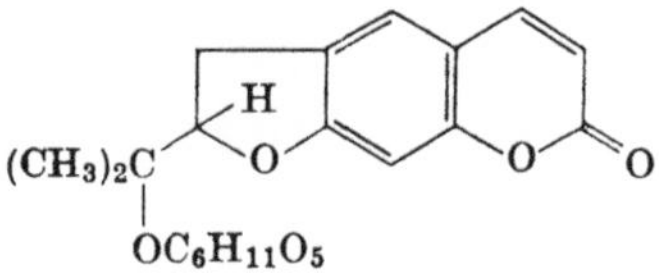

(178)

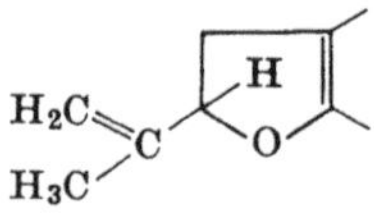

(179)

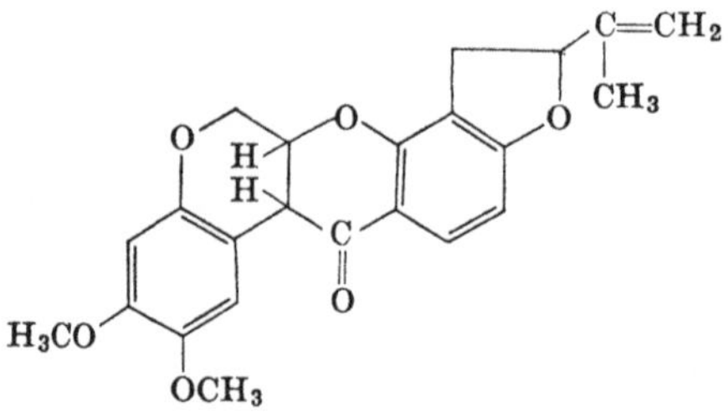

(180)

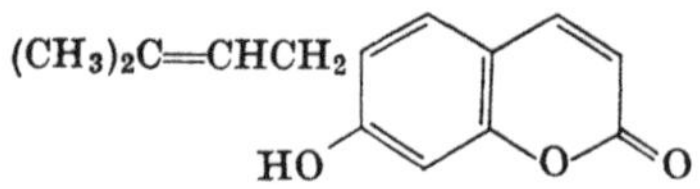

(181)

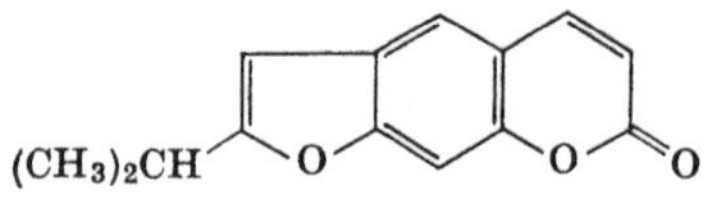

(182)

B. α-Isopropyl-β-hydroxyfurans and Relations

These are similar to the above type, and oreoselone (**183**),[186] and its methyl ether, peucedanin (**184**)[186] are derived directly from the common precursor **173**. For their formation, it is suggested[55] that **173** is oxidized to the epoxide (**185**) or the triol (**186**). Rearrangement of the epoxide or dehydration of the triol yields a ketone (**187**), which can be cyclized to **183**. On the other hand, the epoxide (**185**) or the triol (**186**) can cyclize to form the dihydroxydihydrofuran derivative (**188**) which is found in athamantin (**189**)[187, 188] as the diester of isovaleric acid. On acid hydrolysis athamantin is dehydrated to form oroselone (**190**)[156, 187] (Eq. 40).

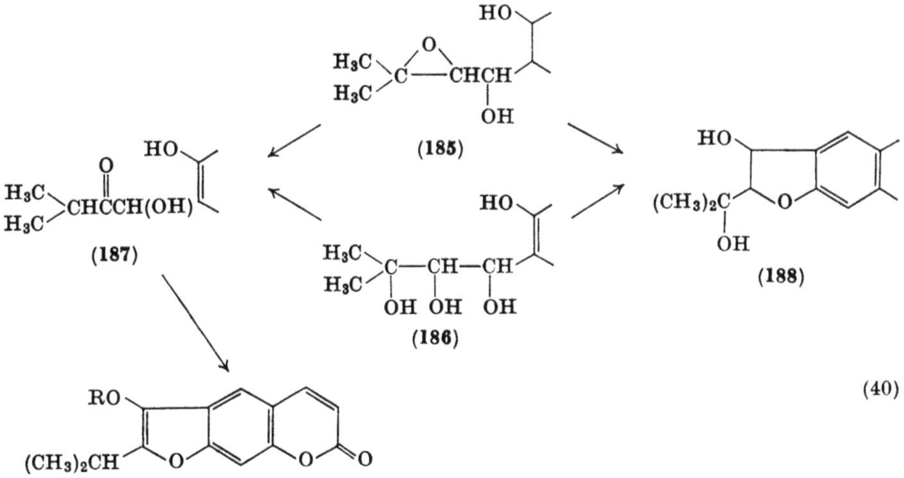

(40)

(**183**) R=H
(**184**) R=CH_3

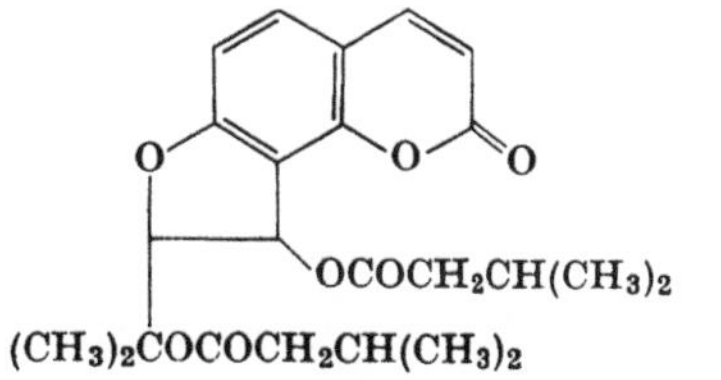

(**189**)

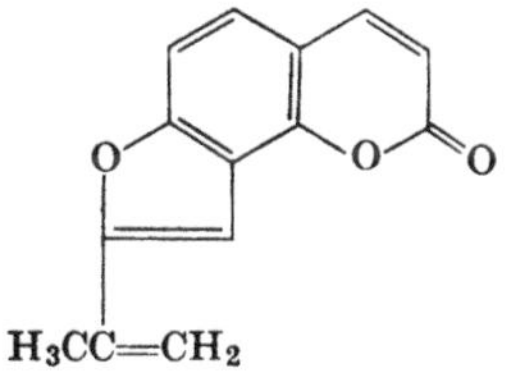

(**190**)

C. Simple Furans

The biogenesis of unsubstituted furan structures has been far more difficult to understand. Furocoumarins are the earliest known, however, furo derivatives of 2-methylchromones and flavones are now also well represented.

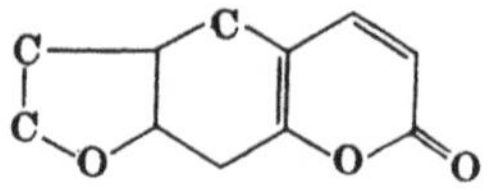

(191)

Späth[1] considered that the four carbon atoms of the furan ring together with one carbon atom of the central benzene ring could constitute an isopentane unit as shown in **191**. However, this is not consistent with his view of the origin of the benzene ring in simple coumarins and those carrying isopentane substituents; in these cases Späth preferred a carbohydrate origin for the benzene ring. Haworth[189] suggested that theoretically the unsubstituted furan rings of these natural coumarins could be derived by elimination of propane from a hypothetical α-isopropyldihydrofuran (**192**). Geissman and Hinreiner[181] also examined this possibility. A two carbon phosphorylated keto alcohol moiety (**193**) was suggested as the precursor, and this was considered to cyclize to furan-3-one (**194**), which subsequently yielded a furan ring by reduction and dehydration (Eq. 41).

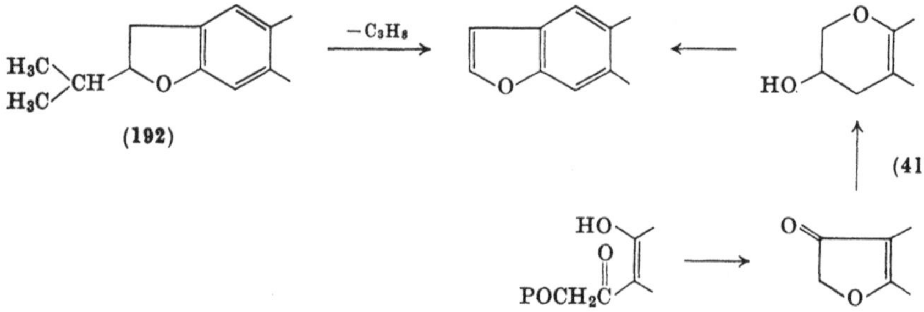

(41)

A review of natural products reveals the remarkable association of compounds having unsubstituted furan rings with those having isopentane units. A more important feature is that, frequently, definite isopentane units and unsubstituted furan rings are found incorporated together in one compound. Based on this close association, Seshadri and coworkers[55] suggested that structure **173**, from which all known types of isopentane structures can be derived, is also the precursor of simple furan rings. The transformation of structure **173** into the glycol (**195**), and oxidative cleavage of the latter, would result in the loss of three carbons, leaving a residue of two as an acetaldehyde (**196**) which on cyclization and dehydration would form unsubstituted furans.[56, 57]

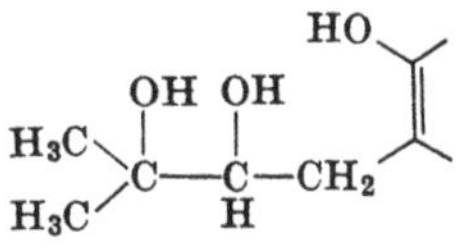
(195)

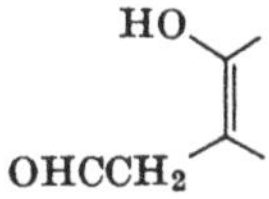
(196)

4. Physiological Activity of Furocoumarins

The subject has been well covered by many investigators. Apart from mentioning the reviews of Soine,[190] Musajo and Rodighiero,[191] Musajo,[192] Seshadri,[193] Pathak and coworkers,[194] Sen,[195] only a brief outline will be presented here.

Natural furocoumarins are not mere metabolic products of the living cell, but they possess varied and often remarkable physiological activities. The role of certain plant juices and extractions as dermal photosensitizing agents has been known for many years. Juices of various parts of these plants, e.g. parsley, celery, figs, and parsnip,[191] after contact with the skin and exposure to sunlight cause changes on mammalian skin manifested by erythema and increased pigmentation. The discovery of this unique activity of the furocoumarin,[196, 197] stimulated much more activity among research workers, mainly Musajo and coworkers.[81, 194, 198–201]

A study of the effect of structural alterations on the erythemal activity of furocoumarins[191, 194] indicates that maximum photosensitizing activity lies in the parent compound, psoralen (**16**), and that the various structurally related compounds have more or less reduced activity, depending on the ring system and the nature of substituents (Table 1b). The presence of a linear unreduced furocoumarin ring system is required for significant activity to be seen. Free phenolic groups inactivate the molecule (xanthotoxol (**54**), bergaptol (**31**)), but the methyl ethers of the two possible phenols are both active (xanthotoxin (**47**), bergapten (**25**)). However, the dimethyl ether, isopimpinellin (**82**), is inactive. Etherifying groups larger than methyl result in progressive reduction of activity as the size of the group increases. Nuclear substitution with methyl groups can cause the loss of activity to be retained depending on the position of the group. Thus, a methyl at the 4,4′,5′, or 8–positions may or may not inhibit the activity, but a methyl group at the 3–position invariably does so. Little success has been noted with the introduction of nitro, amino, or acetylamino group (Table 1b).

In an effort to determine the mechanism of the photodynamic effect of furocoumarins, Fowlks and coworkers[202–204] have studied the effect

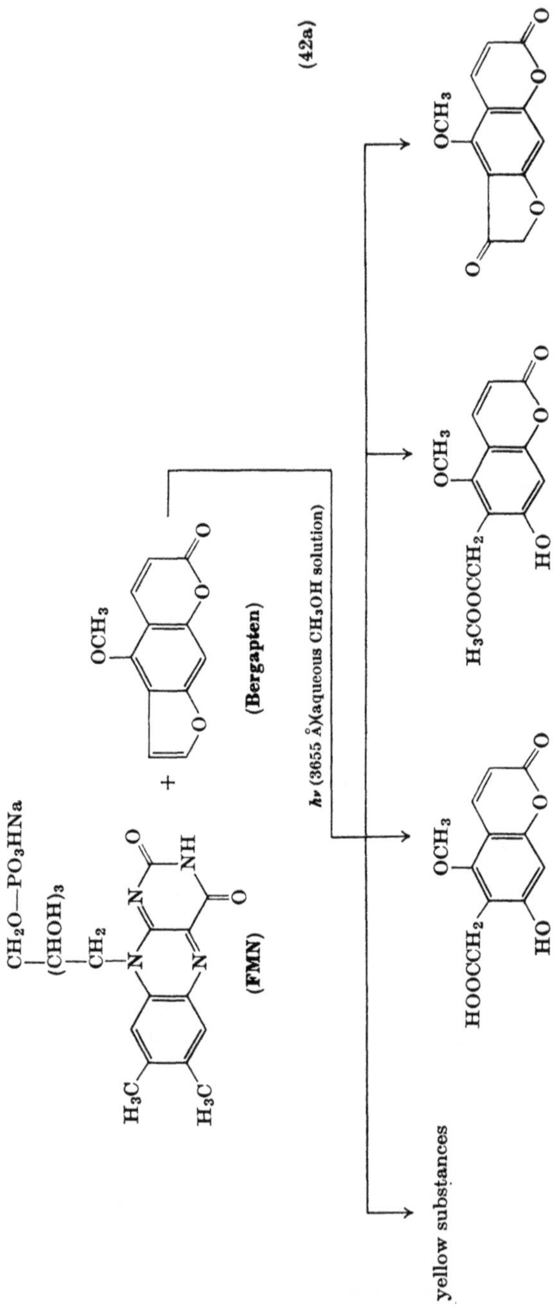
(FMN)
(Bergapten)
hν (3655 Å)(aqueous CH₃OH solution)
yellow substances
(197)
(198)
(199)

(42a)

of furocoumarins on bacteria. It seems that furocoumarins kill bacterial cells. Pathak and Fellman[33] have noted that all biologically active furocoumarins inducing photosensitization, possess absorption and fluorescence peaks of 320–360, and 420–460 mμ, respectively. The long wavelength ultraviolet lamp (3200 Å) is needed to record the response of the skin (erythema and sun tanning) to the coumarins. Furocoumarins have been shown not to act by a photoöxidative mechanism, although other photosensitizing molecules, such as the hematoporphyrins, are known to act in this way. Rodighiero and Capellina[110] have shown that although furocoumarins are dimerized under the influence of irradiation, the dimers are biologically inactive. Pathak and coworkers[205] have explored the possibility that free radicals may be generated from excited furocoumarin molecules under ultraviolet irradiation, but allowing for the biological changes that could occur in such an irradiated system, this probably does not adequately explain the marked activity of these compounds.[191] Recently, Musajo and coworkers[192, 206] have observed that flavinmononucleotide (FMN) will react only with the furocoumarins that are photodynamically active and that the reaction products appear to have been modified mainly in the furan ring. Furthermore, they have demonstrated that FMN in large amounts acts against erythema expected from the psoralen–type molecule. Three new coumarin derivatives have been isolated in the bergapten photoreaction, namely, 7-hydroxy-5-methoxycoumarin-6-acetic acid (**197**), its methyl or ethyl ester (**198**) according to the presence of methyl or ethyl alcohol in the irradiated solution, and probably, 4′,5′-dihydro-4′-oxo-5-methoxy-furocoumarin (**199**) (Eq. 42a).

Two substances have been obtained by the photoreaction of FMN and psoralen in a water-methyl alcohol solution, namely, the methyl ester of 7-hydroxycoumarin-6-acetic acid (**200**), and 6-formyl-7-hydroxy-coumarin (**201**). No new compounds are formed in the photoreaction of FMN and xanthotoxin.[192]

In spite of these findings, the precise mechanism by which furocoumarins function in the treatment of leucoderma is unknown. A full knowledge of the photoreaction, noticed *in vitro*, may give some insight into the mechanism of the sensitizing effect of furocoumarins (Eq. 42b).

Rodighiero and coworkers[207] found that psoralen and xanthotoxin significantly inhibited the growth of the tubercle bacillus. In Souton medium tuberculostatic activity has been tested for pimpinellin (**94**), isopimpinellin (**82**), and aminobergapten (**27**).[208]

The antifungal activity of furocoumarins has been studied by Chakraborty and coworkers[209] who reported that psoralen and

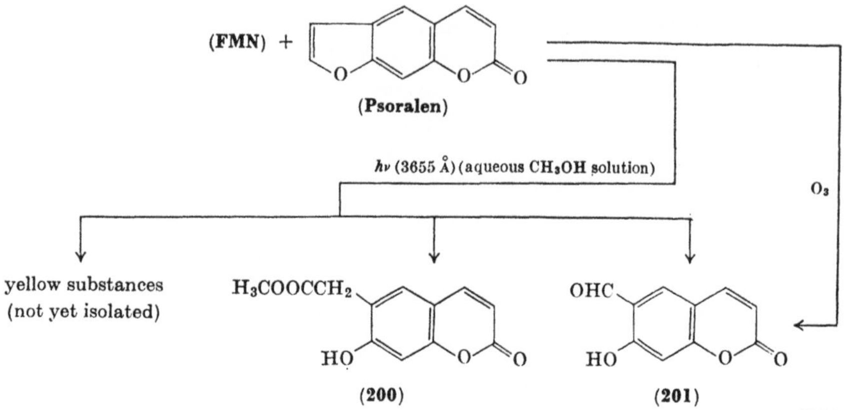

(42b)

imperatorin (**76**) were the most effective antifungal agents tested. A similar result has been confirmed recently.[210]

The relation between structure and antibiotic properties of some natural furocoumrains has been discussed.[211]

Bergapten, isopimpinellin, and less effectively, xanthotoxin have been shown to exert a molluscacidal activity.[212]

Späth and Kuffner[213] noted that naturally occurring furocoumarins had appreciable toxic action on fresh water fish. α-Methylfuro(2′,3′–7,8)coumarin was fairly toxic, but its dihydro derivative was considerably less toxic. Substitution of a methyl group at the 4–position of the latter compound, enhanced the toxicity, whereas, the introduction of a methoxyl group at the 6–position again reduced it. A 3–phenyl group enhanced the toxicity considerably, but the total effect did not exceed the toxicity shown by a simpler compound, 7-methoxy-3-phenylcoumarin. Seshadri and coworkers[193] have concluded that the introduction of a costlier furan ring does not seem to be particularly advantageous.

V. References

1. E. Späth, *Chem. Ber.*, **70A**, 83 (1937).
2. F. M. Dean, in *Fortschritte der Chemie Organischer Naturstoffe*, Vol. 9, (Ed. L. Zechmeister), Wien Springer, Verlag, Austria, 1952, p. 225.
3. F. M. Dean, *Naturally Occurring Oxygen Ring Compounds*, Butterworth, London, 1963, pp. 176–220.
4. L. Reppel, *Pharmazie*, **9**, 278 (1954).
5. W. Karrer, *Konstitution und Vorkommen Der Organischen Pflanzenstoffe*, Birkenhäuser Verlag Basel und Stuttgart, 1958, p. 531.

6. T. Noguchi and M. Kawanami, *J. Pharm. Soc. Japan*, **60**, 57 (1940); *Chem. Abstr.*, **34**, 3717 (1940).
7. A. B. Svendsen, *Pharm. Acta Helv.*, **27**, 44 (1952); *Chem. Abstr.*, **46**, 9255 (1952).
8. A. Chatterjee and A. Choudhury, *Naturwiss.*, **43**, 535 (1955); *Chem. Abstr.*, **51**, 11339 (1957).
9. M. Calvarano, *Essenze Deriv. Agrumari*, **27**, 173 (1957); *Chem. Abstr.*, **52**, 10507 (1958).
10. W. L. Stanley and S. H. Vannier, *U.S. Pat.*, 2,839,337 (1959); *Chem. Abstr.*, **53**, 220 (1959).
11. E. Stähl and P. J. Schorn, *Z. Physiol. Chem.*, **325**, 263 (1961); *Chem. Abstr.*, **57**, 4761 (1962).
12. L. Peyron, *Compt. Rend.*, **257**, 235 (1963); *Chem. Abstr.*, **59**, 9082 (1963).
13. N. S. Vaul'fson, V. I. Zeretskii and L. S. Chetverikova, *Izv. Akad. Nauk SSSR, Ser. Khim.*, **8**, 1903 (1963); *Chem. Abstr.*, **59**, 15584 (1963).
14. E. Späth and F. Vierhapper, *Monatsh.*, **72**, 179 (1939); *Chem. Abstr.*, **33**, 2505 (1939).
15. K. Riedl and L. Neuberger, *Monatsh.*, **83**, 1083 (1952); *Chem. Abstr.*, **47**, 9967 (1953).
16. D. P. Chakraborty, B. K. Barman and K. R. Guha, *Trans. Bose Res. Inst. (Calcutta)*, **24**, 211 (1961); *Chem. Abstr.*, **58**, 12850 (1963).
17. L. Fabbrini, *Sperimentale Sez. Chim. Biol.*, **6**, 7 (1955); *Chem. Abstr.*, **50**, 10342 (1956).
18. F. Wessely and J. Kotlan, *Monatsh.*, **86**, 430 (1955); *Chem. Abstr.*, **50**, 11330 (1956).
19. D. P. Chakraborty and H. C. Chakraborty, *Sci. Cult.* (*Calcutta*), **22**, 117 (1956); *Chem. Abstr.*, **53**, 4019 (1959).
20. S. Berlingozzi and V. Parrini, *Sperimentale Sez. Chim. Biol.*, **6**, 59 (1956); *Chem. Abstr.*, **51**, 4651 (1957).
21. W. L. Stanley and S. H. Vannier, *J. Am. Chem. Soc.*, **79**, 3488 (1957).
22. A. B. Svendsen, *Zur Chemie Norwegischer Umbellifern*, Johan Grandt Tanum Forlag, Oslo, Norway, 1954, p. 144; *Chem. Abstr.*, **52**, 2173 (1958).
23. J. M. Miller and J. G. Kirchner, *Anal. Chem.*, **26**, 2002 (1954).
24. J. Dutta, H. C. Chakraborty and B. K. Barman, *Sci. Cult.* (*Calcutta*), **28**, 84 (1962); *Chem. Abstr.*, **58**, 10763 (1963).
25. S. A. Brown and J. P. Shyluk, *Anal. Chem.*, **34**, 1058 (1962); *Chem. Abstr.*, **57**, 10514 (1962).
26. Y. Ichimura, *Yakugaku Zasshi*, **80**, 775 (1960); *Chem. Abstr.*, **54**, 21646 (1960).
27. R. H. Goodwin and F. Kavanagh, *Arch. Biochem.*, **27**, 152 (1950); *Chem. Abstr.*, **44**, 10519 (1950).
28. R. H. Goodwin and F. Kavanagh, *Arch. Biochem. Biophys.*, **36**, 442 (1952); *Chem. Abstr.*, **47**, 10519 (1953).
29. F. Feigl, H. E. Feigl and D. Goldstein, *J. Am. Chem. Soc.*, **77**, 4162 (1955).
30. T. Nakabayashi, T. Tokoroyama, H. Miyazaki and S. Isono, *J. Pharm. Soc. Japan*, **73**, 669 (1953).
31. R. Andrisano and G. Pappalardo, *Gazz. Chim. Ital.*, **83**, 108 (1953); *Chem. Abstr.*, **47**, 11981 (1953).
32. D. K. Chatterjee, R. M. Chatterjee and K. Sen, *J. Org. Chem.*, **29**, 2467 (1964).
33. M. A. Pathak and J. H. Fellman, *Nature*, **185**, 382 (1960).

34. G. Caporale and E. Cinzolani, *Rend. Ist. Super. Sanita'*, **21**, 943 (1958); *Chem. Abstr.*, **53**, 9811 (1959).
35. K. Sen and P. Bagchi, *J. Org. Chem.*, **24**, 316 (1959).
36. D. P. Chakraborty and S. K. Chakraborty, *Trans. Bose Res. Inst. (Calcutta)*, **24**, 15 (1961); *Chem. Abstr.*, **55**, 24228 (1961).
37. M. E. Brokke and B. E. Christensen, *J. Org. Chem.*, **23**, 589 (1958).
38. S. A. Böhme and T. Sverin, *Arch. Pharm.*, **290**, 486 (1957); *Chem. Abstr.*, **52**, 2536 (1958).
39. L. H. Briggs and L. D. Colebrook, *J. Chem. Soc.*, 2458 (1960).
40. I. P. Kovalev, A. P. Prokopenko and E. V. Titov, *Ukr. Khim. Zh.*, **29**, 740 (1963); *Chem. Abstr.*, **59**, 13480 (1963).
41. K. D. Kaufman, J. F. W. Keana, R. C. Kelly, D. W. McBride and G. Slomp, *J. Org. Chem.*, **27**, 2567 (1962).
42. H. S. Gutowsky, G. H. Holm, A. Saika and G. A. Williams, *J. Am. Chem. Soc.*, **79**, 4596 (1957).
43. M. Karplus, *J. Chem. Phys.*, **30**, 11 (1959).
44. D. R. Davis, R. P. Lutz and J. D. Roberts, *J. Am. Chem. Soc.*, **83**, 246 (1961).
45. A. Mustafa, W. Asker, O. H. Hishmat, M. I. Ali, A. K. E. Mansour, N. M. Abed, K. M. A. Khalil and S. M. Samy, *Tetrahedron*, **21**, 849 (1965).
46. T. Asao, G. Büchi, M. M. Abdel-Kader, S. B. Chang, E. L. Wick and G. N. Wogan, *J. Am. Chem. Soc.*, **85**, 1706 (1963).
47. A. P. Prokopenko and C. O. Tarasenko, *Farmatsevt. Zh. (Kiev)*, **17**, 18 (1962); *Chem. Abstr.*, **59**, 10766 (1963).
48. A. Schönberg, N. Badran and N. A. Starkowsky, *J. Am. Chem. Soc.*, **77**, 5438 (1955).
49. G. N. Nikonov, N. I. Rodina and L. A. Sapunova, *Med. Prom. SSSR*, **17**, 46 (1963); *Chem. Abstr.*, **59**, 13113 (1963).
50. A. C. Curtius, *J. Invest. Dermatol.*, **32**, 133 (1959).
51. E. Späth and O. Pesta, *Chem. Ber.*, **67**, 853 (1934).
52. H. S. Jois and B. L. Manjunath, *Chem. Ber.*, **69**, 964 (1936).
53. B. Krishnaswamy and T. R. Seshadri, *Proc. Indian Acad. Sci.*, **10A**, 151 (1942); *Chem. Abstr.*, **37**, 1430 (1943).
54. E. Späth and M. Pailer, *Chem. Ber.*, **67**, 1212 (1934).
55. R. Aneja, S. K. Mukerjee and T. R. Seshadri, *Tetrahedron*, **4**, 256 (1958).
56. R. Aneja, S. K. Mukerjee and T.R. Seshadri, *Tetrahedron*, **2**, 203 (1958).
57. R. Aneja, S. K. Mukerjee and T. R. Seshadri, *Tetrahedron*, **3**, 230 (1958).
58. K. S. Raizada, P. S. Sarin and T. R. Seshadri, *J. Sci. Ind. Res.*, **19**B, 76 (1960); *Chem. Abstr.*, **55**, 22303 (1961).
59. E. Späth and M. Pailer, *Chem. Ber.*, **68**, 940 (1935).
60. R. M. Naik and V. M. Thakor, *J. Org. Chem.*, **22**, 1696 (1957).
61. S. Tanaka, *J. Am. Chem. Soc.*, **73**, 872 (1951).
62. Y. Kawase, M. Nakayama and H. Tamatsukuri, *Bull. Chem. Soc. Japan*, **35**, 149 (1962).
63. D. B. Limaye, *Rasayanam*, **1**, 1 (1936).
64. K. Okahara, *Bull. Chem. Soc. Japan*, **11**, 389 (1936); *Chem. Abstr.*, **30**, 7575 (1936).
65. E. Späth, K. Okahara and F. Kuffner, *Chem. Ber.*, **70**, 73 (1937).
66. E. Späth, B. L. Manjunath, M. Pailer and H. S. Jois, *Chem. Ber.*, **69**, 1087 (1936).

67. K. D. Kaufman, *J. Org. Chem.*, **26**, 117 (1961).
68. E. C. Horning and D. E. Reisner, *J. Am. Chem. Soc.*, **70**, 3619 (1948).
69. R. T. Foster, A. Robertson and A. Bushra, *J. Chem. Soc.*, 2254 (1948).
70. R. C. Esse and B. E. Christensen, *J. Org. Chem.*, **25**, 1565 (1960).
71. N. Ray, S. S. Silooja and V. R. Vaid, *J. Chem. Soc.*, 812 (1935).
72. G. Caporale and C. Antonello, *Farmaco (Pavia) Ed. Sci.*, **13**, 363 (1958); *Chem. Abstr.*, **53**, 21917 (1959).
73. K. D. Kaufman, F. J. Gaiser, T. D. Leth and L. R. Worden, *J. Org. Chem.*, **26**, 2443 (1961).
74. K. D. Kaufman and L. R. Worden, *J. Org. Chem.*, **26**, 4721 (1961).
75. K. D. Kaufman, W. E. Russey and L. R. Worden, *J. Org. Chem.*, **27**, 875 (1962).
76. Pomeranze, *Monatsh.*, **12**, 379 (1891); **14**, 28 (1893).
77. H. Thoms and E. Baetcke, *Chem. Ber.*, **45**, 3705 (1912).
78. E. Späth, F. Wessely and G. Kubiczek, *Chem. Ber.*, **70**, 478 (1937).
79. W. N. Howell and A. Robertson, *J. Chem. Soc.*, 293 (1937).
80. G. Caporale, *Ann. Chim. (Rome)*, **48**, 650 (1958).
81. G. Caporale, *Farmaco (Pavia) Ed. Sci.*, **13**, 784 (1958).
82. A. Schönberg, N. Badran and N. A. Starkowsky, *J. Am. Chem. Soc.*, **77**, 1019 (1955).
83. A. Schönberg, N. Badran and N. A. Starkowsky, *J. Am. Chem. Soc.*, **77**, 5390 (1955).
84. A. Schönberg and G. Aziz, *J. Am. Chem. Soc.*, **77**, 2563 (1955).
85. E. Späth and G. Kubiczek, *Chem. Ber.*, **70**, 1253 (1937).
86. G. Rodighiero and G. Caporale, *Farmaco (Pavia) Ed. Sci.*, **10**, 760 (1955); *Chem. Abstr.*, **50**, 10091 (1956).
87. E. Späth, F. Wessely and P. Kainrath, *Chem. Ber.*, **70**, 243 (1937).
88. R. T. Foster, W. H. Howell and A. Robertson, *J. Chem. Soc.*, 930 (1939).
89. F. Wessely and E. Nadler, *Monatsh.*, **60**, 141 (1932); *Chem. Abstr.*, **26**, 4603 (1932).
90. E. Späth and A. F. J. Simon, *Chem. Ber.*, **67**, 344 (1934); *Chem. Abstr.*, **28**, 3802 (1934).
91. F. Wessely and F. Kallab, *Monatsh.*, **59**, 161 (1932); *Chem. Abstr.*, **26**, 2441 (1932).
92. G. Rodighiero and C. Antonello, *Farmaco (Pavia) Ed. Sci.*, **10**, 3 (1955); *Chem. Abstr.*, **50**, 12037 (1956).
93. E. Späth and L. Socias, *Chem. Ber.*, **67**, 59 (1934).
94. H. V. Soden and W. Rojahn, *Pharm. Ztg.*, **46**, 778 (1901); *Chem. Zentr.* **II**, 930 (1901).
95. E. Späth and P. Kainrath, *Chem. Ber.*, **70**, 2272 (1937); *Chem. Abstr.*, **32**, 550 (1938).
96. A. Chatterjee and B. Chaudhury, *J. Chem. Soc.*, 2246 (1961).
97. H. Thoms, *Chem. Ber.*, **44**, 3325 (1911).
98. (a) A. Schönberg and A. Sina, *J. Am. Chem. Soc.*, **72**, 4826 (1950). (b) A. Schönberg and A. Aziz, *J. Org. Chem.*, **23**, 590 (1958).
99. B. B. Dey and K. K. Row, *J. Chem. Soc.*, 125, 554 (1924).
100. M. E. Brokke and B. E. Christensen, *J. Org. Chem.*, **24**, 523 (1959).
101. E. Späth and M. Pailer, *Chem. Ber.*, **69**, 767 (1936).
102. G. Langercrantz, *Acta Chem. Scand.*, **10**, 647 (1956); *Chem. Abstr.*, **51**, 6615 (1957).

103. G. Rodighiero, and C. Antonello, *Ann. Chim. (Rome)*, **46**, 960 (1956); *Chem. Abstr.*, **51**, 6616 (1957).
104. C. Antonello, *Gazz. Chim. Ital.*, **88**, 415 (1958); *Chem. Abstr.*, **53**, 20046 (1959).
105. J. R. Merchant and R. C. Shah, *J. Org. Chem.*, **22**, 884 (1957).
106. W. J. Horton and E. G. Paul, *J. Org. Chem.*, **24**, 2000 (1959).
107. K. D. Kaufman and L. R. Worden, *J. Org. Chem.*, **25**, 2222 (1960).
108. W. Asker, A. F. A. M. Shalaby and S. M. A. Zayed, *J. Org. Chem.*, **23**, 1781 (1958).
109. A Darapsky, H. Berger and A. Neuhaus, *J. Prakt. Chem.*, **147**, 145 (1936).
110. G. Rodighiero and V. Capellina, *Gazz. Chim. Ital.*, **91**, 103 (1961); *Chem. Abstr.*, **56**, 8203 (1962).
111. A. Mustafa, *Chem. Revs.*, **51**, 1 (1952).
112. A. Mustafa, M. Kamel and M. A. Allam, *J. Org. Chem.*, **22**, 888 (1957).
113. F. Wessely and K. Dinjaski, *Monatsh.*, **64**, 131 (1934); *Chem. Abstr.*, **28**, 6435 (1934).
114. (a) A. Schönberg, N. Latif, R. Moubasher and A. Sina, *J. Chem. Soc.*, 1364 (1951), (b) C. H. Krauch, S. Farid and G. O. Schenck, *Chem. Ber.*, **98**, 3102 (1965).
115. E. Späth and F. Vierhapper, *Chem. Ber.*, **70**, 248 (1937).
116. G. Rodighiero, *Atti. Ist. Veneto Sci., Lettere Arti, Sci. Mat. Nat.*, **119**, 201 (1960–1961); *Chem. Abstr.*, **58**, 2419 (1963).
117. A. Mustafa, M. M. Sidky and M. R. Mahran, *Ann. Chem.*, **684**, 187 (1965).
118. E. Späth and L. Kahovec, *Chem. Ber.*, **66**, 1146 (1933).
119. E. Späth and E. Dobrovolny, *Chem. Ber.*, **71**, 1831 (1938).
120. A. Butenandt and A. Marten, *Ann. Chem.*, **495**, 187 (1932).
121. E. Späth and K. Klager, *Chem. Ber.*, **66**, 749 (1933).
122. E. Späth and K. Klager, *Chem. Ber.*, **66**, 914 (1933).
123. (a) B. E. Nielsen and J. Lemmich, *Acta Chem. Scand.*, **18**, 1379 (1964); (b) B. E. Nielsen and J. Lemmich, *Acta Chem. Scand.*, **19**, 601 (1965); (c) B. E. Nielsen and J. Lemmich, *Acta Chem. Scand.*, **19**, 1810 (1965); (d) B. E. Nielsen and J. Lemmich, *Acta Chem. Scand.*, **18**, 2111 (1964).
124. C. R. Ghoshal, S. S. Gupta and A. Chatterjee, *Chem. Ind. (London)*, 1430 (1963); *Chem. Abstr.*, **59**, 11459 (1963).
125. E. Späth and A. v. Christiani, *Chem. Ber.*, **66**, 1150 (1933).
126. E. Späth and H. Holzen, *Chem. Ber.*, **66**, 1137 (1933).
127. E. Späth and H. Holzen, *Chem. Ber.*, **68**, 1123 (1935).
128. S. K. Saha and A. Chatterjee, *J. Indian Chem. Soc.*, **34**, 228 (1957).
129. C. Bhar, *J. Indian Chem. Soc.*, **25**, 139 (1948).
130. G. A. Kuznetsova, *Zh. Obshch. Khim.*, **31**, 3818 (1961); *Chem. Abstr.*, **57**, 12889 (1962).
131. G. A. Kuznetsova and G. V. Pigulevskii, *Zh. Obshch. Khim.*, **31**, 323 (1961); *Chem. Abstr.*, **55**, 2228 (1961).
132. Y. N. Sharma, A. Zaman and A. R. Kidawi, *Tetrahedron.* **20**, 87 (1964).
133. T. Noguchi and M. Kawanami, *Chem. Ber.*, **71**, 344, 1428 (1938).
134. L. H. Briggs and R. C. Cambie, *Tetrahedron*, **2**, 256 (1958); *Chem. Abstr.*, **52**, 18487 (1958).
135. T. Noguchi and M. Kawanami, *Chem. Ber.*, **72**, 483 (1939).
136. P. K. Bose and J. C. Chaudhuri, *Ann. Biochem. Exp. Med.*, **6**, 1 (1946); *Chem. Abstr.*, **41**, 4472 (1947).

137. A. Chatterjee, P. K. Bose and S. K. Saha, *Arch. Pharm.*, **295**, 248 (1962); *Chem. Abstr.*, **58**, 519 (1963).
138. T. Noguchi and M. Kawanami, *J. Pharm. Soc., Japan*, **61**, 77 (1941); *Chem. Abstr.*, **36**, 464 (1942).
139. E. Späth and H. Schmid, *Chem. Ber.*, **74**, 595 (1941).
140. T. R. Seshadri and M. S. Sood, *Indian J. Chem.*, **1**, 291 (1963).
141. M. P. Hegarty and P. N. Lahey, *Australian J. Chem.*, **9**, 120 (1956).
142. W. Baker, *J. Chem. Soc.*, 1684 (1934).
143. J. Arina, *J. Chem. Soc.*, 1684 (1934).
144. A. Chatterjee and S. S. Mitra, *J. Am. Chem. Soc.*, **71**, 606 (1949).
145. E. A. Abu-Mustafa and M. B. E. Fayez, *J. Org. Chem.*, **26**, 161 (1961).
146. J. Arima, *Bull. Chem. Soc., Japan*, **4**, 113 (1929).
147. E. Späth and P. Kainrath, *Chem. Ber.*, **69**, 2062 (1936).
148. E. Späth and E. Tyray, *Chem. Ber.*, **72**, 2089 (1939).
149. B. I. Nurunnabi, *Pakistan J. Sci. Ind. Res.*, **3**, 108 (1960); *Chem. Abstr.*, **55**, 29503 (1961).
150. F. E. King, J. R. Housley and T. J. King, *J. Chem. Soc.*, 1392, (1954).
151. M. Nakajima, J. Oda and H. Fukemi, *Agr. Biol. Chem.* (*Tokyo*), **27**, 695 (1963); *Chem. Abstr.*, **60**, 4121 (1963).
152. N. A. Starkowsky and N. Badran, *J. Org. Chem.*, **23**, 1818 (1958).
153. H. Schmid and A. Ebnöther, *Helv. Chim. Acta*, **34**, 1982 (1951).
154. K. N. Gaind, T. S. Gupta, J. N. Ray and K. N. Sareen, *J. Indian Chem. Soc.*, **23**, 370 (1946).
155. G. N. Nikonov, *Zh. Obshch. Khim.*, **31**, 305 (1961); *Chem. Abstr.*, **55**, 22307 (1961).
156. E. Späth, N. Platzer and H. Schmid, *Chem. Ber.*, **73**, 1309 (1940).
157. B. E. Nielsen and J. Lemmich, *Acta Chem. Scand.*, **18**, 932 (1964).
158. O. Halpern, P. Waser and H. Schmid, *Helv. Chim. Acta*, **40**, 758 (1957).
159. W. Bottomley, *Australian J. Chem.*, **16**, 143 (1963); *Chem. Abstr.*, **59**, 5143 (1963).
160. H. Mitsuhashi and T. Itch, *Chem. Pharm. Bull.* (*Japan*), **10**, 514 (1962); *Chem. Abstr.*, **57**, 16744 (1962).
161. R. E. Wilette and T. O. Soine, *J. Pharm. Sci.*, **55**, 275 (1964); *Chem. Abstr.*, **60**, 15847 (1964).
162. M. E. Perel'son, G. Nikonov, G. Yu. Pek and Yu. N. Skeinker, *Dokl. Akad. Nauk SSSR*, **159**, 154 (1964); *Chem. Abstr.*, **62**, 3999 (1965).
163. S. M. Kupchan and A. Afonso, *J. Org. Chem.*, **25**, 2217 (1960).
164. R. E. Buckles, G. V. Mock and L. Locatell, *Chem. Revs.*, **55**, 659 (1955).
165. S. M. Kupchan and A. Afonso, *Pharm. Sci.*, **48**, 731 (1959).
166. A. Chatterjee and S. S. Gupta, *Tetrahedron Letters*, **29**, 1961 (1964); *Chem. Abstr.*, **61**, 10667 (1964).
167. J. H. Birkinshaw, P. Chaplen, A. H. Manchanda and R. Raino-Martin, *Tetrahedron Letters*, **1**, 29 (1965).
168. L. D. Scheel, V. B. Perone, R. L. Larkin and R. E. Kuppl, *Biochemistry*, **2**, 1127 (1963); *Chem. Abstr.*, **59**, 10480 (1965).
169. R. Allcroft and R. B. A. Carmagham, *Chem. Ind.* (*London*), 50 (1963).
170. K. Sergeant, R. B. A. Carmagham and R. Allcroft, *Chem. Ind.* (*London*), 53 (1963).
171. B. F. Nesbitt, K. Sergeant and A. Sheridan, *Nature*, **195**, 1062 (1962).

172. A. S. M. Van der Zijden, W. A. A. Blanche Koelensmid, J. Boldingh, C. B. Barrett, W. O. Ord and J. Philp, *Nature*, **195**, 1060 (1962).
173. H. De-Iongh, R. K. Beerthius, R. O. Vees, C. B. Barrett, and W. O. Ord, *Biochem. Biophys. Acta*, **65**, 548 (1962).
174. E. L. Wick, G. N. Wogan, T. Asuo, G. Büchi, M. M. Abdel-Kader, and S. B. Chang, *J. Am. Chem. Soc.* **87**, 882 (1965).
175. B. E. Nielsen and J. Lemmich, *Acta Chem. Scand.*, **18**, 2111 (1964).
176. S. Takie and M. Koide, *Chem. Ber.*, **62**, 3030 (1929).
177. G. Büchi, L. Crombie, P. J. Godin, J. S. Kaltenbronn, K. S. Siddalingaiah and D. A. Whiting, *J. Chem. Soc.*, 2843 (1961).
178. W. Bencze, J. Eisenbeiss and H. Schmid, *Helv. Chim. Acta*, **39**, 923 (1956).
179. W. D. Ollis and I. O. Sutherland in *Recent Developments in the Chemistry of Natural Phenolic Products* (Ed. D. W. Ollis), Pergamon Press, London, 1961, p. 74.
180. A. C. Jain and T. R. Seshadri, *Quart. Revs.*, **10**, 169 (1956).
181. T. A. Geissman and E. Hinreiner, *Botan. Rev.*, **18**, 229 (1952).
182. R. Robinson, *The Structural Relations of Natural Products*, Clarendon Press, Oxford, 1955, p. 14.
183. D. E. Wolff, C. H. Hoffmann, P. E. Aldrich, H. R. Skeggs, L. D. Wright and K. Folkers, *J. Am. Chem. Soc.*, **78**, 4499 (1956).
184. A. J. Birch, R. J. English, R. A. Massy-Westropp and H. Smith, *J. Chem. Soc.*, 369 (1958).
185. F. B. LaForge, H. L. Haller and L. E. Smith, *Chem. Revs.*, **12**, 189 (1933).
186. E. Späth, K. Klager and C. Schlösser, *Chem. Ber.*, **64**, 2203 (1931).
187. E. Späth and H. Schmid, *Chem. Ber.*, **73**, 709 (1940).
188. H. Schmid, *Sci. Proc. Roy. Dublin Soc.*, **27**, 145 (1939).
189. R. D. Haworth, *Ann. Rep.*, 344 (1937).
190. T. O. Soine, *J. Pharm. Sci.*, **53**, 231 (1964).
191. L. Musajo and G. Rodighiero, *Experientia*, **18**, 153 (1962).
192. L. Musajo, 'Int. Symposium on Pharmaceutical Chemistry, Florence, Italy, 17–19 Sept. 1962', in *The Official J. of the Int. Union of Pure and Appl. Chemistry*, Butterworths, London, 1963, p. 369.
193. R. B. Arora, T. R. Seshadri and N. R. Khrishnaswamy, *Arch. Intern. Pharmacodyn.*, **124**, 150 (1960).
194. M. A. Pathak, J. H. Fellman and K. D. Kaufman, *J. Invest. Dermatol.*, **35**, 165 (1960).
195. A. K. Sen, *J. Sci. Ind. Res. (India)*, **22**, 88 (1963).
196. H. S. Jois, B. L. Manjunath and S. V. Rao, *J. Indian Chem. Soc.*, **10**, 41 (1933).
197. H. Kuske, *Archiv. Dermatol. Syphil.*, **178**, 112 (1938); *Dermatol.*, **82**, 273 (1940).
198. L. Musajo, G. Rodighiero and G. Caporale, *Chem. Ind.*, **35**, 13 (1953).
199. L. Musajo, G. Caporale and C. Antonello, *Farmaco (Pavia) Ed. Sci.*, **13**, 355 (1958).
200. G. Caporale, *Ann. Chim. (Rome)*, **50**, 1135 (1960); *Chem. Abstr.*, **55**, 21106 (1961).
201. M. A. Pathak and T. B. Fitzpatrick, *J. Invest. Dermatol.*, **32**, 255, 509 (1959).
202. E. L. Oginsky, G. S. Green, D. G. Griffith and W. L. Fowlks, *J. Bact.*, **78**, 821 (1959).
203. W. L. Fowlks, D. G. Griffith and E. L. Oginsky, *Nature*, **181**, 571 (1958).

204. W. L. Fowlks, *J. Invest. Dermatol.*, **32**, 233 (1959).
205. M. A. Pathak, B. Allen, D. J. E. Ingram and J. H. Fellman, *Biochim. Biophys. Acta*, **54**, 506 (1961).
206. L. Musajo and G. Rodighiero, *Nature*, **190**, 1109 (1961).
207. G. Rodighiero, B. Perissinotto and G. Caporale, *Atti Ist. Veneto Sci. Lettere Arti Classe Sci. Mat. Nat.*, **114**, 1 (1955–1956); *Chem. Abstr.*, **51**, 10736 (1957).
208. H. W. Bersch and W. Döpp, *Arzneimittel. Forsch.*, **5**, 116 (1955); *Chem. Abstr.*, **49**, 8502 (1955).
209. D. P. Chakraborty, A. D. Gupta and P. K. Bose, *Ann. Biochem. Exp. Med. (Calcutta)*, **17**, 59 (1957); *Chem. Abstr.*, **52**, 1352 (1958).
210. V. E. Mikkelson, E. W. Fowlks, D. G. Griffith, *Arch. Phys. Med. Rehabil.*, **42**, 609 (1961); *Chem. Abstr.*, **56**, 7941 (1962).
211. D. P. Chakraborty and P. K. Bose, *Trans. Bose Res. Inst.*, **24**, 31 (1961).
212. A. Schönberg and N. Latif, *J. Am. Chem. Soc.*, **76**, 6208 (1954).
213. E. Späth, and F. Kuffner, *Monatsh.*, **69**, 75 (1936).
214. C. Ohme, *Ann. Chem.*, **31**, 316 (1839).
215. A. G. Caldwell and E. R. H. Jones, *J. Chem. Soc.*, 40 (1945).
216. H. Priess, *Chem. Ber.*, **21**, 22 (1910); *Chem. Zentr.* **II**, 94 (1911).
217. E. Späth and E. Dobrovolny, *Chem. Ber.*, **72**, 52 (1939).
218. A. Erdmann, *Ann. Chem.*, **32**, 309 (1839).
219. O. Dischendorfer and W. Limonstchew, *Monatsh.*, **80**, 741 (1949).
220. C. Bhar, *Chem. Abstr.*, **41**, 5879 (1947).
221. T. Noguchi and M. Kawanami, *J. Pharm. Soc. Japan*, **59**, 755 (1939).
222. K. Hata, A. Nitta and I. Ogiso, *Yakugaku Zasshi*, **80**, 1800 (1960); *Chem. Abstr.*, **55**, 10802 (1961).
223. A. Mukerjee, *Current Sci. India*, **12**, 209 (1943); *Chem. Abstr.*, **38**, 5640 (1944).
224. C. H. Schlatter, *Ann. Chem.*, **5**, 201 (1833).
225. G. Schnedermann and F. L. Winckler, *Ann. Chem.*, **51**, 315 (1844).
226. E. Späth, K. P. Bose, W. Gruber and N. C. Guha, *Chem. Ber.*, **70**, 1021 (1937).
227. B. Ackacic, D. Kustrak and B. Poje, *Planta Med.*, **9**, 70 (1961); *Chem. Abstr.*, **55**, 15835 (1961).
228. J. Trojanek, J. Hodkova and Z. Cekan, *Planta Med.*, **9**, 200 (1961); *Chem. Abstr.*, **55**, 21481 (1961).
229. I. R. Fahmy, H. Abu-Shady, A. Schönberg and A. Sina, *Nature*, **160**, 468 (1947).
230. I. R. Fahmy and H. Abu-Shady, *Quart. J. Pharm. Pharmacol.*, **20**, 281 (1947).
231. K. Hata, M. Kozowa, K. Yen and Y. Kimura, *Yakugaku Zasshi*, **83**, 611 (1963); *Chem. Abstr.*, **59**, 7318 (1963).
232. Yu. A. Dranitsyma, *Trudy Botan. Inst. im. V. L. Komarova, Akad. Nauk SSSR*, **5**, 43 (1961); *Chem. Abstr.*, **56**, 9134 (1962).
233. Yu. A. Dranitsyma and G. V. Pigulevskii, *USSR Pat.*, 130,896 (1960); *Chem. Abstr.*, **55**, 5878 (1961).
234. Yu, A. Dranitsyma, *Zh. Prikl. Khim.*, **33**, 984 (1960). *Chem. Abstr.*, **54**, 16442 (1960).
235. K. Hata, M. Kozawa and K. Yen, *Yakugaku Zasshi*, **83**, 606 (1963); *Chem. Abstr.*, **59**, 7318 (1963).
236. G. K. Nickonov, R. K. Veremi and M. G. Pimenov, *Zh. Obsch. Khim.*, **34**, 1353 (1964); *Chem. Abstr.*, **61**, 1821 (1964).

237. K. Hata, S. Nomura and T. Takano, *Yakugaku Zasshi*, **80**, 892 (1960); *Chem. Abstr.*, **54**, 23188 (1960).
238. T. Noguchi, *Rept. Japan Assoc. Advancement Sci.*, **17**, 234 (1943); *Chem. Abstr.*, **44**, 3990 (1950).
239. T. Noguchi, S. Fujita and M. Kawanami, *Chem. Ber.*, **71**, 344 (1938).
240. T. Noguchi and M. Kawanami, *J. Pharm. Soc. Japan*, **58**, 1052 (1938); *Chem. Abstr.*, **33**, 2513 (1939).
241. K. Hata and A. Nitta, *Yakugaku Zasshi*, **80**, 742 (1960); *Chem. Abstr.*, **55**, 10802 (1961).
242. A. Nitta, *Yakugaku Zasshi*, **85**, 173 (1965); *Chem. Abstr.* **62**, 14615 (1965).
243. K. Hata and M. Kozawa, *Yakugaku Zasshi*, **81**, 1647 (1961); *Chem. Abstr.*, **56**, 10281 (1962).
244. Y. Tanaka, *Kinki Daigaku Yakugakubu Kiyo*, **4**, 27 (1963); *Chem. Abstr.*, **59**, 14295 (1963).
245. A. B. Svendsen, *Blyttia*, **11**, 96 (1953); *Chem. Abstr.*, **50**, 7963 (1956).
246. T. Nakabayashi, *Nippon Kagaku Zasshi*, **83**, 182 (1962); *Chem. Abstr.*, **59**, 1578 (1963).
247. G. Caporale and G. Rodighiero, *Ric. Sci. Rend. Sec. B*, **1** (2) 127 (1961); *Chem. Abstr.*, **57**, 2595 (1962).
248. G. Rodighiero and G. Allegri, *Farmaco (Pavia) Ed. Sci.*, **14**, 727 (1959); *Chem. Abstr.*, **54**, 19879 (1960).
249. L. Musajo, G. Caporale and G. Rodighiero, *Gazz. Chim. Ital.*, **84**, 870 (1954); *Chem. Abstr.*, **49**, 7188 (1955).
250. J. Iriarte, F. A. Kincl, G. Rosenkranz and F. Sondheimer, *J. Chem. Soc.*, 4170 (1956).
251. F. A. Kincl, J. Romo, G. Rosenkranz and F. Sondheimer, *J. Chem. Soc.*, 4163 (1956).
252. H. N. Khastgir, *J. Indian Chem. Soc.*, **24**, 421 (1947); *Chem. Abstr.*, **42**, 6060 (1948).
253. D. Nomura, *Nagaku no Roiki*, **4**, 561 (1950); *Chem. Abstr.*, **45**, 7112 (1951).
254. M. Calvarano, *Essenze Deriv. Agrumari*, **31**, 167 (1961); *Chem. Abstr.*, **57**, 2355 (1962).
255. G. Rodighiero and G. Caporale, *Atti Ist. Veneto Sci., Lettere Arti Classe Sci. Mat. Nat.*, **112**, 97 (1954); *Chem. Abstr.*, **50**, 5989 (1956).
256. R. A. Berhard, *Nature*, **182**, 1171 (1958); *Chem. Abstr.*, **53**, 5428 (1959).
257. F. D. Dodge, *Chem. Zentr.*, **I**, 2688 (1939).
258. A. Soll, A. Pereira and J. Renz, *Helv. Chim. Acta*, **33**, 1637 (1950).
259. T. Nakaoki and N. Morita, *J. Pharm. Soc. Japan*, **73**, 770 (1953); *Chem. Abstr.*, **48**, 7008 (1954).
260. D. P. Chakraborty, *J. Sci. Ind. Res. (India)*, **18B**, 90 (1959); *Chem. Abstr.*, **53**, 18196 (1959).
261. A. K. Athansios, T. El. S. El-Kholy, G. Soliman and M. A. M. Shaban, *J. Chem. Soc.*, 4253 (1962).
262. S. Fukushi and H. Horitsu, *Nippon Nogai-Kagaku Kaishi*, **32**, 646 (1958); *Chem. Abstr.*, **53**, 5422 (1959).
263. S. Fukushi, M. Saimen and H. Horitsu, *Nippon Nogai-Kagaku Kaishi*, **31**, 593 (1957); *Chem. Abstr.*, **52**, 13016 (1958).

264. Y. Hatsuda, S. Murao, N. Terahina and T. Yokota, *Nippon Nogei-Kagaku Kaishi*, **34**, 484 (1960); *Chem. Abstr.*, **58**, 11689 (1963).
265. Y. Obata and S. Fukushi, *Nippon Nogei-Kagaku Kaishi*, **29**, 451 (1955); *Chem. Abstr.*, **52**, 9323 (1958).
266. G. Rodighiero, G. Caporale and E. Ragazzi, *Atti Ist. Veneto Sci. Lettere Arti. Classe Sci. Mat. Nat.*, **111**, 125 (1952–53); *Chem. Abstr.*, **48**, 14116 (1954).
267. M. N. Galbraith, E. Richtie and W. C. Taylor, *Australian J. Chem.*, **13**, 427 (1960); *Chem. Abstr.*, **55**, 8450 (1961).
268. A. K. Banerjee and P. K. Bose, *Ann. Biochem. Exptl. Med. (Calcutta)*, **19**, 181 (1959); *Chem. Abstr.*, **54**, 13291 (1960).
269. V. Benesova, *Collection Czech. Chem. Commun.*, **27**, 2714 (1962); *Chem. Abstr.*, **58**, 6781 (1963).
270. N. F. Komissarenko, V. T. Chernobai and D. G. Kolesnikov, *Med. Prom SSSR*, **16**, 25 (1962); *Chem. Abstr.*, **58**, 12858 (1963).
271. H. Gutzeit, *Chem. Ber.*, **12**, 2016 (1879).
272. H. Litsuhashi, T. Nomura, U. Nagai, T. Murramatsu and I. Tokuda, *Yakugaku Zasshi*, **81**, 464 (1961); *Chem. Abstr.*, **55**, 1583 (1961).
273. M. Fujita and T. Furuya, *J. Pharm. Soc. Japan*, **76**, 538 (1956); *Chem. Abstr.*, **50**, 12999 (1956).
274. M. Fujita and T. Furuya, *J. Pharm. Soc. Japan*, **74**, 795 (1954); *Chem. Abstr.*, **49**, 4241 (1954).
275. A. B. Svendsen and M. Blyberg, *Pharm. Acta Helv.*, **34**, 33 (1959); *Chem. Abstr.*, **53**, 13503 (1959).
276. A. B. Svendsen and E. Ottestad, *Pharm. Acta Helv.*, **32**, 457 (1957); *Chem. Abstr.*, **52**, 8456 (1958).
277. A. B. Svendsen, E. Ottestad and M. Blyberg, *Planta Med.*, **7**, 113 (1959); *Chem. Abstr.*, **53**, 22739 (1959).
278. D. G. Kolesnikov, N. F. Komissarenko and V. T. Chernobai, *Med. Prom. SSSR*, **15**, 32 (1961); *Chem. Abstr.*, **55**, 27551 (1961).
279. E. Späth and S. Raschka, *Chem. Ber.*, **67**, 62 (1934).
280. T. P. Hilditch and E. E. Jones, *Biochem. J.*, **22**, 326 (1928); *Chem. Abstr.*, **22**, 2766 (1928).
281. A. Skerm, *Deut. Apotheker. Ztg.*, **99**, 511 (1959); *Chem. Abstr.*, **56**, 2511 (1962).
282. G. Heut, *Ann. Chem.*, **176**, 70, 78 (1875).
283. J. Herzog and D. Krohn, *Arch. Pharm.*, **247**, 553 (1909); *Chem. Zentr.*, **II**, 1768 (1909).
284. P. K. Gupta and T. O. Soine, *J. Pharm. Sciences*, **53**, 1543 (1964).
285. Y. R. Naves, *Helv. Chim. Acta*, **26**, 1281 (1943).
286. T. Kariyone and M. Kanno, *J. Pharm. Soc. Japan*, **56**, 662 (1936); *Chem. Abstr.*, **31**, 2583 (1937).
287. T. Kariyone, M. Kanno and R. Sugino, *Chem. Zentr.*, **I**, 1456 (1937).
288. P. K. Bose and A. Mukerjee, *J. Indian Chem. Soc.*, **21**, 181 (1944); *Chem. Abstr.*, **39**, 2498 (1945).
289. E. Späth, K. P. Box, H. Schmid, E. Dobrovolny and A. Mukerjee, *Chem. Ber.*, **73**, 1361 (1940).
290. N. P. Maksyutina and D. G. Kolesnikov, *Med. Prom. SSSR*, **16**, 11 (1962); *Chem. Abstr.*, **57**, 3564 (1962).
291. N. P. Maksyutina and D. G. Kolesnikov, *Dokl. Akad. Nauk SSSR*, **124**, 1335 (1959); *Chem. Abstr.*, **53**, 16292 (1959).

292. T. O. Soine, H. Abu-Shady, and F. E. Digangi, *J. Am. Pharm. Assoc.*, **45**, 426 (1956); *Chem. Abstr.*, **50**, 11616 (1956).
293. I. R. Fahmy, A. H. Saber and E. A. E. Kadir, *J. Pharm. Pharmacol.*, **8**, 653 (1956); *Chem. Abstr.*, **51**, 1536 (1957).
294. N. P. Maksyutina and D. G. Kolesnikov, *Zh. Obshch. Khim.*, **31**, 1386 (1961); *Chem. Abstr.*, **55**, 23516 (1961).
295. J. Arima, *J. Chem. Soc., Japan*, **50**, 205 (1929); *Chem. Abstr.*, **26**, 148 (1932).
296. Y. Akahori, *Chem. Pharm. Bull. (Tokyo)*, **9**, 921 (1961); *Chem. Abstr.*, **57**, 16541 (1962).
297. G. N. Nikonov and A. A. Ivanshenko, *Zh. Obshch. Khim.*, **33**, 2740 (1963); *Chem. Abstr.*, **60**, 1725 (1964).
298. E. Späth and C. Schlösser, *Chem. Ber.*, **64**, 2203 (1931).
299. P. K. Bose and H. H. Filayson, *J. Indian Chem. Soc.*, **15**, 516 (1938); *Chem. Abstr.*, **33**, 2281 (1939).
300. C. H. Yang and S. A. Brown, *Can. J. Chem.*, **40**, 383 (1962).
301. K. A. Thaker, *J. Indian Chem. Soc.*, **41**, 641 (1964).
302. F. Wessely and L. Naigebauer, *Monatsh.*, **84**, 217 (1953); *Chem. Abstr.*, **47**, 5997 (1953).
303. G. V. Pigulevskii and G. A. Kuznetsova, *Chem. Abstr.*, **47**, 12341 (1953).
304. G. V. Pigulevskii and G. A. Kuznetsova, *Chem. Abstr.*, **43**, 3416 (1949).
305. K. K. Chakravarti, A. K. Bose and S. Siddiqui, *J. Sci. Ind. Res. (India)*, **7B**, 24 (1948); *Chem. Abstr.*, **42**, 7492 (1948).
306. H. N. Khastgir, P. C. Duttagupta and P. Sengupta, *Indian J. Appl. Chem.*, **22**, 82 (1959); *Chem. Abstr.*, **54**, 12275 (1960).
307. W. Brandt, *Chem. Zentr.*, **II**, 1199 (1915).
308. H. Mühlemann, *Chem. Zentr.*, **II**, 3950 (1938).
309. G. Rodighiero, G. Caporale,and G. Albiero, *Chem. Abstr.*, **49**, 7188 (1955).
310. M. D. Sayed, F. M. Hashim and M. A. El-Keiy, *Egyptian Pharm. Bull.*, **40**, 203 (1958); *Chem. Abstr.*, **54**, 22879 (1960).
311. J. M. V. Lobo, J. Sanchez and C. Fuertes, *Anales Real. Soc. Espan. Fis. Quim. (Madrid)*, **55B**, 103 (1959); *Chem. Abstr.*, **53**, 17247 (1959).
312. A. St. Pfau, *Helv. Chim. Acta*, **22**, 382 (1939).
313. E. L. Bennett and J. Bonner, *Am. J. Botany*, **40**, 29 (1953); *Chem. Abstr.*, **47**, 3937 (1953).
314. V. T. Chernobai and D. G. Kolesnikov, *Ukr. Khim. Zh.*, **25**, 111 (1959); *Chem. Abstr.*, **53**, 17432 (1959).
315. E. Späth, P. K. Bose, J. Matzke and N. C. Guha, *Chem. Ber.*, **72**, 821 (1939).
316. A. Chatterjee and A. Bhattacharya, *J. Indian Chem. Soc.*, **30**, 33 (1953); *Chem. Abstr.*, **47**, 10811 (1953).
317. E. C. Horning and D. B. Reisner, *J. Am. Chem. Soc.*, **72**, 1514 (1950).
318. B. Krishnaswamy, K. R. Rao and T. R. Seshadri, *Proc. Indian Acad. Sci.*, **19A**, 5 (1944); *Chem. Abstr.*, **39**, 1153 (1945).
319. D. B. Limaye, R. H. Munje, G. S. Shenolikar and S. S. Talwalker, *Rasayanam*, **1**, 187 (1939); *Chem. Abstr.*, **34**, 5071 (1940).
320. V. K. Bhagwat and R. Y. Shahane, *Rasayanam*, **1**, 190 (1939); *Chem. Abstr.*, **34**, 5071 (1940).
321. E. Späth, *Chem. Ber.*, **66**, 1137 (1933).
322. E. Späth, *Monatsh.*, **72**, 179 (1938).
323. J. Arima, *J. Chem. Soc. Japan*, **48**, 88 (1927); *Chem. Abstr.*, **21**, 1817 (1927).

324. V. D. N. Sastri, N. Narasimhachari, P. Rajogopalan, T. R. Seshadri and T. R. Thiruvengadam, *Proc. Indian Acad. Sci.*, **37A**, 681 (1953); *Chem. Abstr.*, **48**, 8227 (1954).
325. C. Antonello, *Gazz. Chim. Ital.*, **88**, 430 (1958); *Chem. Abstr.*, **53**, 20046 (1959).
326. D. N. Shah and N. M. Shah, *J. Org. Chem.*, **19**, 1938 (1954).
327. D. B. Limaye, *Chem. Ber.*, **67**, 12 (1934).
328. T. R. Seshadri and M. S. Sood, *J. Indian Chem. Soc.*, **39**, 539 (1962).
329. A. N. Nesmeyanov, A. F. Vompe, T. S. Zarevich and D. D. Smolin, *J. Gen. Chem. (USSR)*, **7**, 2767 (1937); *Chem. Abstr.*, **32**, 2915 (1938).
330. L. R. Row and T. R. Seshadri, *Proc. Indian Acad. Sci.*, **11A**, 206 (1939); *Chem. Abstr.*, **34**, 5446 (1940).
331. K. B. Doifode and M. G. Marathy, *J. Org. Chem.*, **29**, 2025 (1964).
332. O. Dischendorfer and W. Limonstchew, *Monatsh.*, **80**, 58 (1949); *Chem. Abstr.*, **43**, 7017 (1949).
333. H. A. Shah and R. C. Shah, *J. Indian Chem. Soc.*, **17**, 41 (1940).
334. M. C. Chudgar and N. M. Shah, *J. Univ. Bombay*, **13**, (3), 15 (1944); *Chem. Abstr.*, **39**, 2286 (1945).
335. K. A. Thaker, *J. Indian Chem. Soc.*, **41**, 721 (1964).
336. K. D. Kaufman and W. E. Russey, *J. Org. Chem.* **27**, 670 (1962).
337. R. H. Mehta, *Indian J. Chem.*, **1**, 186 (1965).

CHAPTER III

Furochromones

Natural furochromones, whose structures are similar to those of the linear furocoumarins, are known. Of these, the following deserve special mention: khellin, visnagin, khellinin, ammiol, khellinol and visamminol. They were discovered in *Ammi visnaga* L., a perennial herbaceous plant found in the Eastern Mediterranean, and belonging to the Umbelliferae family.

Comprehensive reviews about the botany and chemistry of the naturally occurring furochromones have been written.[1–9]

I. Isolation

Natural furochromones are isolated from the seeds and fruits of *Ammi visnaga* in several ways known as (a) the ether method,[10, 11] (b) the alcohol method,[12, 13] (c) the hexane or heptane-rich petroleum ether method,[14] (d) the skellylsolve-B (n-hexane fraction, b.p. 63–70°) method,[15] and (e) by formation of oxonium complexes with mineral acids, e.g. H_3PO_4.[16] The furochromone content of *Ammi majus* L. differs with its geographical location,[14, 17–19] and the selection of the correct plant parts may be of great importance.[20, 21]

Purification of crude furochromone fractions may be carried out by fractional crystallization, and/or preparative chromatographic procedures.[22] Commercially prepared (10–30 ν) khellin can be subjected to two dimensional paper chromatography followed by elution of the fluorescent spots with methanol, and measurement of the extinction at 243 mμ.[23] The concentration is calculated by using khellin as standard substance.[24, 25] Chromatography on polyamide–treated paper has been reported to effect sharp separation of visnagin and khellin.[26, 27]

Gas–liquid chromatography has recently been used in the separation of some furochromones.[28]

II. Physical Properties

The main substances isolated from *Ammi visnaga* have been assayed by: (a) ultraviolet absorption in ethanol, (b) reduction at a dropping mercury electrode (polarograms), and (c) infrared absorption.[29] The spectrophotometric method has been successfully applied in the determination of khellin in blood[30] and in urine.[31] Samaan and coworkers[32] found, that by using the potassium hydroxide color test, the ultraviolet spectrum of khellin had two maxima at 250 and 338 mμ and that khellinin had two maxima at 246 and 334 mμ. The latter method was used to determine khellin (3–5 ν) and khellinin (3–5 ν). Furthermore, khellin can be estimated in up to 20 ν khellin/ml amounts by determination of chromium in the dried khellin reineckate ($C_{46}H_{46}O_{15}N_7S_4Cr$), obtained by addition of Reinecke salt in hydrochloric to the test solution.[33]

The absorption bands shown by furochromones and their dihydro derivatives in the ultraviolet region have been studied in the hope that these spectra might have structural significance,[29, 34, 35] as exemplified in the case of visamminol.[36] The unsubstituted 5*H*-furo(3,2-g)–1–benzopyran-5-one (**A**) or 4′,5′-dihydrofuro(3′,2′-6,7)chromone (**B**)

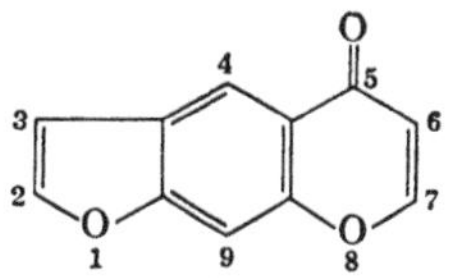

(**A**)

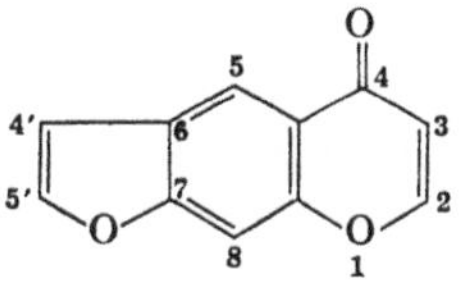

(**B**)

has an absorption maximum at 310 mμ and another band of lower intensity near 280 mμ. The introduction of a 2-methyl group causes a slight shift of the main band towards a shorter wavelength. With the introduction of methoxyl groups into the benzene ring, this main band becomes broader or merges with the smaller band, the maximum varying between 297 and 305 mμ. The replacement of the dihydrofuro ring by the furo ring, as in khellin, 8-methoxy-2-methyl-, and 2-methylfuro-(3′,2′-6,7)chromone causes a displacement of the 310 mμ band of the 4′,5′-dihydro isomers by 10 to 30 mμ, towards longer wavelengths, which conforms to the effect usually found on addition of one conjugated linkage to a conjugated system.

Infrared analysis also has a place in the structural characterization of furochromones.[29] The particular value of such analysis, apart from the detection of groups not fundamentally related to the chromone nucleus, lies in its ability to assign a lactone function.[37]

Molecular structure,[38] and crystallographic studies[39] of khellin have been reported.

III. Nomenclature

Several natural furochromone nomenclature systems are in use; the literature is full of trivial names used for brevity. Naturally occurring furochromones are of the linear type and are limited in number. Many other furochromones are obtained synthetically and they are of the linear and angular types. *The Ring Index* naming system is not commonly used, and the numbering system **B** will be used throughout the text (cf. footnote 1, Ref. 65).

IV. Naturally Occurring Furochromones

1. Chemical Properties

A. Khellin

Khellin (kellin, visammin, 4,9-dimethoxy-7-methyl-5*H*-furo(3,2-g)–1–benzopyran-5-one or 5,8-dimethoxy-2-methylfuro(3′,2′–6,7)chromone) (**1**), $C_{14}H_{12}O_5$, is obtained from the fruits and seeds of *Ammi visnaga* L., which contain a remarkable series of oxygen heterocycles, whose structures have not all been allocated. A close relative of this plant, and often mistaken for it, is *Ammi majus* L., of the Umbelliferae family, but it differs morphologically and also in its constituents. Thus, whereas all the known constituents of *Ammi visnaga* L. are furochromones, with the exception of khellactone, those of *Ammi majus* L. are furocoumarins.[4]

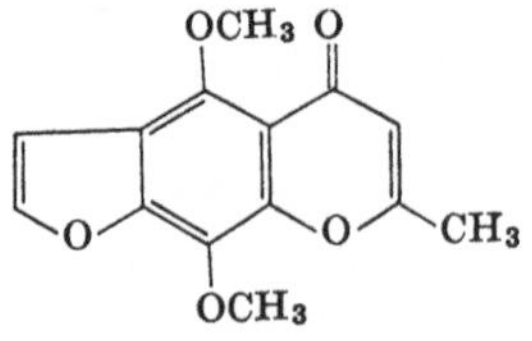

(**1**)

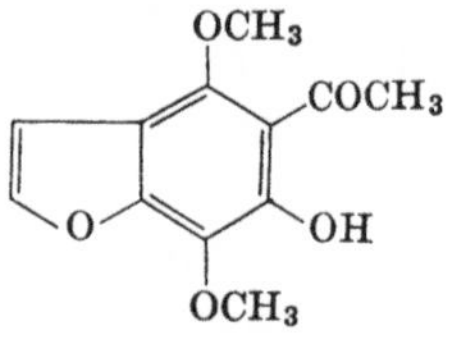

(**2**)

Khellin (**1**) was first diagnosed as a furochromone by Späth and Gruber[10] who later offered a rigid proof. Hot alkali degrades **1** giving acetic acid, but not acetone, together with khellinone, 5-acetyl-4,7-dimethoxy-6-benzofuranol (**2**), and this had misled earlier workers into believing that a coumarin ring was present. Regeneration of **1**, either by conversion of **2** into a β-diketone and cyclization of this with acids,[40] or by subjection of **2** to a Kostanecki synthesis[10], proved that a 2-methylchromone ring is present (Eq. 1).

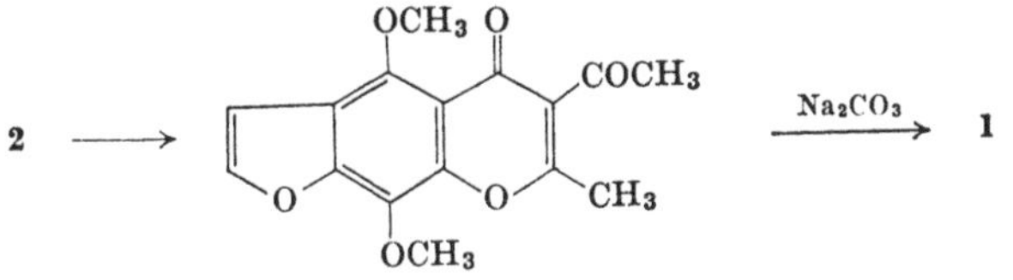

(1)

Khellinone (**2**) can be oxidized to furan-2,3-dicarboxylic acid with hydrogen peroxide, showing that the furo ring is connected with the rest of the structure in one of two ways **1** or **1a**. However, controlled oxidation of **1** with hydrogen peroxide yields 6-hydroxy-4,7-dimethoxybenzofuran-5-carboxylic acid (**3**).[41]

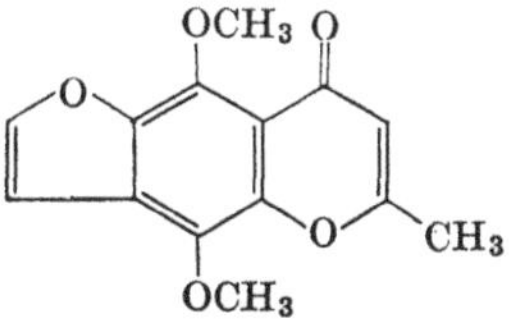

(**1a**)

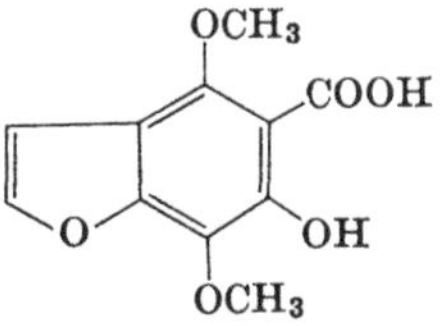

(**3**)

Structure **1**, was chosen for the following reasons. Khellinone ethyl ether, tentatively assumed to have structure **4** gave **5** on ozonization (Eq. 2).

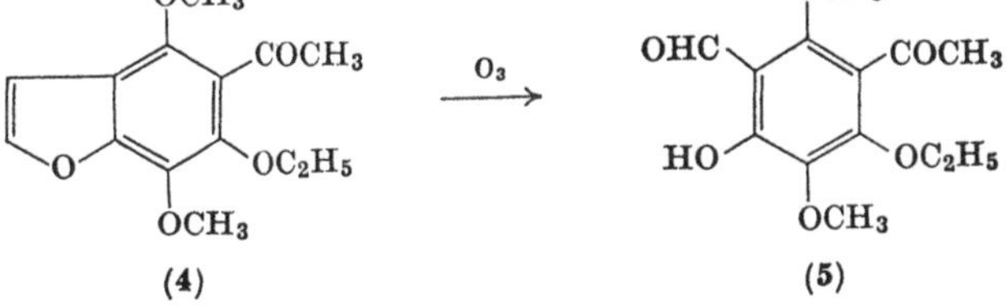

(2)

Since the hydroxyaldehyde (**5**) could not be oxidized directly, the ethyl ether (**6**) was prepared and oxidized to **7** (Eq. 3).
Decarboxylation of **7** gave **8** which is in favor of structure **1** for khellin, since **9**, would result from formula **1a**.

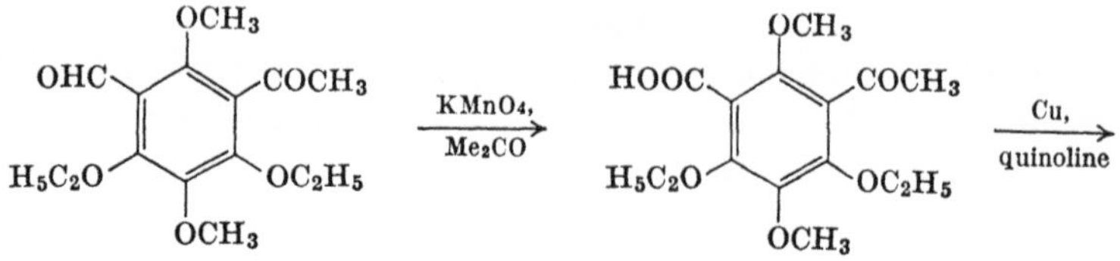

(6) (7)

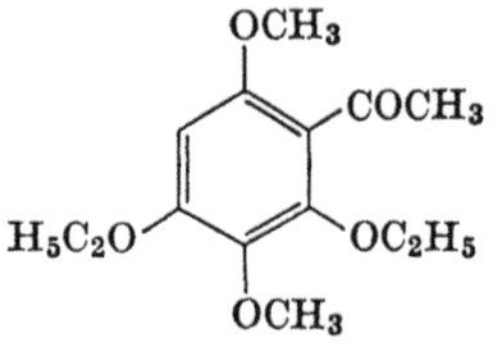

(8)

(3)

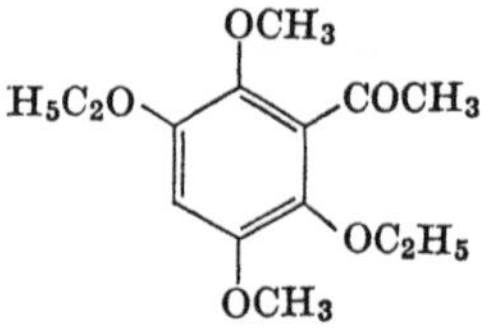

(9)

Wessely and Moser,[42] however, had prepared the ketone (**10**), which with ethyl sulfate and potassium hydroxide, gave **8** (Eq. 4).

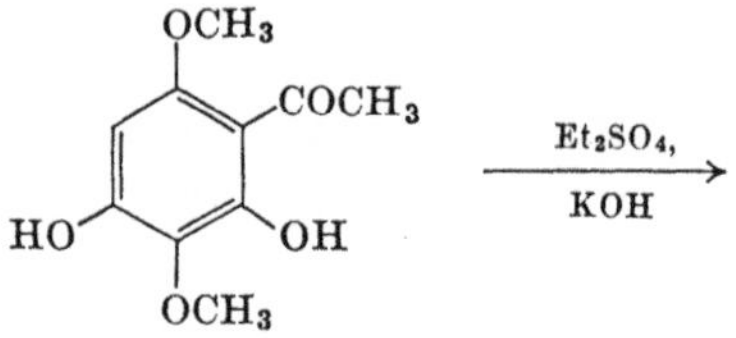

8

(4)

(10)

Oxidation of khellinone (**2**) with nitric acid is less drastic, and merely resulted in the formation of a 1,4-quinone, namely khellinquinone (**11**); **1** itself could also be oxidized to **11** in the same way, thus proving that the two methoxyl groups are in the *para* position.

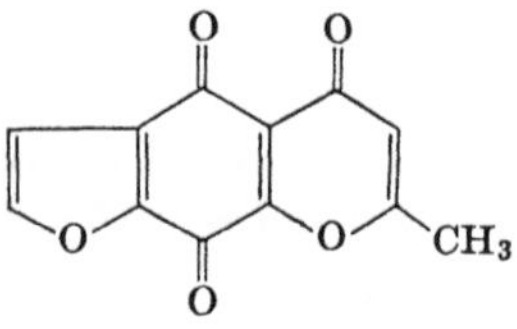

(11)

Späth and Gruber's[10] conclusion was correlated by the synthesis of khellin by many groups of investigators.

(1) Synthesis of khellin. Spath and Gruber[10] achieved a partial synthesis of khellin using the Kostanecki–Robinson method, but this

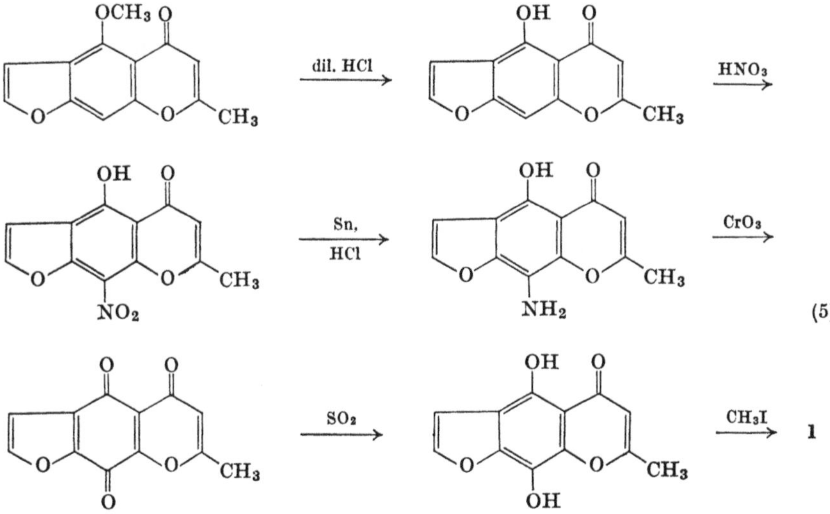

(11) (5)

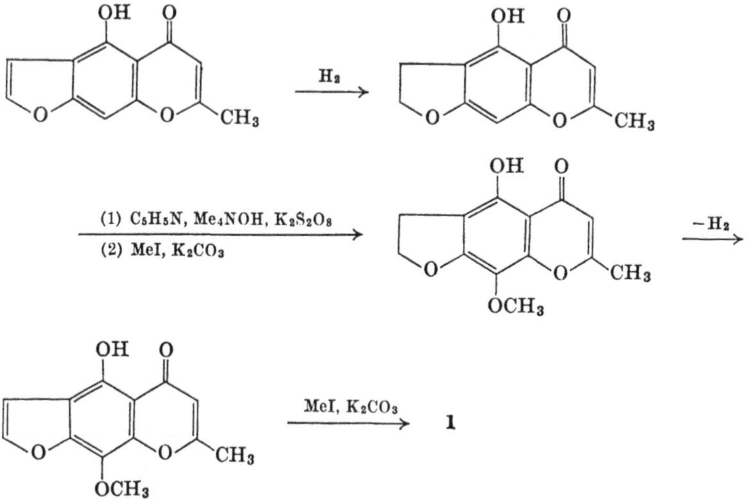

(6)

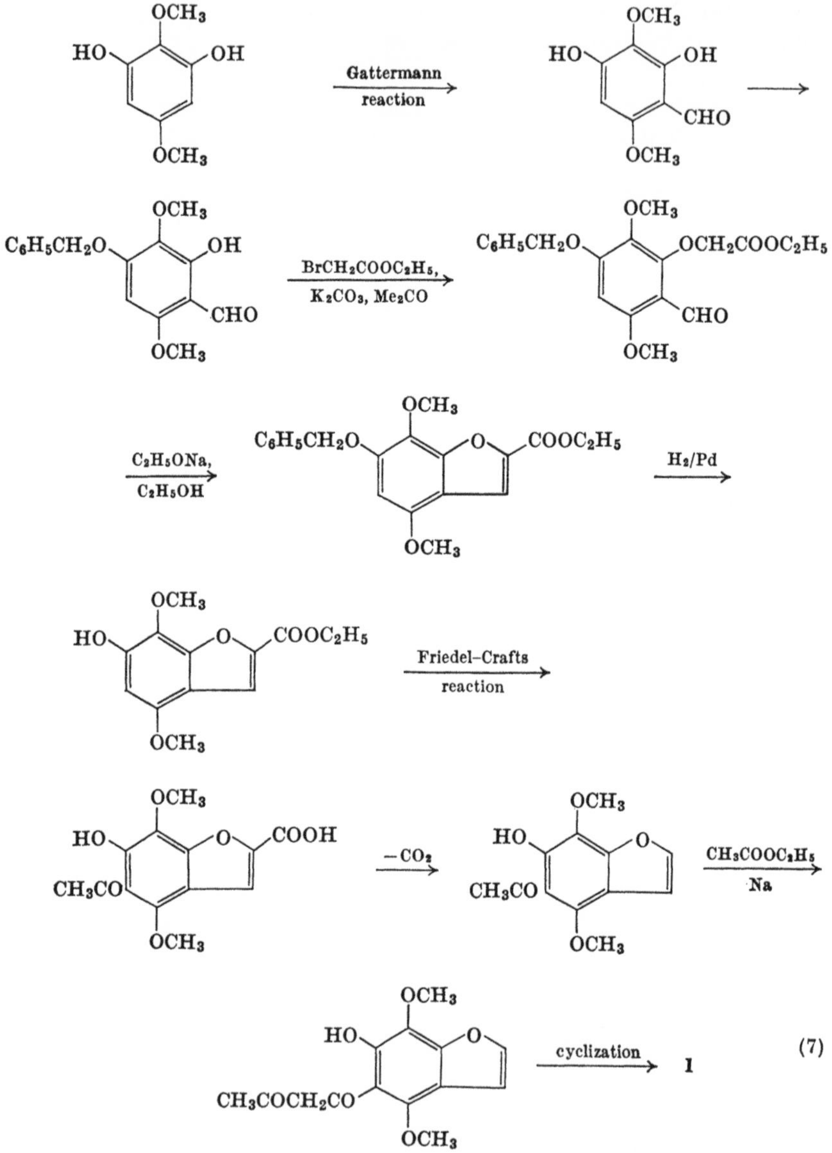

(7)

synthesis was not entirely unambiguous, since it yields a mixture of the desired chromone and the isomeric 4-methylcoumarin. Geissman and Halsall,[43] using a modified Wittig method, treated khellinone (**2**) and khellinone acetate, with sodium hydride in ethyl acetate, followed by trituration of the reaction product with cold hydrochloric acid to obtain **1**.

Starting with visnagin, a by-product in the manufacture of khellin, and with little biological activity, **1** has been obtained in an overall yield not exceeding 3 %[44,45] (Eq. 5). A 9 % yield[44] was obtained by the method shown in equation 6.

Total synthesis of khellin (**1**), starting from 2,5-dimethoxyresorcinol, has been described by Clarke and Robertson[46] (Eq. 7).

Baxter, Ramage and Timson[47] have also achieved a total synthesis of **1** from pyrogallol by the following sequence of reactions (Eq. 8).

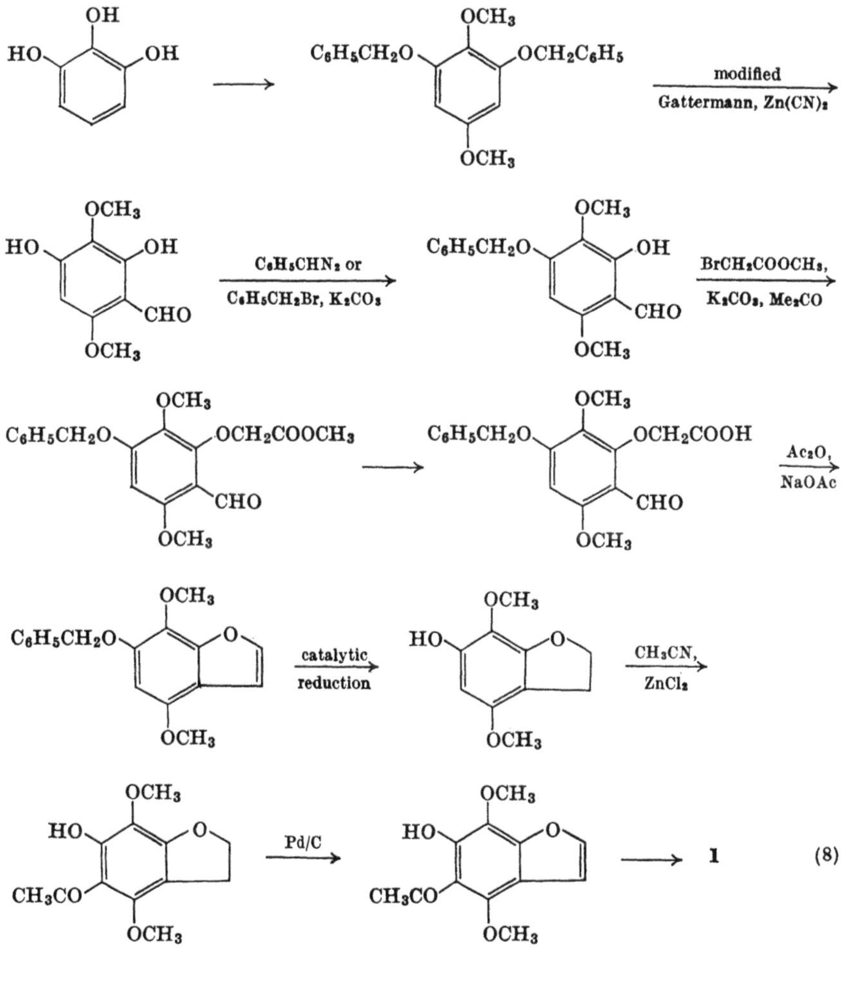

(8)

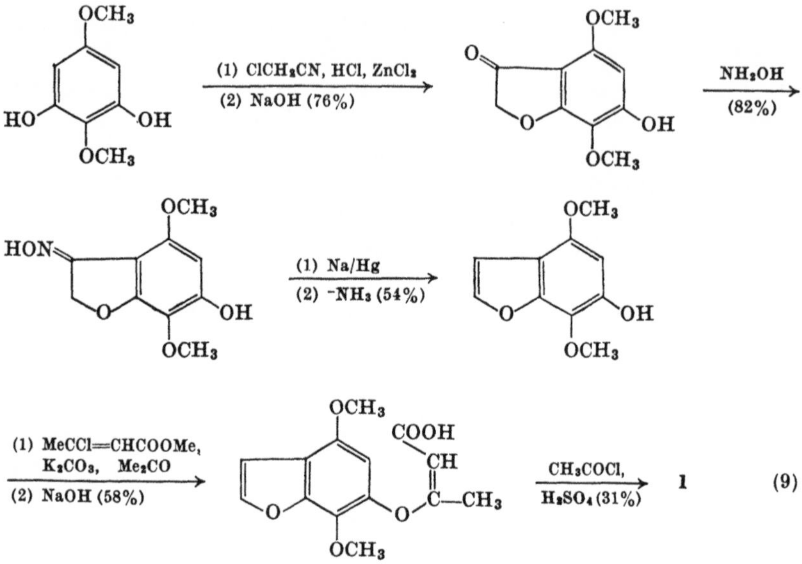

OCH₃
HO
OH
OCH₃
(1) ClCH₂CN, HCl, ZnCl₂
(2) NaOH (76%)
NH₂OH
(82%)
HON
(1) Na/Hg
(2) ⁻NH₃ (54%)
(1) MeCCl=CHCOOMe,
K₂CO₃, Me₂CO
(2) NaOH (58%)
COOH
CH
C—CH₃
CH₃COCl,
H₂SO₄ (31%)
1

(9)

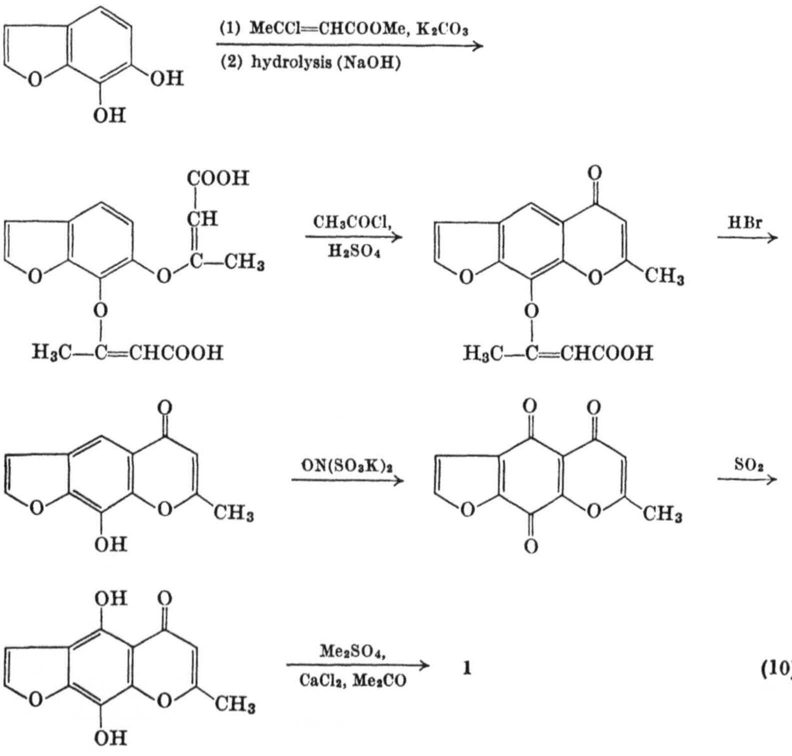

(1) MeCCl=CHCOOMe, K₂CO₃
(2) hydrolysis (NaOH)
H₃C—C=CHCOOH
CH₃COCl,
H₂SO₄
HBr
ON(SO₃K)₂
SO₂
Me₂SO₄,
CaCl₂, Me₂CO
1

(10)

In a similar manner, Dann and Illing[48] have synthesized khellin by building a pyrone ring onto the appropriate coumarone, namely, 6-hydroxy-4,7-dimethoxycourmarone (Eq. 9), and 6,7-dihydroxycoumarone[49] (Eq. 10).

An elegant synthesis by Murti and Seshadri,[50, 51] involved building up the furan ring on to the chromone. They started with 5,7-dihydroxy-2-methylchromone, having the 5-hydroxyl group protected by chelation, thus allowing the condensation reaction to proceed with the 7-hydroxyl group (Eq. 11).

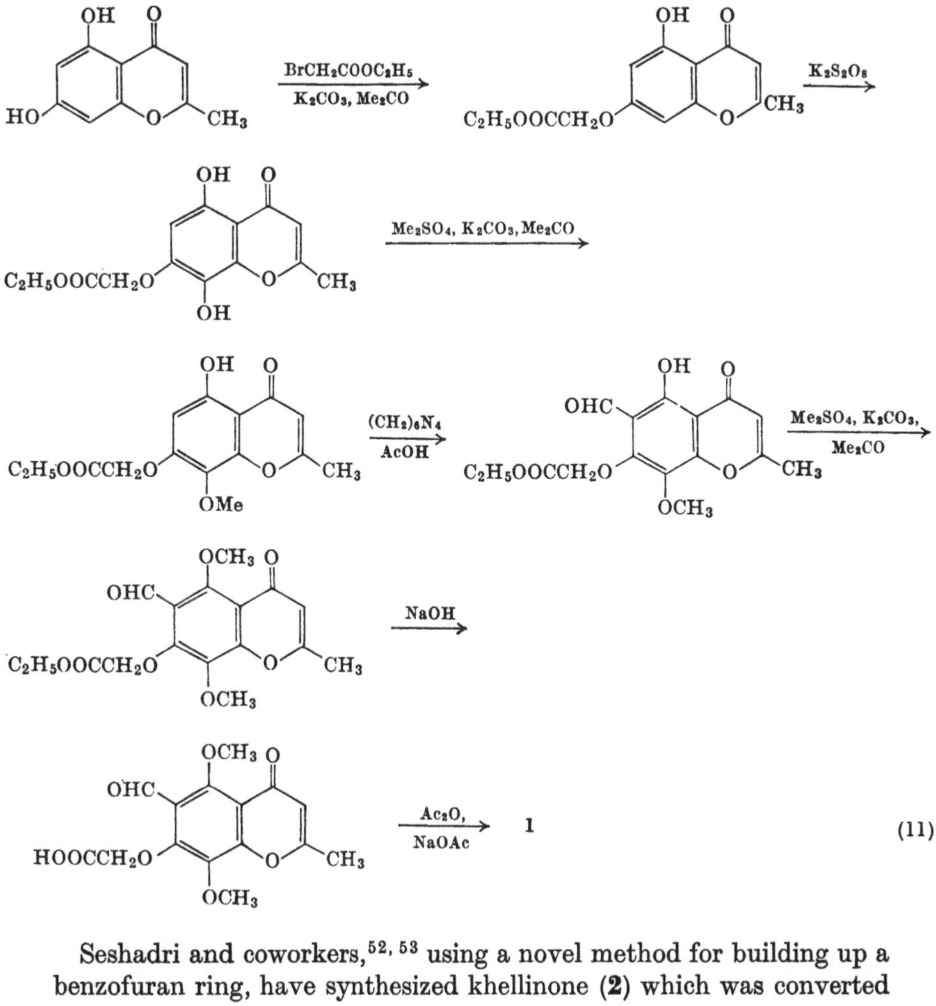

(11)

Seshadri and coworkers,[52, 53] using a novel method for building up a benzofuran ring, have synthesized khellinone (**2**) which was converted into khellin (**1**) by the standard procedure (Eq. 12). The overall yield of **1** from 2,4-dihydroxy-3,6-dimethoxyacetophenone is about 30%.

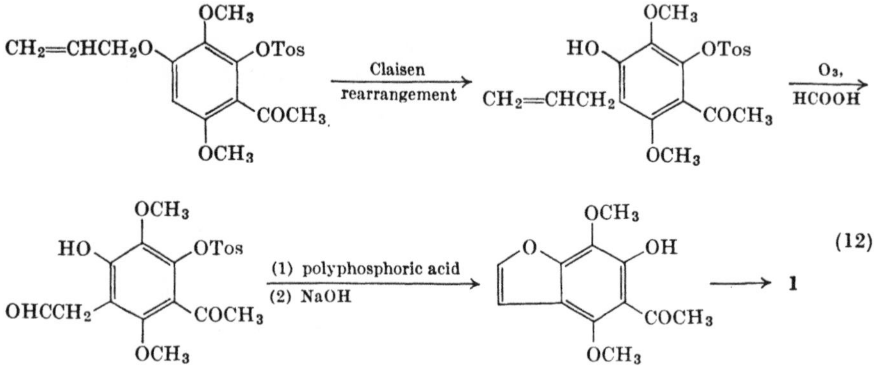

(12)

(2) Synthesis of khellin analogs. Interest has been centered during the last decade in synthesizing a large number of khellin analogs (Table 1), making use of the condensation reaction of khellinone (**2**) with esters in presence of metallic sodium to yield the corresponding diketones, followed by cyclization to the requisite analog (**C**).

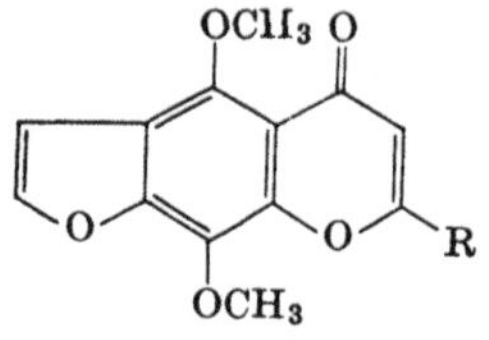

(**C**)

R=H; R=alkyl; R=

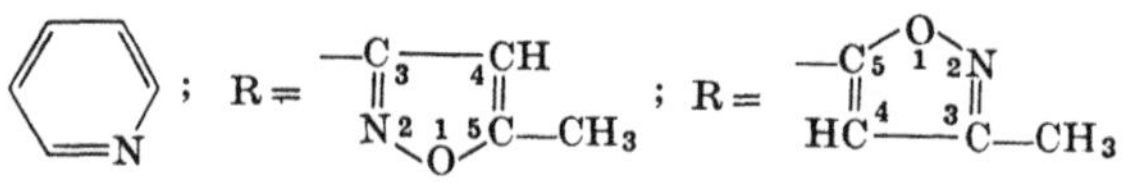

2-Alkyl-,[54] 2-(3′-pyridyl)-,[55] and 2-isoxazolyl-5,8-dimethoxyfuro(3′,2′–6,7)chromone[56] have been synthesized, and the parent compound, the 2-unsubstituted khellin analog, is described.[54]

(3) Reactions. Special attention has been given to the problem of khellin demethylation, since this reaction leads to different products, according to the reagent and the conditions used. Partial demethylation of khellin (**1**) to 5-hydroxy-8-methoxy-2-methylfuro(3′,2′–6,7)chromone

TABLE 1. Furochromones

Substituents	M.p. (°C)	Solvent for crystallization[a]; appearance	Color test with[1]	Derivatives	References
A. *Furo*(3′,2′–6,7)*chromones*					
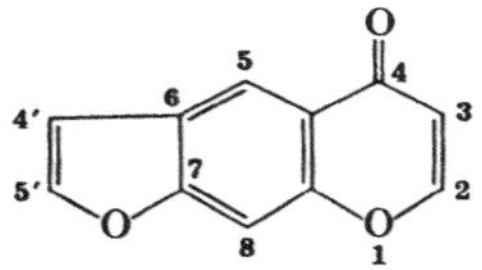					
(a) Alkyl-substituted derivatives					
4′,5′-Dihydro	184–185	A, G			34, 166
2-Me	186	G			34, 166, 167
2-Me, 4′,5′-dihydro	164, 166	G, O			34, 48, 168, 169
2-Et, 4′,5′-dihydro	124–125	N			34, 168
2-n-Pr, 4′,5′-dihydro	103	G			34, 168
2-COOH, 4′,5′-dihydro	283(dec.)	A			34
2,4′-Di-Me,	158	M/P		2-Piperonylidene (m.p. 239°)	170
2,4′-Di-Me, 4′,5′-dihydro	131	M		2-Piperonylidene (m.p. 188°)	170
(b) Alkyl monohydroxy-substituted derivatives					
2-Me, 5-OH, (5-norvisnagin)	155–159	A, G; pale yellow	(c) green; (f) red; scarlet red		45, 53, 63, 99, 102, 172
2-Me, 8-OH,	296–300	A	(g) blue		166, 173
2-Me, 5-OH, 4′,5′-dihydro	165–168	N/P; lemon yellow	(c) dark red	Acetyl (m.p. 169–170°)	35, 53
2-Me, 8-OH, 4′,5′-dihydro	230	O	(b) pale yellow→ apple green		174
2-CH_2OH, 5-OH, (norkhellol)	198–200	G; yellow	(b) orange; (c) green; (f) red		96, 102
2-Me, 5-OH, 8-$CH_2CH{=}CH_2$	162–163	G; yellow			171
2-Me, 5-OH, 8-NH_2	225	B; yellow			45
2-Me, 5-OH, 8-NO_2	260–262(dec.)	B; yellow	(c) brown→red		45
(c) Alkyl monoalkoxy-substituted derivatives					
5-OMe	152	N	(b) yellow		62
2-Me, 5-OMe (visnagin)	139–140	M/P	(a) red-violet; (d) red-violet; (e) red-violet	Urea adduct (m.p. 198°); thiourea adduct (193°); styryl (176°)	12, 26, 54, 84, 85, 90, 95, 99, 121, 126

(Table continued)

TABLE 1 (*continued*)

Substituents	M.p. (°C)	Solvent for crystallization[a]; appearance	Color test with[1]	Derivatives	References
2-Me, 5-OMe, 4′,5′-dihydro	113.5–114	N	(b) pale yellow→ greenish brown		35
2-Me, 8-OMe	196–197 204–205	N; pale yellow	(b) golden yellow→red-brown		166, 173, 174
2-Me, 8-OMe, 4′,5′-dihydro	159	C/M, O	(b) colorless→golden yellow		174
2-Et, 5-OMe	139	N	(a) weak brownish red; (b) yellow		62
2-Me, 8-OEt, 4′,5′-dihydro	130	O	(b) yellow→brown		174
2-(3-C_5H_4N), 5-OMe	226	G	(b) orange; (h) red	HCl (m.p. above 230° (dec.))	95
2-CH_2OH, 5-OMe	175; 179–180	N, O	(a) violet-red	Chloroacetyl (138°); iodoacetyl (150°); benzoyl (145°); *p*-nitrobenzoyl (211°); *p*-toluenesulfonyl (151–152°); piperidinomethyl (214.5°)	61,95,96,115,172
2-CH_2OMe, 5-OMe	124–125	M/P			95
2-CH_2Cl, 5-OMe	140	G	(a) green→violet -red		96
2-CH_2I, 5-OMe	158–159	H, P; pale yellow	(a) bluish green		61, 73, 95, 172
2-CH_2NHPh, 5-OMe	211–212	A			95
2-CHO, 5-OMe	173–174	C/M; yellow	(a) pale brown	Oxime (pale yellow; m.p. 245°)	61
2-COOH, 5-OMe	256	A; pale yellow	(b) red		61
2-COOEt, 5-OMe	167–168	G; yellow	(b) deep red		61
2-Me, 5-OCH_2COOH	246–248(dec.)	A	(a) violet		102
2-Me, 5-OCH_2COOEt	134–135	G			102
2-Me, 8-OCH_2Ph, 4′,5′-dihydro	110	G/O	(b) yellow→brown		174
2-Me, 5-$OCH_2CH_2NEt_2$	ca. 50	G/I		HCl (m.p. 208°)	63
2-Me, 8-$O(CH_3)C{=}CHCOOH$	258–260	D; pale yellow			166, 173
2-$N{=}CHC_6H_4NMe_2$-*p*, 5-OMe	206	F, Q; yellow	(a) brown		61
5-$OPO(OCH_3)_2$, 2-Me	130–132	M			112
5-$OPO(OC_2H_5)_2$, 2-Me	140–142	M			112
5-$OPS(OC_2H_5)_2$, 2-Me	151–153	M			112
5-$OPO(OC_3H_7$-iso), 2-Me	129–131	M			112
(d) Alkyl dihydroxy-substituted derivatives					
2-Me, 5,8-di-OH (5,8-bisnorkhellin)	278, 280–282: 287	A, G; yellow	(b) orange-green; (c) green-brown; (f) violet-red		44, 45, 50, 62, 102, 166, 173 175
2-Me, 5,8-di-OH, 4′,5′-dihydro	260–262	N	(c) blue green→ brown with excess	Acetyl (m.p. 174–175°)	53

Substituents	M.p. (°C)	Solvent for crystallization[a]; appearance	Color test with[1]	Derivatives	References
(e) Alkyl hydroxyalkoxy-substituted derivatives					
5-OH, 8-OMe	204–206	G		Acetyl (m.p. 173)	57, 77, 176
2-Me, 5-OH, 8-OMe, (desmethylkhellin)	196; 201–202; 205	C; yellow	(c) green; (f) red, violet	Styryl (m.p. 197°); *p*-nitrostyryl (260°)	41, 44, 66, 72, 77, 102, 177
2-Me, 5-OH, 8-OMe, 4′,5′-dihydro	176–178	E/K			44, 53
2-Me, 5-OH, 8-(OCH_2CH_2OH)	171	—			63
2-Me, 5-OH, 8-$OCH(OH)CH(OH)CH_3$	138	C			63
2-Me, 5-OH, 8-OCH——$CHCH_3$ (—O— bridge)	160–161	—			63
2-Me, 5-OH, 8-OCH_2COOH	220–222	G			102
2-Me, 5-OH, 8-OCH_2COOEt	129–130	G			102
2-Me, 5-OH, 8-$OCH_2CH_2NMe_2$	124–125	J, G	(c) green; (f) red	HCl (m.p. 250°); picrate (208°); methiodide (166°); ethiodide (274°)	63, 102, 179
2-Me, 5-OH, 8-$OCH_2CH_2NEt_2$ (amikhellin)	93–94; 96–98	H/M		HCl (m.p. 240°); picrate (197°); methiodide (244°); ethiodide (230°)	57, 63, 177, 178
2-Me, 5-OH, 8-(2-piperidinoethoxy)	108–109	J		HCl (m.p. 243–244°)	178, 179
2-Me, 5-OH, 8-(2-OH-3-piperidylpropoxy)	214–215	G/P			63
2-Me, 5-OH, 8-(2-morpholinoethoxy)	131–132	G		HCl (m.p. 251–252°)	178, 179
2-(3-Methylisoxazol-5-yl), 5-OH, 8-OMe	203–204	—			56
2-(5-Methylisoxazol-3-yl), 5-OH, 8-OMe	189–190	—			56
2-Me, 5-$OPO(OCH_3)_2$, 8-OH	180–182	M			112
2-Me, 5-OH, 8-$OPO(OCH_3)_2$	164–166	G; yellow	(c) deep green; (f) orange -red		112
2-Me, 5-OH, 8-$OPO(OC_2H_5)_2$	148–150	M; yellow	(c) deep green; (f) orange-red		112
2-Me, 5-$OPO(OC_2H_5)_2$, 8-OH	166–168	M			112
2-Me, 5-$OPO(OC_3H_7\text{-iso})_2$, 8-OH	134–136	M			112
2-Me, 5-OH, 8-$OPO(OC_3H_7\text{-iso})_2$	132–134	M; yellow	(c) deep green; (f) orange-red		112
(f) Alkyl dialkoxy-substituted derivatives					
5,8-Di-OMe	170, 182	G	(b) yellow		54, 176
2-Me, 5,8-di-OMe (khellin)	153	G	(a) red-violet; (d) deep red-violet; (e) violet-red	Styryl (m.p. 190–192°); *p*-tolylidene (184°); *p*-methoxy benzylidene (171°); piperonylidene (235°); furfurylidene (200°); cinnamylidene (161–163°); thiourea adduct (190° (dec.))	12, 44, 46, 47, 52, 54, 77, 84, 85, 121, 166, 173, 180, 181

(*Table continued*)

TABLE 1 (*continued*)

Substituents	M.p. (°C)	Solvent for crystallization[a]; appearance	Color test with[1]	Derivatives	References
2-Me, 5,8-di-OMe, 2,3-dihydro	136–138	M			72
2-Me, 5,8-di-OMe, 2,3-di-I, 2,3-dihydro	147–148	A			76
2-Me, 5,8-di-OMe, 2,3,4′,5′-tetrahydro	114–114.5	L		Oxime (m.p. 202–202.5), azine (187–188°)	73a
2-Me, 5,8-di-OMe, 4-thione (thiokhellin)	132.5	M; deep red		*p*-Nitrobenzylidene (m.p. 207° (dec.))	76
2-Et, 5,8-di-OMe	126	N	(a) weak brownish red; (b) orange→yellow		54
2-n-Pr, 5,8,-di-OMe	93–94	M	(a) weak brownish red; (b) orange→yellow		54
2-CH_2Ph, 5,8-di-OMe	148	C, G	(b) orange		121
2-Ph, 5,8-di-OMe	176–177	G	(b) orange→yellow		54
2-(3-Pyridyl), 5,8-di-OMe	213–214	G; pale yellow	(b) red; (h) red	HCl (m.p. 233° (dec.)); H_2SO_4 (250°)	55
2-N=$CHC_6H_4NMe_2$-*p*, 5,8-di-OMe	212(dec.)	G; golden yellow	(a) red-brown; (b) red→orange		61
2-(3-Methylisoxazol-5-yl), 5,8-di-OMe	218	—			56
2-(5-Methylisoxazol-3-yl), 5,8-di-OMe	207–208	—			56
2-Me, 5,8-di-OEt	94	G/M	(a) red-violet; (b) orange-yellow		12, 62, 121
2-Me, 5-OEt, 8-OMe	109–110	G	(a) red	Styryl (m.p. 151°)	57, 77
2-Me, 5-OEt, 8-OMe, 4-thione	112–114	M; red-brown		*p*-Nitrostyryl (m.p. 210° (dec.))	76
2-Me, 5,8-di-n-OPr	81	M	(a) orange-yellow		62
2-Me, 5-iso-OPr, 8-OMe	93	C/M	(a) red		57
2-Me, 5-(ethyl-α-butoxy), 8-OMe	134–138	—			182
2-Me, 5-phenacyl, 8-OMe	202	C/M			57
2-Me, 5-n-OBu, 8-OMe	99–100	C/M	(a) red		57
2-Me, 5-$OCH_2C_6H_5$, 8-OMe	146	G			66
2-Me, 5-OCH_2CH=CH_2, 8-OMe	104.5	C/M	(a) red		57
2-Me, 5,8-di-OCH_2CH=CH_2	124	M	(b) orange-brown		62
2-Me, 3-OH, 5,8-di-OMe	150–151	N			73a
2-Me, 3-OH, 5,8-di-OMe, 4′,5′-dihydro	199–201	L			73a
2-Me,3-OH, 5,8-di-OMe, 2,3,4′,5′-tetrahydro	137–138	N			73a
2-Me, 3-OAc,5,8-di-OMe	164–165	N			73a
2-Me, 3-OAc, 5,8-di-OMe, 4′,5′-dihydro	135–137	C/E_1, N			73a
2-Me, 3-$OCH_2C_6H_5$, 5,8-di-OMe	100–102	N			73a
2-Me, 3-$OCH_2C_6H_5$, 5,8-di-OMe, 4′,5′-dihydro	105–106.5	N			73a

5

Substituents	M.p. (°C)	Solvent for crystallization[a]; appearance	Color test with[1]	Derivatives	References
2-Me,3-OAc,5,8-di-OMe, 2,3,4′,5′-tetrahydro	128–129	L, N			73a
2-CH_2OH, 5,8-di-OMe	211	G; yellow	(a) bright red; (b) yellow	Acetyl (m.p. 102–103°, yellow)	61
2-Me, 5,8-di-OMe, 3-(2-piperidinomethyl)	267 (dec.)	G; yellow			184
2-Me, 5,8-di-OMe, 3-Ac, 4′-OH	145–146	H/M		Acetyl (m.p. 118–120°)	183
2-CH_2CN, 5,8,-di-OMe	273.5	C, G			176
2-CHO, 5,8-di-OMe	182–184	M/Q; orange	(a) brown	Oxime (m.p. 253–255°); semicarbazone (245–246° (dec.)); phenylhydrazone (218–220°); 2,4-dinitrophenylhydrazone (320° (dec.)); anil (187°)	61
2-COOH, 5,8-di-OMe	277–279 (dec.)	A; yellow	(b) red		61
2-COOEt, 5,8-di-OMe	132	G/K; yellow	(b) red		54
2-Me, 5-OCH_2COOH,8-OMe	278 (dec.)	A			57, 182
2-Me, 5-OCH_2COOEt, 8-OMe	159	G; light yellow	(a) red		57
2-Me, 5,8-di-OCH_2COOH	276	G	(a) red-violet; (b) orange-yellow		12, 62, 102, 121
2-Me, 5,8-di-OCH_2COOMe	125	G	(b) orange-yellow		62
2-Me, 5,8-di-OCH_2COOEt	125–127	G			62, 102
2-Me, (5)8-OCH_2COOEt, (8)5-OCH_2COOH	198–200	G	(a) crimson red		102
2-Me, 5,8-di-$OCH_2CH_2NMe_2$	123.5	—		HCI (m.p. 244–246°); dipicrate (197°)	175, 177
2-Me, 5,8-di-$OCH_2CH_2NEt_2$	—	—		HCl (m.p. 241–243°); dipicrate (160°); dimethiodide (224°); diethiodide (235°)	175, 177
2-Me, 5-$OCH_2CH_2NEt_2$, 8-OMe	93–94	H/M; pale yellow			57
2-Me, 5-$OCH_2CH_2NEt_2$, 8-$OCH_2CH_2NMe_2$	214	—		Dipicrate (m.p. 189°): dimethiodide (126°); diethiodide (215°)	177
2-Me, 5-$OCH_2CH_2NMe_2$, 8-$OCH_2CH_2NEt_2$	232	—		Dipicrate (m.p. 130°); dimethiodide (248°); dimethopicrate (218°)	177
2-Me, 3-NEt_2, 5,8-di-OMe	270 (dec.)	F, G; pale yellow			184
2-Me, 5,8-di-$OCH_2CH_2NC_5H_{10}$	108–110	—			175
2-(5-Methylisoxazol-3-yl), 5-$OCH_2CH_2NEt_2$, 8-OMe	179–180				56
2-Me, 5-OCH_2CONMe_2, 8-OMe	289	—; light yellow	(a) red		57

(Table continued)

TABLE 1 (*continued*)

Substituents	M.p. (°C)	Solvent for crystallization[a]; appearance	Color test with[1]	Derivatives	References
2-Me, 5-OCH_2CONEt_2, 8-OMe	140; 143	G		Styryl (m.p. 131–132°)	77, 177
2-Me, 5-$OPO(OMe)_2$, 8-OMe	166–168	M			112
2-Me, 5-$OPO(OEt)_2$, 8-OMe	114–116	M			112
2-Me, 5-$OPS(OEt)_2$, 8-OMe	103–105	M			112
2-Me, 5-$OPO(OPr\text{-}iso)_2$, 8-OMe	120–122	M			112
2-Me, 5-(diethyl α-succinoxy), 8-OMe	216 (dec.)	—			182
B. *Furo*(2′,3′–6,7) *chromones*					
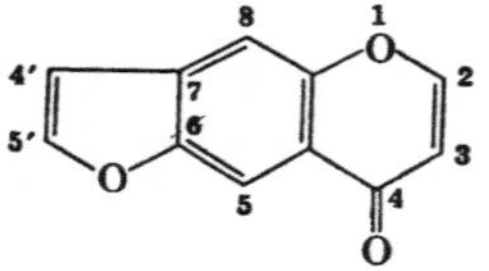					
Unsubstituted	149	Sublimation			166, 170
4′,5′-Dihydro	170	Sublimation			166, 170
2-Me	169	O			166
2-Me, 4′,5′-dihydro	139	O			166, 170
2-COOH	341 (dec.)	A			166, 170
2-COOH, 4′,5′-dihydro	307	A			166, 170
C. *Furo*(2′,3′–7,8)*chromones*					
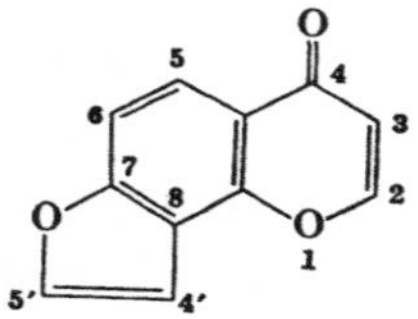					
2-Me	105–106 (dec.)	N; pale yellow			184
2,6-Di-Me	140	G			170

Substituents	M.p. (°C)	Solvent for crystallization[a]; appearance	Color test with[1]	Derivatives	References
2,4′-Di-Me	198	G; pale yellow		Piperonylidene (m.p. 242°)	170, 186
2,3-Di-Me, 4′-Ph	153–155	—			186
2-Me, 3-OH	240–242	—	(b) brown		185
2-Me, 3-CH_2CH_2Ph, 5-OH	154–156	—	(a) deep yellow; (b) yellow		185
2-Me, 3-OMe	153–154	N			99, 185
2-Me, 5-OMe (allovisnagin)	162	—	(b) green	Piperonylidene (m.p. 264°)	170
2-Me, 5-OMe, 4′,5′-dihydro	193–194	G	(b) yellow→green→bluish		35
2-Me, 6-OMe, 5-OH (desmethylisokhellin)	177	B; bright yellow	(a) red; (b) yellow→green→blue	Acetyl (m.p. 147°)	65
2-Me, 3,5-di-OMe	132–133	G, P; pale yellow			187
2-Me, 5,6-di-OMe (isokhellin)	176–180	N; pale yellow	(a) red; (b) yellow; →green→blue; (d) violet-red; (e) violet-red	Piperonylidene (m.p. 226°)	46, 64, 66, 90, 126
2-Me, 5-OCH_2Ph, 6-OMe	140	N; pale yellow			68
2-Me, 3-Ac	118–119	N; pale yellow			184
2-Me, 3-OMe, 5′-COOH	309–311 (dec.)	G; yellow			99

D. *Furo*(3′,2′–5,6)*chromones*

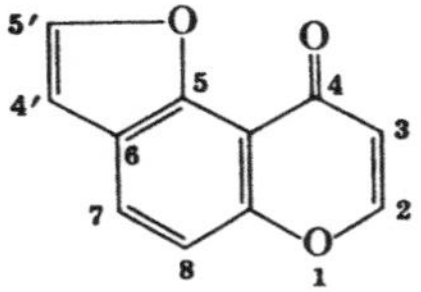

Substituents	M.p. (°C)	Solvent for crystallization[a]; appearance	Color test with[1]	Derivatives	References
2-Me	225–228	N			48, 167
2,5′-Di-Me	195–196	G; yellow	(b) pale green		188
2,3-5′-Tri-Me	216	A; yellow			188
2-Me, 7-OH	318 (dec.)	A			92
2-Me, 7-OMe (isovisnagin)	247	E, G		Piperonylidene (m.p. 244°)	92, 106

(*Table continued*)

TABLE 1 (*continued*)

Substituents	M.p. (°C)	Solvent for crystallization[a]; appearance	Color test with[1]	Derivatives	References
2-Me, 7-OMe, 4′,5′-dihydro	183–184	N			35, 92
2-Me, 8-OMe	112–113	—			189
2,5′-Di-Me, 8-Cl	228–230	G; pale yellow			188
2-Me, 3-Ac, 7-OAc	—	—			190
2-Me, 3-Ac, 7-OMe	218–220	—			106
E. *Furo(2′,3′–5,6)chromones*					
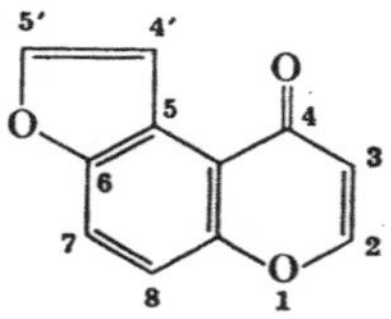					
2-Me	172.4–173.6	G/O			167
F. *Furo(2′,3′–6,7)chromones*					
2-Me, 5,8-di-OMe, 4′,5′-dihydro	82—84	E_1			73a
2-Me, 3,4-di-OH, 5,8-di-OMe, 4′,5′-dihydro	145–147	N			73a
2-Me, 3,4-di-OAc, 5,8-di-OMe,4′.5′-dihydro	153–154	N			73a

[a] A, acetic acid; B, acetone; C, benzene; D, n-butyl alcohol; E, chloroform, E_1, cyclohexane; F, dioxan; G, ethyl alcohol; H, ether; I, methyl ethyl ketone; J, heptane; K, hexane; L, isopropyl alcohol; M, benzine; N, methyl alcohol; O, water; P, ethyl acetate; Q, toluene.

[1] (a) alkali hydroxide; (b) sulfuric acid; (c) ferric chloride; (d) *m*-dinitrobenzene in alkaline medium; (e) sodium nitroprusside; (f) uranyl acetate, followed by separation of the colored complex; (g) 4′-antipyrine; (h) magnesium and hydrochloric acid.

(khellinol, 5-norkhellin, desmethylkhellin) (**12**) is effected by boiling with hydrochloric acid,[44, 57, 58] dilute hydrobromic acid (sp. gr. 1.38),[57] thiophenol in the presence of piperidine,[59] or aniline hydrochloride.[58] The efficiency of the latter demethylating reagent has been questioned.[60] Prolonged treatment of **1** with aluminum isopropoxide brings about a similar result.[61]

Complete demethylation of **1** to 5,8-dihydroxy-2-methylfuro(3′,2′–6,7)chromone (5,8-dinorkhellin) (**13**) can be brought about either by treatment with magnesium iodide in the absence of solvent,[62] or by refluxing with pyridine hydrochloride.[63]

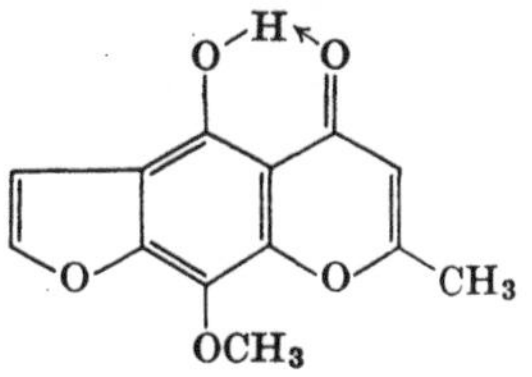

(**12**)

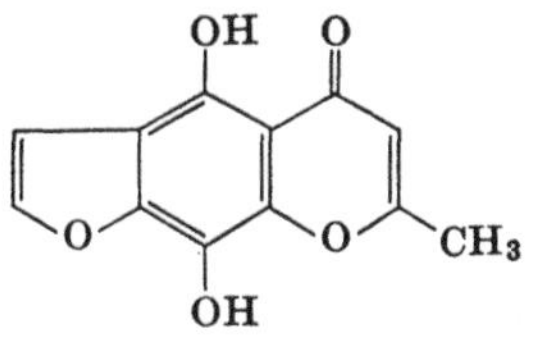

(**13**)

Demethylation is believed to be connected with the methoxyl group in the 5-position, since the partially demethylated product (**12**) is

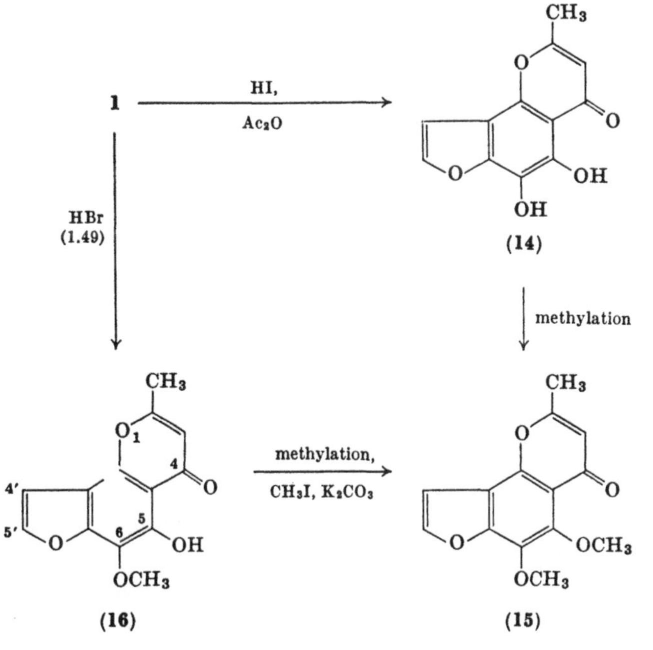

(13)

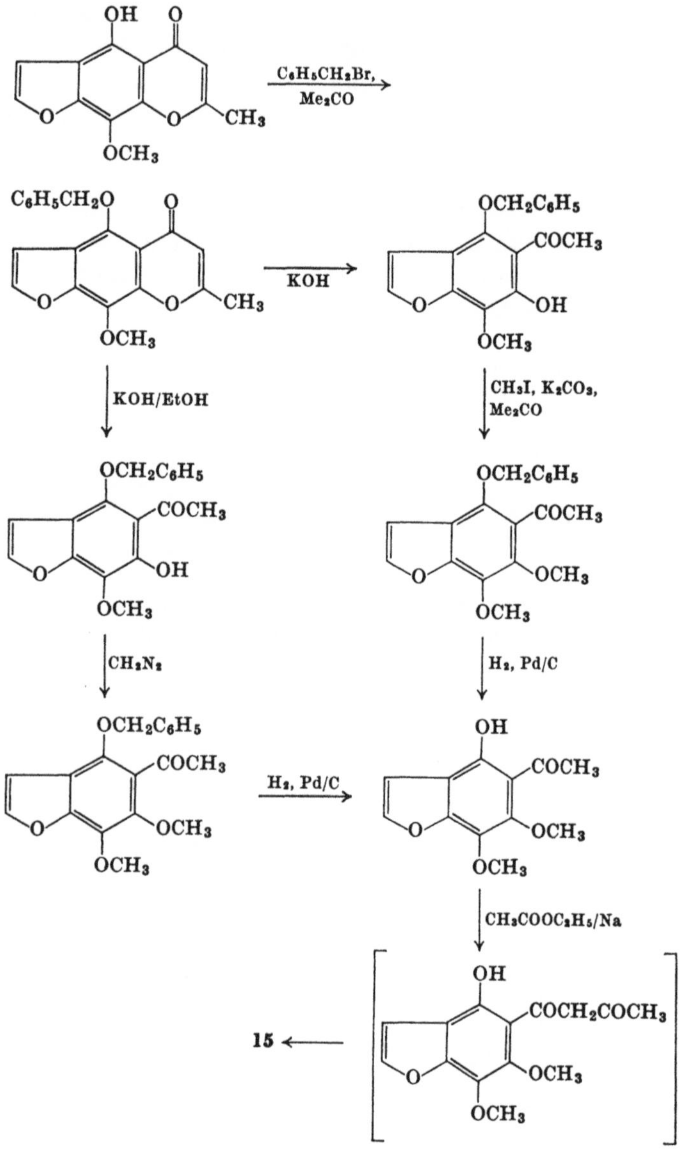

(14)

insoluble in alkali and stable towards diazomethane, characteristics thought to be due to chelation.

The hydrogen iodide demethylation of **1**, followed by remethylation,[44, 64] gave an isomer, isokhellin (5,6-dimethoxy-2-methylfuro-

(2′,3′–7,8)chromone or 5,6-dimethoxy-2-methyl-4*H*-furo(2,3-h)-1–benzopyran-4-one) (**15**), which is still a 2-methylchromone since like khellin it yields a piperonylidene derivative. Similar rearrangement, accompanied by partial demethylation resulting in the formation of 5-norisokhellin (**16**), but not 5,6-dinorisokhellin (**14**), is achieved by using moderately concentrated hydrobromic acid (sp. gr. 1.49)[65] (Eq. 13).

Clarke and Robertson[46] have suggested that in such a rearrangement, it is the furan ring which opens, and recyclizes at an alternative site.

The structure of isokhellin (**15**) has been confirmed by its identity with a synthesized product[66] (Eq. 14), and its difference from the second possible structure,[46, 66] allokhellin (**17**), which has been synthesized recently by Horton and Stout[67] (Eq. 15).

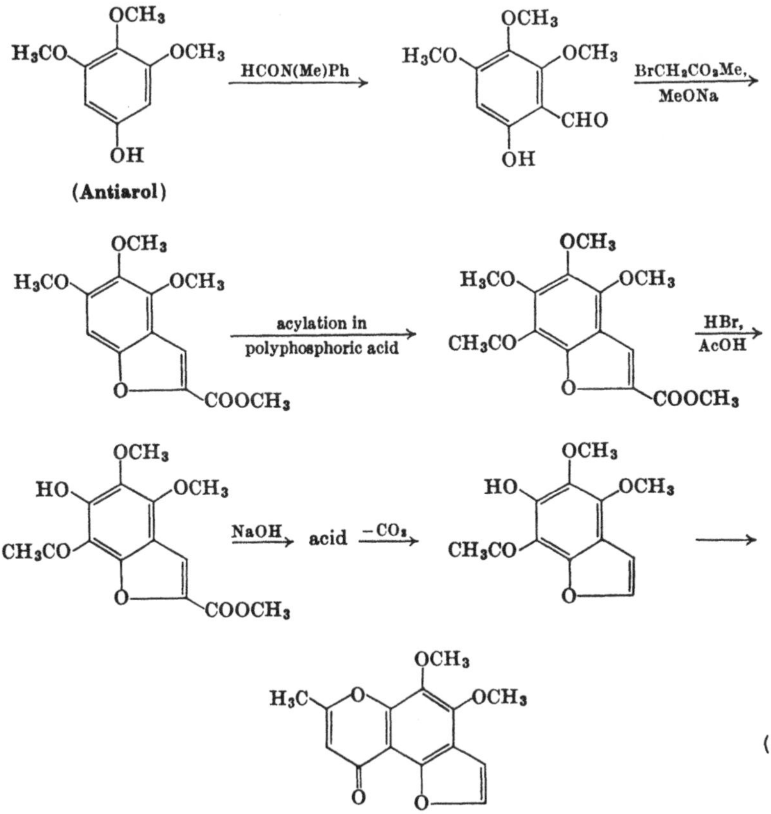

(**17**)

(15)

5-Norisokhellin (desmethylisokhellin) (**16**) cannot be methylated with diazomethane, but treatment with a methyl iodide–potassium carbonate mixture gave isokhellin (**15**).[65] The infrared spectrum of **16** showed a band at 3100 cm.$^{-1}$ (due to a bonded OH group). The product of alkaline degradation of 5-norisokhellin benzyl ether, followed by methylation, was found to be identical with authentic khellinone benzyl ether[68] (Eq. 16).

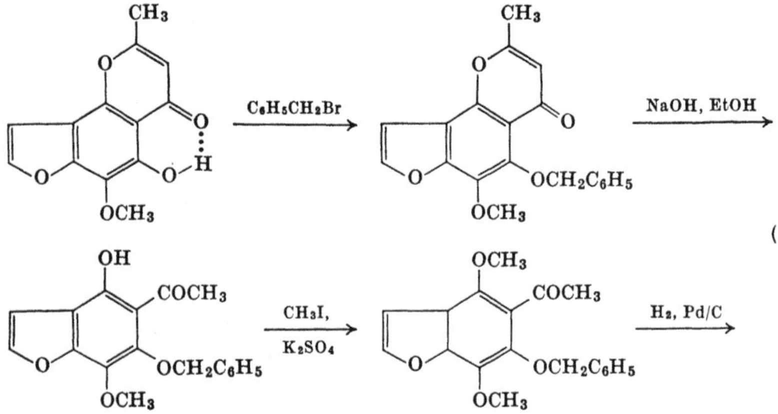

(16)

Khellinone (2)

The maxima of the ultraviolet spectra of khellin (**1**), isokhellin (**15**), and their partially demethylated products, desmethylkhellin (**12**) and desmethylisokhellin (**16**), together with the isomeric allokhellin (**17**) are presented in (Table 2). It is interesting to note that the 8 mμ shift, toward the methyl ether of the phenol, in each case is in the first maximum.[29, 67, 68]

TABLE 2. Ultraviolet Absorption Data of Khellins

Compound	λ_{max}^{EtOH} (95%) (mμ) (log ϵ)
Khellin (**1**)	247 (4.57); 281 (3.67); 331 (3.67)
Isokhellin (**15**)	243 (4.47); 319 (3.66)
Allokhellin (**17**)	255.5 (4.37); 294 (3.93); ~244 (4.29)i; ~261 (4.33)i
Desmethylkhellin (**12**)	255 (4.52); ~287.5 (–); 353 (3.49)
Desmethylisokhellin (**16**)	251 (4.45); ~344 (3.45)

i, inflexion.

The easy degradation of khellin to khellinone (**2**) could be explained by regarding khellin as vinylog (**D**) of the ester (CH_3OCOR),[69] but this view has been questioned.[70] Alkaline degradation in a similar manner to that of isokhellin (**15**), leads to the formation of isokhellinone (**18**).

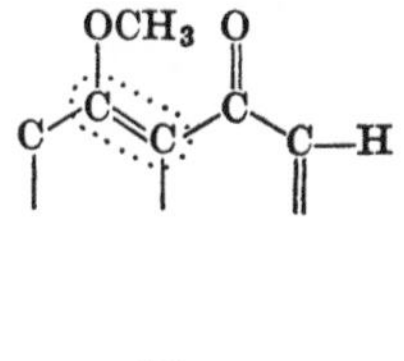

(**D**)

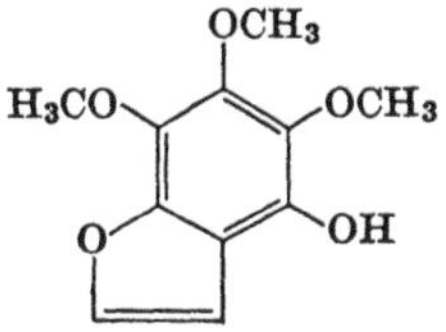

(**18**)

Aqueous sodium hydroxide solution converted a number of furochromone derivatives to the corresponding dibenzofurans[71] (Eq. 17).

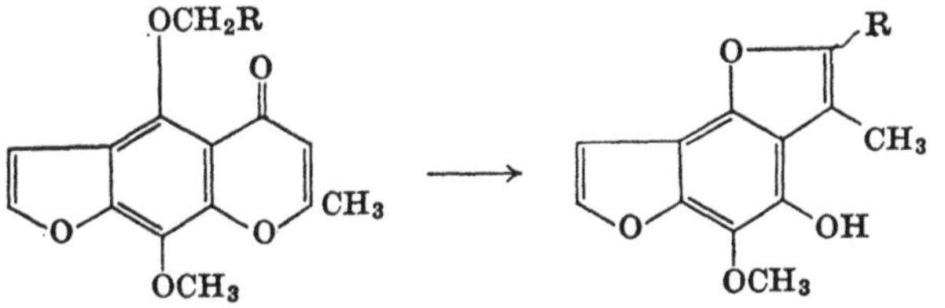

(17)

Khellinones, **2** and **18**, are useful in direct syntheses. Thus, upon treatment with ethyl carbonate in the presence of sodium, they produce the 4-hydroxyfurocoumarins, **19** and **20**, respectively. The latter compounds condense with formaldehyde to give the dicoumarols, **21** and **22**, respectively.[55]

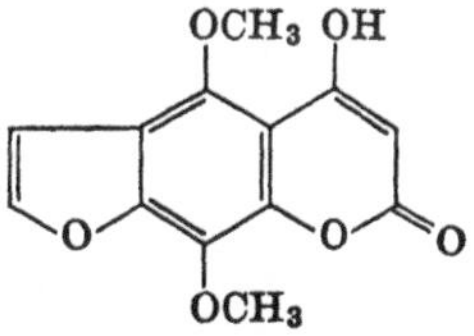

(**19**)

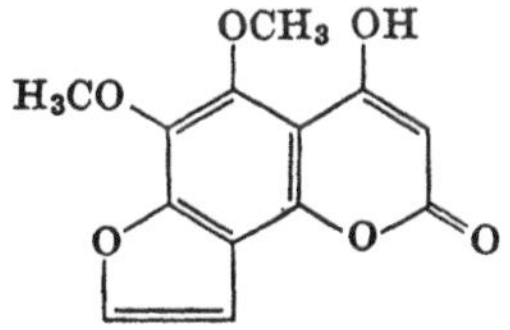

(**20**)

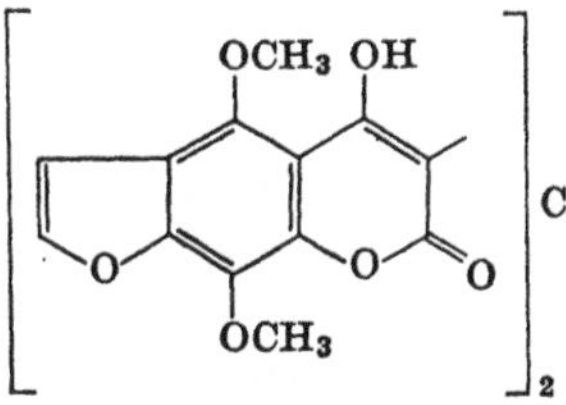

(**21**)

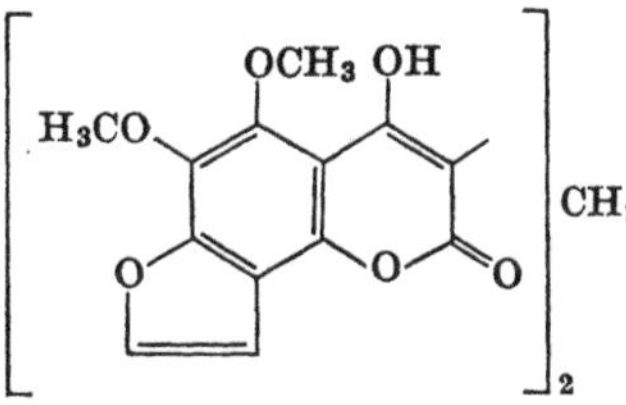

(**22**)

Condensation of **2** with aromatic aldehydes, followed by oxidation of the chalkones formed, established a new route to 2-arylidene-4,7-dimethoxyfuro(5,6–4′,5′)coumaran-3-ones[41] (Eq. 18).

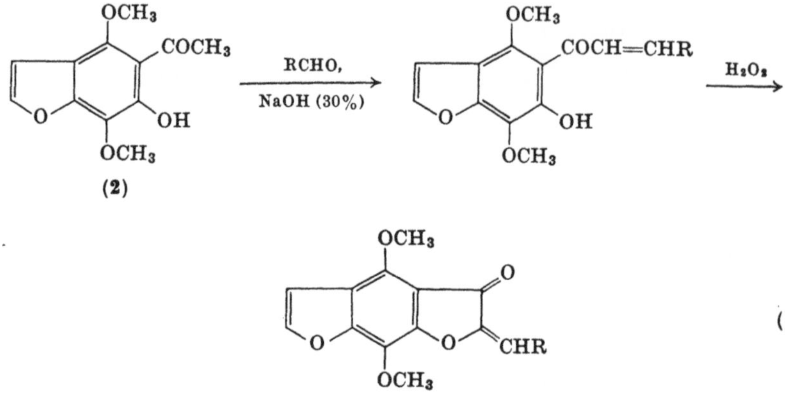

(18)

Hydrogenolysis of khellin with lithium aluminum hydride is accompanied by γ-pyrone ring opening with the formation of 5-butyryl-6-hydroxy-4,7-dimethoxycoumarone (**23**). On the other hand, catalytic hydrogenation of **1** in the presence of $PtO_2.H_2O$, at room temperature, avoids the heterocyclic ring opening, with the formation of 5,8-dimethoxy-2-methylfuro(3′,2′–6,7)chromanone (**24**).[72] The survival of the carbonyl group in **23** is explained as the result of a catinoid–ethenoid grouping.[73] The ultraviolet absorption spectrum of **24** shows a maximum at 290 mμ and that of khellin at 282 and 330 mμ; thus showing that the maximum at 330 mμ may be connected with the double bond in khellin.

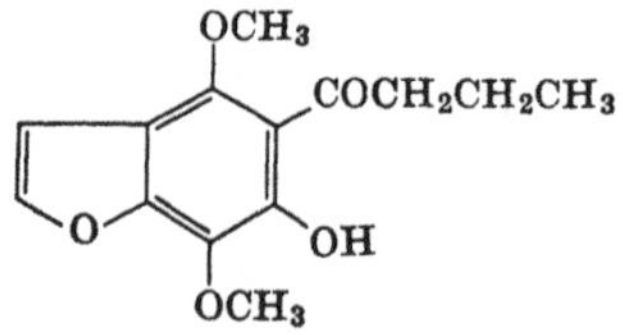

(23)

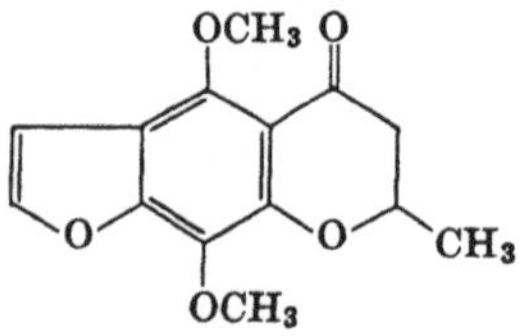

(24)

In contradiction to the above results Dann and Volz,[73a] have recently reported that khellin is catalytically reduced, in the presence of platinum oxide at room temperature and atmospheric pressure, to give an alkali soluble component, which on acidification yields **25a** (R = H). In the presence of a $Pd(OH)_2$–$BaSO_4$ catalyst two equivalents of hydrogen are absorbed, and treatment of the resulting crystalline residue with a chromic trioxide–pyridine complex yields **25b** ($RR^1 = O$). The latter is readily reduced by amalgamated zinc and hydrochloric acid to **25b** ($R = R^1 = H$) which yields **25c** ($R = H$) when allowed to react with isoamylnitrite. The latter compound is catalytically reduced in the presence of $Pd(OH)_2$–$BaSO_4$, under pressure, to give **25d** ($R = R^1 = H$), which upon treatment with a chromic trioxide–pyridine complex yields **25e** ($R = Ac$). Khellin, in tetrahydrofuran at $-80°$ is reduced by lithium aluminum hydride to **24**, which is readily reconverted to khellin by the action of chloranil in boiling xylene.

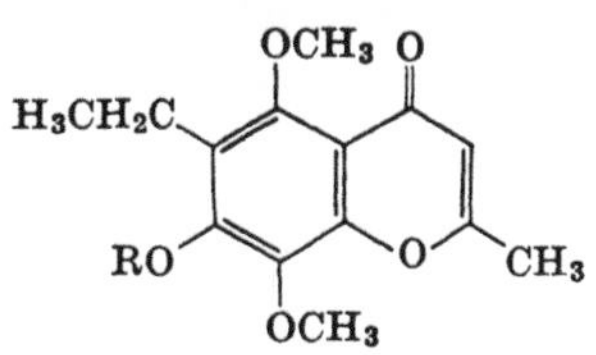

(25a)

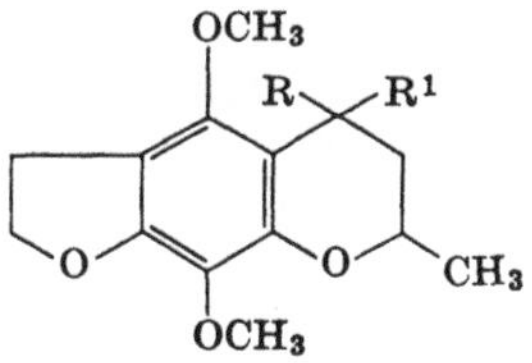

(25b)

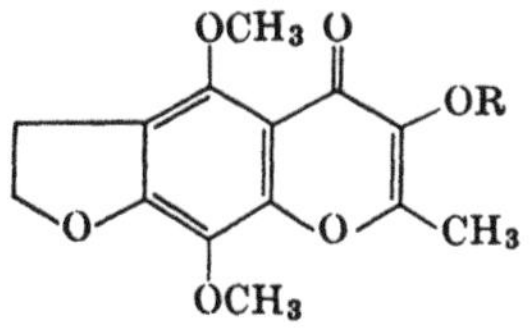

(25c)

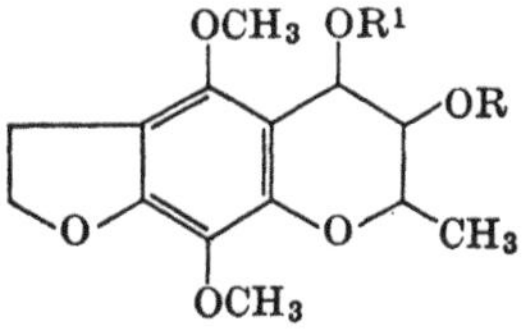

(25d)

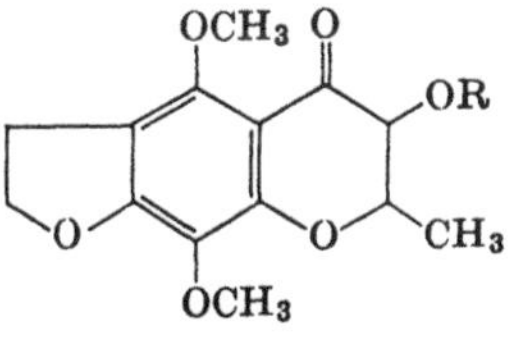

(25e)

Oxidation of khellin by various reagents has been the subject of extensive studies. 5,8-Bisnorkhellin (khellinquinol) (**13**) is produced during the oxidative demethylation of khellin with nitric acid to khellinquinone **26**,[50] which can be reduced to **13** by sulfur dioxide.[50, 63]

Controlled oxidation, with alkaline hydrogen peroxide, leads to the formation of a salicylic acid derivative (**27**), but under drastic conditions, furan-2,3-dicarboxylic acid is produced.[10, 41]

The γ-pyrone ring of khellin is opened by the action of hydrazine and

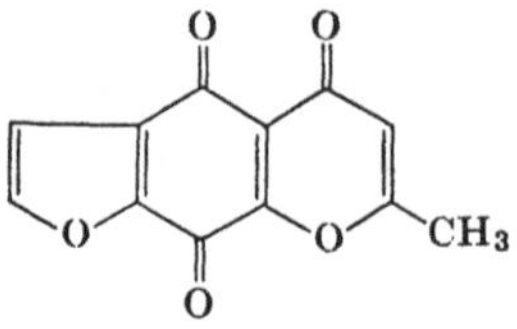

(26)

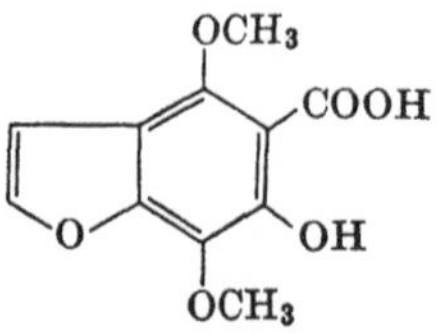

(27)

its derivatives with the formation of pyrazole compounds, containing the 6-hydroxy-4,7-dimethoxybenzofuran group in the 3- or 5–position of the pyrazole ring **28a, b**.[74, 75] The pyrazoline derivative (**29**)[72] is obtained by treatment of the chromanone (**24**) with hydrazine in ethanol. Thiokhellin (**40**) reacts with hydrazine, in a similar manner to khellin, yielding **28a** or **28b** (R = H).[76] The γ-pyrone ring in 2-styryl derivatives of khellin is opened by hydrazine in ethanol to yield **30**.[77]

Ruggieri[78] has shown that the condensation of khellin with hydrazines and hydrazides, in refluxing ethanol, gave **31**, **32**, and/or **33** (cf. Table 3).

TABLE 3. Reaction Products of Khellin with Hydrazines and/or Hydrazides

Starting compound	Molar ratio of khellin to starting compound in the product	Product	M.p. (°C)	Solvent for crystallization[a]
1-Hydrazinophthalazine	1	**32** (R = 1-phthalazinyl-amine)	119–120	A
1,4-Dihydrazinophthal-azine	2	**33** (R = 1,4-phthalazinyl-enediamino)	107–108	A
Isonicotinylhydrazine	1	**33** (R = iso-nicotenoyl)	135	B
Cyanoacetylhydrazine	1	**31** (R = cyano-acetamido)	234–235	B
Benzidine	2	**32** (R = 4,4′-biphenylene)	87	B
α-Naphthylamine	1	**31** (R = α-naphthyl)	145	B
2,4-Dinitrophenyl-hydrazine	1	**33** (R = 2,4-dinitrophenyl)	126–127	B

[a] A, acetone; B, ethanol.

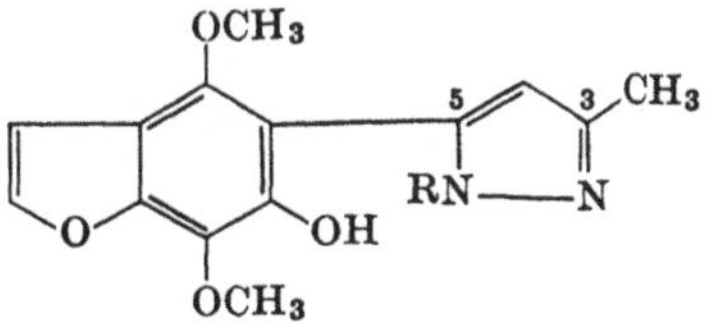

(28a)

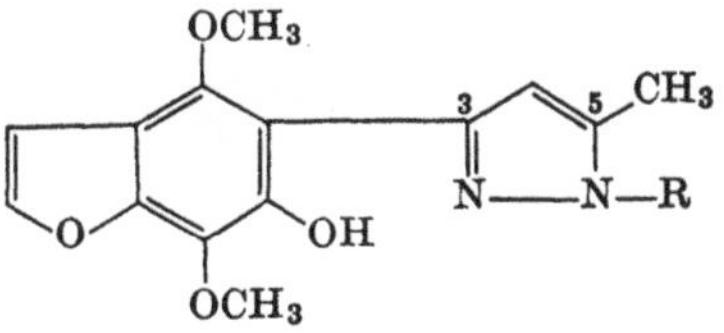

(28b)

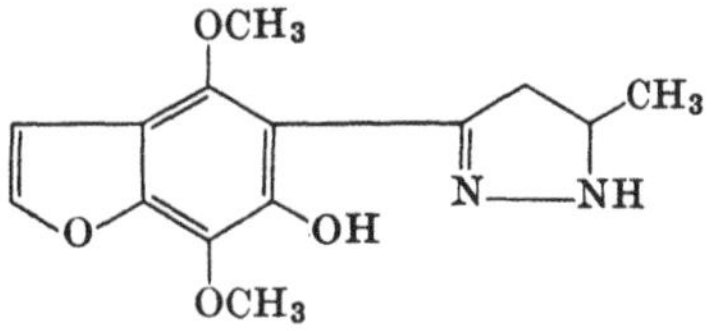

(29)

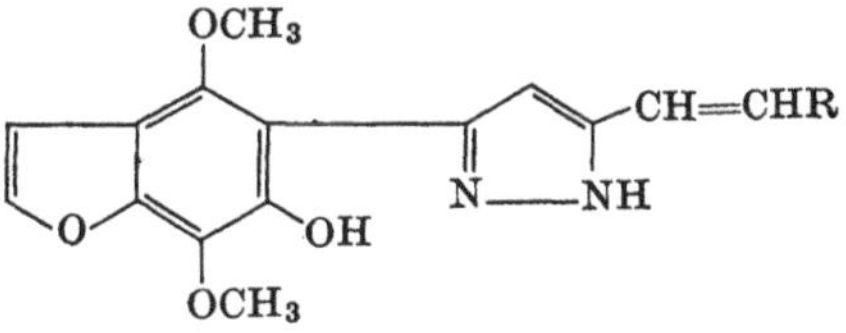

(30)

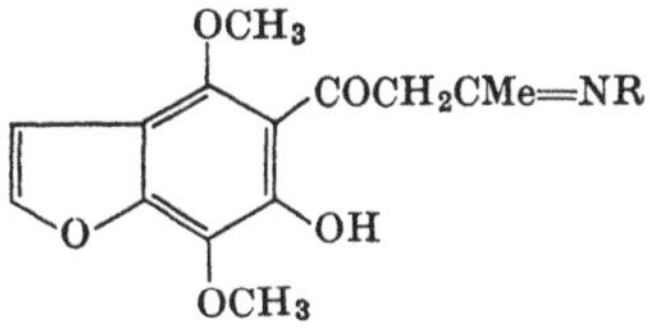

(31)

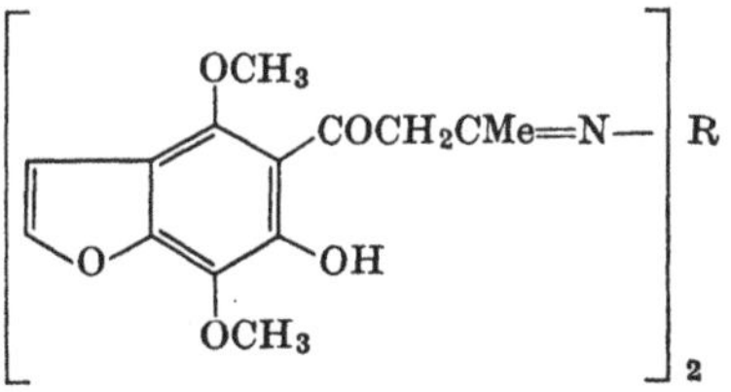

(32)

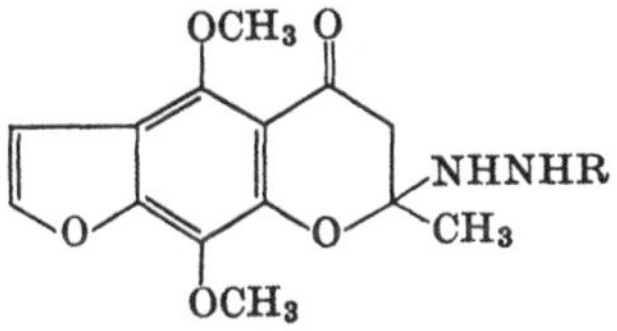

(33)

The opening of the heterocyclic ring of khellin has also been effected by the action of hydroxylamine hydrochloride in pyridine,[74] and of guanidine,[77] yielding the isoxazole derivative (**34a**) or (**34b**) and (**35**) respectively.

It has been reported that the benzisoxazole derivative (**36**), is readily obtained by cyclization of khellinone oxime, obtained by treatment of khellinone (**2**) with hydroxylamine hydrochloride in the presence of sodium bicarbonate, with acetic anhydride.[79]

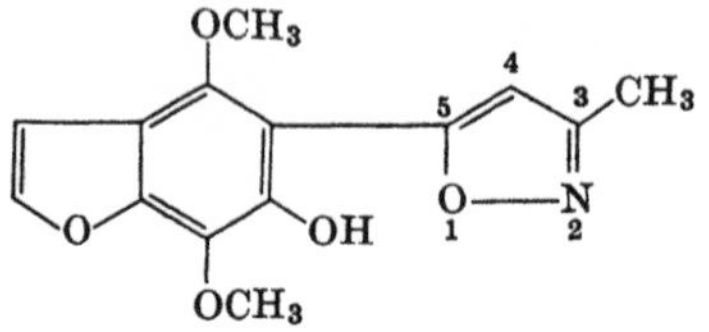
(34a)

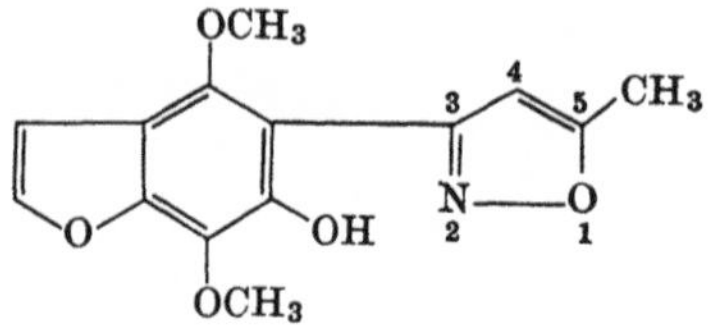
(34b)

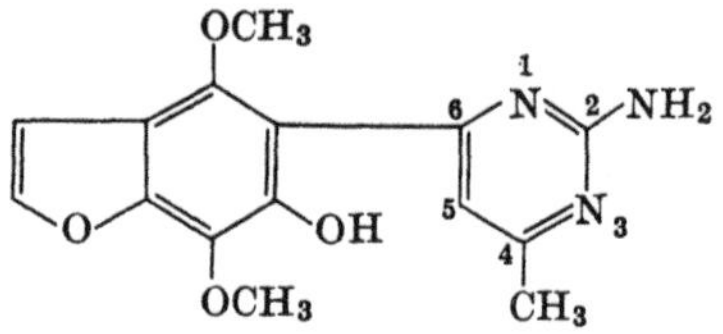
(35)

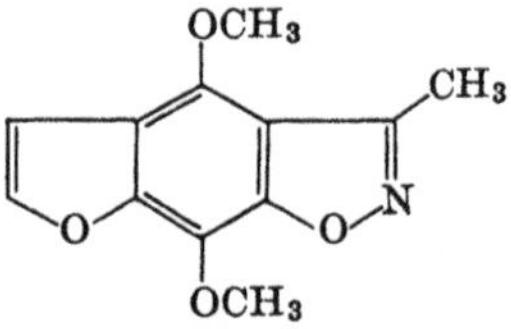
(36)

Aliphatic primary amines, but not secondary amines, react with khellin with the opening of the γ-pyrone ring, and formation of derivatives of 6-hydroxy-4,7-dimethoxycoumarone (**37**).[80] Similarly, **37** ($R = CH_3$) has been obtained by treatment of khellin or thiokhellin with methylamine.[76] Analogous derivatives (**38**) are obtained by treatment of khellin with other aliphatic amines.[80]

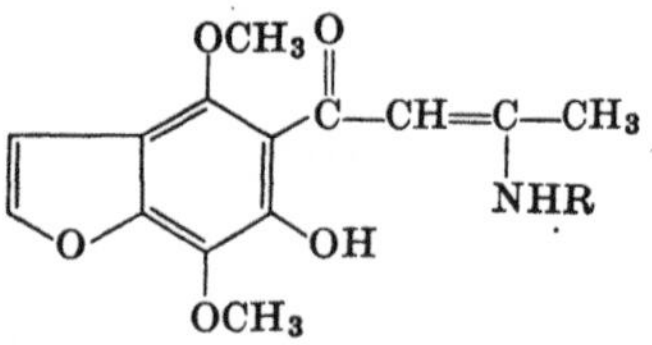
(37)

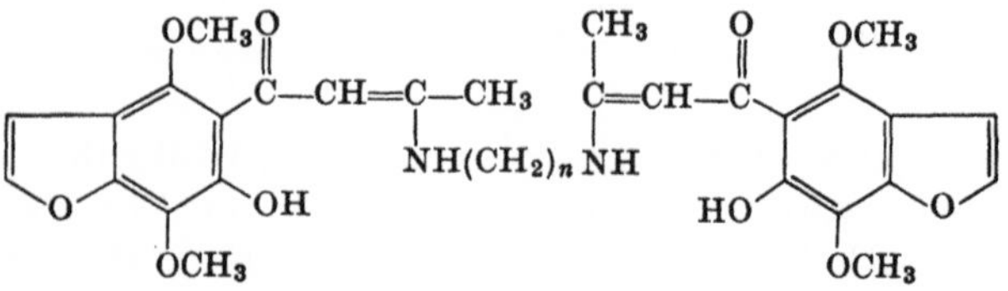
(38)

The behavior of khellin in the Mannich reaction has been investigated by Reichert[81] in an attempt to obtain water soluble derivatives. Dimethylamine, piperidine, and/or morpholine led to the formation of the 3-*N*-substituted aminomethyl derivative (**39**). 3-Piperidinomethylkhellin (**39**, R = piperidyl) has been reported to show double spasmolytic effect as in khellin.[81]

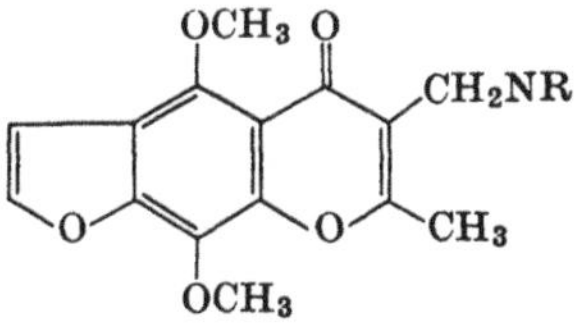

(**39**)

Thiokhellin (**40**) is obtained as a dark–red microcrystalline powder[72] by treatment of khellin with phosphorus pentasulfide in boiling xylene.

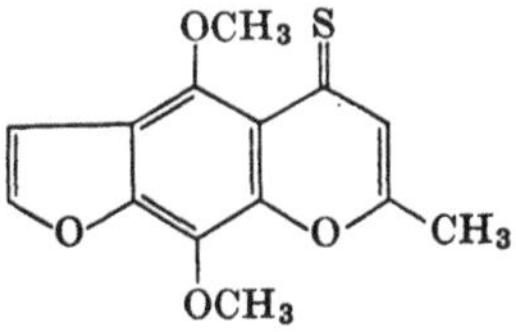

(**40**)

Khellin, being a chromone derivative with a methyl group in the 2-position, is expected to condense with aromatic aldehydes as do other 2-methylchromones.[82] In the presence of sodium alcoholate, such condensation brings about the formation of 2-styrylkhellins.[54, 77] Norkhellin (**12**) reacts similarly with benzaldehyde to give 2-styrylnorkhellin,[83] and thiokhellin (**40**) reacts with *p*-nitrobenzaldehyde to give the corresponding 2-*p*-nitrostyrylthiokhellin.[72] 2-Styrylkhellins, having conjugated double bonds, one of which is part of the heterocyclic ring, behave in a similar manner to 2-styrylchromones, and take part in the Diels–Alder reaction leading to derivatives of 5*H*-furo(3,2-b)xanthene[84] (Eq. 19).

Khellin condenses, via its pyridinium iodide, with *p*-nitrosodimethylaniline in the presence of sodium carbonate, to give the corresponding azomethine derivative[61] (Eq. 20).

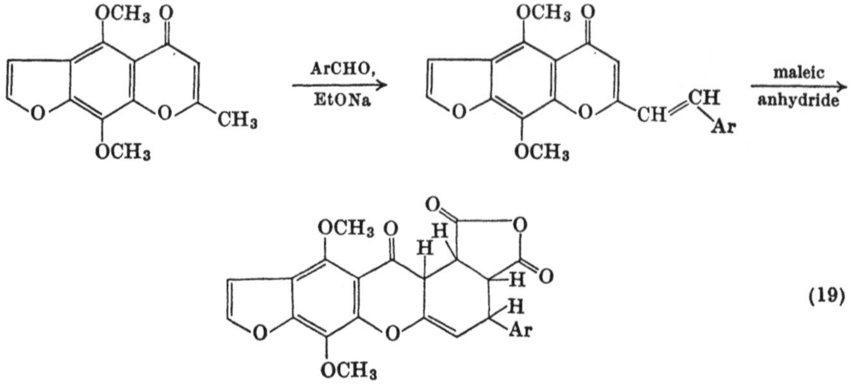

(19)

Compounds, believed to be oxonium salts, have been obtained by direct treatment of **1** with sulfuric and hydrochloric acid. Assay of **1** in its medical preparations is based on the photoelectric colorimetric determination of the intensity of the yellow color imparted by khellin to sulfuric acid of known concentration.[20] A perchlorate is formed with perchloric acid (60 %).[54]

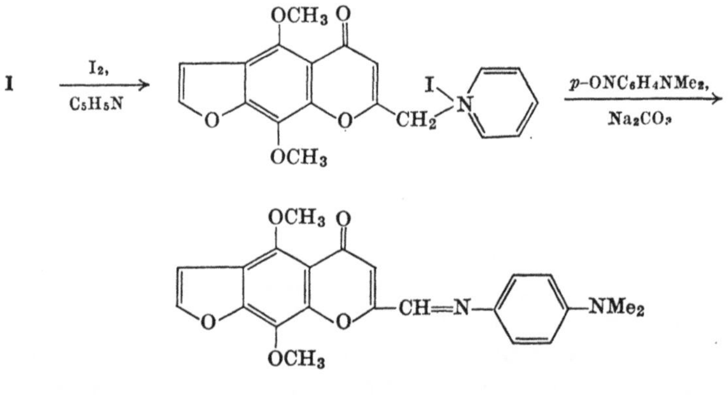

(20)

The iodochlorohydrate oxonium salt of **41a** is obtained by addition of iodine monochloride in hydrochloric acid to an acetic acid solution of **1**; **41b** is obtained during the addition of iodine to a chloroform solution of **40**.[72]

Khellin forms a crystalline adduct with thiourea (1:1), but not with urea; a deeply colored crystalline addition compound is formed when **1** is treated with iodine.[85]

A photochemical addition reaction is known to take place between olefins and *o*-quinones in sunlight,[86–88] for example the photoreaction of

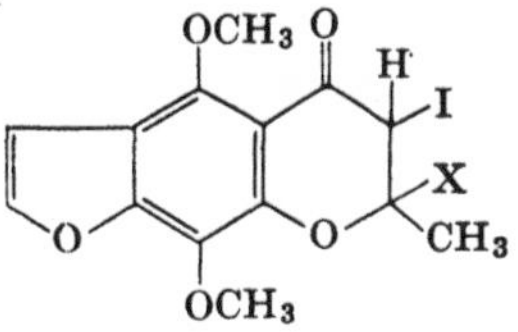

(**41a**) X = Cl;
(**41b**) X = I

2,3-diphenylbenzofuran and phenanthraquinone.[89] The same is true of khellinone (**2**), which forms the photoproduct **42**.[90]

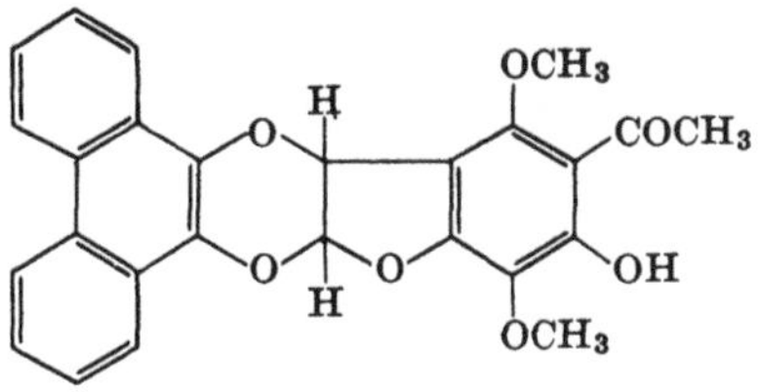

(**42**)

Photochemical addition of benzaldehyde to khellinquinone (**26**) in sunlight, with the formation of 5-benzyloxy-8-hydroxy-2-methylfuro-(3′,2′–6,7)chromone (**43**) has been reported.[90a]

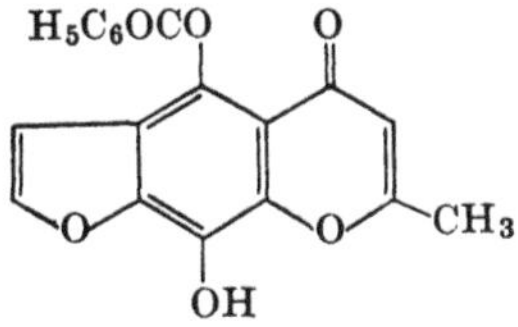

(**43**)

B. Visnagin

Visnagin (visnagidin, 4-methoxy-7-methyl-5*H*-furo(3,2-g)–1–benzopyran-5-one or 4-methoxy-2-methylfuro(3′,2′–6,7)chromone) (**44**), $C_{13}H_{11}O_4$, is an important companion of khellin found in the seeds of *Ammi visnaga* L.[91] It is a furochromone, containing one methoxyl group, parallel to the furocoumarin bergapten (**25**, Chapter 2), and on solution in fuming nitric acid forms a bright green crystalline oxonium nitrate which regenerates visnagin by hydration. The analytical proof of structure **44**, adduced by Späth and Gruber,[91] is closely similar to that given for khellin (**1**).

(1) Synthesis of visnagin. Partial synthesis of visnagin (**44**) has been achieved by condensation of visnagone (5-acetyl-4-methoxy-6-benzofuranol) (**45**), the alkaline degradation product of **44**, with ethyl acetate

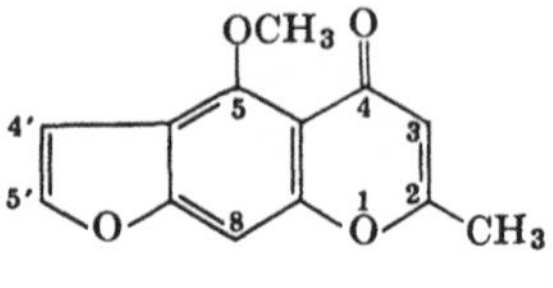

(44)

in the presence of sodium, followed by cyclization of the resultant β-diketone (**46**)[92] with hydrochloric acid, or by drastic treatment with a sodium acetate-acetic anhydride mixture[93, 94] (Eq. 21).

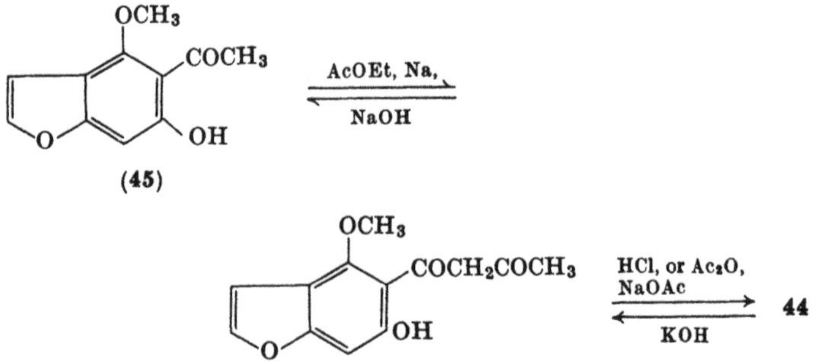

(45) **(46)** (21)

Reduction of **47c**, obtained by treatment of khellol tosylate (**47b**) with sodium iodide,[95] or by treatment of **47d**, prepared by the action of thionyl chloride on khellol (**47a**), with zinc and acetic acid,[96] yielded **44**. This method established a route to visnagin and its 2-methyl substituted derivatives.

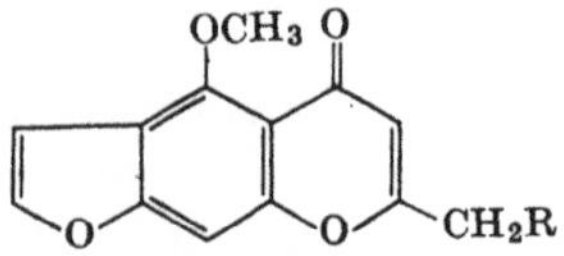

(47a) R=OH;
(47b) R=$OSO_2C_6H_4CH_3$-p;
(47c) R=I;
(47d) R=Cl

Total synthesis of visnagin has been achieved independently, by Gruber and Horváth,[97] Davies and Norris,[35] and Geissman and

Hinreiner,[98] via the total synthesis of visnagone (**45**). This synthesis involved a large number of steps and the final yield was small, yet it solved orientation problems by the use of coumarans rather than coumarones. 4,6-Dihydroxycoumaran (**48a**) supplied both possible acetophenones in the Hoesch reaction. The requisite acetophenone (**48b**) was selected and monomethylated to give the possible monomethyl ethers. Again, the requisite material was selected, converted into the β-diketone and cyclized to 4′,5′-dihydrovisnagin (**49**) which was dehydrogenated most successfully by *N*-bromosuccinimide.

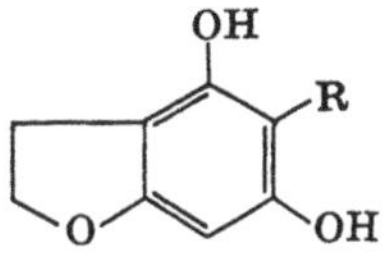

(**48a**) R = H;
(**48b**) R = $COCH_3$

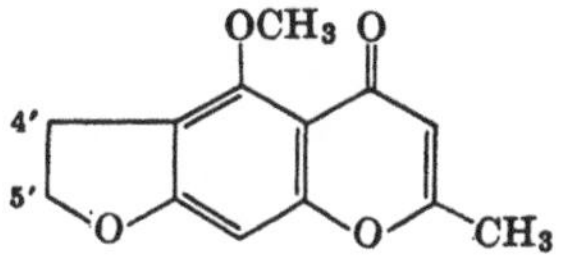

(**49**)

The synthesis, in which the furan ring is built up last, may be closer to the biosynthetic process in plants. Seshadri and coworkers,[99] starting with 5,7-dihydroxy-2-methylchromone, have introduced an allyl group into the 6-position by Claisen migration of a 5-allyl ether. The initial protection of the 7-hydroxyl group is best effected by tosylation. If the tosyl group is removed before Claisen migration and the *C*–allyl compound subjected to ozonolysis and ring closure, norvisnagin is obtained in a poor yield and can be methylated to visnagin (Eq. 22).

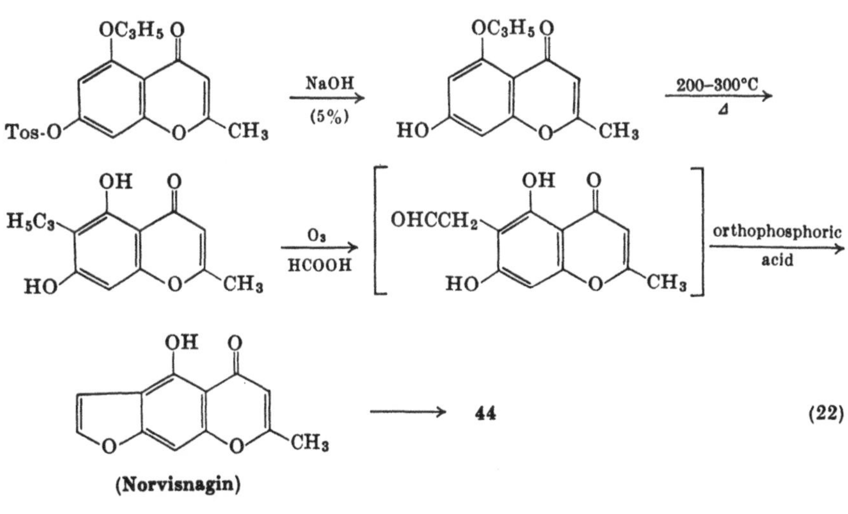

(22)

(**Norvisnagin**)

In an alternative method, the tosyl group is removed just before ozonolysis, and previous methylation of the 5-position is advantageous. This route gives visnagin directly in good yield (Eq. 23).

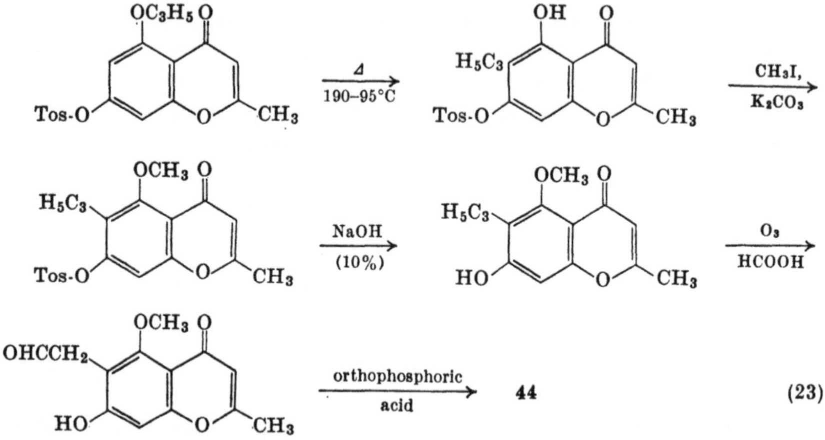

(23)

(2) Synthesis of visnagin analogs and related transformations. Due to the marked interest in the pharmacological properties of visnagin and related compounds, a great number of reactions has been described in the literature, (Table 1). Condensation of visnagone (**45**) with esters in the presence of sodium has led to the production of 2-ethyl-[62] (**50b**), 2-(3′-pyridyl)visnagin analog[55] (**50c**), and the parent 2-unsubstituted analog[62] (**50a**). The pyridyl analog (**50c**) is easily oxidized with chromic acid to 6-formyl-7-hydroxy-5-methoxy-2-(3′-pyridyl)chromone (**51**).[55]

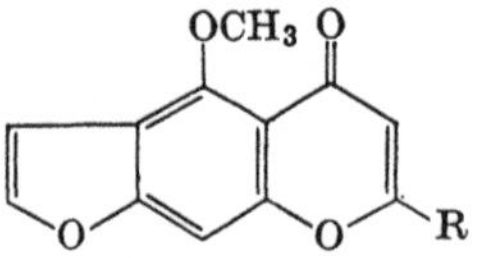

(**50a**) R = H;
(**50b**) R = C_2H_5;
(**50c**) R =

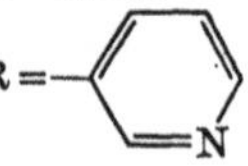

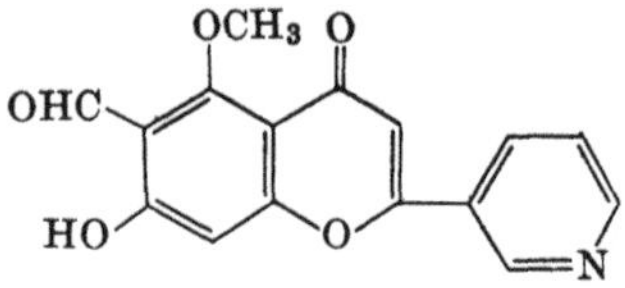

(**51**)

Synthesis of baicelin from visnagin in a six step process has been reported[100] (Eq. 24).

6-Formyl-7-hydroxy-5-methoxy-2-methylchromone (**52**), the oxidation product of visnagin, can replace the furan ring in **44** by an α-pyrone ring.[101] Thus, condensation of **52** with ethyl malonate in the presence of piperidine gave the ester (**53b**), which upon hydrolysis, produced the

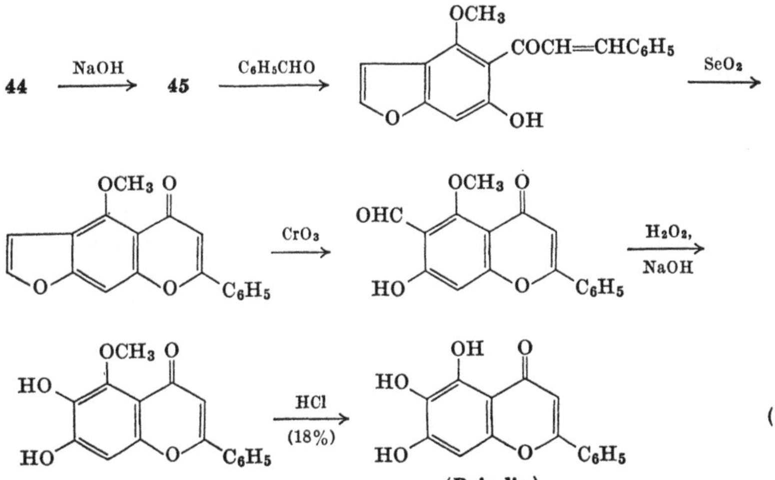

(Baicelin)

(24)

acid (**53c**), which decarboxylated to **53a**. **53a** and **53c** can be readily demethylated with hydrochloric acid to yield **53e** and **53d** respectively.[102] Moreover, alkali treatment of **53a** produced **54**, which on alkaline hydrogen peroxide oxidation gave **55**. **53a**, a hybrid between visnagin and bergapten, can also be conveniently obtained by treatment of the aldehyde (**52**) with acetic anhydride and sodium acetate.

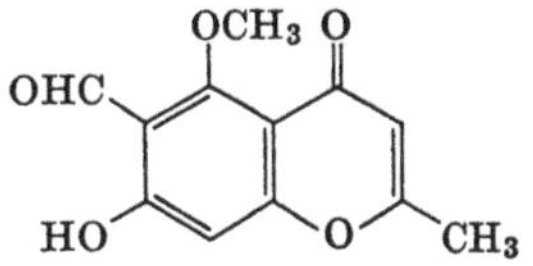

(**52**)

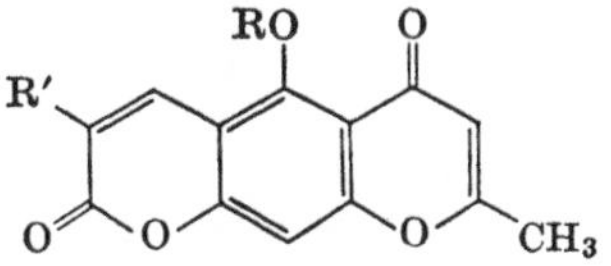

(**53a**) $R=CH_3$, $R'=H$;
(**53b**) $R=CH_3$, $R'=COOC_2H_5$;
(**53c**) $R=CH_3$, $R'=COOH$;
(**53d**) $R=H$, $R'=COOH$;
(**53e**) $R=R'=H$

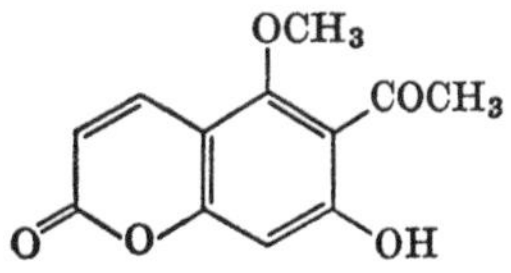

(**54**)

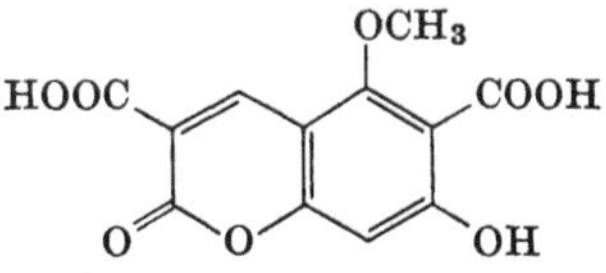

(**55**)

(3) Reactions. When warmed with dilute hydrochloric, visnagin behaves as khellin, forming the phenol, 5-norvisnagin (**56**). Thiophenol, in the presence of piperidine,[59] and/or prolonged treatment with aluminum isopropoxide brings about a similar demethylation reaction.[61] Demethylation of chromones, substituted with methoxy groups in positions 3 and 5 has been reported.[103, 104]

5-Norvisnagin reacts with α-acetobromoglucose in quinoline and silver oxide to yield 5-norvisnagin tetra-*O*-acetylglucoside, which readily hydrolyzed to 5-norvisnagin and glucose.[105] Furthermore, **56** can be nitrated; the nitro derivative can be reduced to the corresponding amino derivative, which upon oxidation gives khellinquinone (**26**).[45] **56**, being a 2-methylchromone derivative, condenses with benzaldehyde, to give the 2-styryl derivative.[83]

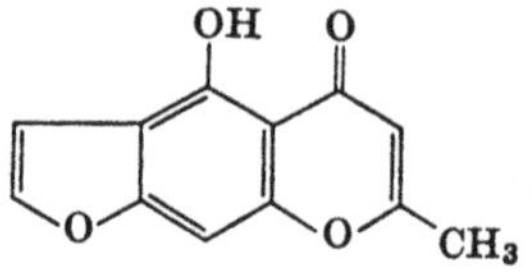

(**56**)

As with khellin (**1**), hydrogen iodide demethylations entail a rearrangement, so that remethylation of the resulting phenol, 7-norisovisnagin

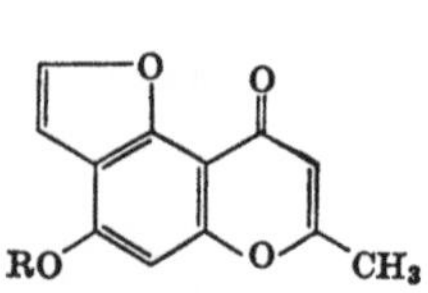

(**57a**) R = H;
(**57b**) R = CH_3

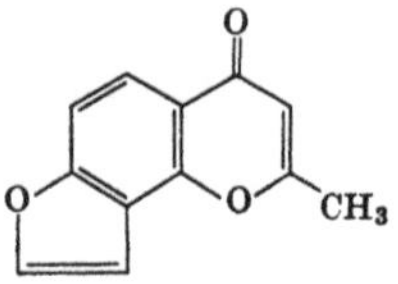

(**58**)

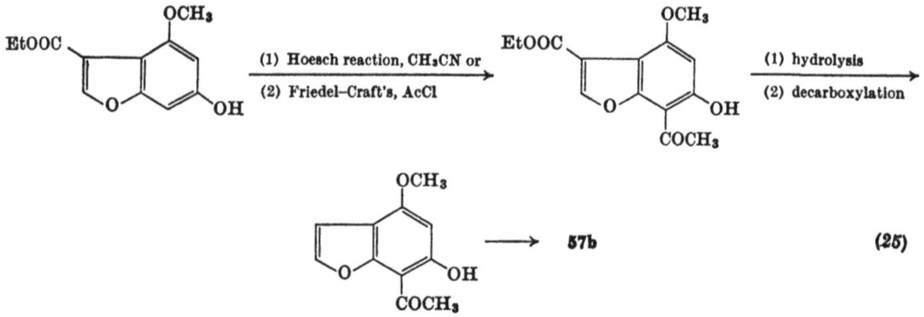

(**Isovisnagone**)

(**25**)

(**57a**), gives a new compound, isovisnagin (**57b**) whose structure has been confirmed by synthesis[92, 106] (Eq. 25).

The rearrangement clearly shows that the furan ring of visnagin (**44**) is opened during demethylation (presumably to a derivative of phenylacetaldehyde) with subsequent closure to **57b**, since the opening and subsequent recyclization of the chromone ring would give the possible isomeric allovisnagin (**58**), which has been synthesized [107] (Eq. 26).

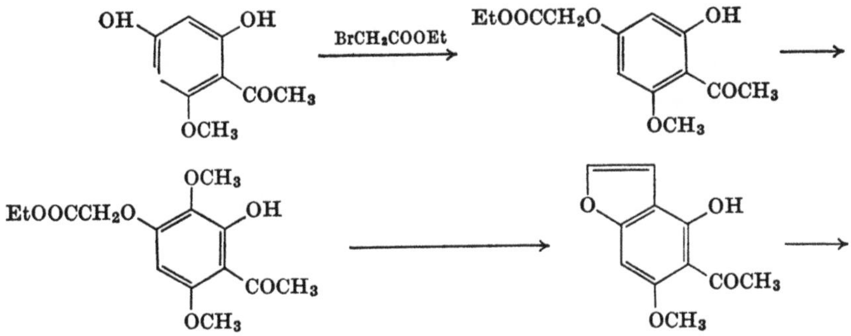

(26)

58

(Allovisnagone)

The great stability to boiling hydriodic acid achieved by khellin (**1**) and visnagin (**44**) during demethylation is exceptional for α,β-unsubstituted benzofurans. A similar exceptional stability to strong mineral acids has been noticed in *o*-ethylvisnagone, which is reduced by the Clemmensen method to the corresponding ethylcoumarone (**59**); and by khellinone (**2**) which is oxidized with nitric acid to the quinone (**60**) (Eqs. 26a, 26b).

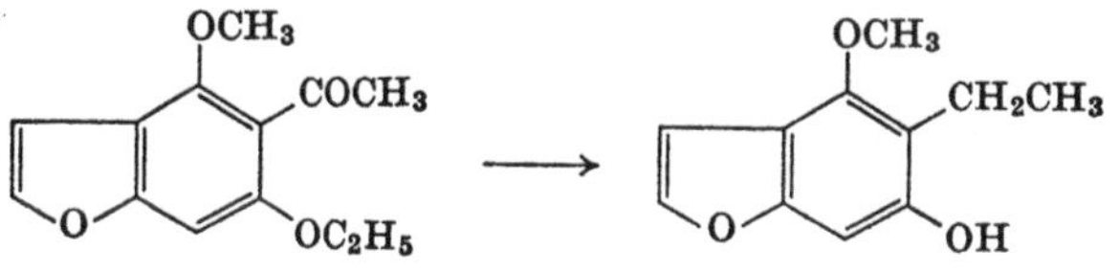

(26a)

(59)

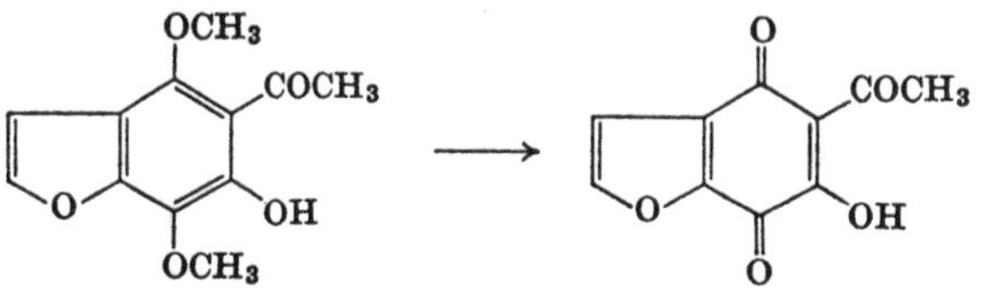

(26b)

(Khellinone) **(60)**

Dilute alkali (1%) cleaves the chromone ring in visnagin[91] to form acetic acid and visnagone (**45**) together with 6-hydroxy-4-methoxycoumarone-5-carboxylic acid.[108] The presence of a phloroglucinol nucleus in **45**, and hence in visnagin (**44**), is shown by fusion of **45** with caustic potash. Ozonolysis of *O*-ethylcoumarone (**59**) cleaves the furan ring, with the formation of the liquid *o*-hydroxy aldehyde. Ethylation of the latter gives the *O*-ethyl ether derivative, which forms a semicarbazone, identical with a synthetic specimen, and which is converted to the corresponding acid on oxidation (Eq. 27). Thus the linear structure of visnagin is established since the aldehyde has been successfully synthesized from phloroacetophenone, and was found to be identical with the natural material.[109]

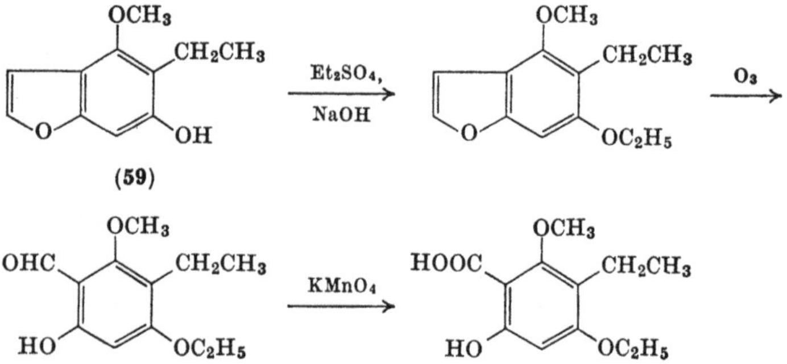

(27)

Visnagone (**45**) reacts with ethyl carbonate in the presence of sodium to give the corresponding 4-hydroxycoumarin, which condenses with formaldehyde to yield the dicoumarol[55] (Eq. 28).

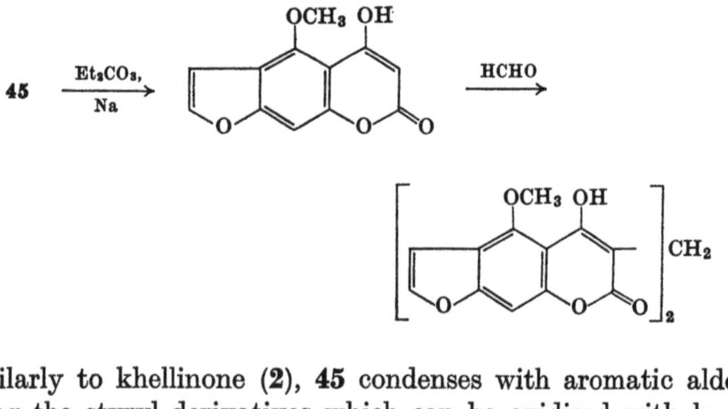

(28)

Similarly to khellinone (**2**), **45** condenses with aromatic aldehydes yielding the styryl derivatives which can be oxidized with hydrogen peroxide to give 2-arylidene-4-methoxyfuro(5,6–4′,5′)coumaran-3-ones[41] (Eq. 29).

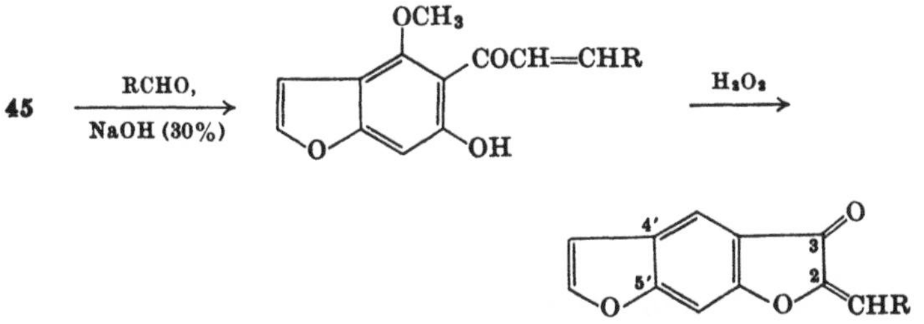

(29)

Aryldiazonium chloride couples with **45** to give 5-acetyl-7-arylazo-6-hydroxy-4-methoxycoumarones (**61a**) which can be reduced to 7-aminovisnagone (**61b**),[108] which has been proved to be identical with the authentic product, obtained by reduction of 7-nitrovisnagone (**61c**).

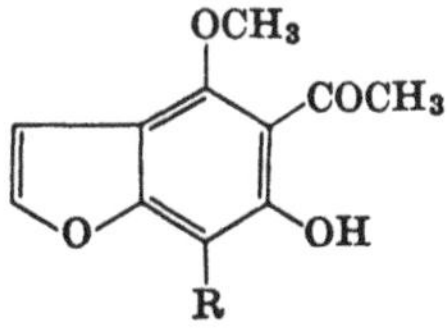

(**61a**) R = N=NAr;
(**61b**) R = NH_2;
(**61c**) R = NO_2

Visnagone reacts photochemically with phenanthraquinone to give the photoadduct (**62**).[90]

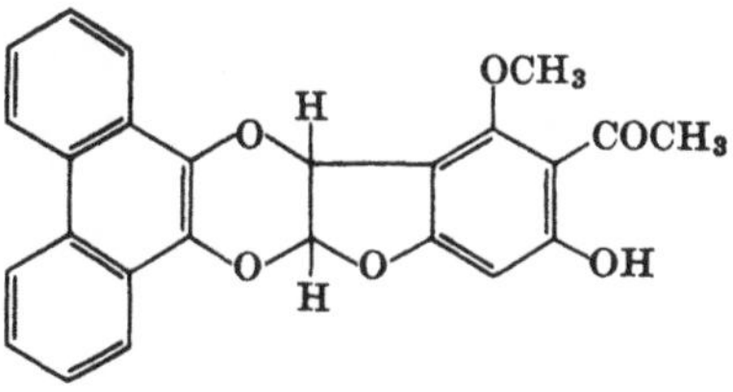

(**62**)

Oxidation of visnagin with hydrogen peroxide to furan-2,3-dicarboxylic acid has been described by Späth and Gruber.[10, 91] Controlled oxidation with the same reagent yields 6-hydroxy-4-methoxycoumarone-5-carboxylic acid (**63**).[41] This reaction does not seem to proceed via visnagone (**45**) since the latter does not yield **43** under the same conditions. Norvisnagin (**56**) and khellol (**47a**) also give **63** under similar conditions.

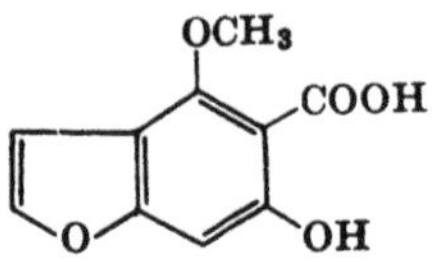

(63)

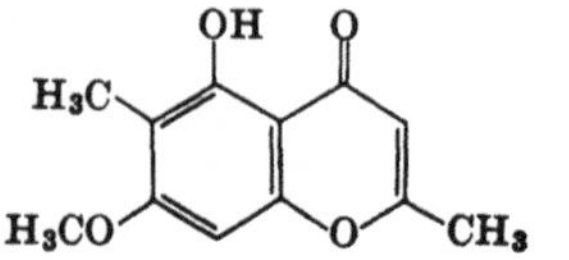

(64)

Selective oxidation of visnagin with chromic acid to 6-formyl-7-hydroxy-5-methoxy-2-methylchromone (**52**)[41, 96] provides an important alternative to ozonolysis. Demethylation, reduction of the formyl group, and selective methylation produce eugenitin (**64**).[41]

Visnagin behaves in a similar manner toward hydrazine hydrate forming the pyrazole derivative (**65a**) or (**65b**)[74].

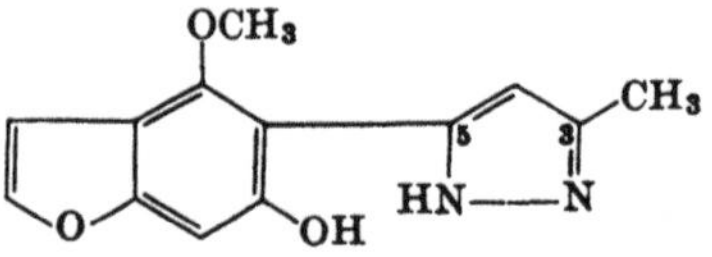

(65a)

or

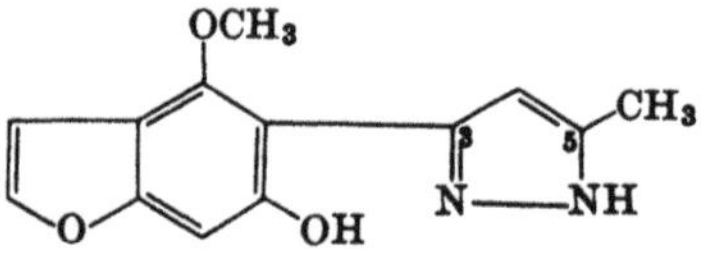

(65b)

In contrast to its stability to hydroxylamine hydrochloride in acetic acid at room temperature, visnagin reacts in boiling pyridine, to give the isoxazole derivative (**66a**) or (**66b**).[74]

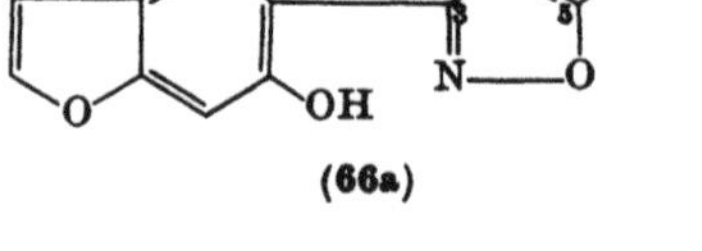

(66a)

or

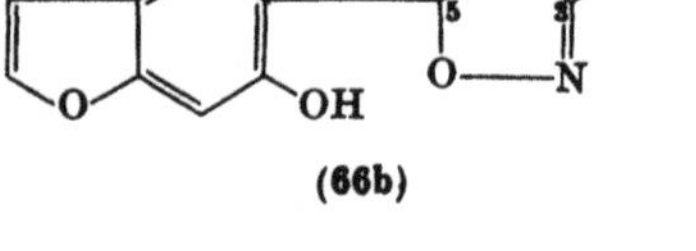

(66b)

Visnagin reacts with bromine[85] to give 8-bromovisnagin (**67a**), whose structure was proved by alkaline degradation to 5-acetyl-7-bromo-6-hydroxy-4-methoxycoumarone (**68**) and oxidation to 8-bromo-6-formyl-7-hydroxy-5-methoxy-2-methylchromone (**69**).[41]

Nitration of 5-norvisnagin (**56**) with nitric acid, followed by methylation of the 8-nitro-5-norvisnagin formed, produces 8-nitrovisnagin (**67b**).[108]

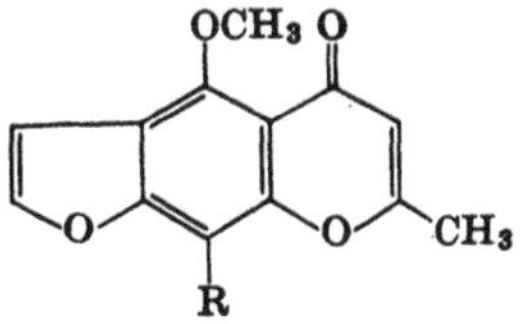

(**67a**) R = Br;
(**67b**) R = NO_2

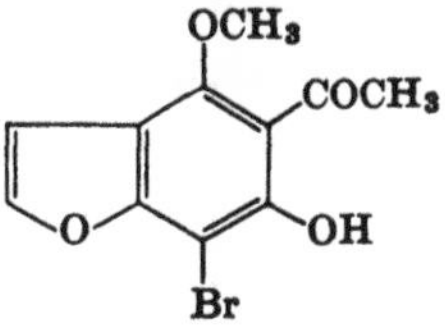

(**68**)

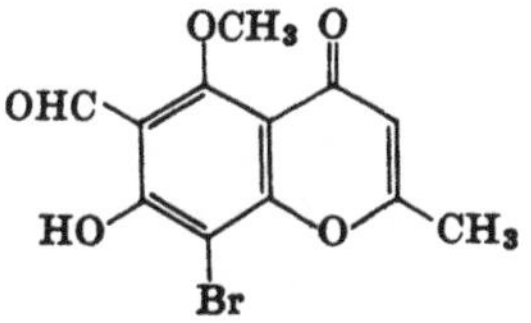

(**69**)

Visnagin condenses with aromatic aldehydes to give the corresponding 2-styrylvisnagins,[83, 84] which have been found to act as diene components in the Diels–Alder reaction, and add maleic anhydride to yield the xanthone derivatives[84] (Eq. 30). Mustafa and coworkers have shown that 2-styrylchromones and 4-styrylcoumarins can also act as diene components in the Diels–Alder reaction.[110]

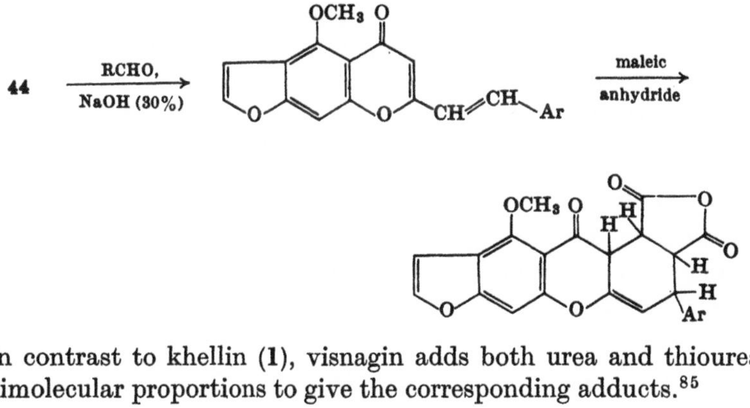

(30)

In contrast to khellin (**1**), visnagin adds both urea and thiourea in equimolecular proportions to give the corresponding adducts.[85]

C. Khellinol

Khellinol (5-norkhellin, 4-hydroxy-9-methoxy-7-methyl-5*H*-furo(3,2-g)-1-benzopyran-5-one) (**12**), $C_{13}H_{10}O_5$, is a minor congenor of khellin,[65,111] from which it can be prepared by partial demethylation with acids.[58] It forms yellow crystals, gives a green ferric chloride reaction, forms a complex with aluminum chloride, and is insoluble in dilute alkali solution.

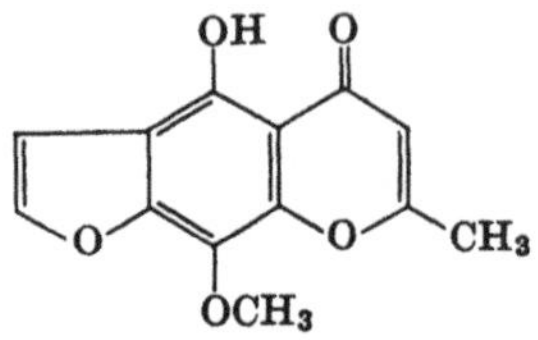

(12)

Khellinol reacts with α-acetobromoglucose, suspended in quinoline and silver oxide, to give 5-tetra-*O*-acetyl-α-glucosyloxy-8-methoxy-2-methylfuro(3′,2′–6,7)chromone, which is readily hydrolyzed to the 5-α-glucosyloxy-8-methoxy derivative by the action of sodium methoxide.[105] Khellinol, 5-norvisnagin (**56**), and khellinquinol (**13**) are readily phosphorylated with dialkyl chloridophosphoridates and dialkyl chloridophosphorothioates to yield the corresponding phosphoric esters (Table 1).[112]

D. Khellinin

Khellinin (khellol glucoside, 7-(glucosyloxymethyl)-4-methoxy-5*H*-furo(3,2-g)–1–benzopyran–5–one), $C_{19}H_{20}O_{10}.2H_2O$, is also found with khellin in *Ammi visnaga* L.[13,113] Alkaline degradation of khellinin (**70a**) yields visnagone (**45**) together with the β-glucoside of glycollic acid, which is best hydrolyzed to khellol (**70b**) by hydrobromic acid under definite conditions, as there is danger of concomitant demethylation[58,113] to 5-norkhellol (**71**). The compound is shown to be glucopyranoside since it consumes two molecules of periodic acid in titrations.[114] The infrared spectrum of **70a** is consistent with the proposed structure.

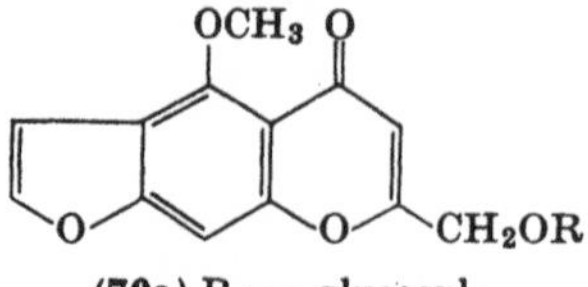

(70a) R = α-glucosyl;
(70b) R = H;
(70c) R = CH_3

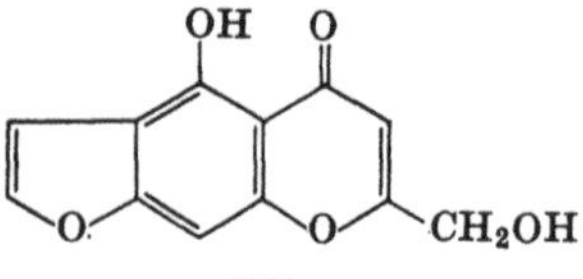

(71)

E. Khellol

Khellol (chellol, 7- (hydroxymethyl) - 4 - methoxy - 5*H* - furo(3,2-g) –1–benzopyran-5-one), $C_{13}H_{10}O_5$, is the aglycone of khellinin, and is obtained from it.[13] Geissman and Bolger[115] have synthesized khellol (**70b**) by submitting visnagone benzyloxyacetate (**72a**) to a Baker–Venkataraman rearrangement to give the diketone (**73**), which was simultaneously cyclized, and debenzylated with hydrochloric acid to **70b**. Visnagone acetoxyacetate has also been used to build a chromone ring containing a 2-hydroxymethyl substituent. Moreover, *C*-alkylation of visnagone with methyl methoxyacetate in the presence of sodium, followed by ring closure of the diketone formed, yields khellol methyl ether (**70c**).[116]

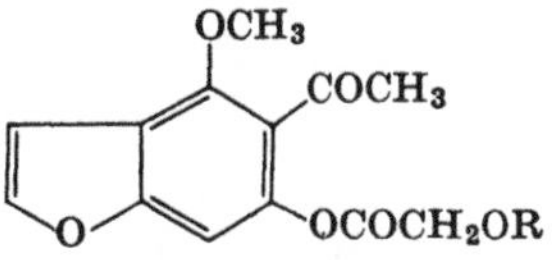

(**72a**) R=$CH_2C_6H_5$;
(**72b**) R=$COCH_3$;
(**72c**) R=OCH_3

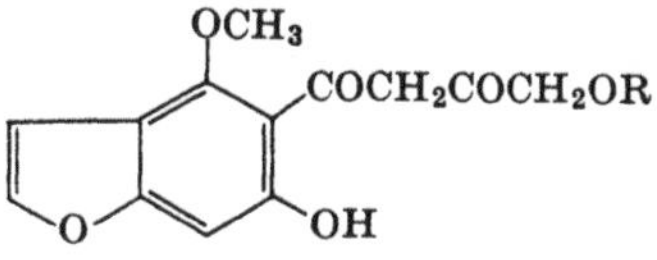

(**73**)

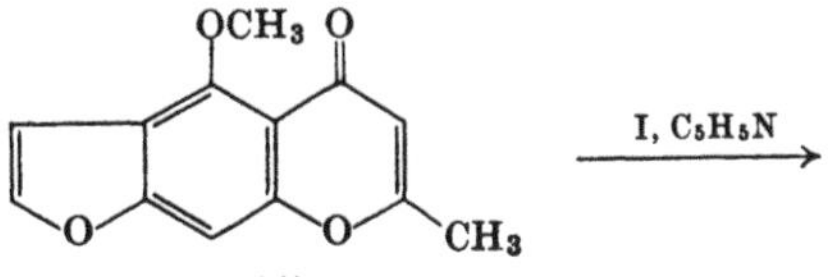

(**44**)

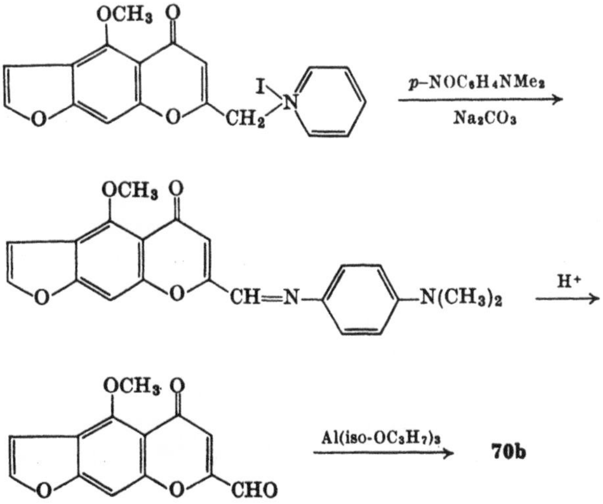

(31)

Mustafa and coworkers[61] have recently reported the synthesis of khellol by a different route, making use of 2-formyl-5-methoxyfuro-(3′,2′–6,7)chromone according to the following sequence of reactions (Eq. 31).

Moreover, treatment of the formyl derivative with ammoniacal silver nitrate gave 5-methoxyfuro(3′,2′–6,7)chromone-2-carboxylic acid (**74**).[61]

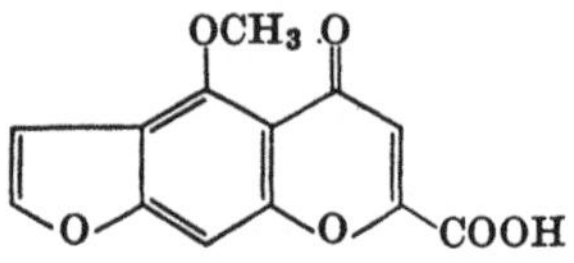

(**74**)

Khellol, as can be seen from its structural formula **70b**, is closely related to visnagin (**44**). It can be converted into the corresponding 2-methylpyrone, visnagin, by treatment with thionyl chloride and reduction of the product (**76a**) with zinc. It is interesting to note that **70b**, an allylic alcohol, can be oxidized with chromic acid to the aldehyde (**75**) without affecting the alcohol grouping.[96] It seems that the 2,3-double bond is too closely implicated in the γ-pyrone structure to provide allylic activation.

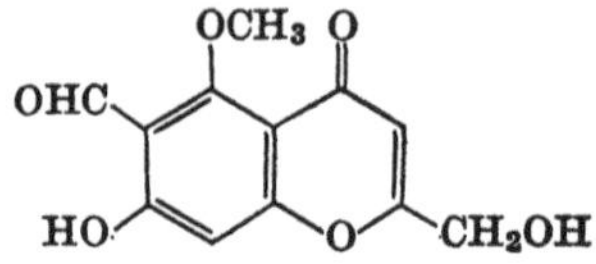

(**75**)

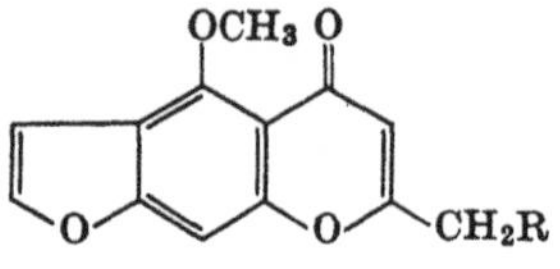

(**76a**) R=Cl;
(**76b**) R=I;
(**76c**) R=NHC_6H_5;
(**76d**) R=$OSO_2C_6H_4CH_3$-*p*

Under mild conditions, alkaline hydrogen peroxide oxidizes khellol to the salicylic acid derivative (**63**) also obtained from visnagin.[41]

Khellol tosylate (**76d**) is of particular interest. Its reaction with sodium iodide, and with aniline, leads to the formation of **76b** and **76c**, respectively. Treatment of **76b** with zinc dust in acetic acid, or **76c** with sodium iodide in acetic anhydride gave visnagin (**44**).[95]

Khellol behaves similarly towards hydrazine hydrate, and hydroxylamine hydrochloride in pyridine, to yield the pyrazole derivative (**77a**) or (**77b**) and the isoxazole derivative (**78a**) or (**78b**) respectively.[74]

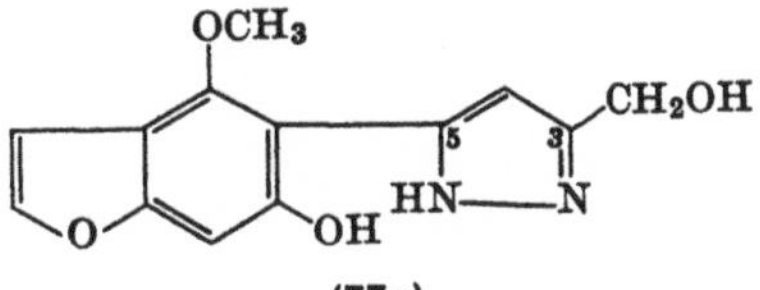
(77a)

or

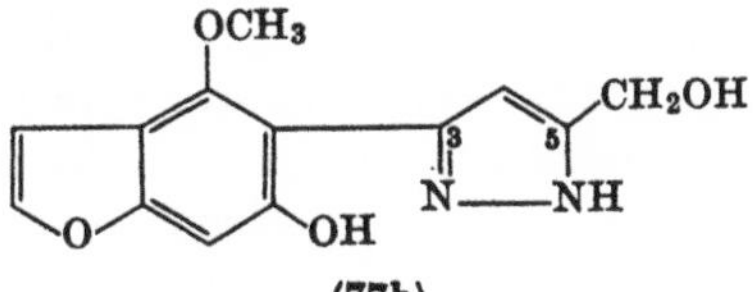
(77b)

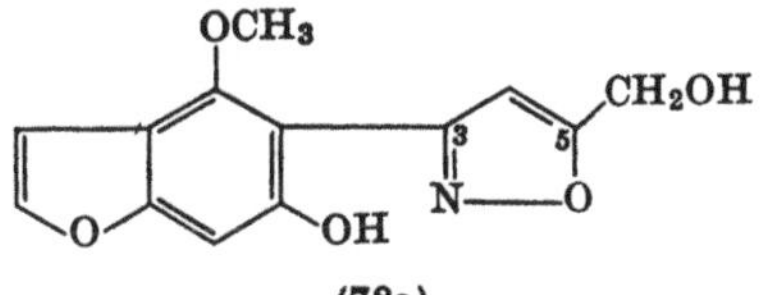
(78a)

or

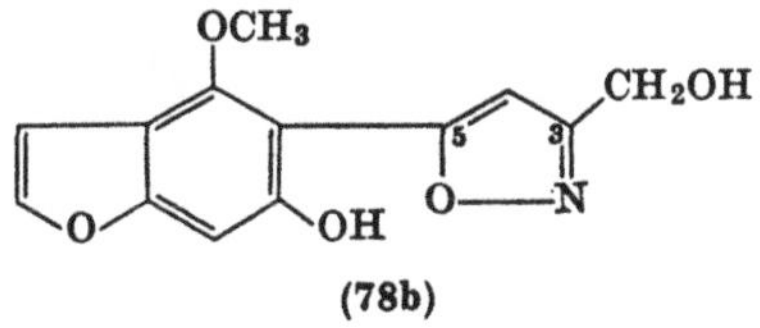
(78b)

F. Ammiol

Ammiol (8-methoxykhellol, 7-(hydroxymethyl)-4,9-dimethoxy-5*H*-furo(3,2-g)–1–benzopyran-5-one), $C_{14}H_{12}O_6$, another constituent of *Ammi visnaga* L., is best purified through its perchlorate. Alkaline hydrolysis of ammiol (**79**)[117] gave khellinone (**2**) and glycollic acid. The proposed structure has been recently confirmed by its synthesis (Eq. 32), achieved by Mustafa and coworkers,[61] following the route applied to 2-methylchromone for the synthesis of 2-oxochromones.[118]

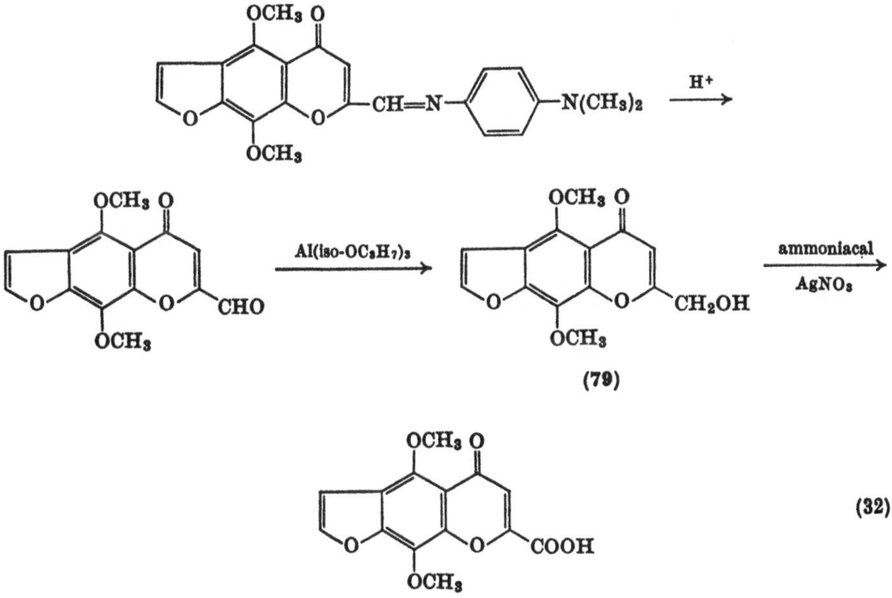

(79)

(80)

(32)

Oxidation of 2-formyl-5,8-dimethoxyfuro(3′,2′–6,7)chromone with ammoniacal silver nitrate establishes a route for the preparation of 5,8-dimethoxyfuro(3′,2′–6,7)chromone-2-carboxylic acid (**80**),[61] which has also been synthesized from khellinone (**2**), thus establishing the linear structure for ammiol (Eq. 33).

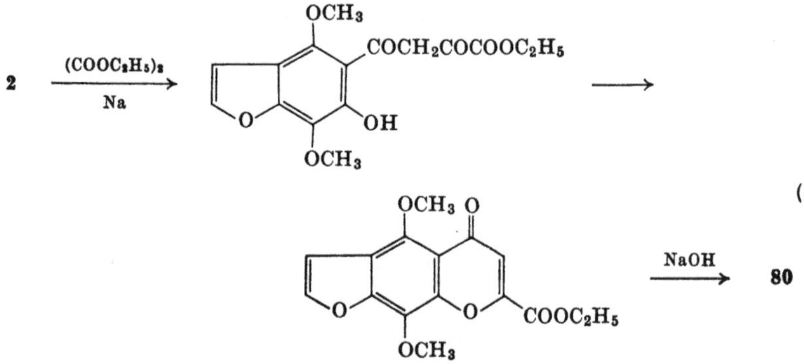

(33)

G. Visamminol

Visamminol (2,3-dihydro-4-hydroxy-2-(1-hydroxy-1-methylethyl)-7-methyl-5*H*-furo(3,2-g)–1–benzopyran-5-one), $C_{15}H_{16}O_5$, was discovered first, amongst the furochromones of *Ammi visnaga* L., by Smith and coworkers[36] and Cavallito and Rockwell.[119] A dihydrofurochromone structure was first proposed based on the similarity of its ultraviolet spectrum[36] to the spectra reported by Davies and co-

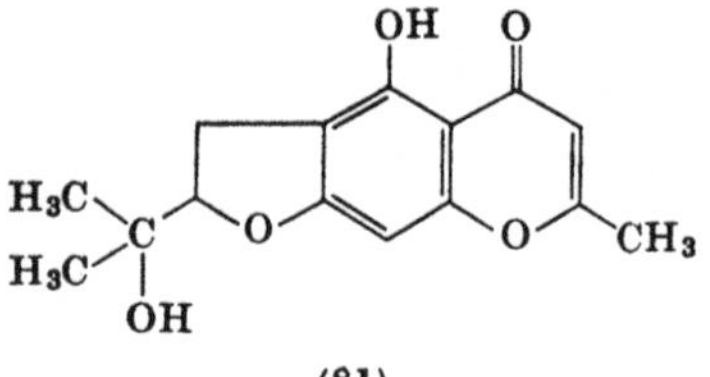

(**81**)

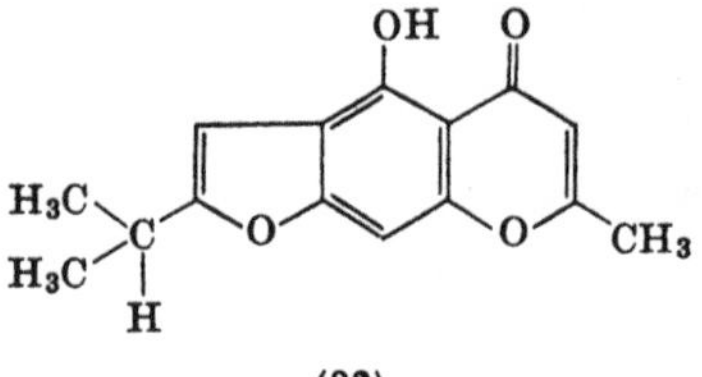

(**82**)

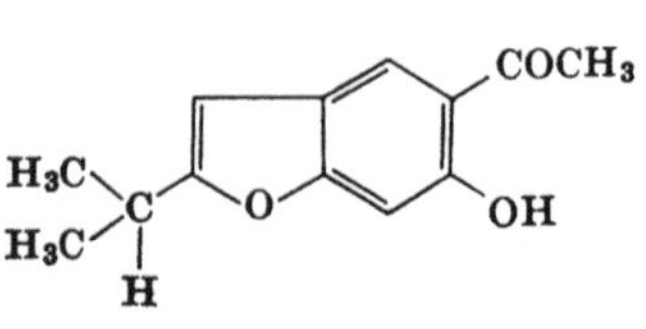

(**83**)

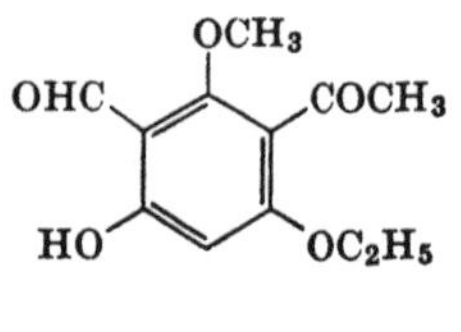

(**84**)

workers[34, 35] for dihydrofurochromones. Schmid and coworkers[111] have confirmed the proposed structure **81**.

Visamminol behaves as a 5-hydroxychromone. It gives a red ferric reaction, and the main ultraviolet absorption band underwent a bathochromic shift when aluminum chloride was added to its solution in ethanol. A methyl ether can be prepared, which with potassium hydroxide gives the red color typical of 2-methylchromones. An alcoholic hydroxyl group is present which can be acetylated, but not methylated; while dehydration with *p*-toluenesulfonic acid as a catalyst leads to the furan, anhydrovisamminol (**82**), by means of the double bond migration. This acid treatment is vigorous, but did not result in isomerization, of the ring opening and recyclization type, since the methyl ether of **82** can also be obtained by first methylating, and then dehydrating **81**.

Anhydrovisamminol (**82**) decomposes to acetone when heated with alkali, but its methyl ether yields acetic acid together with a phenolic ketone (**83**). The latter ketone gives a green ferric reaction, and by ethylation and ozonolysis, yields a compound (**84**) together with isobutyric acid; therefore its structure is **83**. Compound **84** can be easily obtained by ozonolysis of the ethyl ether of visnagone. It was felt that structure **82** did not adequately describe anhydrovisamminol since chromic oxidation of **81** gives acetone but not isobutyric acid; it seems likely that **81** describes the natural product.

Schmid and coworkers[111] have also eliminated structure **85**, which might possibly have contracted to the furan (**82**). Alkaline hydrolysis of the synthetic epoxyacetate (**86**) liberated the phenolic hydroxyl group, which then attacked the epoxide grouping giving **87**; cyclization to the

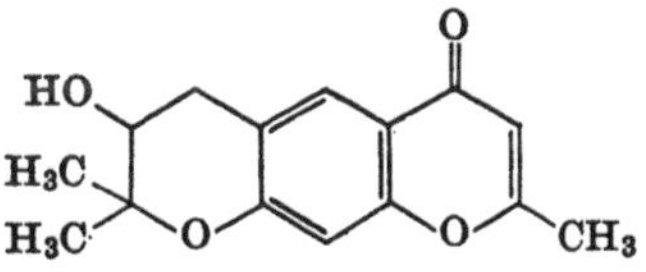

(85)

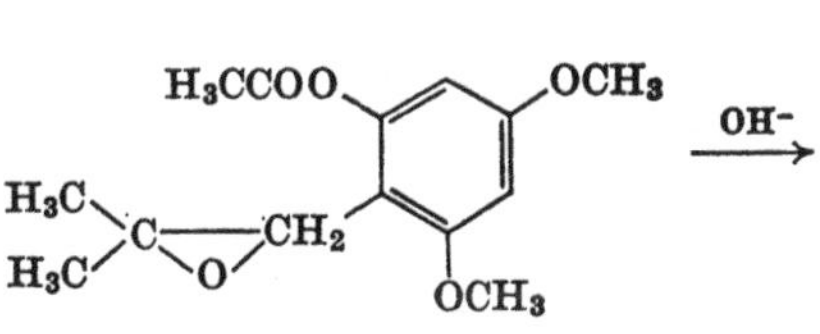

(86)

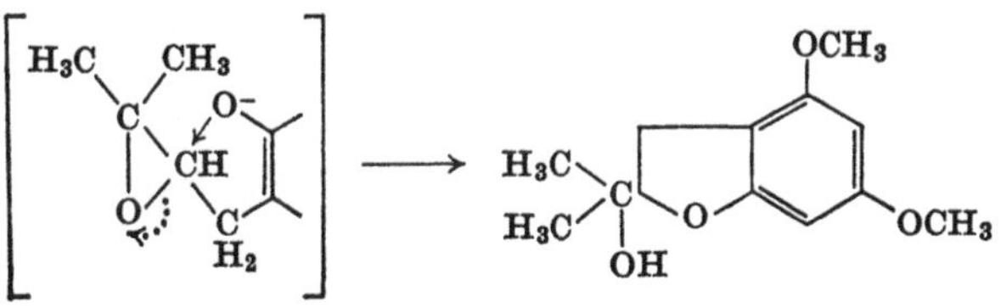

(87)

five-membered ring rather than to the six-membered ring being favored both by steric and electronic factors. Späth and coworkers[120] have shown that this approach through the epoxide gives furan derivatives, and that it is an effective complement to the acid-catalysed cyclization of phenolic olefins which usually produce 2,2-dimethylchroman derivatives.

Apart from its racemic nature, **87** was found to be identical with a compound $(\alpha)_D + 52°$, made by alkaline hydrolysis of the pyrone ring in visamminol (**81**), and methylation of the resulting dihydric phenol; thus structure **81** for the natural product is confirmed.

2. Color Reactions of Furochromones

Furochromones, having a methyl group in position 2, develop a reddish-violet color with potassium hydroxide,[20, 54] in contrast to those having no substituent in this position. The importance of the 2-methyl group, and the absence of a phenolic hydroxyl group has been stressed.[62, 121]

The violet substance, obtained by the reaction of khellin (**1**) with

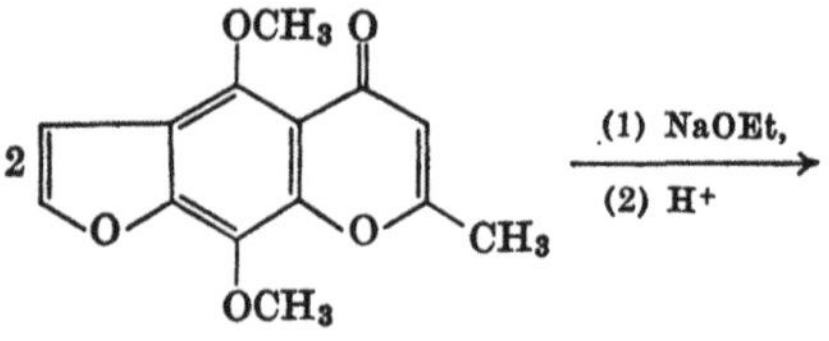

1

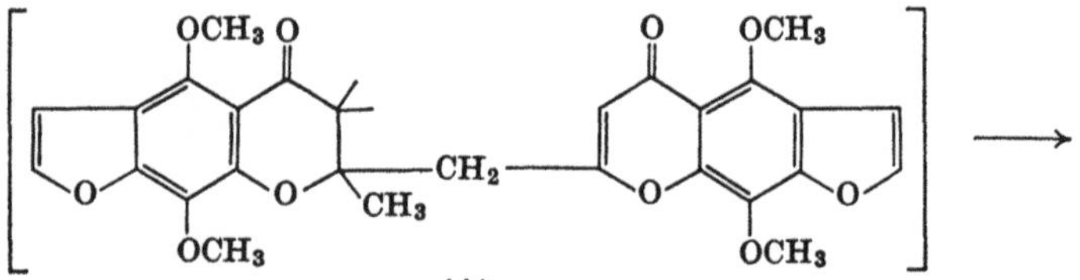

(A)

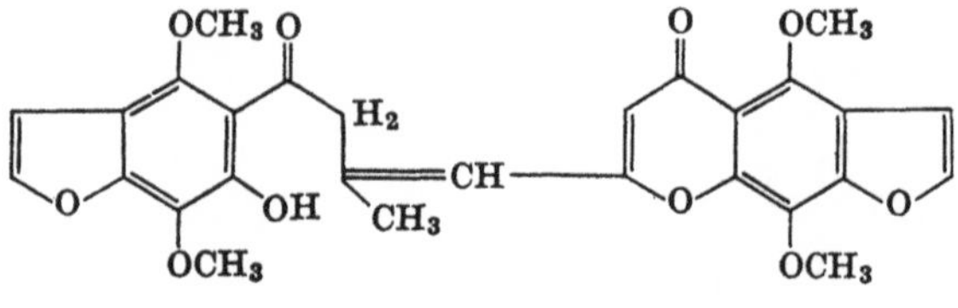

(88)

(34)

alkali, has been studied[122] although difficult to isolate[121] and structure **88** is proposed for the yellow crystalline compound, formed by treatment of the violet product with acetic acid. The yellow compound develops red–violet color with aqueous potassium hydroxide (1%). Its structure is based on its infrared and nuclear magnetic resonance spectra, the formation of two molecules of khellin on pyrolysis, and its similarity to the dimeric product of 2-methylchromone.[123] It is believed that **88** could be formed by a Michael condensation reaction mechanism, involving 1,4-addition of the carbanion to the α,β-unsaturated system of another molecule, followed by ring opening of the intermediate adduct (**A**). This mechanism is supported by the activation of the methyl group in position 2 in alkaline medium, allowing condensation reactions[54] to take place as found with other 2-methylchromones (Eq. 34).[82,118]

The principle of vinylogy, which has been applied to 2-methylchromones, considered as vinylogs of acetophenone,[70] which develops color with *m*-dinitrobenzene[124] and with sodium nitroprusside,[125] has been used successfully with furochromones containing a methyl group in position 2, and having no free phenolic substituent.[90a,126]

A stable red–violet color has been reported upon the treatment of khellin (**1**) with 2,4-dinitrophenylhydrazine followed by potassium hydroxide solution (30%).[127,128] Khellin and other furochromones, having OH or OCH_3 groups in the 5– and 8–positions, on treatment with nitric acid, followed by destruction of the oxonium salt formed with alkali, give a color believed to be due to a quinone formed by oxidation with the reagent.[129]

Mustafa and coworkers[102] have reported a color reaction for furochromones, containing a free hydroxyl group in the 5- and/or 8–position, with aqueous uranyl acetate solution, accompanied by a separation of the metal complex upon dilution with water, in some cases. The color developed is destroyed by mineral acids or acetic acid. The detection limit for norvisnagin (**56**) is 5 γ/ml, and that for uranium in uranyl acetate with **56** is 5.6 γ/ml. The importance of a free OH in the 5–position is stressed as a protected hydroxyl group, in this position, gives no color.

3. Physiological Activity

The active constituents of *Ammi visnaga* L. (khellah plant) were known by the Egyptians to be effective antispasmodic agents; khellin has been used particularly for the relief of spasms of the ureter, bile duct, and gall bladder, and also in bronchial asthma, the anginal syndrome (increasing coronary flow), and against whooping cough. Moreover, it has been found to be a potent coronary vasodilator. During the last decade, khellin has been known, in medical circles, as the only useful substance of these active constituents,[130] and exhaustive pharmacological and clinical investigations have been carried out. For a complete review and bibliography of the subject, reference should be made to Hutterer and Dale,[4] Schmid,[3] Pointet,[5] Illing,[8] Uhlenbroock and coworkers,[1] Schnidler,[131] Kern,[6] Aubertin,[132] and Lesser.[133]

Khellin is effective in uretral spasms; it has a selective antispasmodic effect upon the ureter,[134–136] gall bladder,[137–138] and bile duct.[139] A bronchodilating action of khellin has been reported;[140–142] thus relaxation of the bronchial musculature relieves bronchial asthma and whooping cough.

Khellin may be considered to have specific relaxing effect on the coronary vessels.[143–145] It was found to be a potent coronary vasodilator.[146–154] Contrary to Samaan's[155] claim, khellin has been found to be capable of increasing the contractable force, and frequency of the heart.[156, 157] The inhibitory action of khellin on gastric ulcers,[158–160] human tumors,[161] and intestinal activity[162] has been studied.

Synergism was observed between khellin and barbiturate, since it shortened the interval before the hypnotic effect of the barbiturate appeared. Its antispasmodic effect was reinforced by small doses of papaverine. Khellin combined with barbiturate and papaverine produced a sustained hypotensive action.[163]

Khellol glucoside, khellinin, possesses a persistent and rather selective stimulating action on the heart, producing a more complete systol and diastole, with corresponding increase in cardiac output. It increases the coronary flow, and is not cumulative. Contrary to khellin and visnagin, it afforded no protection against poisoning by histamine aerosol. It is not converted into khellin in the digestive tract or in the body tissues.[4] A procedure for the determination of khellin content in blood organs, and tissues is reported.[164]

By studying the chemical constitution and the antispasmodic activity of a number of furochromones, Schönberg and coworkers[54] formed the generalization, which applies to compounds of the khellin and of the visnagin type, that loss of the furan ring leads to a 70–90 % diminution in activity. Opening of the γ-pyrone ring results in a 75 % loss of activity. Removal of the 8-methoxyl group results in a 30–50 % loss of activity and removal of both 5-, and 8–methoxyl groups reduces the activity by 90 %. Simultaneous replacement of both 5-, and 8–methoxyl groups by hydroxyl groups reduces the activity by 90 % but replacement by ethoxyl causes only 25 % loss of activity, whereas higher alkoxyl groups reduce the activity by 80–90%. Replacement of the 2-methyl group by hydroxymethyl group reduces the activity more than 50% but, on the other hand, replacement of the 2-methyl group by the glucose radical renders the compound almost inactive. Replacement of the 2-methyl group by hydrogen or by ethyl results in 45% and 25% loss of activity respectively.

Jongebreur,[165] while studying the chemical constitution and pharmacological action of 40 pyrone derivatives in comparison with khellin, found that spasmolytic effect of the molecule is increased if, the hydrogen in the 2-position is replaced by an alkyl, 2-furyl, 3-pyridyl, 3-aminophenyl, and/or phenyl group, increasing in this order; the *O* in the ring is replaced by *S*; the hydrogen in the 3-position is replaced by methyl; methoxyl groups are introduced into the benzene nucleus; and alkoxy groups are introduced into the 7-position. Saturation of the 2,3-double bond, or transformation of the chromone to the coumarin, have no effect. By their action on a perfused surviving rat heart, it is claimed that 7-isopropoxy-2-methylchromone, 2-(3-pyridyl)chromone, 2-(2-furyl)-chromone, 3-aminoflavone, flavone, 5,7,8-trimethoxyflavone, 5,8- dimethoxyflavone, and 2-isopropyl-5,8-dimethoxychromone were all more active than khellin.

V. References

1. K. Uhlenbroock, K. Mulli and O. Schmid, *Arzneimittel–Forsch.*, **3**, 130, 177, 219, 407 (1953); *Chem. Abstr.*, **48**, 13619 (1954).
2. H. Schmidler, *Pharmazie*, **8**, 176 (1953); *Chem. Abstr.*, **47**, 11666 (1953).
3. H. Schmid in *Fortschritte der Chemie Organischer Naturstoffe*, Vol. 11, (Ed. L. Zechmeister), Wein, Springler-Verlag, 1954, p. 124
4. P. C. Hutterer and E. Dale, *Chem. Revs.*, **48**, 543 (1951).
5. M. Pointet, *Contribution á l'etude de L'Ammi visnaga* L., (Ed. Jouve), Paris (1954).
6. W. Kern, *Pharm. Ztg.-Nachr.*, **89**, 275 (1951); *Chem. Abstr.*, **47**, 12341 (1953).
7. J. E. Molfino, *Chacra* (*Buenos Aires*), **16**, 78 (1946).
8. G. Illing, *Arzneimittel–Forsch.*, **7**, 497 (1957); *Chem. Abstr.*, **52**, 660 (1958).
9. M. W. Quimby, *Econ. Botany*, **7**, 89 (1953); *Chem. Abstr.*, **47**, 4554 (1953).
10. E. Späth and W. Gruber, *Chem. Ber.*, **71**, 106 (1938).
11. K. Samaan, *Quart. J. Pharm. Pharmacol.*, **4**, 14 (1931).
12. I. R. Fahmy and M. El-Keiy, *Rept. Pharm. Soc. Cairo*, **3**, 36 (1931).
13. P. Fantl and S. I. Salem, *Z.*, **226**, 166 (1930).
14. H. Supniewska, *Dissertationes Pharm.*, **9**, 61 (1957); *Chem. Abstr.*, **51**, 13076 (1957).
15. H. Abu-Shady and T. O. Soine, *J. Am. Pharm. Assoc. Sci. Ed.*, **41**, 481 (1952).
16. J. Baisse (Promidel), *Fr. Pat.*, 1,031,549 (1953); *Chem. Abstr.*, **52**, 13801 (1958).
17. H. Supniewska, *Bull. Acad. Polon. Sci. Ser. Biol.*, **5**, 6 (1957); *Chem. Abstr.*, **54**, 6881 (1960).

18. Ö. T. Baytop, *Bull. Fac. Med. Istanbul*, **23**, 703 (1960); *Chem. Abstr.*, **55**, 16703 (1961).
19. M. A. T. Rainerie, *Univ. Chile, Fac. Quim. y Farm. Tesis Quim. Farm.*, **5**, 436 (1953); *Chem. Abstr.*, **50**, 4457 (1956).
20. I. R. Fahmy, N. Badran and M. F. Messeid, *J. Pharm. Pharmacol.*, **1**, 529, 535 (1949).
21. D. N. Grindley, *J. Sci. Food Agr.* (*England*), **1**, 53 (1950).
22. O. Dann and G. Illing, *Arch. Pharm.*, **289**, 718 (1956); *Chem. Abstr.*, **51**, 6841 (1957).
23. G. Illing, *Arch. Pharm.*, **290**, 291 (1957); *Chem. Abstr.*, **51**, 17099 (1957).
24. J. Delphaut and C. Hedel, *Bull. Soc. Chim. Biol.*, **35**, 368 (1953); *Chem. Abstr.*, **47**, 11296 (1953).
25. T. Kawatami, T. Ohno and M. Ito, *Eisei Skikenjo Hôkoku*, (74), 73 (1956); *Chem. Abstr.*, **51**, 7478 (1957).
26. H. Friedrich and C. Hortsmann, *Pharmazie*, **16**, 319 (1961); *Chem. Abstr.*, **56**, 14393 (1962).
27. C. Hortsmann, *Kulturpflanze*, **10**, 132 (1962); *Chem. Abstr.* **59**, 6708 (1963).
28. N. Narasimhachari and E. V. Rudolff, *Can. J. Chem.*, **40**, 1123 (1962); *Chem. Abstr.*, **57**, 6588 (1962).
29. S. D. Bailey, P. A. Geary and A. E. DeWald, *J. Am. Pharm. Assoc. Sci. Ed.*, **40**, 280 (1951); *Chem. Abstr.*, **45**, 7306 (1951).
30. S. D. Solani and J. F. Márquez, *J. Am. Pharm. Assoc. Sci. Ed.*, **42**, 20 (1953); *Chem. Abstr.*, **47**, 3391 (1953).
31. M. S. El-Ridi, S. N. El-Din, K. Samaan and A. M. Hossein, *Proc. Pharm. Soc. Egypt Sci. Ed.*, **35**, 21 (1953); *Chem. Abstr.*, **51**, 8189 (1957).
32. K. Samaan, A. H. Hossein and M. S. El-Ridi, *Quart. J. Pharm. Pharmacol.*, **20**, 502 (1947).
33. E. E. Vonech and O. A. Guagnine, *Anales Assoc. Quim. Argentina*, **41**, 23 (1953); *Chem. Abstr.*, **47**, 9555 (1953).
34. J. S. H. Davies, P. A. McCrea, W. L. Norris and G. R. Ramage, *J. Chem. Soc.*, 3206 (1950).
35. J. S. H. Davies and W. L. Norris, *J. Chem. Soc.*, 3195 (1950).
36. E. Smith, L. A. Pucci and W. C. Bywater, *Science*, **115**, 520 (1952).
37. P. K. Grover, G. D. Shah and R. C. Shah, *Chem. Ind.* (*London*), 62 (1955).
38. G. Gabrielli and A. Ficalbi, *Ann. Chim.* (*Rome*), **47**, 1013 (1957); *Chem. Abstr.*, **52**, 834 (1958).
39. S. Morgante and A. Damiani, *Periodico Mineral.* (*Rome*), **31**, 99 (1962); *Chem. Abstr.* **57**, 11945 (1962).
40. T. A. Geissman, *J. Am. Chem. Soc.*, **71**, 1498 (1949).
41. A. Schönberg, N. Badran and N. A. Starkowsky, *J. Am. Chem. Soc.*, **75**, 4992 (1953).
42. F. Wessely and G. H. Moser, *Monatsh.*, **56**, 97 (1930).
43. T. A. Geissman and T. G. Halsall, *J. Am. Chem. Soc.*, **73**, 1280 (1951).
44. S. K. Mukerjee and T. R. Seshadri, *Proc. Acad. Sci.*, **35A**, 323 (1952); *Chem. Abstr.*, **47**, 8742 (1953).
45. A. Schönberg and N. Badran, *J. Am. Chem. Soc.*, **73**, 2960 (1951).
46. J. R. Clarke and A. Robertson, *J. Chem. Soc.*, 302 (1949).
47. R. A. Baxter, G. R. Ramage and J. A. Timson, *J. Chem. Soc.*, S30 (1949).
48. O. Dann and G. Illing, *Ann. Chem.*, **605**, 146 (1957).

49. O. Dann and H. G. Zeller, *Chem. Ber.*, **93**, 2829 (1960).
50. V. V. S. Murti and T. R. Seshadri, *Proc. Indian Acad. Sci.*, **30A**, 107 (1949).
51. V. V. S. Murti and T. R. Seshadri, *J. Sci. Ind. Res.* (*India*), **8B**, 112 (1949).
52. R. Aneja, S. K. Mukerjee and T. R. Seshadri, *J. Sci. Ind. Res.* (*India*), **17B**, 382 (1958).
53. R. Aneja, S. K. Mukerjee and T. R. Seshadri, *Chem. Ber.*, **93**, 297 (1960).
54. A. Schönberg and A. Sina, *J. Am. Chem. Soc.*, **72**, 1611 (1950).
55. A. Schönberg, N. Badran and N. A. Starkowsky, *J. Am. Chem. Soc.*, **77**, 5438 (1955).
56. C. Musante, *U.S. Pat.*, 3,053,841 (cl. 260-247.5), (1962); *Chem. Abstr.*, **58**, 1463 (1963).
57. H. Abu-Shady and T. O. Soine, *J. Am. Pharm. Assoc. Sci. Ed.*, **41**, 325 (1952); *Chem. Abstr.*, **47**, 5937 (1953).
58. A. Schönberg and G. Aziz, *J. Am. Chem. Soc.*, **75**, 3265 (1953).
59. W. Asker, A. F. A. M. Shalaby and S. M. A. D. Zayed, *J. Org. Chem.*, **23**, 1781 (1958).
60. M. E. Brokke and B. E. Christensen, *J. Org. Chem.*, **23**, 589 (1958).
61. A. Mustafa, N. A. Starkowsky and T. I. Salama, *J. Org. Chem.*, **26**, 886 (1961).
62. A. Schönberg and A. Sina, *J. Am. Chem. Soc.*, **72**, 3396 (1950).
63. J. P. Fourneau, *Ann. Pharm. Franc.*, **11**, 685 (1953); *Chem. Abstr.*, **49**, 1027 (1955).
64. S. K. Mukerjee and T. R. Seshadri, *J. Sci. Ind. Res.* (*India*), **13B**, 400 (1954); *Chem. Abstr.*, **49**, 12454 (1955).
65. H. Abu-Shady and T. O. Soine, *J. Am. Pharm. Assoc. Sci. Ed.*, **41**, 429 (1952); *Chem. Abstr.*, **47**, 11188 (1953).
66. H. Abu-Shady and T. O. Soine, *J. Am. Pharm. Assoc. Sci. Ed.*, **41**, 403 (1952); *Chem. Abstr.*, **47**, 11188 (1953).
67. W. J. Horton and M. G. Stout, *J. Org. Chem.*, **26**, 1221 (1961).
68. H. Abu-Shady and T. O. Soine, *J. Am. Pharm. Assoc. Sci. Ed.*, **42**, 573 (1953); *Chem. Abstr.*, **48**, 10015 (1954).
69. H. Abu-Shady and T. O. Soine, *J. Am. Pharm. Assoc. Sci. Ed.*, **41**, 395 (1952).
70. A. Schönberg, M. M. Sidky and G. Aziz, *J. Am. Chem. Soc.*, **76**, 5115 (1954).
71. C. Musante, *Ann. Chim.* (*Rome*), **46**, 768 (1956).
72. L. Fabbrini, *Ann. Chim.* (*Rome*), **46**, 130 (1956); *Chem. Abstr.*, **50**, 13887 (1956).
73. R. Robinson, *Endeavour*, **13**, 173 (1954); *Chem. Abstr.*, **49**, 684 (1955).
73a. O. Dann and G. Volz, *Ann. Chem.*, **685**, 167 (1965); *Chem. Abstr.*, **63**, 9926 (1965).
74. A. Schönberg and M. M. Sidky, *J. Am. Chem. Soc.*, **75**, 5128 (1953).
75. C. Musante and S. Fatutta, *Farmaco* (*Pavia*), **9**, 328 (1954); *Chem. Abstr.*, **49**, 9634 (1955).
76. L. Fabbrini, *Ann. Chim.* (*Rome*), **46**, 137 (1956); *Chem. Abstr.*, **50**, 15524 (1956).
77. C. Musante and S. Fatutta, *Ann. Chim.* (*Rome*), **45**, 918 (1955); *Chem. Abstr.*, **51**, 1946 (1957).
78. R. Ruggieri, *Gazz. Chim. Ital.*, **93**, 329 (1963); *Chem. Abstr.*, **59**, 6402 (1963).
79. C. Musante, *Gazz. Chim. Ital.*, **88**, 910 (1958); *Chem. Abstr.*, **53**, 18939 (1959).
80. C. Musante and A. Sterner, *Gazz. Chim. Ital.*, **86**, 297 (1956); *Chem. Abstr.*, **52**, 352 (1958).

81. B. Reichert, *Arch. Pharm.*, **293**, 111 (1960); *Chem. Abstr.*, **54**, 18492 (1960).
82. I. M. Heilbron, H. Barnes and R. A. Morton, *J. Chem. Soc.*, 123, 2559 (1923).
83. A. J. C. Rahla, *Rev. Port. Farm.*, **2**, 121 (1952); *Chem. Abstr.*, **48**, 3358 (1954).
84. A. Schönberg, A. Mustafa and G. Aziz, *J. Am. Chem. Soc.*, **76**, 4576 (1954).
85. N. A. Starkovsky, *Egypt. J. Chem.*, **2**, 111 (1959); *Chem. Abstr.*, **54**, 8802 (1960).
86. A. Mustafa, *Nature*, **175**, 992 (1955).
87. A. Mustafa, A. K. E. Mansour and H. A. A. Zaher, *J. Org. Chem.*, **25**, 949 (1960).
88. A. Mustafa, in *Advances in Photochemistry*, (Ed. W. A. Noyes Jr., G. S. Hammond and J. N. Pitts Jr.,) Interscience, 1964, p. 104.
89. A. Mustafa and M. Islam, *J. Chem. Soc.*, S81 (1949).
90a. A. Schönberg and M. M. Sidky, *J. Org. Chem.*, **21**, 476 (1956).
90b. C. H. Krauch, S. Farid and G. O. Schenck, *Chem. Ber.*, **98**, 3102 (1965).
91. E. Späth and W. Gruber, *Chem. Ber.*, **74**, 1492 (1941).
92. J. R. Clarke, G. Glaser and A. Robertson, *J. Chem. Soc.*, 2260 (1948).
93. W. Gruber and F. E. Hoyos, *Monatsh.*, **78**, 417 (1948).
94. W. Gruber and F. Traub, *Monatsh.*, **77**, 414 (1947).
95. T. A. Geissman, *J. Am. Chem. Soc.*, **73**, 3355 (1951).
96. A. Schönberg, N. Badran and N. A. Starkowsky, *J. Am. Chem. Soc.*, **77**, 1019 (1955).
97. W. Gruber and K. Horváth, *Monatsh.*, **81**, 819 (1950).
98. T. A. Geissman and E. Hinreiner, *J. Am. Chem. Soc.*, **73**, 782 (1951).
99. R. Aneja, S. K. Mukerjee and T. R. Seshadri, *Tetrahedron*, **3**, 230 (1958).
100. A. Schönberg, N. Badran and N. A. Starkowsky, *J. Am. Chem. Soc.*, **77**, 5390 (1955).
101. A. Mustafa, N. A. Starkovsky and E. Zaki, *J. Org. Chem.*, **26**, 523 (1961).
102. A. Mustafa, N. A. Starkovsky and E. Zaki, *J. Org. Chem.*, **25**, 794 (1960).
103. B. Krishnaswamy and T. R. Seshadri, *Proc. Indian Acad. Sci.*, **15A**, 437 (1942).
104. K. V. Rao and T. R. Seshadri, *Proc. Indian Acad. Sci.*, **22A**, 383 (1945).
105. C. Musante and S. Fatutta, *Farmaco (Pavia) Ed. Sci.*, **16**, 343 (1961); *Chem. Abstr.*, **56**, 463 (1962).
106. W. Gruber and K. Horváth, *Monatsh.*, **80**, 563 (1949).
107. G. H. Phillips, A. Robertson and W. B. Whalley, *J. Chem. Soc.*, 4951 (1952).
108. C. Musante, L. Fabbrini and A. Franchi, *Farmaco (Pavia) Ed. Sci.*, **17**, 520 (1962); *Chem. Abstr.*, **58**, 2427 (1963).
109. F. H. Curd and A. Robertson, *J. Chem. Soc.*, 437 (1933).
110. (a) A. Mustafa, M. Kamel and M. A. Allam, *J. Am. Chem. Soc.*, **78**, 4692 (1956); (b) A. Mustafa and M. I. Ali, *J. Org. Chem.*, **21**, 849 (1959).
111. W. Bencze, J. Eisenbeiss and H. Schmid, *Helv. Chim. Acta*, **39**, 923 (1956).
112. A. Mustafa, M. M. Sidky and R. Mohran, *Ann. Chem.*, **684**, 187 (1965).
113. E. Späth and W. Gruber, *Chem. Ber.*, **74**, 1549 (1941).
114. L. Fabbrini and A. Franchi, *Ann. Chim. (Rome)*, **49**, 894 (1959); *Chem. Abstr.*, **54**, 4558 (1960).
115. T. A. Geissman and J. W. Bolger, *J. Am. Chem. Soc.*, **73**, 5875 (1951).
116. T. A. Geissman, *J. Am. Chem. Soc.*, **73**, 3355 (1951).
117. G. Seitz, *Arch. Pharm.*, **287**, 79 (1954); *Chem. Abstr.*, **51**, 14705 (1957).
118. J. Schmutz, R. Hirt and H. Lauener, *Helv. Chim. Acta*, **35**, 1168 (1952).
119. O. J. Cavallito and H. E. Rockwell, *J. Org. Chem.*, **15**, 820 (1950).

120. E. Späth, K. Eiter and T. Meinhard, *Chem. Ber.*, **75**, 1623 (1942).
121. A. Schönberg and A. Sina, *J. Chem. Soc.*, 3344 (1950).
122. M. M. Sidky and M. R. Mahran, *J. Org. Chem.*, **27**, 4112 (1962).
123. A. Schönberg, E. Singer and M. M. Sidky, *Chem. Ber.*, **94**, 660 (1961).
124. B. V. Bitto, *Ann. Chem.*, **269**, 377 (1892).
125. B. V. Bitto, *Ann. Chem.*, **267**, 372 (1892).
126. M. M. Sidky and M. R. Mahran, *Arch. Pharm.*, **296**, 569 (1963).
127. L. Strassberger and E. E. Vonesch, *Anales Assoc. Quim. Argentina*, **40**, 203 (1952); *Chem. Abstr.*, **47**, 3520 (1953).
128. C. J. C. Tovar, *Annales Fac. Farm. Bioquim. Univ. Nacl. Mayor Sen Marcos (Lima Peru)*, **4**, 436 (1953); *Chem. Abstr.*, **49**, 13282 (1955).
129. A. C. Ralha, *Rev. Port. Farm.*, **6**, 109 (1956); *Chem. Abstr.*, **51**, 6091 (1957).
130. Schwartz, Mintz and Plainfield, *J. Med. Soc. New Jersey*, **49**, 8 (1952).
131. A. W. H. Schnidler, *Arch. Pharm.*, **286**, 523 (1953); *Chem. Abstr.*, **48**, 11003 (1954).
132. E. Aubertin, *J. Med. Bordeaux*, **127**, 819 (1951); *Chem. Abstr.*, **45**, 2587 (1951).
133. M. A. Lesser, *Drug and Cosmetic Ind.*, **67**, 480, 556 (1950); *Chem. Abstr.*, **45**, 2152 (1951).
134. G. Colombo, W. Montorsi and S. Salvaneschi, *Arch. Sci. Med.*, **97**, 71 (1954); *Chem. Abstr.*, **48**, 7193 (1954).
135. G. Colombo and S. Salvaneschi, *Atti Lombarda Sci. Med. Biol.*, **8**, 340 (1953); *Chem. Abstr.*, **48**, 7783 (1954).
136. W. Montorsi, S. Salvaneschi and G. Colombo, *Presse Med.*, **63**, 81 (1955); *Chem. Abstr.*, **50**, 4382 (1956).
137. J. Delphaut, M. Lanza and H. Sarles, *Compt. Rend. Soc. Biol.*, **150**, 1407 (1956); *Chem. Abstr.*, **51**, 5989 (1957).
138. J. Paris, J. Vanlerenberghe and R. Godchaux, *Compt. Rend. Soc. Biol.*, **151**, 1184 (1957); *Chem. Abstr.*, **52**, 12196 (1958).
139. K. Uhlenbroock and K. Mulli, *Arzneimittel-Forsch*, **7**, 166 (1957); *Chem. Abstr.*, **51**, 9942 (1957).
140. J. Sobral, *Compt. Rend. Soc. Biol.*, **151**, 810 (1957); *Chem. Abstr.*, **52**, 8358 (1958).
141. R. Gölng and H. Kempe, *Arch. Intern. Pharmacodyn.*, **107**, 255 (1956); *Chem. Abstr.*, **51**, 6841 (1957).
142. P. I. Halonen and J. Hakkila, *Ann. Med. Exp. Biol. Fenniae*, **30**, 118 (1952); *Chem. Abstr.*, **47**, 3463 (1953).
143. G. Caccia, L. Chiampo and G. Peracchia, *Arch. Ital. Sci. Farmacol.*, **11**, 45 (1961); *Chem. Abstr.*, **56**, 6605 (1962).
144. N. G. Polyakov, *Farmakol i Toksikol*, **24**, 172 (1961); *Chem. Abstr.*, **56**, 5345 (1962).
145. P. Crepax and A. Volta, *Studi Urbinati, Fac. Farm.*, **31-4**, (1), 82 (1957); *Chem. Abstr.*, **57**, 1499 (1962).
146. I. Halonen, O. Vartiainen and E. V. Venho, *Ann. Med. Exp. Biol. Fenniae*, **29**, 23 (1951); *Chem. Abstr.*, **46**, 4663 (1952).
147. I. R. Leusen and H. E. Essex, *Am. J. Physiol.*, **172**, 226 (1953); *Chem. Abstr.*, **47**, 5027 (1953)..
148. H. J. Kuschke, *Arch. Intern. Pharmacodyn.*, **106**, 100 (1956); *Chem. Abstr.*, **50**, 13266 (1956).
149. H. Jourdan, *Arzneimittel-Forsch.*, **7**, 82 (1957); *Chem. Abstr.*, **51**, 7584 (1957).

150. H. Jourdan, *Arzneimittel-Forsch*, **8**, 141 (1958); *Chem. Abstr.*, **53**, 581 (1959).
151. F. Jourdan and G. Faucon, *Therapie*, **12**, 927 (1957); *Chem. Abstr.*, **53**, 11660 (1959).
152. L. Schmidt, *Klin. Wochschr.*, **36**, 127 (1958); *Chem. Abstr.*, **53**, 5507 (1959).
153. W. Heinmann and W. Wilbrandt, *Helv. Physiol. Pharmacol. Acta*, **12**, 230 (1954); *Chem. Abstr.*, **49**, 1954 (1955).
154. R. L. Farrand and S. M. Horváth, *Am. J. Physiol.*, **196**, 391 (1959); *Chem. Abstr.*, **53**, 10571 (1959).
155. K. Samaan, A. M. Hossein and I. Fahim, *J. Pharm. Pharmacol.*, **1**, 538 (1949).
156. F. Jourdan, G. Faucon and G. Schoff, *Compt. Rend. Soc. Biol.*, **156**, 1855 (1962); *Chem. Abstr.*, **59**, 2084 (1963).
157. J. Mercier, M. Gavend, M. R. Gavend and F. Mercier, *Arch. Intern. Pharmacodyn.*, **122**, 394 (1959); *Chem. Abstr.*, **54**, 14472 (1960).
158. M. J. Hans and J. J. Desmarez, *Compt. Rend. Soc. Biol.*, **150**, 598 (1956); *Chem. Abstr.*, **50**, 17157 (1956); *Compt. Rend. Soc. Biol.*, **150**, 1806, 1820 (1956); *Chem. Abstr.*, **51**, 11556 (1957).
159. J. J. Desmarez, *Compt. Rend. Soc. Biol.*, **149**, 1291 (1955); *Chem. Abstr.*, **50**, 8017 (1956).
160. Ya. I. Khadzhai and G. V. Obolentseva, *Farmakol. i Toksikol.*, **25**, 450 (1962); *Chem. Abstr.*, **58**, 1831 (1963).
161. C. A. Appfel, *Deut. Med. Wochschr.*, **80**, 414 (1955); *Chem. Abstr.*, **49**, 9151 (1955).
162. Raymond-Hamet, *Compt. Rend.*, **238**, 1624 (1954); *Chem. Abstr.*, **48**, 12303 (1954).
163. K. Merz, F. Monroy and B. Uribe, *Annales Soc. Biol. Bogotá*, **5**, 124 (1952); *Chem. Abstr.*, **48**, 12295 (1954).
164. K. Uhlenbroock and G. Möller, *Med. Welt*, 253, (1953); *Chem. Abstr.*, **47**, 7670 (1953).
165. G. Jongebreur, *Arch. Intern. Pharmacodyn.*, **90**, 384 (1952); *Chem. Abstr.*, **47**, 760 (1953).
166. British Schering Research Laboratories Ltd. and C. V. Stead, *Brit. Pat.*, 748,223 (1956).
167. P. K. Ramachandran, A. T. Tefteller, G. O. Paulson, T. Cheng, C. T. Lin and W. J. Horton, *J. Org. Chem.*, **28**, 398 (1963).
168. R. A. Baxter, G. R. Ramage and J. A. Timson, *Brit. Pat.*, 663,418 (1951); *Chem. Abstr.*, **47**, 7550 (1953).
169. W. Gruber and K. Horváth, *Monatsh. Chem.*, **81**, 828 (1950).
170. G. R. Ramage and C. V. Stead, *J. Chem. Soc.*, 3602 (1953).
171. H. Abu-Shady, and T. O. Soine, *J. Am. Pharm. Assoc. Sci. Ed.*, **42**, 351 (1953); *Chem. Abstr.*, **48**, 3969 (1954).
172. T. A. Geissman, *U.S. Pat.*, 2,657,218 (1953); *Chem. Abstr.*, **48**, 10779 (1954).
173. O. Dann, *U.S. Pat.*, 3,099,600 (1963); *Chem. Abstr.*, **60**, 1758 (1964).
174. J. S. H. Davies and T. Deagan, *J. Chem. Soc.*, 3202 (1950).
175. J. Klosa, *Pharmazie*, **10**, 62 (1955); *Chem. Abstr.*, **50**, 4132 (1956).
176. P. L. G. Eymard (To Laboratories Berthier, J. S. A.), *Fr.* M1859 (1963); *Chem. Abstr.*, **60**, 1757 (1964).
177. R. Selleri and G. D. Paco, *Ann. Chim.* (*Rome*), **48**, 1205 (1958); *Chem. Abstr.*, **53**, 11362 (1959).

178. J. M. R. A. Delourne-Houde, *Brit. Pat.*, 735,729 (1955); *Chem. Abstr.*, **50,** 3503 (1956).
179. J. M. R. A. Delourne-Houde, (Jean-Pierre Fourneau, inventor), *Fr. Addn.*, 63,346 (1955); *Chem. Abstr.*, **53,** 2702 (1959).
180. O. Dann, *U.S. Pat.*, 2,855,406 (1958).
181. O. Dann, *Ger. Pat.*, 956.047 (1957); *Chem. Abstr.*, **53,** 6257 (1959).
182. G.m.b.H. Hefa, Chem. Pharm. Fabrik (Ludwig Ritter and H. Kunsch, inventors), *Ger. Pat.*, 952,899 (1956); *Chem. Abstr.*, **53,** 2258 (1959).
183. T. S. Gardner, E. Wenis and J. Lee, *J. Org. Chem.*, **15,** 841 (1950).
184. C. B. Rao, G. Subramanyan and V. Venkateswarlu, *J. Org. Chem.*, **24,** 685 (1959).
185. B. L. Manjunath and A. Seetharamiah, *Chem. Ber.*, **72,** 97 (1939).
186. G. R. Kelkar and D. B. Limaye, *Rasayanam*, **1,** 228 (1941); *Chem. Abstr.*, **36,** 1037 (1942).
187. V. K. Ahluwali, S. K. Mukerjee and T. R. Seshadri, *J. Sci. Ind. Res. (India)*, **12B,** 283 (1953); *Chem. Abstr.*, **49,** 13980 (1955).
188. N. M. Shah and P. M. Shah, *Chem. Ber.*, **93,** 24 (1960).
189. C. B. Rao, G. Subramanyan and V. Venkateswarlu, *J. Org. Chem.*, **24,** 685 (1959).
190. W. Gruber and F. E. Hoyos, *Monatsh. Chem.*, **80,** 303 (1949).

CHAPTER IV

Furoxanthones

I. Naturally Occurring Furoxanthones

While xanthones have been found in plants and in fungi, our knowledge of the naturally occurring furoxanthones has been gained by the isolation of sterigmatocystin, which has a unique structure. The parent pyrans, the furoxanthenes, are not known to occur in nature, and synthetic furoxanthones are not numerous (Table 1).

1. Sterigmatocystin

Some strains of the *Aspergillus versicolor* (Vuill) produce a crystalline metabolite named sterigmatocystin (3a,12a-dihydro-8-hydroxy-6-methoxy-7*H*-furo(3′,2′–4,5)furo(2,3-c)xanthen-7-one) (**4**),[1–3] which may be isolated from the dried and powdered mycelium by successive solvent extraction, chromatography, crystallization, and sublimation. It forms pale yellow needles, and has molecular formula $C_{18}H_{12}O_6$. The molecular weight, found by X–ray crystallography and other methods, has shown that the value previously accepted had been too low, but nevertheless, it was possible to confirm the important degradative results obtained by Hatsuda and coworkers,[1] namely the conversion of the pigment into 1,3,8-trihydroxyxanthone by aluminum chloride.

Sterigmatocystin is strongly *laevo* rotatory $(\alpha)_D^{20.5} - 387°$ ($CHCl_3$). Hot ethanolic potassium hydroxide solution slowly converts it into an optically inactive isomer, namely, isosterigmatocystin. It yields a green ferric reaction in ethanol, and gives a deep–yellow color with aqueous sodium hydroxide, in which it is insoluble. The molecule contains one methoxyl and one free hydroxyl group, and does not contain a methylenedioxy group or *C*-methyl group.

Davies, Kirkaldy, and Roberts[4] found that many of the properties of sterigmatocystin were those of 1-hydroxyxanthone. This view was confirmed by the following evidence: (a) the ultraviolet light absorption agrees with those recorded for many hydroxylated and/or methoxylated

TABLE 1. Furoxanthones

Substituents	M.p. (°C)	Solvent for crystallization[a]; appearance	Remarks (u.v. spectrum mμ (log ε))	References
A. *Sterigmatocystin and Isosterigmatocystin*				
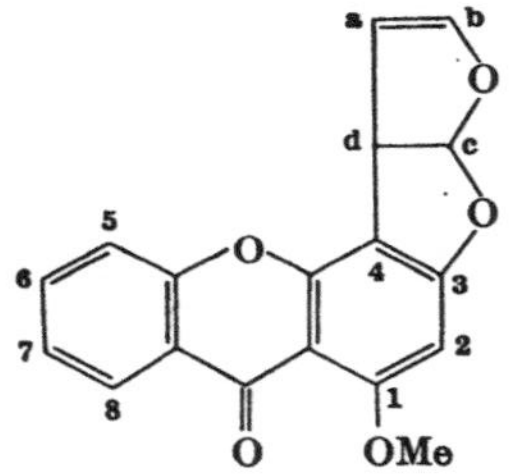				
8-Hydroxy (sterigmatocystin)	241–242; 246 (subl.)	C; pale yellow	208(4.28), 235(4.39), 249(4.44), 329(4.12)	4, 6
8-OMe (*O*-methylsterigmatocystin)	265–267	J; pale yellow	236(4.61), 309(4.23)	4
8-$OCOC_6H_5$ (*O*-benzoyloxysterigmatocystin)	258–260	G		4
8-Hydroxy, a,b-dihydro (dihydrosterigmatocystin)	229–230	G; yellow	208(4.31), 232(4.42), 247(4.49), 325(4.21)	4
8-OMe, a,b-dihydro	282–283	J	203(4.42), 237(4.59), 311(4.24)	4
8-$OCOC_6H_5$, a,b-dihydro	256–258	—		4
8-OAc, b-OAc, a,b-dihydro	227–228	G		4
8-OAc, a,b-dihydro	215–216	G		4
8-OH, b-OMe, a,b-dihydro	296–297; 298–299 (subl.)	G; yellow		6
8-OH, b-OEt, a,b-dihydro	252–254	J; yellow		6
8-OMe, b-OAc, a,b-dihydro	176–177 (subl).			6
8-OH, 6-OMe	223 (subl.)	G; yellow	232(–), 248(–), 279(–), 331(–)	7

(*Table continued*)

TABLE 1 (*continued*)

Substituents	M.p. (°C)	Solvent for crystallization[a]; appearance	Remarks (u.v. spectrum mμ (log ε))	References
8-OAc, 6-OMe	228	G		7
8-OH, 6-OMe, a,b-dihydro	248	G; pale yellow	red color with H_2SO_4	7
8-OAc, 6-OMe, a,b-dihydro	231	G		7
5,8-Di-OH, a,b-dihydro	260–262; 264(subl.)	D/G; yellow		7
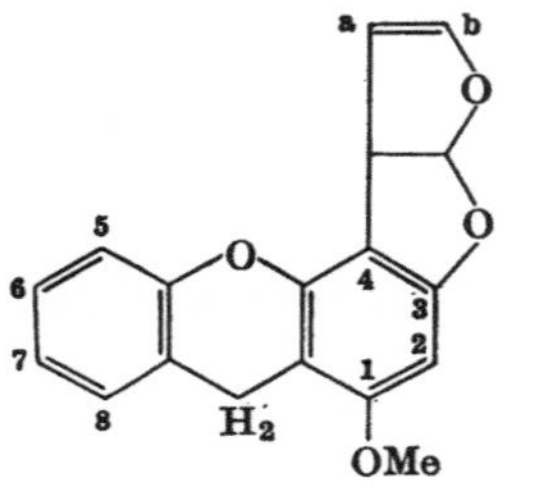				
8-OMe	220–222	J	221(4.62), 275(3.62)	4
8-OMe, a,b-dihydro	233–236	G		4
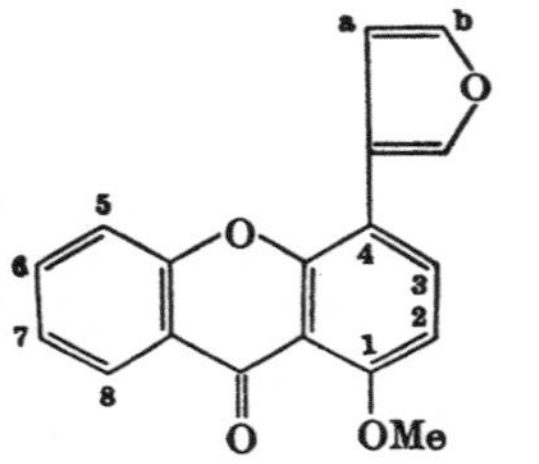				

Substituents	M.p. (°C)	Solvent for crystallization[a]; appearance	Remarks (u.v. spectrum mμ (log ε))	References
3,8-Di-OH (isosterigmatocystin)	233–234	G; pale yellow	212(4.48), 238(4.52), 253(4.59), 340(4.24)	4, 6
3,8-Di-OMe	206–208	G	209(4.34), 242(4.53), 301(3.86), 350(3.51)	6
3-OMe, 8-OH	243	J; pale yellow	green-brown with $FeCl_3$	6
3-OMe, 8-OAc	268	G		6
3-OH, 8-OMe	230	G; pale yellow	211(4.41), 241(4.56), 260(4.37), 321(4.17)	6
3-OAc, 8-OMe	189–190	G		6
3,8-Di-OMe	173–174	E/I	216(4.40), 243(4.55), 320(4.16); adduct with maleic anhydride (m.p. 168°)	6
B. *Furo(2′,3′–1,2)xanthones*				
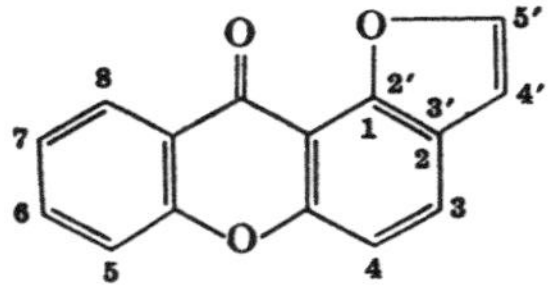				
4′-OH	253–255	G		12
4′-OAc	230–231	B		12, 13
4′-OH, 5′-Me	238	G		13
4′-OH, 5′-Ph	315–317	F		13
5′-Me, 4′,5′-dihydro	181	I	234(ε,31,300), 254(ε,39,600), 364(ε,7,200)	14
3,5,6-Tri-OMe	245	H	254(4.58), 263(4.60), 281(4.14), 315(4.25)	21
3,5,6-Tri-OMe, 4,4′-di-Me, 8-$CH_2CH{=}CMe_2$, 4′,5′-dihydro	238–239(dec.)	K/I; pale yellow	257(4.69), 276(4.06), 314(4.29)	21
3,5,6-Tri-OMe, 4′,4′,5′-tri-Me, 4′,5′-dihydro	155–156	I	248(4.66), 285s(4.04), 299(4.39)	21

(Table continued)

TABLE 1 (*continued*)

Substituents	M.p. (°C)	Solvent for crystallization[a]; appearance	Remarks (u.v. spectrum mμ (log ε))	References
C. *Furo*(3′2′–1,2)*xanthones*				
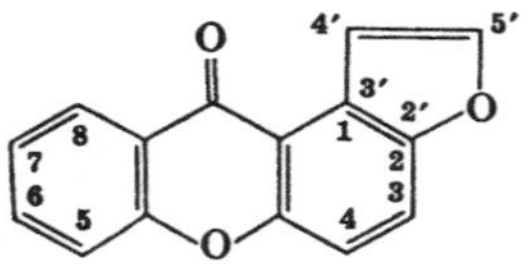				
Unsubstituted	144	A	233(ε,25,800), 250(ε,23,100), 305(ε,11,00), 351(ε,9,800)	9, 11
5′-Me, 4′,5′-dihydro	158–160	I; pale yellow		13
4′-OH, 4′, 5′-dihydro	208	G	229(ε,29,500), 250(ε,30,900), 300(ε,3,000), 358(ε,9,500)	11
4′-Me	137	G	235(ε,27,300), 318(ε,9,600), 354(ε,10,600)	11
4′-COOH	332(dec.)	A		9
4′-CCOEt	207	E		9
D. *Furo*(3′,2′–4,3)*xanthones*				
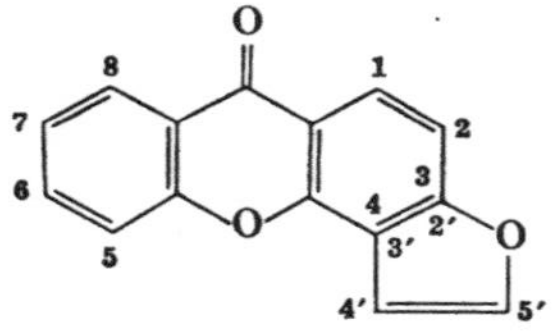				
Unsubstituted	192	G		10
4′-Me	216	G		18
5′-Me	170	G; yellow		17

Substituents	M.p. (°C)	Solvent for crystallization[a]; appearance	Remarks (u.v. spectrum mμ (log ε))	References
5′-Me, 4′,5′-dihydro	180	G; pale yellow		17
5′-COOH	332(dec.)	A		10
5′-COOEt	205	G		10
4′,6-Di-Me	227	G		18
5′,6-Di-Me	255	G		17
E. *Furo*(2′,3′–4,3)*xanthones* 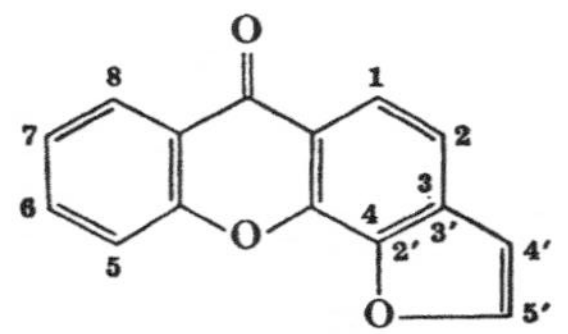				
5′-Me, 4′,5′-dihydro	139	K		13
F. *Furo*(3′,2′–2,3)*xanthones* 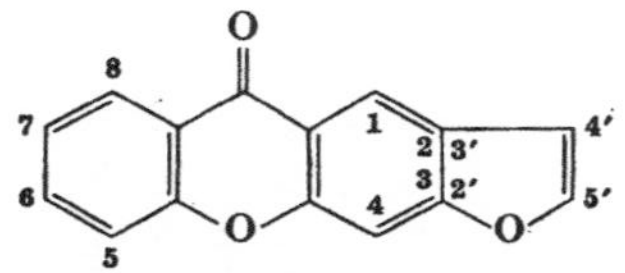				
5′-Me, 4-Ac	187	G; pale yellow		22
5′,7-Di-Me, 4-Ac	241	G; pale yellow		22

[a] A, acetic acid; B, acetic anhydride; C, acetone; D, benzene; E, chloroform; F, dioxan; G, ethyl alcohol; H, ethyl acetate; I, light-petroleum (b.p. 60–80°); J, methyl alcohol; K, petroleum ether (b.p. 110–140°).

xanthones; (b) the infrared absorption spectrum has a strong band at 1650 cm^{-1} which is assigned to the stretching vibration of a xanthone hydrogen–bonded carbonyl group (1-hydroxyxanthone); (c) when reduced with lithium aluminum hydride *O*-methylsterigmatocystin yielded a xanthene derivative; (d) the ultraviolet absorption spectrum of *O*-methylsterigmatocystin was very similar to that of 1,3,8-trimethoxyxanthone.

Oxidation of *O*-methylsterigmatocystin with potassium permanganate gave, among other products, 2-hydroxy-6-methoxybenzoic acid. Similar oxidation of sterigmatocystin gave γ-resorcylic acid, proving the unprotected hydroxyl group in this compound to be in position 8. With a smaller proportion of oxidant, sterigmatocystin gave a crystalline acid (**2**), $C_{15}H_{10}O_7$, which when pyrolyzed and sublimed, gave 3,8-dihydroxy-1-methoxyxanthone. Thus the methoxyl group in sterigmatocystin, occupies position 1.

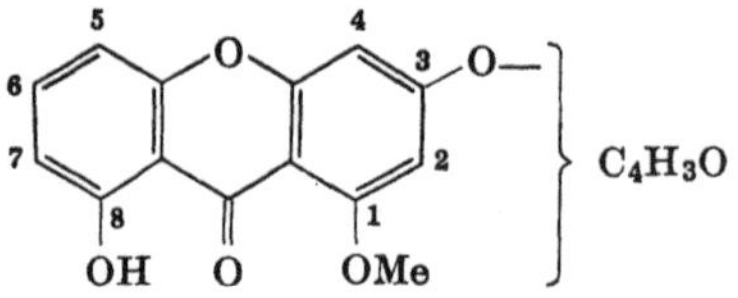

(1)

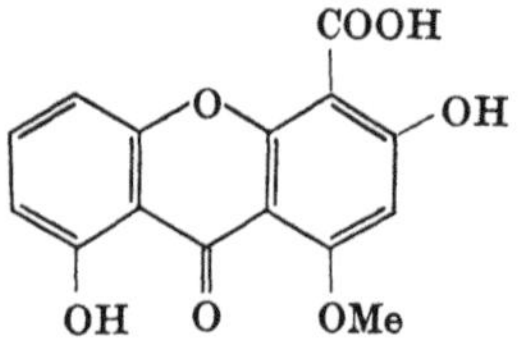

(2)

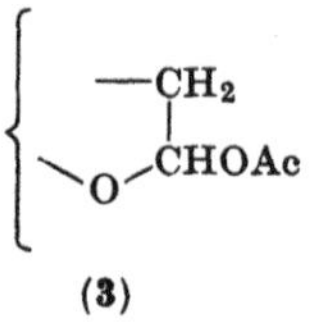

(3)

Quantitative hydrogenation of sterigmatocystin, under mild conditions, established the presence of one ethylenic bond in the molecule. Infrared absorption bands for sterigmatocystin (3099(w), 1610(s), 1067, and 722 cm^{-1}), which were absent from the spectrum of dihydrosterigmatocystin, indicated the presence of a vinyl ether grouping and the bands correspond to those recorded for 2,3-dihydrofuran. Moreover, the presence of a vinyl ether group was confirmed by ozonolysis of *O*-methylsterigmatocystin, followed by hydrolysis of the product and isolation of formic acid, by titration with peracid, and by addition of methanol and/or ethanol to the double bond.[4–6]

One molecule of acetic acid added to the vinyl ether grouping of sterigmatocystin to give acetoxymono-*O*-acetyldihydrosterigmatocystin

(**3**). Dihydrosterigmatocystin behaves normally on acetylation, yielding a mono-*O*-acetyl derivative.

Since the ultraviolet absorption spectrum of dihydrosterigmatocystin is essentially similar to that of sterigmatocystin, the ethylenic link of the vinyl ether sytem cannot be in conjugation with the main chromophore. Valency analysis of dihydrosterigmatocystin indicates the presence of two rings in addition to those in the xanthone nucleus.

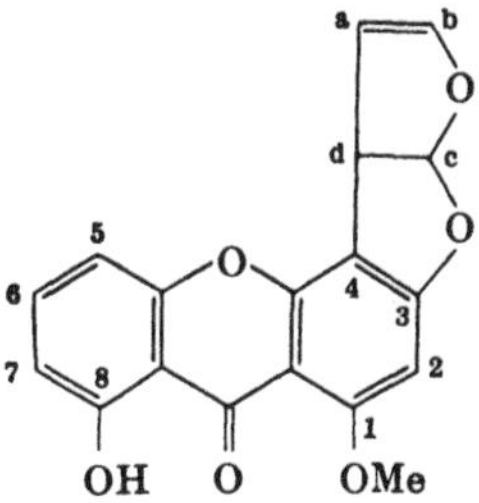

(**4**)

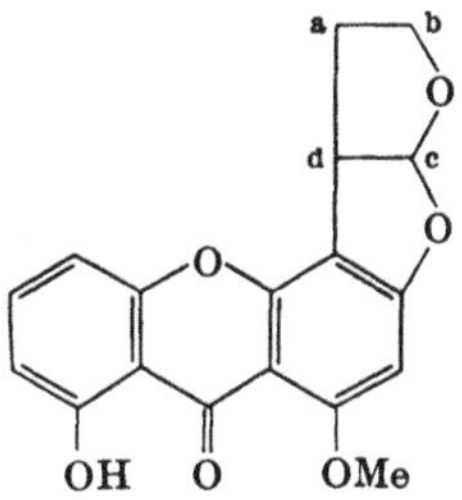

(**5**)

Isosterigmatocystin (**6a**), $C_{18}H_{12}O_6$, is a pale–yellow crystalline compound which gives a ferric reaction. It is xanthonoid in nature and in contrast to sterigmatocystin is insoluble in aqueous sodium carbonate solution; thus indicating the presence of a hydroxyl group in the 3-position of the xanthone nucleus. It contains one methoxyl group and, two free hydroxyl groups,[6] since it forms a di-*O*-acetyl derivative (**6b**).

Ozonolysis of di-*O*-methylisosterigmatocystin (**6c**) yields formic acid (2 mol.) and 1,3,8-trimethoxyxanthone-4-carboxylic acid. This reaction, and the stability of dihydrosterigmatocystin (**5**) to isomerization conditions, establish the retention of an intact xanthone nucleus during isomerization.

A comparison of the ultraviolet spectrum of **4** with that of 3-*O*-methylisosterigmatocystin (**6d**) showed that the bathochromic and hypsochromic effects are of the order expected for the additional conjugation of one ethylenic link with the xanthone nucleus. A comparison of the infrared absorption of **6c** with that of 1,3,8-trimethoxyxanthone indicates the presence of a furan ring in **6c**. The presence of this ring is confirmed by the ready formation of a Diels–Alder adduct (**7**) from **6c** and maleic anhydride. Since ozonolysis of **6c** yields 2 molecules of formic acid, the furan ring is attached to the xanthone nucleus through its β-position.[6]

The structural assignments have been confirmed by investigation of the proton magnetic resonance spectra of sterigmatocystin (**4**), dihydrosterigmatocystin (**5**), and di-*O*-methylisosterigmatocystin **6c**.[6, 7]

The isolation of a new metabolite from the mycelium of a variant strain of *Aspergillus versicolor* (Vuillemin) Tiraboschi has recently been

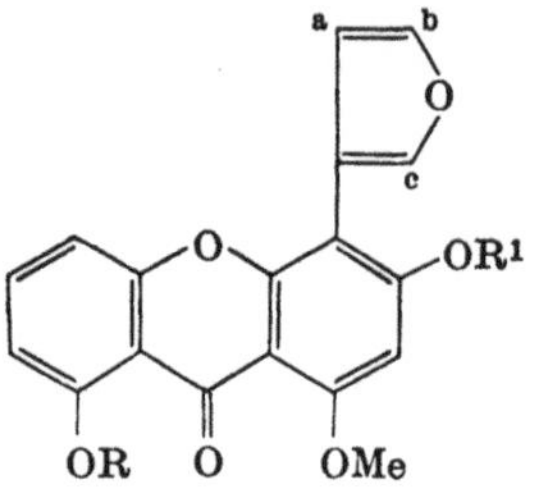

(**6a**) R=R¹=H; (**6b**) R=R¹=Ac; (**6c**) R=R¹=Me; (**6d**) R=H, R¹=Me

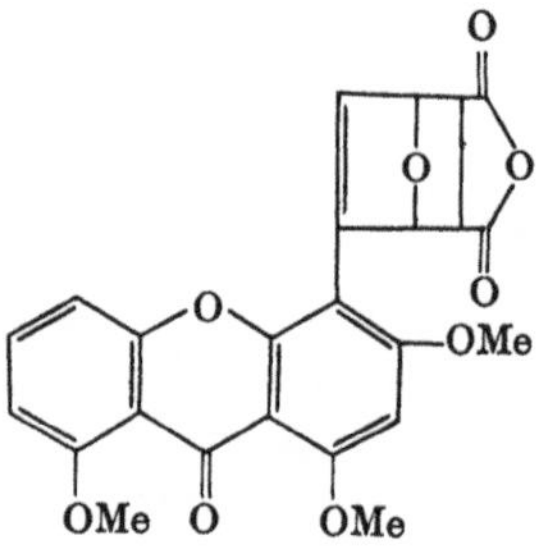

(**7**)

reported.[7] The metabolite, $C_{17}H_8O_5(OMe)_2$, $(\alpha)^{20}_{D} -360°$ ($CHCl_3$), crystallized in yellow needles. It is almost insoluble in aqueous sodium hydroxide, gives a ferric reaction, a positive Gibbs test, and a monoacetate, which is catalytically hydrogenated to a dihydro derivative (**9**).

The infrared absorption of the metabolite (**8**), which closely resembled that of sterigmatocystin (**4**) (except in the region associated with the aromatic substitution pattern), indicated the presence of a hydrogen-bonded hydroxyl group (3447 cm^{-1}), a carbonyl group (1662 cm^{-1}), and a vinyl ether system (3096, 1618, 1050, and 724 cm^{-1}). The bands due to the vinyl ether system were absent from the spectrum of **9**.

The ultraviolet absorption spectra of **8** and its dihydro derivative (**9**) are similar to that of dihydro-5-hydroxysterigmatocystin obtained by Elb's oxidation of dihydrosterigmatocystin (**5**).[6] All these spectra possess an extra peak at about 280 mμ which is not in the spectrum of **4**. The spectra of the dihydro compound (**9**) and its acetyl derivative are sufficiently similar to those of the respective unsaturated materials to indicate that the double bond of the vinyl ether system is not in conjugation with the main chromophore,

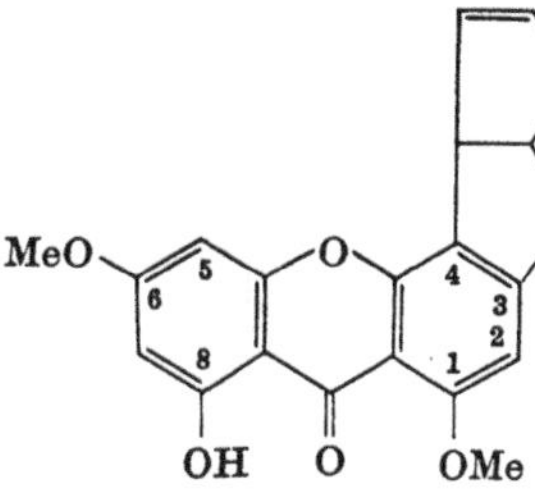

(8)

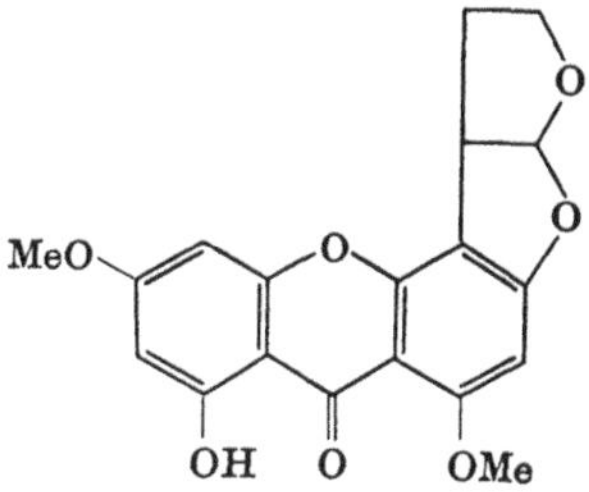

(9)

Proton magnetic resonance studies of the acetyl derivative of **9** gave results which are in favor of structure **8**.

II. Synthetic Furoxanthones

A condensed furan nucleus occurs in a number of related natural products, viz. furocoumarins, –chromones, –flavones, etc., and in an attempt to prepare physiologically active molecules of simple structure, analogous to biologically active khellin[8] the synthesis of furoxanthones has been recently achieved. This synthesis employs one of the typical methods for the preparation of benzofuran derivatives, viz. a mixed Claisen–type condensation or internal aldol condensation using the appropriately substituted aldehyde or ketone and bromoacetic or bromomalonic ester. Thus, cyclization of 1-formyl-2-xanthyloxyacetate, obtained by condensation of 1-formyl-2-hydroxyxanthone and ethyl bromoacetate in ethanol with sodium ethoxide, yields 5′-carbethoxyfuro-(3′,2′–1,2)xanthone. Hydrolysis of this ester, followed by decarboxylation, gave the parent angular furoxanthone (**10**), which was also obtained by ring closure of 1-formyl-2-xanthyloxyacetic acid accompanied by decarboxylation[9] (Eq. 1). Similarly, 4-formyl-3-hydroxyxanthone when condensed with bromomalonic ester, which effected simultaneous esterification and internal aldol condensation followed by hydrolysis and decarboxylation of the 5′-carbethoxy derivative formed, yielded furo(3′,2′–4,3)xanthone (**11**, $R = R^1 = H$).[10]

Treatment of 2-hydroxy-1-xanthylethylene oxide, obtained by the action of diazomethane on 1-formyl-2-hydroxyxanthone, with sulfuric

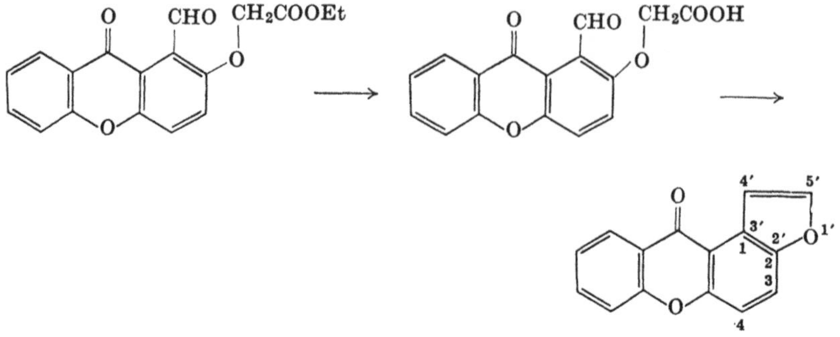

(1)

(10)

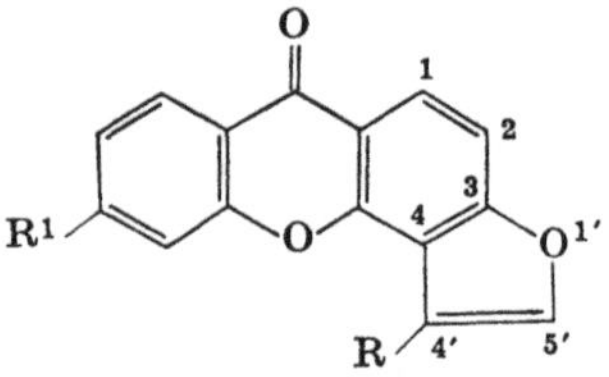

(11)

acid, led to the formation of the 4′-methyl derivative of **10**.[11] On the other hand, the corresponding ketone, viz. 2-acetyl-1-hydroxyxanthone, which can also be submitted to a similar internal Claisen condensation leading to the formation of a furan skeleton, can be more readily prepared and in better yield from the hydroxyxanthones by means of the Friedel–Craft's–Fries reactions. However, while aldehydo esters give rise to furo compounds, unsubstituted in the furan ring, the use of substituted ketones for such internal Claisen condensation results in the formation of furo compounds carrying a methyl substituent in the β-position.

2-ω-Bromoacetyl-1-hydroxyanthone is readily cyclized to 4′,5′-dihydro-4′-oxofuro(2′,3′–1,2)xanthone (**12**).[12,13] Similarly, 4′,5′-dihydro-4′-oxofuro(3′,2′–1,2)xanthone (**13**) is formed upon cyclization of 1-ω-bromoacetyl-2-hydroxyxanthone.[11]

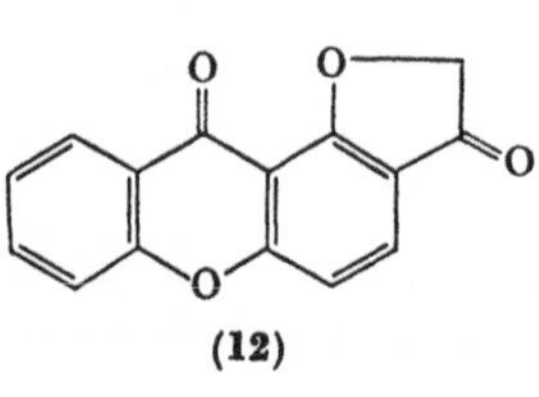

(12)

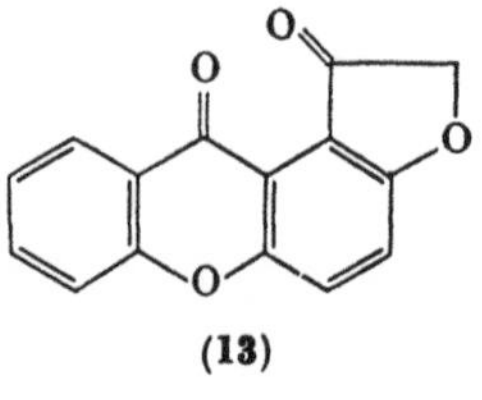

(13)

Cyclization of 2-β-bromopropyl-1-hydroxyxanthone, obtained by treatment of 2-allyl-1-hydroxyxanthone with hydrogen bromide in ethanol with sodium hydroxide, yields 4′,5′-dihydro-5′-methylfuro-(2′,3′–1,2)xanthone (**14**).[14] Similar treatment of 1-β-bromopropyl-2-hydroxyxanthone, and 3-β-bromopropyl-4-hydroxyxanthone led to the formation of 4′,5′-dihydro-5′-methylfuro(2′,3′–2,1)xanthone (**15**), and 4′,5′-dihydro-5′-methylfuro(2′,3′–4,3)xanthone (**16**), respectively.[13]

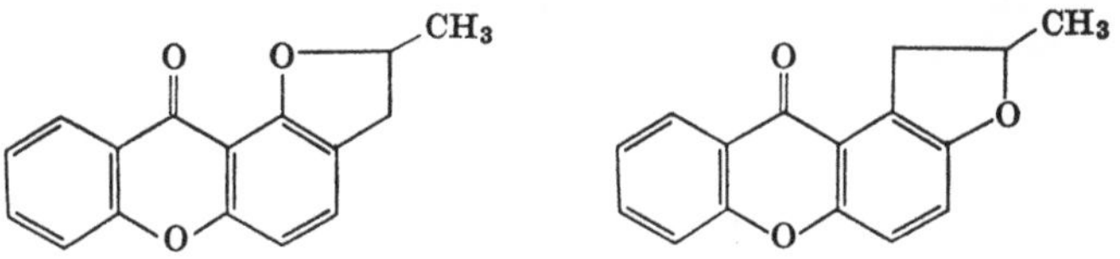

(**14**) (**15**)

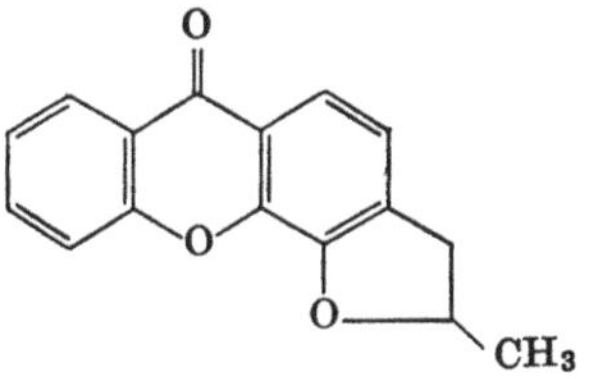

(**16**)

Two methods for the furan ring closure of 4-allyl-3-hydroxyxanthones were adopted. The first based on Scheinmann and Suschitzky's work,[14] gives dihydrofuro compounds which are dehydrogenated by treatment with *N*-bromosuccinimide in the presence of benzoyl peroxide, followed by dehydrobromination with pyridine[15] to α-methylfuroxanthones (**17**). The second route is an adaption of Adams and Rindfusz's[16] method, involving addition of bromine to 3-acetoxy-4-allyl derivatives, followed by cyclization and dehydrobromination[17].

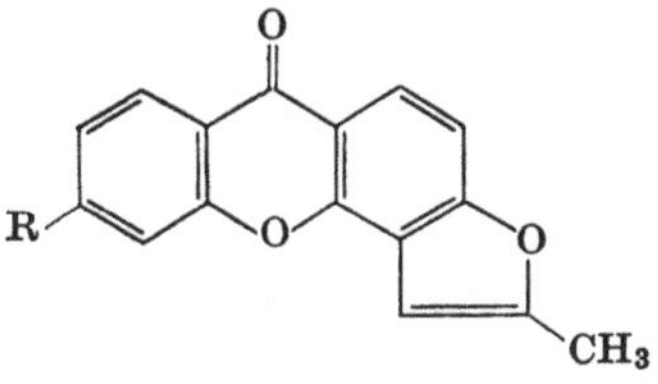

(**17a**) R=H; (**17b**) $R=CH_3$

4-Acetyl derivatives of 3-hydroxy-, and 3-hydroxy-6-methylxanthone condense with bromoacetic ester, in the presence of potassium carbonate, to yield the corresponding 3-*O*-carbethoxy derivatives, which on hydrolysis give the corresponding acids. The internal Claisen condensation of the acids with sodium acetate and acetic anhydride yields 4′-methylfuro(3′,2′–4,3)xanthone (**11**, $R = CH_3$, $R^1 = H$) and 4′,6-dimethylfuro(3′,2′–4,3)xanthone (**11**, $R = R^1 = CH_3$).[18]

Seshadri and coworkers' new method[19] of synthesising benzopyronofurans has recently been adopted by Scheinmann and coworkers[20, 21] for the preparation of the furoxanthones (**19**, **20**) by ozonolysis of the allyl derivatives (**18a**, **18b**), followed by cyclization with polyphosphoric acid.

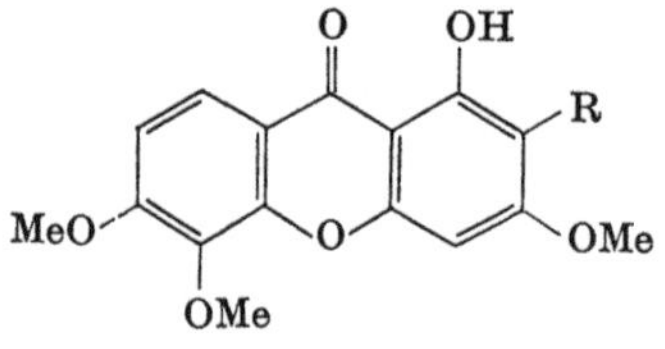

(**18a**) $R = C_3H_5$;
(**18b**) $R = CMe_2CH{=}CH_2$

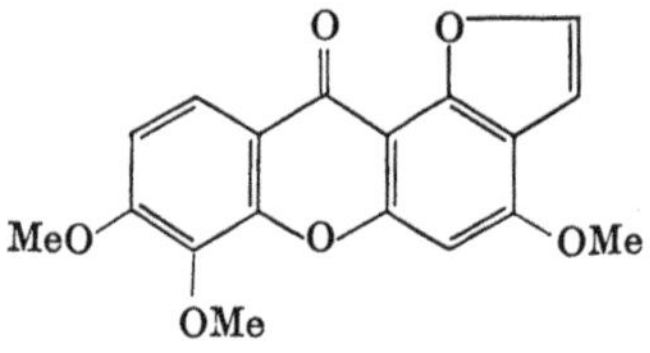

(**19**)

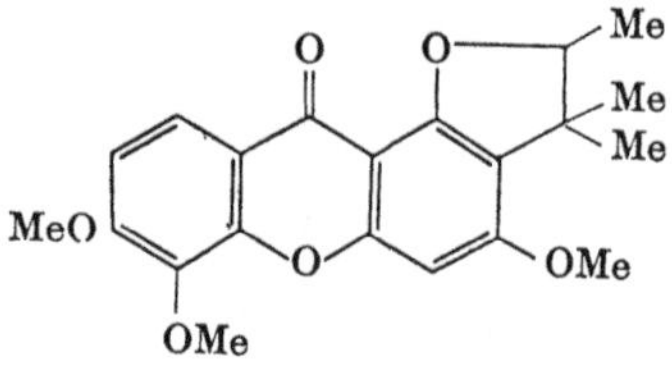

(**20**)

The reaction of 1,5-dihydroxy-3,6-dimethoxyxanthone with 3,3-dimethylallylbromide gave the ether (**21**) which rearranged in dimethylaniline to two xanthones (**22**, **23a**). Migration in the other ring occurs to yield the *ortho* product, 5-hydroxy-3,6-dimethoxy-8-(3′,3′-dimethylallyl)-4′,5′-dihydro-4′,4′,5′-trimethylfuro(2′,3′–1,2)xanthone (**23a**), and in addition the *para* isomer (**22**). Methylation of **23a** gave the trimethyl ether (**23b**).[21]

4-Acetyl derivatives of 3-hydroxy-, and 3-hydroxy-6-methylxanthones, obtained by treatment of the appropriate xanthone derivative with acetyl chloride in the presence of aluminum chloride, have been alkylated, and subjected to Claisen rearrangement to yield the 4-acetyl-2-allyl-3-hydroxyxanthones. The latter, after acetylation and bromination, are treated with potassium hydroxide in ethanol to bring about simultaneous dehalogenation and cyclization with the formation of 4-acetyl-5′-methylfuro(4′,5′–2,3)xanthone (**24**, $R = H$), and 4-acetyl-5′,7-dimethylfuro(4′,5′–2,3)xanthone (**24**, $R = CH_3$)[22] (Eq. 2).

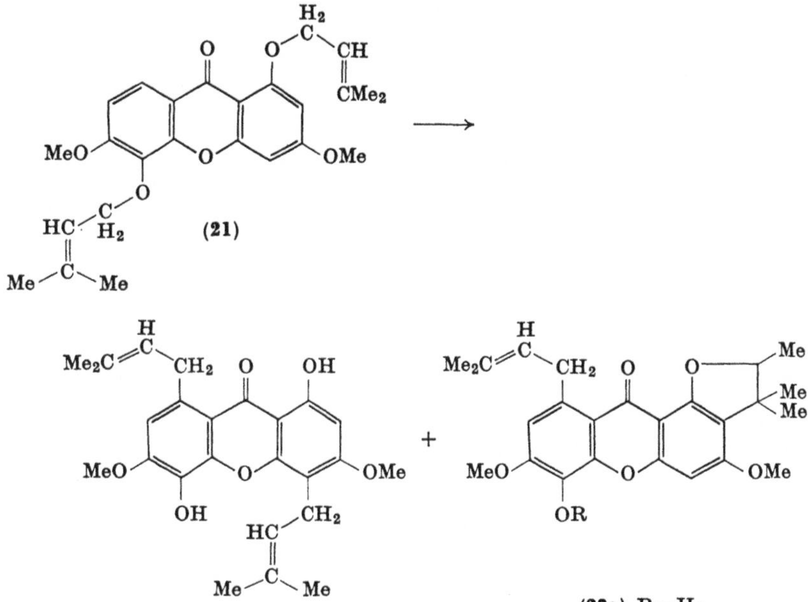

(21)

(22)

(23a) R=H; (23b) R=Me

Comparison of the ultraviolet absorption of xanthone and those of its extended ring systems possessing a conjugated ring in the 1,2-positions shows that the latter compounds have a marked depression in the

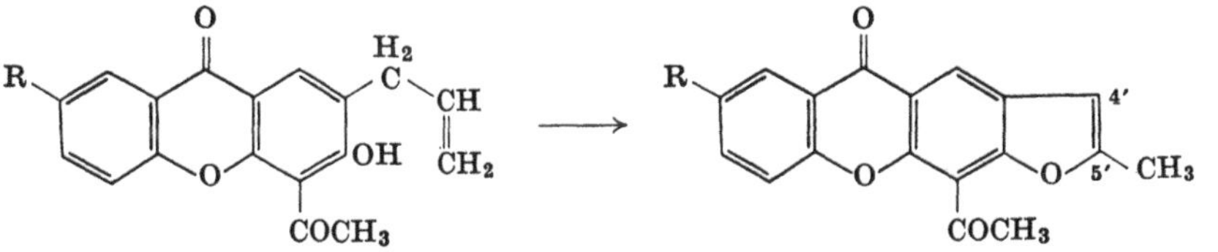 (2)

(24)

intensity of the principal absorption band without appreciable change in the location of the maximal absorption (Table 1), for example xanthone absorbs maximally at 238 mμ, while furo(3′,2′–1,2)xanthone (**10**) absorbs maximally at 233 mμ. Similar spectral behavior is observed in 1-formyl-2-hydroxyxanthone, which owing to chelation, may be regarded as equivalent to xanthone with a fused conjugated angular ring. The intensity depression in the furoxanthone (**13**) suggests predominance of the conjugated enol– over the keto–form.

III. References

1. Y. Hatsuda and S. Kuyoma, *J. Agric. Chem. Soc. Japan*, **28**, 989 (1954); *Chem. Abstr.*, **50**, 15522 (1956).
2. J. H. Birkinshaw and J. M. M. Hammady, *Biochem. J.*, **65**, 162 (1957).
3. J. E. Davies, J. C. Roberts and S. C. Wallwork, *Chem. Ind. (London)*, 178 (1956).
4. J. E. Davies, D. Kirkaldy and J. C. Roberts, *J. Chem. Soc.*, 2169 (1960).
5. J. C. Roberts and J. G. Underwood, *J. Chem. Soc.*, 2060 (1962).
6. (a) E. Bullock, J. C. Roberts and J. G. Underwood, *J. Chem. Soc.*, 4179 (1962); (b) J. A. Knight, J. C. Roberts and J. G. Underwood, *J. Chem. Soc.*, 5784 (1965).
7. E. Bullock, D. Kirkaldy, J. C. Roberts and J. G. Underwood, *J. Chem. Soc.*, 829 (1963).
8. C. P. Huttrer and E. Dale, *Chem. Revs.*, **48**, 543 (1951).
9. J. S. H. Davies, F. Lamb and H. Suschitzky, *J. Chem. Soc.*, 1790 (1958).
10. G. S. Puranik and S. Rajopal, *Chem. Ber.*, **96**, 976 (1963).
11. F. Lamb and H. Suschitzky, *Tetrahedron*, **5**, 1 (1959).
12. J. S. H. Davies, F. Scheinmann and H. Suschitzky, *J. Chem. Soc.*, 2140 (1956).
13. A. Mustafa, M. M. Sidky, S. M. A. Zayed and F. M. Soliman, *Tetrahedron*, **19**, 1335 (1963).
14. F. Scheinmann and H. Suschitzky, *Tetrahedron*, **7**, 31 (1959).
15. T. A. Geissman and T. G. Halsall, *J. Am. Chem. Soc.*, **73**, 1280 (1951).
16. R. Adams and B. Rindfusz, *J. Am. Chem. Soc.*, **41**, 648 (1919).
17. G. S. Puranik and S. Rajopal, *J. Chem. Soc.*, 1522 (1965).
18. G. S. Puranik and S. Rajopal, *J. Org. Chem.*, **29**, 1089 (1964).
19. R. Aneja, S. K. Mukerjee and T. R. Seshadri, *Tetrahedron*, **2**, 203 (1958); **3**, 230 (1958).
20. E. D. Burling, A. Jefferson and F. Scheinmann, *Tetrahedron Letters*, (12), 599 (1964).
21. E. D. Burling, A. Jefferson and F. Scheinmann, *Tetrahedron*, **21**, 2653 (1965).
22. Y. S. Agasimundin and S. Rajopal, *J. Org. Chem.*, **30**, 2084 (1965).

CHAPTER V

Furoflavones

I. Isolation

Eight furoflavones have been successfully isolated from plants. The plant *Pongamia pinnata* (L) Merr. is as fruitful as *Angelica archangelica* is in its own field for it produces at least four furoflavones: karanjin (3-methoxylanceolatin B, 3-methoxyfuro(2″,3″–7,8)flavone), and pongapin (3′,4′-methylenedioxykaranjin) derived from resorcinol[1] and gamatin (5-methoxyfuro(3′,2′–6,7)flavone), and pinnatin (3′,4′-methylenedioxygamatin) derived from phloroglucinol. This is the first recorded instance of four furoflavones of different types occurring together in nature.[2] The seeds of *Pongamia glabra* contain karanjin and 6-methoxy-4-oxo-2-phenylfuro(2,3-h)-1-benzopyran, and pongamol,[3] and the root-bark of *Tephrosia lanceolata* Grab. contains both lanceolatin B and the related furan diketone, pongamol. The latter when demethylated is converted into lanceolatin B.[4]

Isolation of furoflavones depends upon successive extraction with solvents of increasing polarity. Thus, petroleum ether is frequently used, although the oils extracted together, as in the case of the seeds of *Pongamia glabra*, together with the furoflavones, tend to exert a solubilizing effect. Ethanol extraction of the oil provided karanjin and by subjecting the remaining extract to countercurrent, distribution between aqueous acetic and light petroleum was obtained.[3]

Crude furoflavone fractions may be purified by fractional crystallization or by separations employing column chromatography or any other preparative chromatographic procedures. Examples of the use of chromatographic methods in effecting separations has been demonstrated by Seshadri and coworkers[5] by the isolation of the furoflavone constituents of *Pongamia glabra*.

II. Physical Properties

Spectrophotometric measurements have been employed in identifying the natural furoflavones. Since the completely alkylated hydroxyfuroflavones, e.g. gamatin and pinnatin[6] which occur in nature, do not

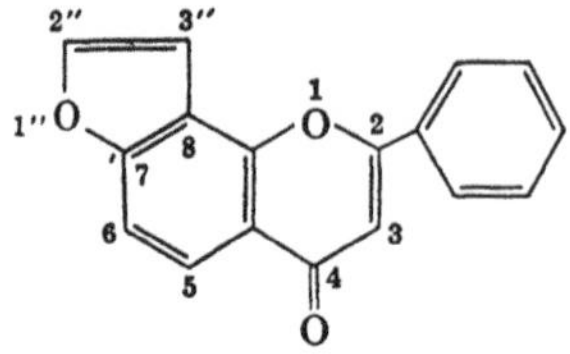

(Lanceolatin B)

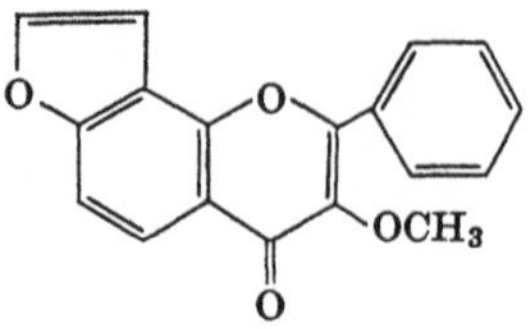

(Karanjin)

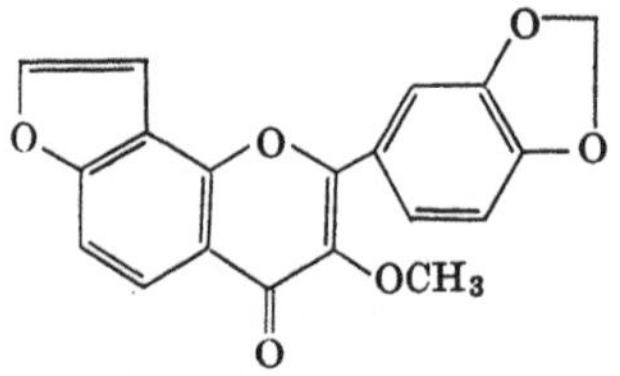

(Pongapin)

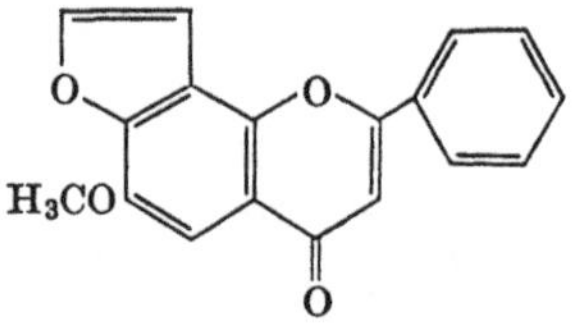

(6-Methoxyfuroflavone)
(Kanjone)

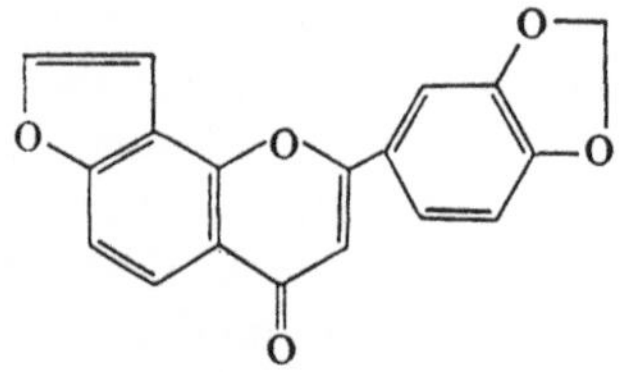

(Pongaglabrone)

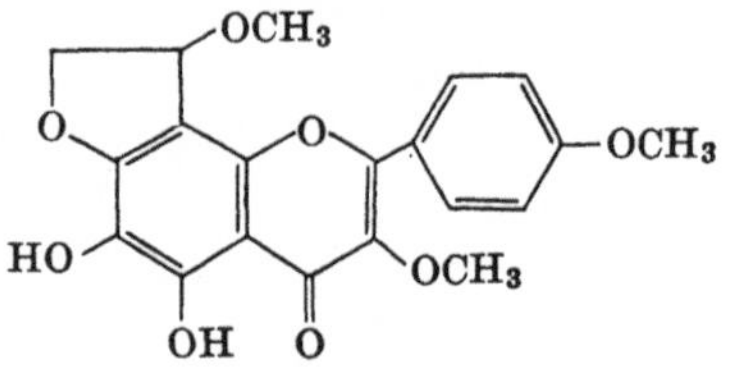

(Atanasin)

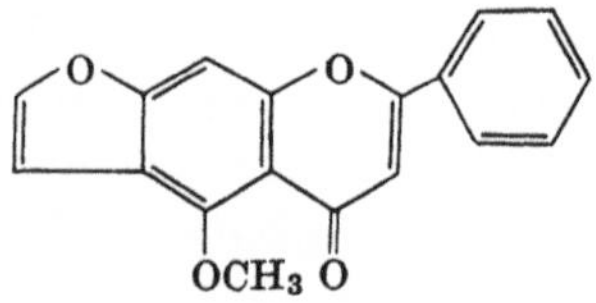

(Gamatin)

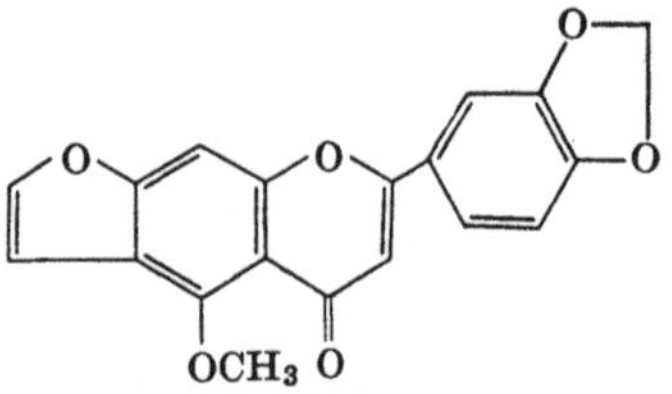

(Pinnatin)

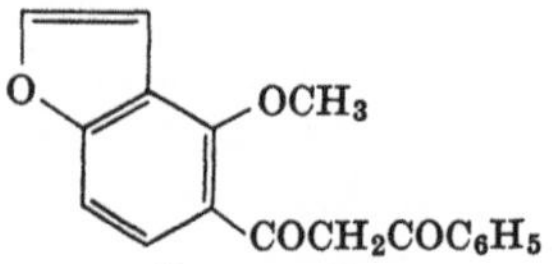

(Pongamol)

contain ionizable or chelating groups, their identification by spectral methods must be based solely on their spectra in neutral solvents. Thus, similarly to flavones,[7] they generally exhibit high absorption intensity in the 320–330 mμ region (Band 1) and the 240–270 mμ region (Band 2).

TABLE 1. Furoflavones

Substituents	M.p. (°C)	Solvent for crystallization[a]; appearance	Remarks	References
A. *Furo(2″,3″–6,7)flavones*				
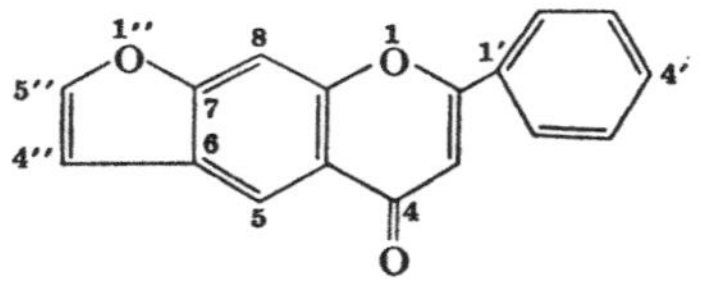				
Unsubstituted	168–169	E	yellow color with H_2SO_4	46
4″,5″-Dihydro	202–203	E		46
5″-Me_2CH, 2,3,4″,5″-tetrahydro	106–109	E		39
3-OH	203–205	H; yellow	brown color with $FeCl_3$; yellow color with Mg + HCl; greenish-blue fluorescence in ethanol	36
3-OH, 4″,5″-dihydro	188–189	E; bright yellow	brownish-violet with $FeCl_3$; orange-yellow with Mg + HCl	36
3-OMe	169–170	E; pale yellow	pale-yellow color with H_2SO_4 with light-blue fluorescence; yellow with Mg + HCl	36
3-OMe, 4″,5″-dihydro	161–162	E	pale yellow with H_2SO_4 with light-blue fluorescence; orange-yellow with Mg + HCl	36
4′-OMe, 5″-Me_2CH, 4″,5″-dihydro	168	E; pale yellow		39
4′-OMe, 5″-Me_2CH, 2, 3,4″,5″-tetrahydro	111	E		39
5-OMe (gamatin)	181–182	—		6
8-OMe	189–190; 190–191	C, E	green color with H_2SO_4; λ_{max} 266, 305 mμ	35, 40
3-OAc, 4″,5″-dihydro	169–170	E		36
3-OH, 4′-OMe, 5″-Me_2CH, 4″,5″-dihydro	192	E or E/C		39

(Table continued)

TABLE 1 (*continued*)

Substituents	M.p. (°C)	Solvent for crystallization[a]; appearance	Remarks	References
3′,4′-Di-OMe	170–172	C		35
3′,4′-Di-OMe, 4″,5″-dihydro	199–200	G; pale brown		35
5,8-Di-OMe	177–178; 180	A/H,E; yellow	red color with H_2SO_4	37, 38
3′,4′-(:O_2CH_2)	266	A/E; pale brown		35
3′,4′-(:O_2CH_2), 4″,5″-dihydro	246–247	E/A		35
3,3′,4′-Tri-OMe	156–160	E; yellow	pink color with Mg + HCl; yellow changing to green on warming with H_2SO_4	2
3,3′,4′-Tri-OMe, 4″,5″-dihydro	175-176	E	yellow color with H_2SO_4 changing to red-orange on heat	2
5,8,4′-Tri-OMe	184	E,G; pale yellow		37, 38
5,8,4′-Tri-OMe, 4″,5″-dihydro	141	G; orange	red color with H_2SO_4	37
3-OMe, 3′,4′-(:O_2CH_2)	180–181	E; yellow	green color changing to blue on warming with gallic acid and H_2SO_4	2
3-OMe, 3′-4′-(:O_2CH_2), 4″,5″-dihydro	215–216	E; pale yellow		2
5-OMe, 3′4′-(:O_2CH_2), (pinnatin)	234	—		6
3″,5-Di-OH, 4′-OMe, 5″-Me	238–240	E; yellow	brown color with $FeCl_3$; Ac. deriv., (m.p. 229–231°)	47
5,8,3′,4′-Tetra-OMe	182	G; yellow		37, 38
5,8-Di-OMe, 3′,4′-(:O_2CH_2), 4″,5″-dihydro	181	B/H; pale yellow	red color with H_2SO_4	37
8-OMe, 3′,4′-(:O_2CH_2)	228–230	E		35
8-OMe, 3′,4′-(:O_2CH_2), 4″,5″-dihydro	220–221	E		35

B. *Furo*(2″,3″–7,8)*flavones*

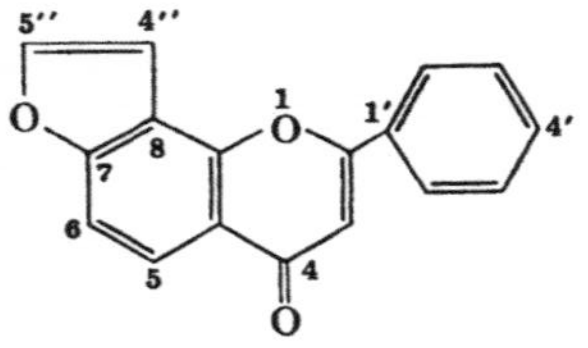

Substituents	M.p. (°C)	Solvent for crystallization[a]; appearance	Remarks	References
Unsubstituted (lanceolatin B)	135–142	—		24
3-OH (karanjonol)	192–193; 200	E	violet color with $FeCl_3$	14, 18
4″-OH	206–207	E	benzylidene deriv. (m.p. 224–225°); acetyl deriv. (m.p. 260–261°)	43
3-OMe (karanjin)	157–158	G		18, 21
5-OMe	182–183	D/F; yellow	green color with H_2SO_4; EtOH λ_{max}, 271,320 mμ	3
6-OMe (kanjone)	190	G; pale yellow	yellow color with H_2SO_4	3
3-OAc	177	E		18
3-OMe, 5″-Me (α-methylkaranjin)	121–122	E		41
3-OMe, 5″-Me, 4″,5″-dihydro	142–143	E		41
3-OMe, 5″-COOH	283(dec.)	E		18
3-OMe, 5″-COOMe	194–195	C		18
3-OMe, 5″-COOEt	183	C		18
3,4′-Di-OH	271 (sinters) 282 (melts)	A; yellow	violet color with $FeCl_3$	11
3,4″-Di-OH	284–286	E	benzylidene deriv. (m.p. 274°); Ac. deriv. (m.p. 192°)	43
3,4′-Di-OMe	166–168	G; yellow		11
3′,4′-(:O_2CH_2) (pongaglabrone)	233	E		5
3,4′-Di-OMe,5″-Me	159–160	E		41
3,4′-Di-OMe, 5″-Me, 4″,5″-dihydro	147–148	E		41
3,4′,5′-Tri-OH	282–299	A; yellow	red solution in alkali	11
3,4′,5′-Tri-OMe	184	G; yellow		11
3,3′,4′-Tri-OMe, 5″-Me	152–153	E; pale yellow		41
3,3′,4′-Tri-OMe, 5″-Me,4″,5″-dihydro	144–145	E		41
3-OMe, 3′,4′-(:O_2CH_2) (pongapin)	275–276	E		21
3-OMe, 3′,4′-(:O_2CH_2), 5″-Me	190–191	E		41
3-OMe, 3′,4′-(:O_2CH_2), 5″-Me, 4″,5″-dihydro	145–146	E		41

(Table continued)

TABLE 1 (*continued*)

Substituents	M.p. (°c)	Solvent for crystallization[a]; appearance	Remarks	References
3,3′,4′,5′-Tetra-OH	ca. 315	A; yellow	brown-red solution in H_2SO_4	11
3,3′,4′,5′-Tetra-OMe	158–159	G		11
3,4′,4″-Tri-OMe, 5,6-di-OH (atanasin)	221	—		30, 31
C. *Furo*(3″,2″–7,8)*flavones*				
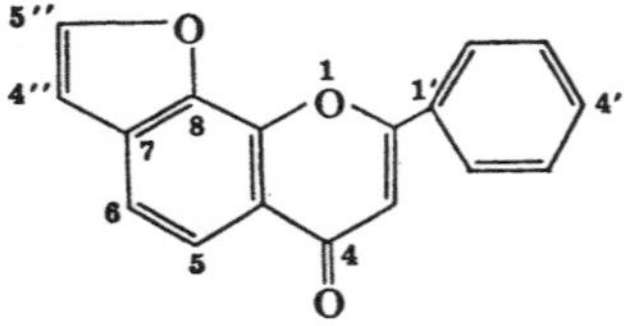				
4′-OMe, 4″-Me	223	A		48
4′-OMe, 5″-Br, 4″-Me	240	A; grey		48
D. *Furo*(3″,2″–5,6)*flavones*				
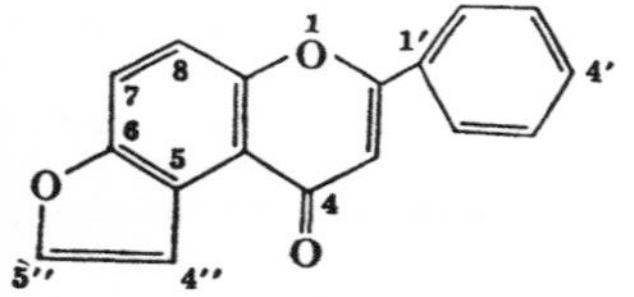				

Substituents	M.p. (°C)	Solvent for crystallization[a]; appearance	Remarks	References
4′-OMe, 4″-Me	241–242	—		48
E. *Furo(2″,3″–5,6)flavones*				

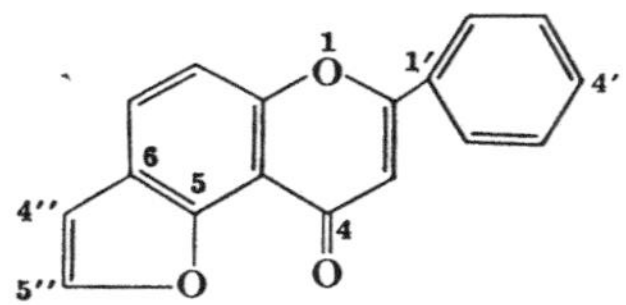

Substituents	M.p. (°C)	Solvent for crystallization[a]; appearance	Remarks	References
7-OMe	233	D/F; pale yellow	green color with H_2SO_4; EtOH λ_{max}, 265, 315 mμ	40
8-OMe	160–162	E; pale yellow	green color with H_2SO_4	40
4′-OMe, 5″-Br, 4″-Me	230	—		45
3′,4′-(:O_2CH_2), 4″-Me, 5″-Br	207	—		45

[a] A, acetic acid; B, acetone; C, benzene; D, chloroform; E, ethanol; F, light petroleum; G, methanol; H, water.

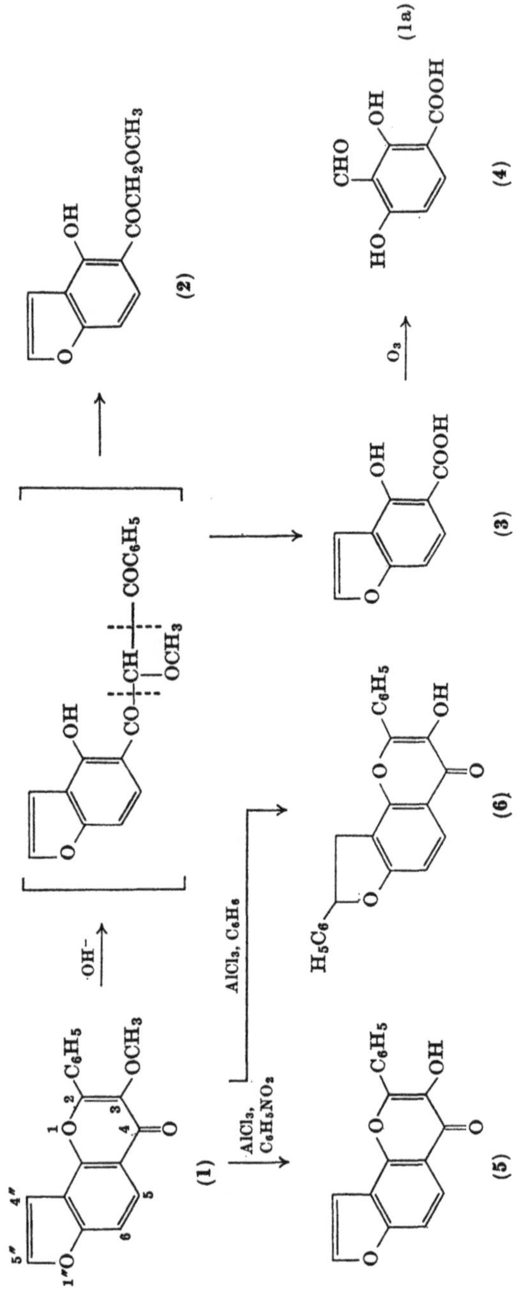
(1)
(2)
(3)
(4)
(5)
(6)
(1a)
OH⁻
O3
AlCl3, C6H5NO2
AlCl3, C6H6
C6H5
OCH3
OH
COCH2OCH3
COC6H5
COOH
CHO
HO
H5C6

Ultraviolet Spectra of Furoflavones

Furoflavone	U.v. spectrum (max (log E))	Reference
Pongapin	251.5 (4.34); 332.5 (4.34)	5
Gamatin	251.5 (4.52); 334 (4.39)	5
Pongaglabrone	249 (4.44); 331 (4.31)	5
Pinnatin	269 (4.52); 305 (4.19)	6
Karanjin	— —	1

III. Naturally Occurring Furoflavones

1. Chemical Properties

A. Karanjin

Karanjin (3-methoxylanceolatin B) (**1**), $C_{18}H_{12}O_4$ was the first member of this group recognized by Limaye.[8] It occurs in the seed oil of *Pongamia glabra*,[9, 10] the root bark of *Pongamia pinnata*,[1] and the roots of *Pongamia glabra*;[11] it is said to be opposed to leucoderma. Alkaline hydrolysis[8] ruptures the chromone ring with production of benzoic acid, a phenolic ketone (**2**), and the related acid, karanjic acid (**3**), by splitting of the intermediate diketone (across the dotted lines).

The flavone nature of karanjin (**1**) is further demonstrated by recyclization of karanjol (**2**) to **1** by vigorous benzoylation with sodium benzoate and benzoic anhydride. Karanjic acid (**3**) is oxidized by hydrogen peroxide to furan-2,3-dicarboxylic acid,[11] which is additional evidence for the structure assigned to **3** and hence for the angular formulation of karanjin, and gives 3-formyl-β-resorcylic acid (**4**) on ozonolysis (Eq. 1a). Decarboxylation of **3** yields 4-hydroxycoumarone.[8]

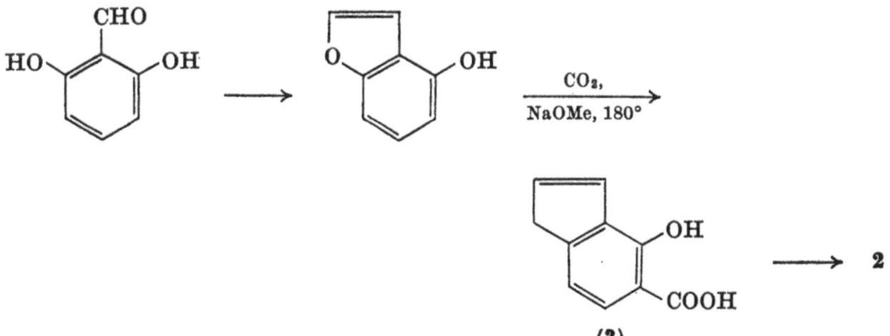

(1b)

(**3**)

Karanjin is demethylated, in the presence of aluminum chloride, to karanjonol (**5**). However, in benzene, addition of the latter to the furan ring double bond takes place with the formation of α-phenyl-α,β-dihydrokaranjonol (**6**).[14]

Synthesis of karanjic acid (**3**)[12, 13] has been achieved and confirms the degradative evidence (Eq. 1b).

The reaction of the aldehydo ester (**7**) with ethyl bromoacetate yields **8**, which on hydrolysis to the corresponding dibasic acid, and subsequent cyclization with sodium acetate–acetic anhydride gave a mixture of *O*-acetylkaranjic acid and karanjic acid (**3**).[15] The latter acid

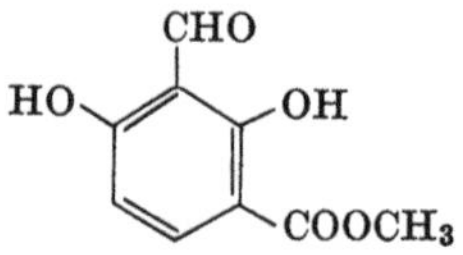
(**7**)

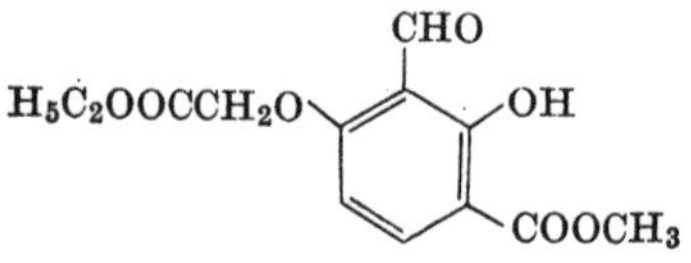
(**8**)

can be readily converted into karanjin ketone (**2**) by the use of α,γ-dimethylacetoacetic ester, according to Robinson's[16] method. Row and Rao[17] have reported a new synthesis of **2**: acetylkaranjoyl chloride (**9**) was treated with diazomethane to yield the diazoketone (**10**), which suffered decomposition, in absolute methanol with copper bronze, to yield 4-hydroxy-5-(ω-methoxyacetyl)coumarone, identical with **2**. The latter was converted into karanjin (**1**) by the Allan–Robinson method (Eq. 2).

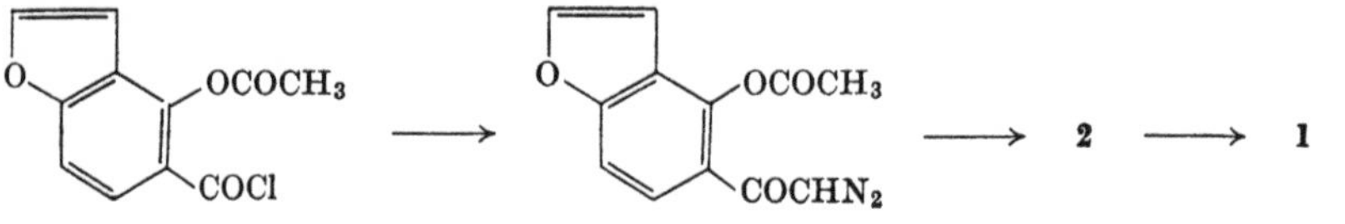
(**9**) (**10**) (2)

Kawase and coworkers[18] explored an alternative synthetic route to karanjin. Starting with 7-hydroxy-3-methoxyflavone-8-aldehyde, as in the method of Seshadri and coworkers,[13, 19] and Limaye,[8] 3-methoxyfuro-(2″,3″–7,8)flavone-5″-carboxylic acid (**11**) was synthesized by Tanaka's[20] method for benzofuran synthesis, and was decarboxylated, in quinoline with copper bronze, to yield **1** (Eq. 3). The synthetic **1** was related to the 3-hydroxy derivative, karanjonol, (**5**).

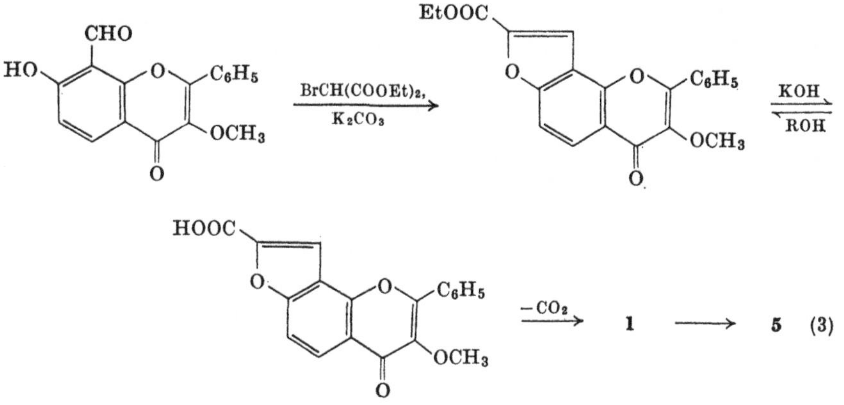

(11)

A new method of preparing benzofuran derivatives using allyl phenols has been worked out. Ozonolysis of 3-allyl-2,4-dihydroxy-(ω-methoxy)-acetophenone and of 8-allyl-7-hydroxy-3-methoxyflavone, followed by cyclization with orthophosphoric acid of the acetaldehyde derivative obtained gave **2** and **1** respectively, in good yield.[21, 22]

Karanjin gives a yellow diacetoxymercurikaranjin upon boiling with mercuric acetate.[23]

B. Lanceolatin B

Extraction of root bark of *Tephrosia lanceolata* gave three crystalline substances: lanceolatin A, lanceolatin B, and lanceolatin C. The latter has been proved to be identical with *O*-methylkaranjoylbenzoylmethane (pongamol) (**12a**), since it forms benzoic acid, acetophenone, and *O*-methylkaranjic acid on alkaline degradation, and yields benzoic acid on oxidation with potassium permanganate.

Lanceolatin B (**13**), $C_{17}H_{10}O_3$, resembles karanjin (**1**), but lacks the 3-methoxyl group.[24] It has been identified as furo(2″,3″–7,8)flavone by the isolation of benzoic acid, acetophenone, karanjic acid (**3**), 5-acetyl-4-hydroxycoumarone, and karanjoylbenzoylmethane (**12b**) on alkaline

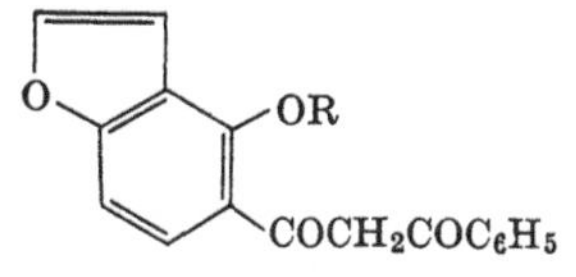

(**12a**) R=CH_3; (**12b**) R=H

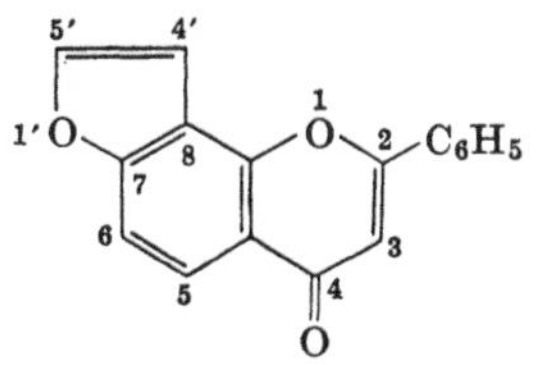

(**13**)

degradation, and by the formation of lanceolatin B (**13**) by demethylation of **12a** with hydrobromic acid in acetic acid, or with magnesium iodide.[25]

C. Pongapin

Pongapin (3′,4′-methylenedioxykaranjin) (**14**), $C_{19}H_{12}O_6$, is a pale yellow substance obtained from the root bark of the Australian *Pongamia pinnata* (L) Merr. Row[1] identified it as 3-methoxy-2-(3′,4′-methylenedioxyphenyl)-4*H*-furo(2,3-h)-1-benzopyran-4-one since on alkaline hydrolysis it yields piperonylic acid, and 4-hydroxy-5-methoxyacetoxycoumarone (**2**) which is identical with karanjin ketone. In an Allan–Robinson synthesis, karanjin ketone and piperonylic acid regenerate pongapin.[1]

A more complex, naturally occurring furoflavone **14**, has also been synthesized by ozonolysis of 8-allyl-7-hydroxy-3-methoxy-3′,4′-methylenedioxyflavone (**15**) in formic acid solution, followed by cyclization of the resulting acetaldehyde derivative (**16**) to yield pongapin[21] (Eq. 4).

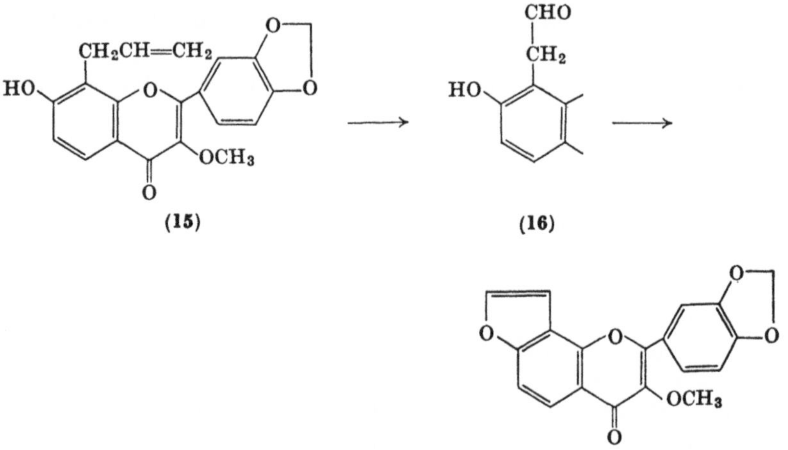

(**15**) (**16**)

(**14**) (**4**)

D. Kanjone

Kanjone (6-methoxy-2-phenylfuro(2,3-h)-1-benzopyran-4-one or 6-methoxyfuro(2″,3″–7,8)flavone) (**17**), $C_{18}H_{12}O_4$, is isolated from *Pongamia glabra* together with pongamol (**12a**).[1] It contains one methoxyl group, and has a prominent flavone carbonyl absorption at 1640 $cm.^{-1}$ With aqueous alcoholic potassium hydroxide, it yielded acetophenone, and *o*-hydroxy acid (**18**), $C_{10}H_8O_5$. When absolute alcoholic potassium hydroxide solution was used, benzoic acid, and *o*-hydroxy ketone (**19**), $C_{11}H_{10}O_4$, were obtained. These reactions are in agreement with the behavior of a flavone derivative without a substituent in the 2-phenyl group. The ultraviolet bands at 271 and 305 mμ are very similar to those of karanjin (**1**),[1] which indicates that it has the furoflavone skeleton with the furan ring attached to the condensed benzene ring. From the results of alkali fission, the methoxyl group should be present in the benzene ring.

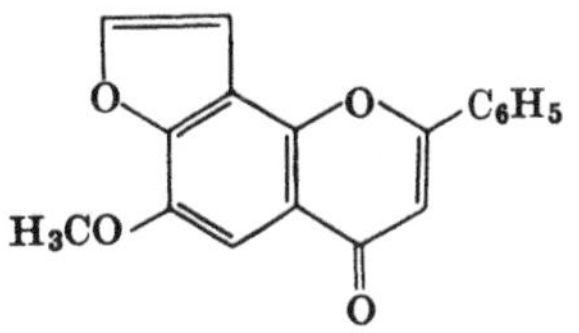

(**17**)

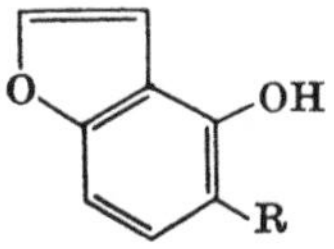

(**18**) R = COOH
(**19**) R = $COCH_3$

The synthesis of kanjone (**17**) has been achieved.[3] 4-Benzyloxy-2-hydroxyacetophenone was subjected to persulfate oxidation, and the dihydroxy compound (**20a**) partially methylated and debenzylated to yield the acetophenone (**20b**). Partial allylation in the 4-position, and Claisen migration yielded 3-allyl-2,4-dihydroxy-5-methoxyacetophenone (**21a**).

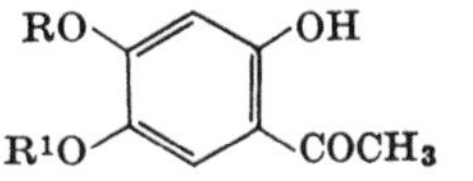

(**20a**) R = C_7H_7, R^1 = H;
(**20b**) R = H, R^1 = CH_3

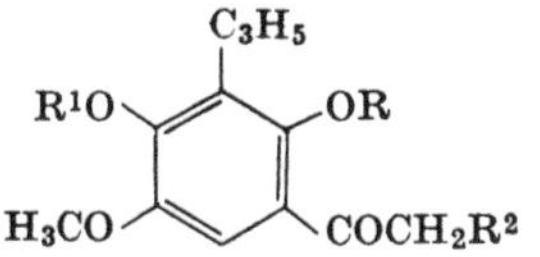

(**21a**) R = R^1 = R^2 = H;
(**21b**) R = R^1 = COC_6H_5, R^2 = H

The next step of the flavone-ring closure was effected by the dibenzoate (**21b**), followed by a Baker–Venkatarman transformation with pyridine and potassium hydroxide, and finally debenzoylation. Final furan-ring closure was brought about by ozonolysis of 8-allyl-7-hydroxy-6-methoxyflavone, followed by cyclization of the resulting acetaldehyde derivative with polyphosphoric acid to yield **17**[21] (Eq. 5).

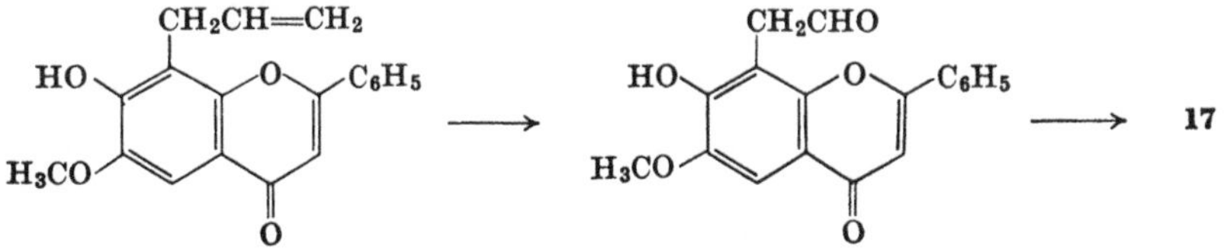

(5)

E. Pongaglabrone

Pongaglabrone (3′,4′ - methylenedioxyfuro(2″,3″–7,8)flavone) (**22**), $C_{18}H_{10}O_5$, isolated from the seeds of *Pongamia glabra*, is a colorless and fat-soluble compound.[5] Its insolubility in aqueous sodium hydroxide solution, and the absence of a ferric reaction indicate that there are no enolic or phenolic hydroxyl groups. The formation of a bright green color with gallic acid and sulfuric acid indicates the presence of a methylenedioxy group. Its infrared spectrum includes a strong absorption at 1640 cm^{-1}, characteristic of a flavone-carbonyl group and a strong band at 1035 cm^{-1} and a weak band at 935 cm^{-1} characteristic of the methylenedioxy group.[26] The ultraviolet absorption spectrum shows two bands at 249 and 331 mμ ,which are similar to those of pongapin (**14**) and gamatin[6] in wavelength and intensities, and indicate a furoflavone skeleton for pongaglabrone.

The position of the methylenedioxy group and the furan ring has been determined by alkaline treatment. Fission of **22** with absolute alcoholic potash yields piperonylic acid and 5-acetyl-4-hydroxycoumarone (**23**). The latter, which was obtained earlier by direct alkali fission of pongamol[27] has now been prepared by a modified method.[5] Demethylation of pongamol gives the furo(2″,3″–7,8)flavone (lanceolatin B) (**13**).[28, 29] Fission of **13**[2] yields the *o*-hydroxy ketone (**19**). Thus, pongaglabrone may be represented as 3′,4′-methylenedioxyfuro(2″,3″–7,8)flavone (**22**).

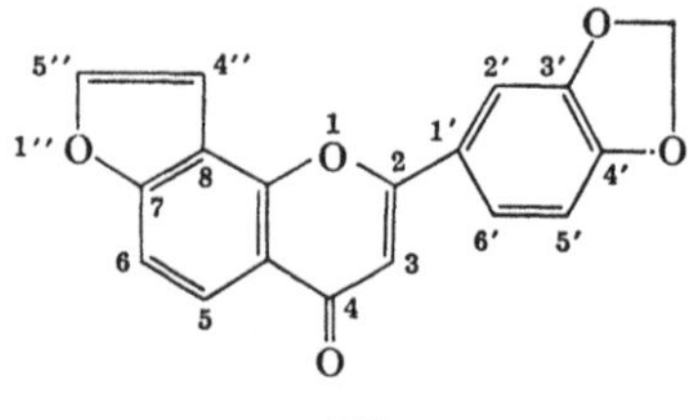

(**22**)

The structure assigned to pongaglabrone (**22**) has been confirmed by synthesis.[5] The diketone (**24**), obtained from the piperonylic ester (**23**), may be converted into **22** by heating with hydrochloric acid in acetic acid (Eq. 6).

A total synthesis of **22** involves 8-allyl-7-hydroxy-3′,4′-methylenedioxyflavone (**29**), which has been obtained from 4-*O*-allylresacetophenone (**25**). Pyridine chloride or acetone, potassium carbonate, and acid chloride converted **25** into the *O*-piperonyl derivative (**26**), which was

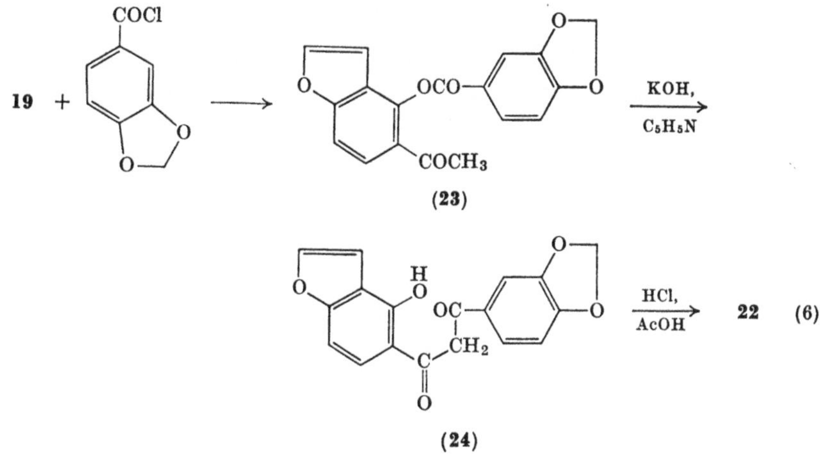

smoothly rearranged to the diketone (**27**). Subsequent cyclization proceeds easily to the flavone (**28**). Application of the Claisen migration to **28** yields **29,** which after ozonolysis, and ring closure of the resulting aldehyde (**30**) with polyphosphoric acid, yields **22** (Eq. 7).

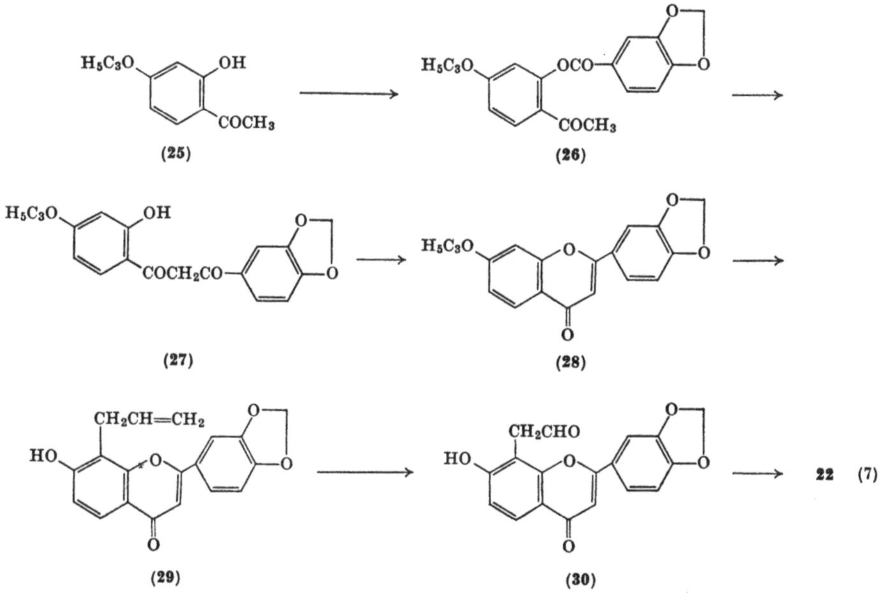

F. Atanasin

Atanasin (**31**), $C_{20}H_{18}O_8$, is isolated, together with penduletin (5,4′-dihydroxy-3,6,7-trimethoxyflavone), from *B. squarrosa*, commonly named *Atansia amarga* (*Brickelia* genus).[30, 31] It contains three methoxyl groups and forms a diacetate which accounts for five of the oxygen atoms. The ultraviolet spectrum, and the effect on this of sodium acetate and sodium methoxide, together with the positive ferric reaction and the formation (in mild conditions) of a monoacetate, strongly suggested the presence of a 5-hydroxyflavone nucleus.[32] Demethylation with hydriodic acid is accompanied by loss of two carbon atoms and an oxygen atom (C_2H_4O, residue), the product being 5,6,7,4′-tetrahydroxyflavone. This reaction confirms the relative positions of the substituents in atansain (**31**). Ozonolysis of **31** yields anisic acid; thus locating one methoxyl group at the 4′-position, while the spectroscopic data indicated the presence of a second methoxyl group at the 3-position. These facts all point to structure **31** for atanasin, which is therefore related on one hand, to penduletin, and on the other hand, to certain coumarins.

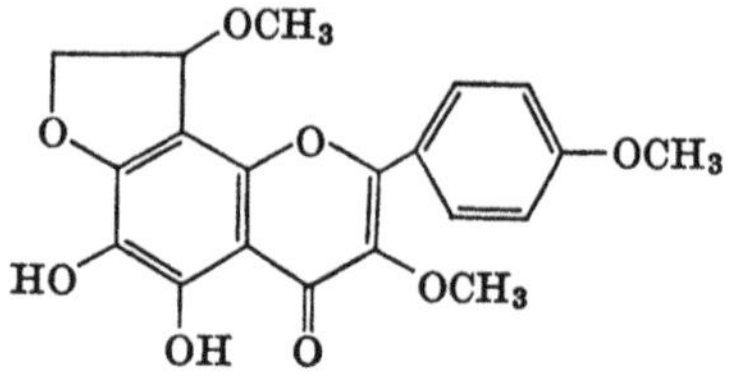

(**31**)

G. Gamatin

Gamatin (5-methoxyfuro(3″,2″–6,7)flavone) (**33**), $C_{18}H_{12}O_4$, was isolated from the root bark of *Pongamia pinnata* (L.), Merr. by extraction with petroleum ether and chromatography on alumina.[1] It contains one methoxyl group. Treatment with alcoholic potash produced benzoic acid, but the phenolic fraction was a mixture. Unlike karanjin (**1**), it belongs to the linear type. It has been synthesized by means of a Baker–Venkataraman reaction conducted on the benzoate of visnaginone (**32**)[6, 33] (Eq. 8).

A more convenient, alternative, synthesis of **33** has recently been reported[3] in which the appropriate flavone is first prepared and the furan built finally. Chrysin was converted into the 7-*O*-allyl-5-*O*-methyl derivative (**34**) by partial allylation of the 7-position, followed by

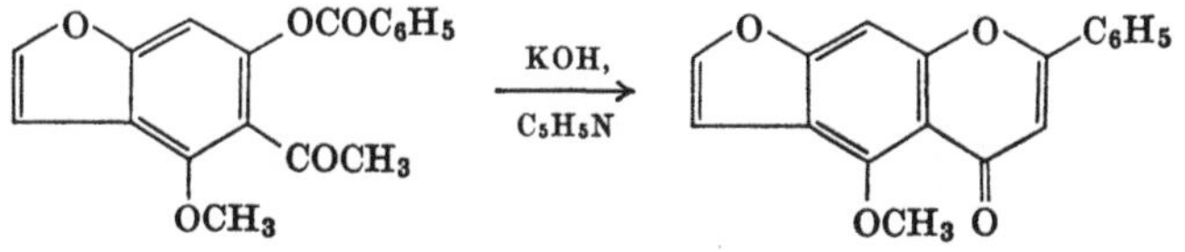

(32) (33) (8)

methylation of the 5-position. It underwent Claisen migration, yielding 8-allyl-7-hydroxy-5-methoxyflavone (**35**). The furan ring was built up after, using the procedure described by Seshadri and coworkers.[21] Ozonolysis of **35** proceeded in formic acid, and the aldehyde (**36**), so obtained, was cyclized with polyphosphoric acid giving 5-methoxy-2-phenylfuro(2,3-h)-1-benzopyran-4-one (5-methoxyfuro(3″,2″–6,7)-flavone) (**33**).

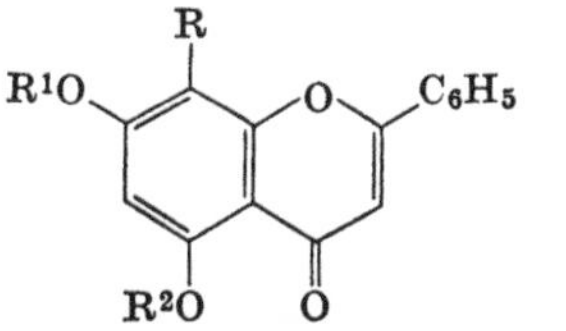

(**34**) $R = H, R^1 = C_3H_5, R^2 = CH_3$
(**35**) $R = C_3H_5, R^1 = H, R^2 = CH_3$
(**36**) $R = CH_2CHO, R^1 = H, R^2 = CH_3$

Selenium dioxide in amyl alcohol effects cyclization of 5-cinnamoyl-6-hydroxy-4-methoxycoumarone[34] (Eq. 9).

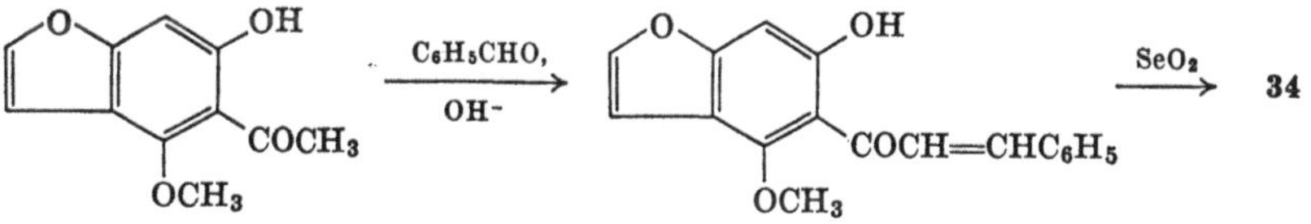

(9)

H. Pinnatin

Pinnatin (3′,4′-methylenedioxygamatin) (**37**), $C_{19}H_{12}O_6$, isolated from the root bark of *Pongamia pinnata* (L.) Merr[1] is colorless, and possesses one methoxyl group. Piperonylic acid could be isolated upon treatment of **37** with alkali. Synthesis has been achieved according to the sequence of reactions in (Eq. 10).[6]

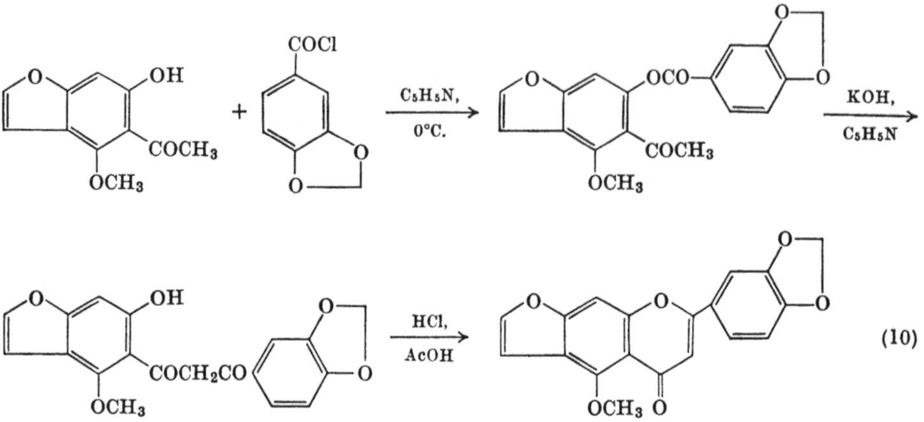

(37)

(10)

IV. Synthetic Furoflavones

1. Linear-type

The furoflavones (**39a–39e**) have been synthesized by subjecting the benzoyl, and piperonyl derivatives of 5-acetyl-6-hydroxy-7-methoxycoumaran to a Baker–Venkataraman migration, followed by cyclization of the resulting diketone, and dehydrogenation of **38** with a Pd/C Linstead catalyst[35] (Eq. 11).

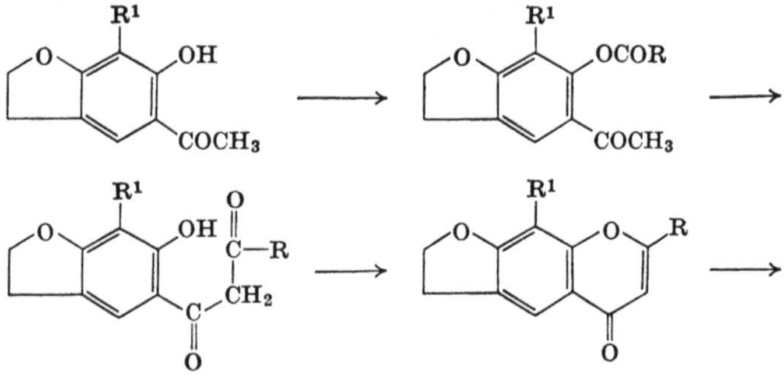

(38)

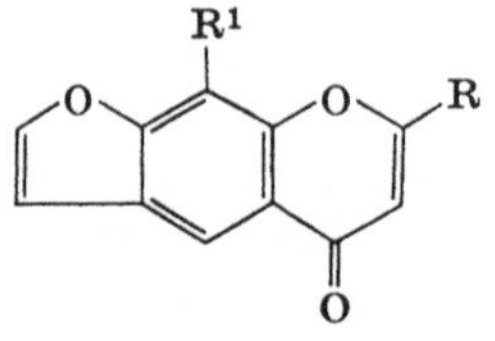

(**39a**) $R = C_6H_5$, $R^1 = OCH_3$;
(**39b**) $R = C_6H_3{=}O_2CH_2(3'{:}4')$, $R^1 = OCH_3$;
(**39c**) $R = C_6H_3{=}O_2CH_2(3'{:}4')$, $R^1 = H$; (11)
(**39d**) $R = C_6H_3(OCH_3)_2(3'{:}4')$, $R^1 = H$;
(**39e**) $R = C_6H_5$, $R^1 = H$

Starting with 6-hydroxycoumaran and methoxyacetonitrile, 6-hydroxy-5-(ω-methoxyacetyl)coumaran was obtained by a Hoesch condensation. The ketone was subjected to an Allan–Robinson condensation with benzoic anhydride and sodium benzoate to yield 3-methoxy-(**40a**), and 3-hydroxy-4″,5″-dihydrofuro(3″,2″–6,7)flavone (**40b**). Dehydrogenation was accompanied by demethylation to the furoflavone (**41b**). Its methyl ether (**41a**) was readily obtained from (**41b**) on methylation[36] (Eq. 12).

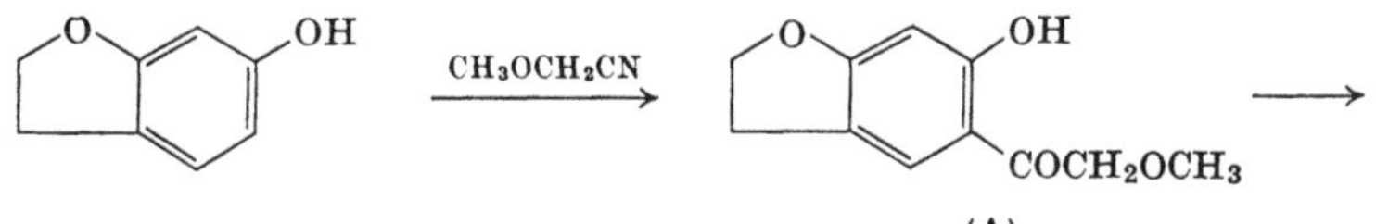

(**A**)

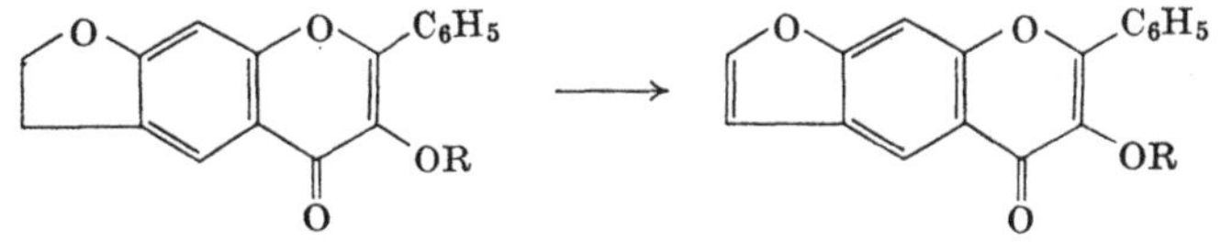

(12)

(**40**) (**41a**) R=CH_3;
(**41b**) R=H

Aroylation of the ketone (**A**), obtained by Row and Rao's[17] method with veratric anhydride and piperonylic anhydride resulted in the formation of the dihydrofuroflavones (**42a** & **42b**) respectively, which were readily dehydrogenated to the corresponding furoflavones (**43a** & **43b**).

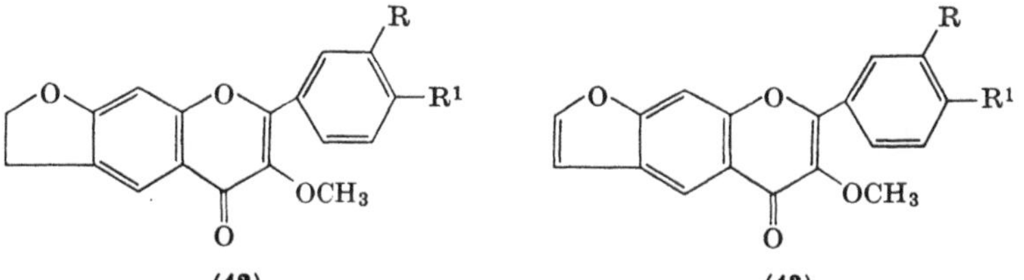

(**42**) (**43**)

(**a**) R=R^1=OCH_3; (**b**) R=R^1==O_2CH_2(3′:4′)

A number of substituted furoflavones (**45a–45c**)[37, 38] has been obtained by cyclization of the appropriate diketones, produced from benzoyl-, *p*-anisoyl-, and veratroyl derivatives of khellinone (**44a–44c**), respectively. On the other hand the furoflavanones (**46a–46c**) are readily obtained by treatment of the chalcones, produced by condensation of khellinone with aromatic aldehydes either in presence of alcoholic phosphoric acid,[37] or of selenium dioxide in amyl alcohol[38] (Eq. 13).

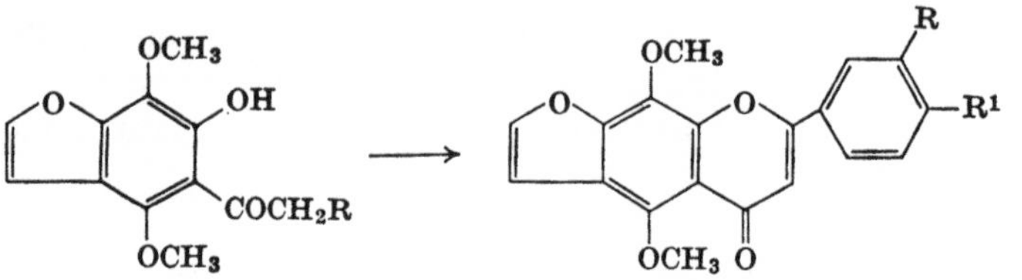

(13)

(44)

(45a) $R = H, R^1 = OCH_3$;
(45b) $R = R^1 = {=}O_2CH_2(3':4')$;
(45c) $R = R^1 = OCH_3$

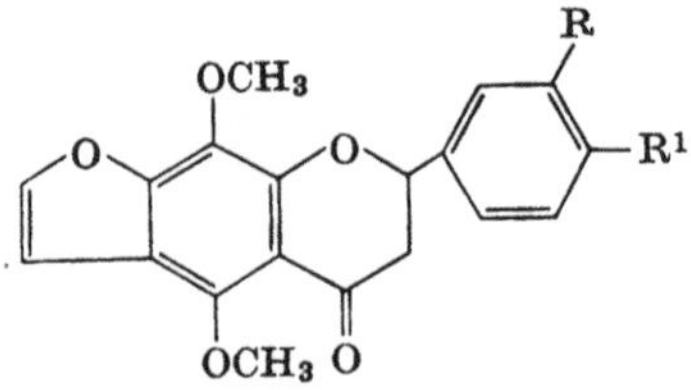

(46a) $R = H, R^1 = OCH_3$;
(46b) $R = R^1 = {=}O_2CH_2(3':4')$;
(46c) $R = R^1 = OCH_3$

Alford and Mentzer[39] have reported cyclization of the styryl ketones (**48**), obtained by condensation of tetrahydroeuparin (**47**) with aromatic aldehydes, to give the corresponding flavanones (**49**), which are converted to the flavones (**50**) (Eq. 14).

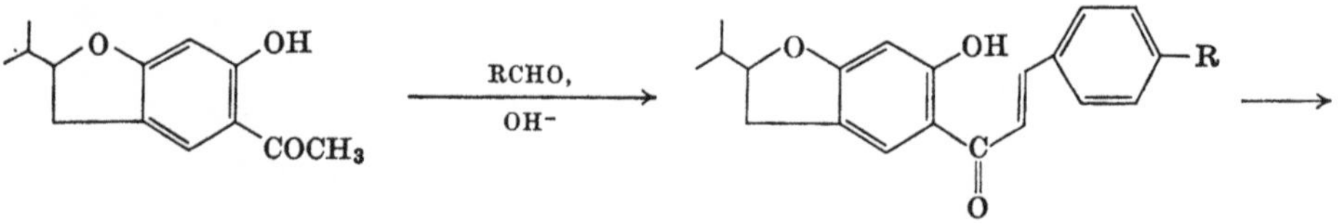

(47) (48)

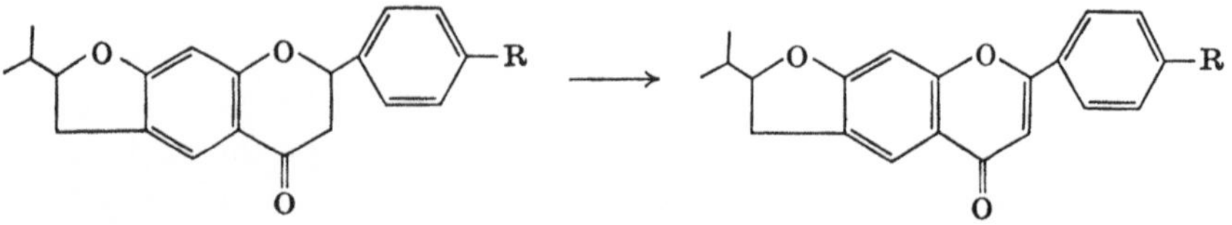

(14)

(49) (50)

Synthesis of 8-methoxyfuro(3″,2″–6,7)flavone (**51**) has recently been achieved.[40] Galloacetophenone is converted to 4-*O*-allyl-3-*O*-methyl-galloacetophenone, which is subjected to a Claisen migration to yield the 5-allyl compound. The flavone ring is built up by the Baker–Venkataraman transformation to produce 6-allyl-7-hydroxy-8-methoxyflavone. After ozonolysis of the latter compound, followed by ring closure of the resulting acetaldehyde derivative, **51** is obtained (Eq. 15).

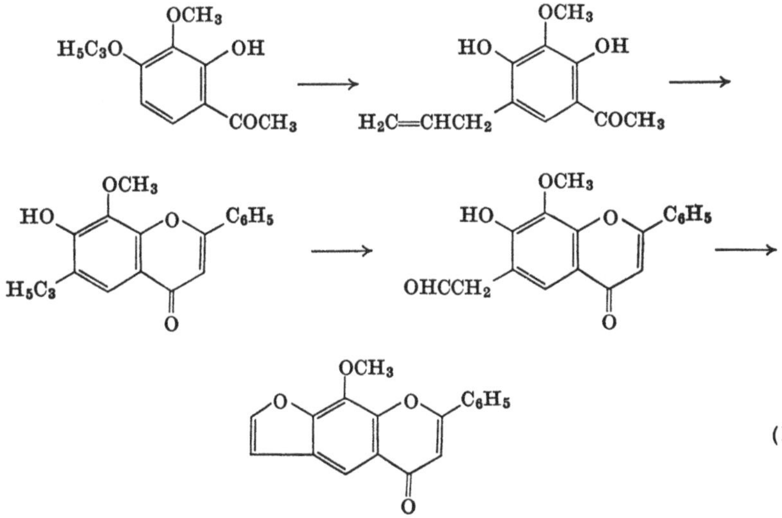

(15)

(51)

2. Angular-type

The furoflavones, having the furan ring condensed in the 5,6-position, have been similarly obtained starting with *O*-allyl ethers of tectochrysin (**52a**) and primetin monomethyl ether (**53a**), followed by ozonolysis of the *C*-allyl phenols (**52b**) and (**53b**). Furan-ring closure effected the formation of 7-methoxyfuro(2″,3″–5,6)-flavone (**54**), and 8-methoxyfuro(2″,3″–5,6) flavone (**55**) in poor yield.

Synthesis of α-methylfuroflavones, and their dihydro derivatives has been reported.[41] Ring closure of 8-allyl-7-hydroxyflavone derivatives is effected either by anhydrous hydrogen bromide, followed by pyridine, and dehydrogenation with *N*-bromosuccinimide, or by treatment with mercuric chloride. In the latter case, the mercury is replaced by iodine by

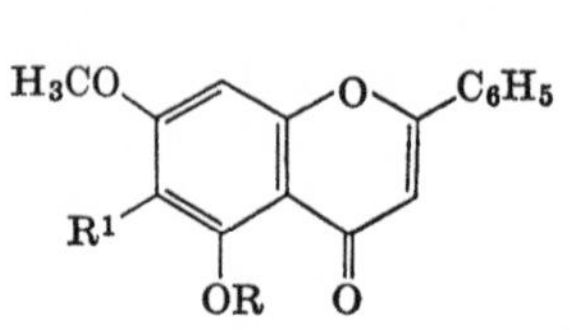

(52)

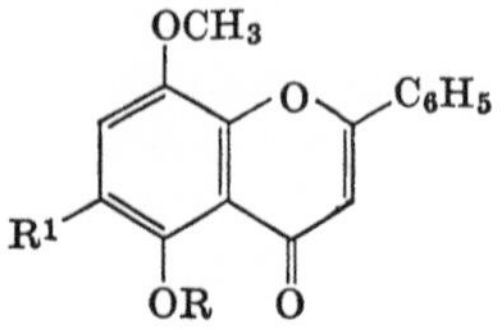

(53)

(a) $R=C_3H_5$, $R^1=H$;
(b) $R=H$, $R^1=C_3H_5$

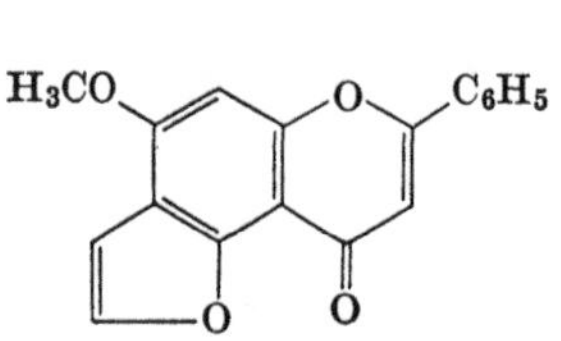

(54)

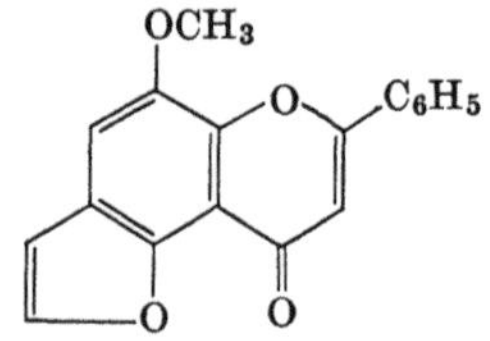

(55)

treatment with potassium iodide, and the iodomethylfuroflavone, so obtained, is then reduced with sodium in ethanol to the α-methyldihydrofuroflavone, namely, α-methyldihydrokaranjin[42] (Eq. 16, 17).

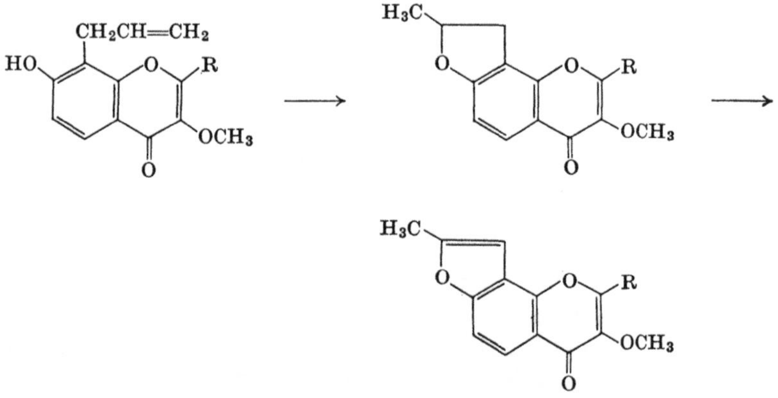

(16)

$R=C_6H_5$; $R=C_6H_4OCH_3$-*p*; $R=C_6H_3(OCH_3)_2(3':4')$, $R=C_6H_3{=}O_2CH_2(3':4')$

An alternative method[43] consists in introducing a chloroacetyl group into a hydroxybenzopyrone by the action of chloroacetyl chloride, followed by a Fries migration, accompanied by simultaneous ring closure (Eq. 18).

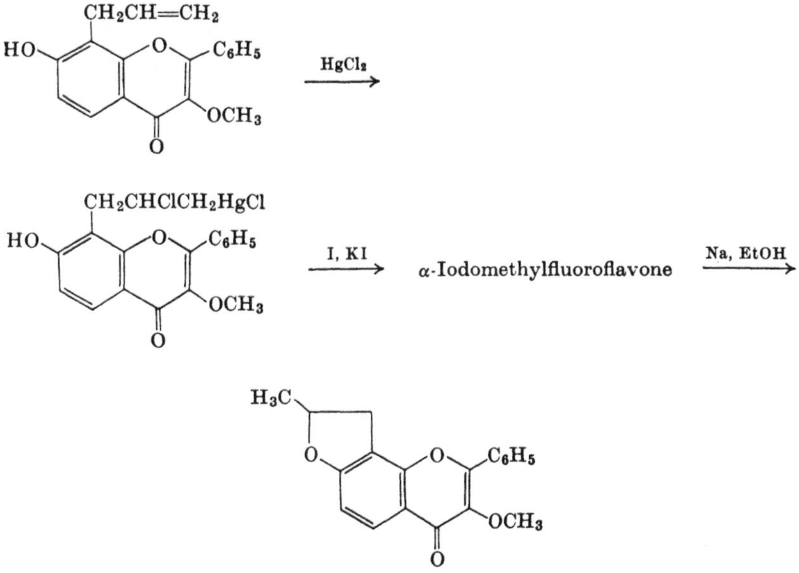

(17)

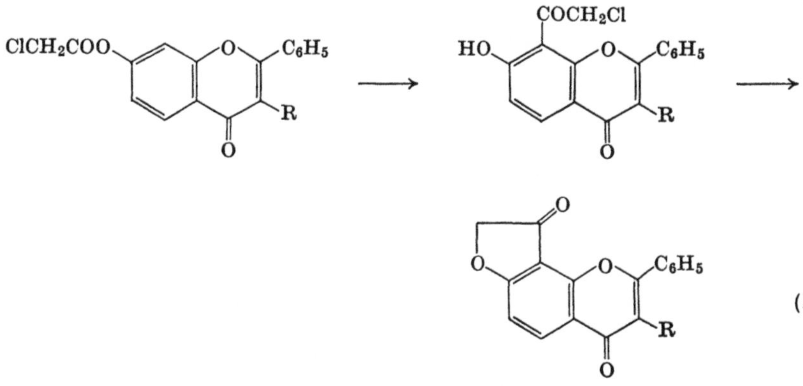

(18)

R = OH; R = H

Synthesis of angular furoflavones has been achieved by treatment of the diketone (**56**) with a glacial acetic and hydrochloric acid mixture to yield **57**,[6] or by cyclization of the styryl ketone (**58**), by treatment with bromine followed by alcoholic sodium hydroxide,[44, 45] to **59** (Eq. 19a, 19b).

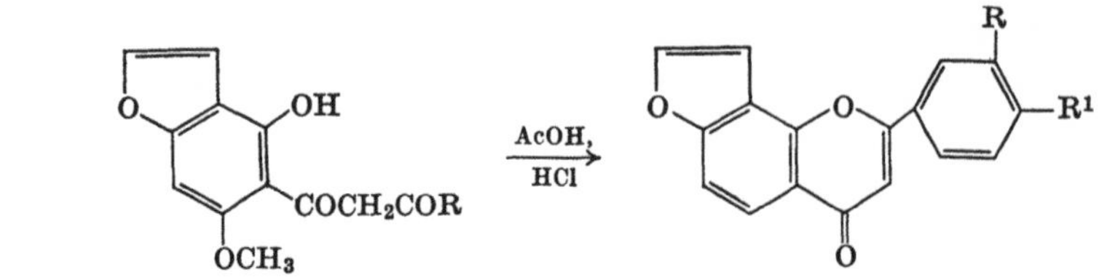

(56) (57)

(19a)

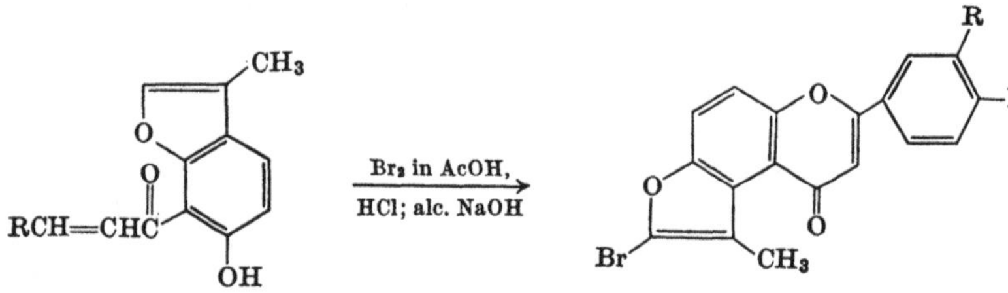

(58) (59)

(19b)

V. References

1. L. R. Row, *Australian J. Sci. Res.*, **5A**, 754 (1952); *Chem. Abstr.*, **47**, 12372 (1953).
2. S. K. Pavanaram, L. R. Row and G. S. R. S. Rao, *J. Sci. Ind. Res. (India)*, **19B**, 57 (1960); *Chem. Abstr.*, **55**, 24732 (1961).
3. R. Aneja, R. N. Khanna and T. R. Seshadri, *J. Chem. Soc.*, 163 (1963).
4. B. C. Rao and V. Venkateswarlu, *Curr. Sci.*, **27**, 482 (1958).
5. R. N. Khanna and T. R. Seshadri, *Tetrahedron*, **19**, 219 (1963); *Chem. Abstr.*, **58**, 11315 (1963).
6. S. K. Pavanaram and L. R. Row, *Australian J. Chem.*, **9**, 132 (1956).
7. T. A. Geissman, in *Modern Methods of Plant Analysis*, Vol. III, (Ed. K. Paech and M. V. Tracey), Julius Springer, Berlin, 1955, p. 485.
8. D. B. Limaye, *Rasayanam*, **1**, 1 (1936); **1**, 119 (1937); **1**, 161 (1939).
9. G. D. Beal and M. C. T. Katti, *J. Am. Pharm. Assoc.*, **14**, 1086 (1926); *Chem. Zentr.*, **II**, 596 (1926).
10. S. K. K. Jatkar and B. N. Mattoo, *J. Indian Chem. Soc. Ind. and News Ed.*, **17**, 39 (1954); *Chem. Abstr.*, **49**, 4307 (1955).
11. B. L. Manjunath, A. Seetharamiah and S. Siddapa, *Chem. Ber.*, **72**, 93, 97 (1939).
12. T. R. Seshadri and V. Venkataraswarlu, *Proc. Indian Acad. Sci.*, **13A**, 414 (1941); *Chem. Abstr.*, **35**, 7961 (1941); **17A**, 16 (1943); *Chem. Abstr.*, **37**, 4398 (1943).
13. S. Rangaswami and T. R. Seshadri, *Proc. Indian Acad. Sci.*, **9A**, 259 (1939); *Chem. Abstr.*, **34**, 103 (1940).
14. B. Krishnaswamy and T. R. Seshadri, *Proc. Indian Acad. Sci.*, **15A**, 437 (1942); *Chem. Abstr.*, **36**, 7024 (1942).

15. R. T. Foster and A. Robertson, *J. Chem. Soc.*, 115 (1948).
16. D. D. Pratt and R. Robinson, *J. Chem. Soc.*, 168 (1925).
17. L. R. Row and D. V. Rao, *J. Sci. Ind. Res. (India)*, **17B**, 199 (1958); *Chem. Abstr.*, **53**, 5255 (1959).
18. Y. Kawase, T. Matsumoto and K. Fukui, *Bull. Chem. Soc. Japan*, **28**, 273 (1955); *Chem. Abstr.*, **52**, 4595 (1958).
19. L. R. Row and T. R. Seshadri, *Proc. Indian Acad. Sci.*, **33A**, 168 (1951).
20. S. Tanaka, *J. Am. Chem. Soc.*, **73**, 872 (1951).
21. R. Aneja, K. Mukerjee and T. R. Seshadri, *Tetrahedron*, **2**, 203 (1958); *Chem. Abstr.*, **52**, 18387 (1958).
22. K. S. Raizada, P. S. Sarin and T. R. Seshadri, *J. Sci. Ind. Res. (India)*, **19B**, 76 (1960).
23. N. V. S. Rao, J. Veerabhadrarao and T. R. Seshadri, *Proc. Indian Acad. Sci.*, **10A**, 65 (1939); *Chem. Abstr.*, **34**, 1663 (1940).
24. S. Rangaswami and B. V. R. Sastry, *Curr. Sci. (India)*, **24**, 13 (1955); *Chem. Abstr.*, **49**, 10278 (1955).
25. B. C. Rao and V. Venkateswarlu, *Curr. Sci. (India)*, **27**, 471 (1958).
26. L. H. Briggs, L. D. Colebrook, H. M. Fales and W. C. Wildman, *Anal. Chem.*, **29**, 904 (1957).
27. B. Rao and V. Venkateswarlu, *Curr. Sci.(India)*, **25**, 357 (1956).
28. S. Rangaswami and T. R. Seshadri, *Proc. Indian Acad. Sci.*, **15A**, 421 (1942).
29. S. Narayanswami, S. Rangaswami and T. R. Seshadri, *J. Chem. Soc.*, 1871 (1954).
30. S. E. Flores and J. Herrán, *Tetrahedron*, **2**, 308 (1958).
31. S. E. Flores and J. Herrán, *Chem. Ind. (London)*, 291 (1960); *Chem. Abstr.*, **54**, 15376 (1960).
32. O. Kubota and A. G. Perkin, *J. Chem. Soc.*, 127, 1189 (1925).
33. S. K. Pavanaram and L. R. Row, *Nature*, **176**, 1177 (1955); *Chem. Abstr.*, **50**, 10716 (1956).
34. A. Schönberg, N. Badran and N. A. Starkowsky, *J. Am. Chem. Soc.*, **77**, 5390 (1955).
35. L. R. Row, C. Rukmini, G. S. R. S. Rao and C. Srinivasulu, *Indian J. Chem.*, **1**, 521 (1963).
36. S. K. Pavanaram and L. R. Row, *J. Sci. Ind. Res. (India)*, **14B**, 157 (1955); *Chem. Abstr.*, **50**, 12040 (1956).
37. J. R. Clarke and A. Robertson, *J. Chem. Soc.*, 302 (1949).
38. A. Stener, *Farmaco (Pavia) Ed. Sci.*, **15**, 642 (1960); *Chem. Abstr.*, **58**, 497 (1963).
39. S. Alford and Ch. Mentzer, *Compt. Rend. Sci. Paris*, **258**, 2854 (1964).
40. R. N. Khanna and T. R. Seshadri, *Indian J. Chem.*, **1**, 385 (1963); *Chem. Abstr.*, **60**, 1724 (1964).
41. S. S. Chibber, A. K. Ganguli, S. K. Mukerjee and T. R. Seshadri, *Proc. Indian Acad. Sci.*, **46A**, 19 (1957); *Chem. Abstr.*, **52**, 4617 (1958).
42. B. Krishnaswamy and T. R. Seshadri, *Proc. Indian Acad. Sci.*, **13A**, 43 (1941).
43. L. R. Row and T. R. Seshadri, *Proc. Indian Acad. Sci.*, **11A**, 206 (1939); *Chem. Abstr.*, **34**, 5446 (1940).
44. S. D. Limaye ahd K. G. Marathe, *Rasayanam*, **2**, 6 (1950); *Chem. Abstr.*, **45**, 7115 (1951).

45. M. G. Marathy, *Sci. Cult.* (*Calcutta*), **17**, 86 (1951); *Chem. Abstr.*, **46**, 11188 (1952).
46. S. K. Pavanaram and L. R. Row, *Curr. Sci.* (*India*), **24**, 301 (1955); *Chem. Abstr.*, **51**, 4365 (1957).
47. A. Kogure and T. Kubota, *J. Inst. Polytech. Osaka City Univ. Ser. C*, **2**, 70 (1952); *Chem. Abstr.*, **47**, 10525 (1953).
48. S. D. Limaye and G. Bhide, *Rasayanam*, **2**, 15 (1950); *Chem. Abstr.*, **45**, 7115 (1951).

CHAPTER VI

The Furoisoflavanoids

I. Introduction

There are many compounds that have structural connections with the true isoflavones. Comparison of the structures shows some striking similarities. The repetition of various oxygenation patterns is particularly interesting, and comparison with the structures of the families of naturally occurring furoisoflavones suggests that they may be regarded as being derived from 3-phenylchroman, and may have a similar or common biosynthetic origin. The natural occurrence of a 3-arylcoumarin has recently been detected by the discovery of pachyrrhizin. Coumestrol, wedelolactone, erosnin and its relatives, are also derivatives of 3-arylcoumarin, but their structures contain a modification which makes them analogous also to the flavan-4-ol derivatives, for example pterocarpin, homopterocarpin, and especially the rotenoids.

II. Furoisoflavanones

1. Naturally Occurring Furoisoflavanones

While the rotenoids, which are complex isoflavanones, have been known for a considerable time, and have been studied thoroughly, only two furoisoflavanones have as yet been found amongst natural products, and those only recently.

Investigations of the root of the leguminous plant, *Neorautanenia pseudopachyrrhiza*, has brought to light two new furoisoflavanones, namely, nepseudin and neotenone.

A. Nepseudin

Nepseudin (**1**), $C_{20}H_{18}O_6$, contains three methoxyl groups, and gave a blue–green Rogers and Calamari test, but a negative Durham reaction. The ultraviolet spectrum resembles that of (±)-neotenone (**2**), and

TABLE 1. The Furoisoflavanoids

Substituents	M.p. (°c)	Solvent for crystallization[a]; appearance	Remarks (u.v. spectrum, λ_{max} (mμ) (log ϵ))	References
A. *Furoisoflavones and Furiosoflavanones*				
(a) Furo(3″,2″–6,7)isoflavones				
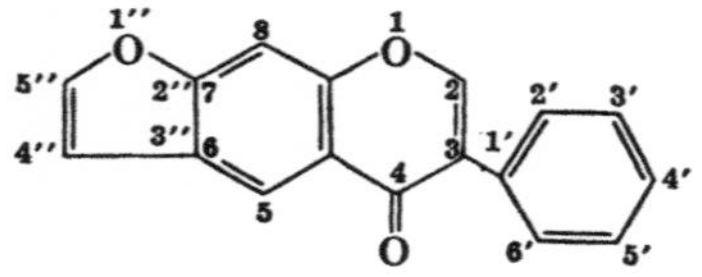				
Unsubstituted	202–203; 212–213	J; pale brownish-yellow	solution in H_2SO_4 acquires light-violet fluorescence; 250(4.60), 325(3.89)	28, 29
4″,5″-Dihydro	192–193	J	solution in H_2SO_4 acquires bright-violet fluorescence; 248(4.43), 311(4.07)	28–30
2,3,4″,5″-Tetrahydro	134	O	bright-yellow solution in H_2SO_4	30
2-Me	230–232	J	weak fluorescence in H_2SO_4	28
2-Me, 4″,5″-dihydro	138–139	J; pale yellow	bright-violet fluorescence in H_2SO_4	28
2-Me, 2′-OMe, 4″,5″-dihydro	152–153	J	intense-violet fluorescence in H_2SO_4	31
2-Me, 3′,4′-di-OMe, 4″,5″-dihydro	193–195	J	weak-violet fluorescence in H_2SO_4	31
2-Me, 3′,4′-(:O_2CH_2)	204–205	J	green-blue color changing when warmed with gallic and sulfuric acid mixture	31
2-Me, 3′,4′-(:O_2CH_2), 4″,5″-dihydro	184–185	J	violet fluorescence in H_2SO_4; green color with gallic and sulfuric acid mixture	31
2′,3′,4′-Tri-OMe (dehydronepseudin)	160; 164–165	O	244(4.59), 324(3.86)	2, 4, 5a
2′,3′,4′-Tri-OMe, 2,3-dihydro (nepseudin)	115–116	E/O	235(4.74), 257i(4.08), 272(3.94), 335(3.61)	2, 5a
2′,3′,4′-Tri-OMe, 4″,5″-dihydro	143–144	—		4
2′,4′,5′-Tri-OMe	190.5–191	J	236(4.44), 304(4.18)	23
2′,4′,5′-Tri-OMe, 4″,5″-dihydro	197–198	J	250i(4.32), 305(4.28)	23

Substituents	M.p. (°C)	Solvent for crystallization[a]; appearance	Remarks (u.v. spectrum, λ_{max} (mμ) (log ϵ))	References
2′-OMe, 4′,5′-(:O_2CH_2) (dehydro-neotenone)	240–241	G/J	239(4.61), 308(4.10)	5a
2′-OMe, 4′,5′-(:O_2CH_2), 2,3-dihydro (neotenone)	180–180.5	G/O	235(4.68), 275(3.83), 300(3.87), 335(3.61)	5a
(b) Furo(2″,3″–7,8)isoflavones				
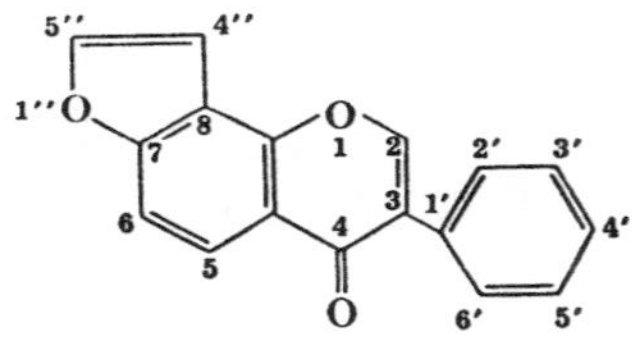				
Unsubstituted	152–152.5	J		11
2-Me	187–188; 95–96	K		8, 10
2′-OMe	182	J/K		12
5″-COOH	283(dec.)	A		11
5″-COOEt	172–173	J		11
2,5″-Di-Me	168–169	K/N		17
2,5″-Di-Me, 4″,5″-dihydro	148–150	K/N		17
2-Me, 2′-OMe	168	J, K		12
2-CH_2Br, 2′-OH	210(dec.)	K		14
2-CH_2OH, 2′-OH	224–225	J		14
2-CH_2OEt, 2′-OMe	119–120	J, K		12
2-Me, 5″-COOH	253–255; 295–297(dec.)	O		8, 10
2-Me, 5″-COOEt	171–172; 225–227	J		8, 10

(Table continued)

TABLE 1 (*continued*)

Substituents	M.p. (°c)	Solvent for crystallization[a]; appearance	Remarks (u.v. spectrum, λ_{max} (mμ) (log ϵ))	References
2′-OMe, 5″-COOH	273.5–274	J		12
2′-OMe, 5″-COOEt	164.5	J		12
2,5″-Di-Me, 5-OH, 4″,5″-dihydro	148–149	K/N		16
2,5″-Di-Me, 4′-OMe	185	O; yellowish		17
2,5″-Di-Me, 2′-OMe, 4″,5″-dihydro	160–161	K		17
2,5″-Di-Me, 4′-OMe, 4″,5″-dihydro	206–207	K		17
3′,4′-(:O_2CH_2)	200–201		227(4.49), 254(4.41), 294(4.05)	13
2-Me, 2′-OMe, 5″-COOH	294	J		12
2-Me, 2′-OMe, 5″-COOEt	161.5–162.5	J		12
2-CH_2OEt, 2′-OMe, 5″-COOH	233–234	J		12
2-CH_2OEt, 2′-OMe, 5″-COOEt	177.5–178.5	J		12
3′,4′-(:O_2CH_2), 5″-COOH	283–284	—	228(4.48), 253(4.60), 287.5(4.39)	13
3′,4′-(:O_2CH_2), 5″-COOEt	185–186	—	224(4.62), 248(4.53), 261(4.47), 295.5(4.37)	13
5″,5″-$(COOEt)_2$, 4″-OH, 3′,4′-(:O_2CH_2), 4″,5″-dihydro	226–227			13
5″,5″-$(COOEt)_2$, 4″-OAc, 3′,4′-(:O_2CH_2), 4″,5″-dihydro	143–144		223(4.51), 267(4.29), 295(4.20)	13
2,5″-Di-Me, 5-OEt, 4″,5″-dihydro	152–154	K/N		16
2,5″-Di-Me, 5-OH, 6-allyl, 4″,5″-dihydro	145–146	K/N; bright yellow	bluish-green color with H_2SO_4	16

B. *Coumaranochromans*

(a) 6a,11a-Dihydro-6H-benzofuro(3,2-c)–1–benzopyran

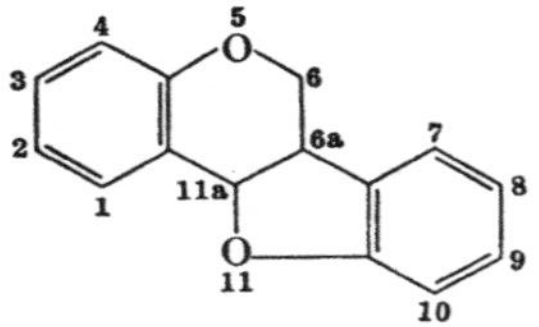

Substituents	M.p. (°C)	Solvent for crystallization[a]; appearance	Remarks (u.v. spectrum, λ_{max} (mμ) (log ϵ))	References
3-OH, 9-OMe(demethylhomopterocarpin)	194–195	N/O	206.5(4.80), 228s(4.24), 282(3.92), 310(2.01)	36, 66
3,9-Di-OMe (homopterocarpin)	82–83; 87	J, O or N/O	$(\alpha)^{20.5}_{5461} -236.6°$	36, 66
3-OAc, 9-OMe	105–106	J	206(4.85), 226s(4.21), 285.5(4.06), 320(3.03)	36
3-OMe, 8,9-(:O_2CH_2) (pterocarpin)	164.5	J or O	$(\alpha)^{20.5}_{5461} -207.5°$	66
3-OH, 8,9-(: O_2CH_2) (*d*-maackiain)	180–181	P	281(3.58), 287(3.65), 310(3.86); $(\alpha)^{26}_{D} +259°$ (C)	48
3-*O*-Glucosyl, 8,9-(:O_2CH_2) (*d*-maackiain-mono-β-D-glucoside or sophojaponicin B_2)	202–204(dec.)	P	279s(3.57), 284(3.63), 309(3.89); $(\alpha)^{17}_{D} -104°$ (A)	47, 48
3-OMe, 8,9-(:O_2CH_2) (*d*-maackiain methyl ether or *d*-pterocarpin)	159–160	P	280(3.58), 286(3.65), 310(3.87); $(\alpha)_D +232°$ (G)	48
3-OMe, 8,9-(:O_2CH_2) (*l*-maackiain methyl ether or *l*-pterocarpin)	158; 159–160	—	$(\alpha)^{20.5}_{5461} -208°$	48, 54, 58
3-*O*-Glucosyl, 8,9-(:O_2CH_2) (*l*-maackiain-mono-β-D-glucoside, sophojaponicin B_1, or trifolirhizin)	145(dec.)	P	280(3.61), 284(3.66), 310(3.86); $(\alpha)_D -185°$ (A)	47, 48, 147
Sophojaponicin tetraacetate (B_2)	165–166	J	278s(3.60), 284(3.65), 310(3.88); $(\alpha)_D -82°$ (A)	48
Sophojaponicin tetraacetate (B_1)	187	J	280(3.60), 284(3.65), 310(3.88); $(\alpha)_D -124°$ (C)	48
3-OH, 8,9-(:O_2CH_2) (*dl*-maackiain)	195–196	P	281(3.58), 287(3.65), 310(3.86)	48
3-OMe, 8,9-(:O_2CH_2) (*dl*-pterocarpin or *dl*-maackiain methyl ether)	185–186	C/P	285(3.58), 286(3.65), 310(3.87)	48
3-OAc, 8,9-(:O_2CH_2) (*dl*-maackiain acetyl deriv.)	159–160	P	279s(3.54), 283(3.59), 309(3.88)	48
3-$OCOC_6H_4NO_2$-*p*, 8,9-(:O_2CH_2)	263–264(dec.)	I; yellow-orange	259(4.19), 307(4.02)	48
3-OMe, 6a-OH, 8,9-(:O_2CH_2) (pisatin)	72	O	213(4.75), 280(3.62), 286(3.68), 309(3.86); $(\alpha)^{20}_{364} +940°$	44, 58
dl-Pisatin	188–190	G/O	285.5(3.56), 309(3.94);	43

(*Table continued*)

TABLE 1 (*continued*)

Substituents	M.p. (°c)	Solvent for crystallization[a]; appearance	Remarks (u.v. spectrum, λ_{max} (mμ) (log ϵ))	References
(b) 6*H*-Benzofuro(3,2-c)–1–benzopyran				
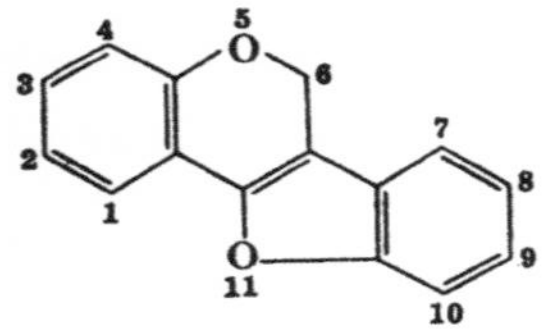				
3-OMe, 8,9-(:O_2CH_2) (anhydropisatin or *O*-methylanhydrosophorol)	72	o	215(4.49), 234(4.30), 244(4.2), 257(4.1), 291(3.8), 339(4.58)	44, 58
(c) Furo(3′,2′–6,7)benzofuro(3″,2″–3,4)chroman				
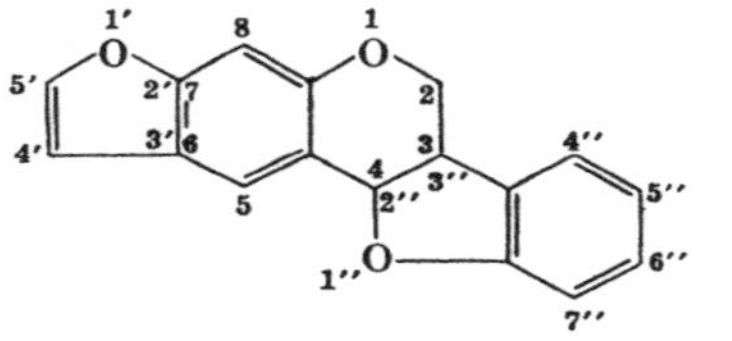				
5″,6″-(:O_2CH_2) (neodulin)	225	o	250(ϵ13,680), 257(ϵ12,500), 305(ϵ12,600)	60

Substituents	M.p. (°c)	Solvent for crystallization[a]; appearance	Remarks (u.v. spectrum, λ_{max} (mμ) (log ϵ))	References
(d) Furo(3,2-c:5,4-f′)benzopyran 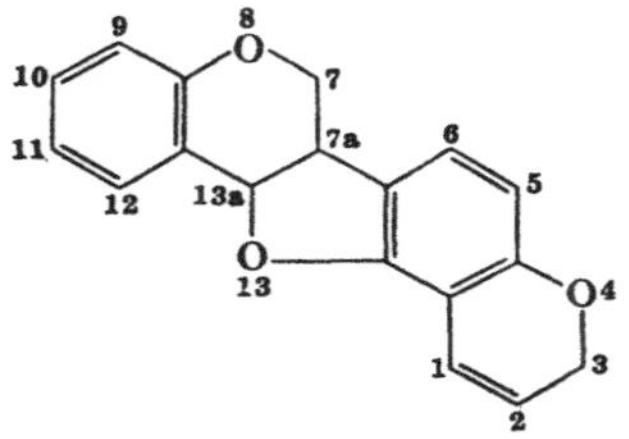				
3,3-Di-Me, 10-OH (phaseollin)	177–178	J	279(3.96), 286(3.90), 315(3.34); $(\alpha)^{20}_{578} - 145°$ (J)	67, 71
(e) 6*H*-Benzofuro(2,3-c)–1–benzopyran 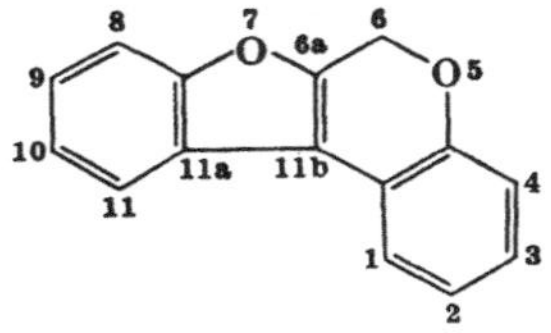				
3-OMe	69–71	J	231(4.41), 265(4.03), 301(3.84); red color with H_2SO_4; pyrylium perchlorate (brownish-yellow, m.p. 220°(dec.))	151
Cis-3-OMe, 6a,11b-dihydro	95–96	J	282(3.58); dark brown color with H_2SO_4; pyrylium perchlorate (m.p. 220° (dec.))	151

(*Table continued*)

TABLE 1 (*continued*)

Substituents	M.p. (°c)	Solvent for crystallization[a]; appearance	Remarks (u.v. spectrum λ_{max} (mμ) (log ϵ))	References
1,2-Di-OMe	105–106	J	220(4.46), 272(4.03); pyrylium perchlorate (orange-red, m.p. 232° (dec.))	151
3,9-Di-OMe	140	J	242(4.44), 248(4.44), 271(3.84), 290(3.98); pink color with H_2SO_4 with green fluorescence; crimson red with HNO_3; pyrylium ferrichloride (light orange, m.p. 215°); pyrylium perchlorate (brown, m.p. 239°(dec.))	151
Cis-3,9-di-OMe, 6a,11b-dihydro	125	J	287(3.84)	151
1,2,9-Tri-OMe	127	J	242(4.40), 288(3.96); pink color with H_2SO_4; pyrylium perchlorate (greyish-brown, m.p. 231°(dec.))	151
Cis-1,2,9-tri-OMe	104	J	286(3.63)	151
C. *Coumaronoflavans*				
(a) 11*H*-Benzofuro(3,2-b)–1–benzopyrans				
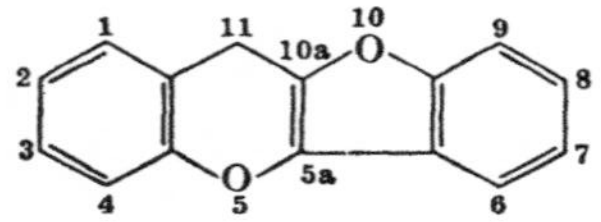				
8-OMe	108	P		108, 148
3,8-Di-OMe	139–140	P		108, 148
1,3,8-Tri-OMe	160–162	P/R		108,148
3,8-Di-OAc	160	G/P	248(4.44), 321(4.35), 370(4.44)	81
1,3,9,11-Tetra-OH, 5a,10a-dihydro (cyanomaclurin)	above 360	N	281(−); $(\alpha)_D^{21}$ +204°; (±)-trimethyl ether (m.p. 145°); (+)-trimethyl ether (m.p. 75°; $(\alpha)_D^{19}$ +135°)	76

Substituents	M.p. (°c)	Solvent for crystallization[a]; appearance	Remarks (u.v. spectrum, λ_{max} (mμ) (log ϵ))	References
(b) 11*H*-Benzofuro(3,2-b)–1–benzopyranyl-5-ium compounds				
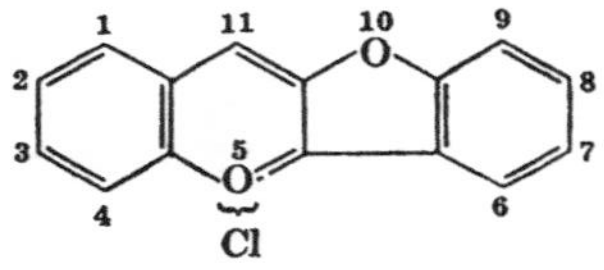				
8-OH	200(dec.)	dull red	231(−), 427(−)	81
3,8-Di-OH	>290	red	260(−), 465(−)	81
3,8-Di-OMe	200	bright yellow	255(−), 420(−)	81
1,3,8-Tri-OH	280	deep red	320(−), 495(−)	81
1,3-Di-OH, 8-OMe	>200	deep red	240(−), 270(−), 310(−), 425(−), 495(−)	81
1,3-Di-OMe, 8-OH	170	deep red	270(−), 320(−), 485(−)	81
1,3,8-Tri-OMe	>270	deep red	270(−), 330(−), 420(−), 485(−)	81
(c) 11*H*-Benzofuro(3,2 b)–1–benzopyran-11-ones				
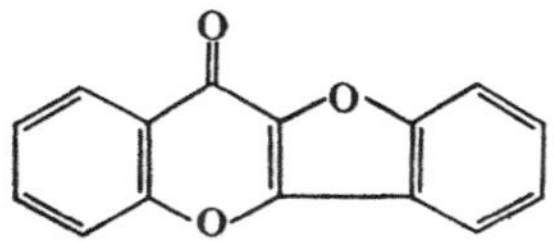				
3-OH	315–317	G/J	252(ϵ14,900), 271(ϵ,7,450), 316(ϵ29,600)	87
6-OH	>340	—	acetate, (m.p. 207–208°); methyl ether, (m.p. 262–263°)	86
3-OMe	213–214	J	255(ϵ,15,750), 268(ϵ,4,980), 318(ϵ,32,000)	87
6,8-Di-OH	340–343	—	diacetate (m.p. 218–220°)	86
1,3-Di-OMe	262–263	G/J; salmon	259(ϵ,25,500), 309(ϵ,20,500), 330(ϵ,17,300)	87

(Table continued)

TABLE 1 (*continued*)

Substituents	M.p (°C)	Solvent for crystalization[a]; appearance	Remarks (u.v. spectrum, λ_{max} (mμ) (log ϵ))	References
D. *Coumaronocoumarins*				
(a) 6*H*-Benzofuro(3,2-c)–1–benzopyran-6-ones				
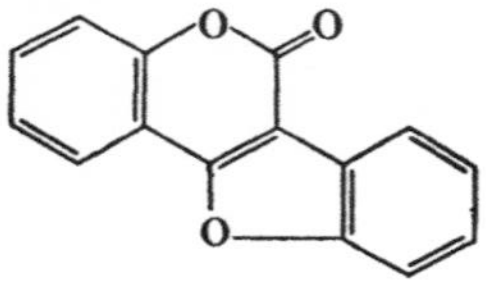				
Unsubstituted	176–177; 180–181; 181–182	A, J or P	233(4.20), 295(4.02), 310(4.10), 323(4.2)	98, 101, 104
3-OH	270–272; 274–276; 285	P; slightly brown	242(–), 297(–), 331(–), 347(–)	88, 101, 102
3-OMe	190; 192–193; 195–196	C/P; cream		88, 101, 102
3-OAc	208–209; 210–211	C/P	236(—), 295(–), 323(–), 340(–)	88, 100, 102
3,9-Di-OH (coumestrol)	> 350	J; yellow		95, 101, 102, 110
3-OH, 9-OMe	337–338	C	243(–), 303(–), 341(–)	93, 102
3-OH, 10-OMe	289	C/P; yellow	252(–), 261(–), 296(–), 330(–)	102
3-OMe, 9-OH	274; 275–276	C/P; pale yellow	243(–), 303(–), 342(–)	93,102
3-$OCH_2C_6H_5$, 9-OH	211	C/P	244(–), 303(–), 343(–)	93
1,3-Di-OMe	254	A		105
3,9-Di-OMe	197	A		105
3,10-Di-OMe	195	P; cream	250(–), 259(–), 294(–), 328(–), 344(–)	102
3-OMe, 9-OAc	208	C/P	240(–), 299(–), 333(–), 348(–)	93

Substituents	M.p. (°C)	Solvent for crystalization[a]; appearance	Remarks (u.v. spectrum, λ_{max} (mμ) log ϵ))	References
3-OAc, 9-OMe	240	C/P	243(−), 302(−), 337(−)	93
3-$OCH_2C_6H_5$, 9-OAc	205	C/P	240(−), 298(−), 333(−), 348(−)	93
3-$OCH_2C_6H_5$, 9-OMe	187	C/P	243(−), 303(−), 341(−)	93
3,9-Di-OAc	235–236; 227–228	D		96, 100
2-$CH_2CH{=}CMe_2$, 3,9-di-OH (psoralidin)	290–292; 315	C	208(ϵ,40,800), 244(ϵ,20,300), 305(ϵ,7,000); 347(ϵ,25,300); λ_{max}^{EtOH} in 0.1 N KOH: 256 (ϵ,14,700), 285(ϵ,18,500), 395(ϵ,26,900); violet fluorescence in dilute ethanol	114, 115
2-$CH_2CH_2CHMe_2$, 3,9-di-OH (dihydropsoralidin)	274–276; 294–295; 295–297	N, P; light yellow	210(ϵ,39,000), 244(ϵ,22,000), 305(ϵ,7,500), 345(ϵ,27,700); diacetate from ethanol, (m.p. 245–247)	32, 114, 120
2-$CH_2CH{=}CMe_2$, 3,9-di-OMe	190–191	C	208(ϵ,42,600), 243(ϵ,24,000), 304(ϵ,8,500), 344(ϵ,28,500); λ_{max}^{EtOH} in 0.1 N KOH: 224(ϵ,27,800), 304(ϵ,7,200), 345(ϵ,11,200); violet fluorescence in dilute ethanol	114, 115
2-$CH_2CH_2CHMe_2$, 3,9-di-OMe	193–195	C	209(ϵ,39,100), 243(ϵ,21,600), 304(ϵ,7,700), 344(ϵ,26,500)	32, 114
3,7,9-Tri-OH (trihydroxycoumestan)	346–348	P; cream	270(−), 303(−), 351(−); λ_{max} in EtOH-NaOAc: 272(−), 380(−); λ_{max} in EtOH-NaOEt: 392(−), 398(−)	109
3,7-Di-OH, 9-OMe (trifoliol)	332(dec.)	H	249(−), 270(−), 303(−); λ_{max} in EtOH-NaOEt: 295(−), 381(−)	109
3-OMe, 8,9-di-OH	>300	C; pale yellow	210(4.80), 250(4.34), 288(3.92), 301(3.94), 310(3.97), 355(4.45); violet color with $FeCl_3$; blue fluorescence in ethyl acetate and in acetone	103
3,9-Di-OH, 7-$OCOC_6H_5$	303(dec.)	C/P	239(−), 303(−), 344(−); λ_{max} in EtOH-EtONa: 282(−), 389(−)	109

(Table continued)

TABLE 1 (*continued*)

Substituents	M.p. (°C)	Solvent for crystallization[a]; appearance	Remarks (u.v. spectrum, λ_{max} (mμ) (log ϵ))	References
3,9-Di-OMe, 7-OH (trifoliol methyl ether)	209–212	P		109
3,7,9-Tri-OMe	255–258	H		109
3,8,9-Tri-OMe	247–248; 254	C, P	211(4.77), 246(4.30), 283.5(3.98), 297(3.89), 310(4.06), 348(4.52); blue fluorescence in ethyl acetate	24, 103
3-OMe, 8,9-(:O_2CH_2)	269–270	C	245(4.25), 283(3.87), 297(3.83), 309(3.99), 347(4.40); green color with gallic acid and blue fluorescence in ethyl acetate and in acetone	146
3,9-Di-OMe, 7-$OCOC_6H_5$	248	C/P	238(—), 300(—), 341(—)	109
3,9-Di-OAc, 7-OMe (trifoliol acetate)	243	C	245(—), 301(—), 336(—)	109
3-OMe, 8,9-di-OAc	225–226	C	240(4.34), 266(3.90), 288.5i(3.90), 299(4.12), 335(4.54), 342i(4.44), 351(4.45)	146
3,9-Di-OAc, 7-$OCOC_6H_5$	258	C/P	235(—), 329(—), 343(—)	109
3,8,9-Tri-OAc	255–257	G	236(3.34), 298(3.12), 328(3.42), 344(3.32)	111a, 103
3-OH, 8,9-(:O_2CH_2) (medicagol)	324–325	L/Q	245(4.22), 309(4.01), 347(4.43)	111a
3-OAc, 8,9-(:O_2CH_2)	262–263	B	239(4.25), 298(4.13), 333(4.52), 349(4.45)	111a
1,8,9-Tri-OH, 3-OMe (wedelolactone)	327–330	P; greenish yellow	211(ϵ,37,400), 250(ϵ,19,200), 300(ϵ,8,200), 350(ϵ,22,700)	94, 114
1,3,8,9-Tetra-OMe (tri-*O*-methyl-wedelolactone)	245–246; 247; 247–248	A, N	247(4.59), 300(4.15), 350(4.71)	94, 103–106, 108, 114
1,8,9-Tri-OEt, 3-OMe	200–201	A/J or C/J		103
1,8,9-Tri-OAc, 3-OMe	235–237; 243–244	A		95, 106
1,8,9-Tri-$OCOC_6H_5$, 3-OMe	267–269	M		94

8

Substituents	M.p. (°C)	Solvent for crystallization[a]; appearance	Remarks (u.v.spectrum, λ_{max} (mμ) (log ϵ))	References
(b) 6*H*-Benzofuro(2,3-c)–1–benzopyran-6-ones				
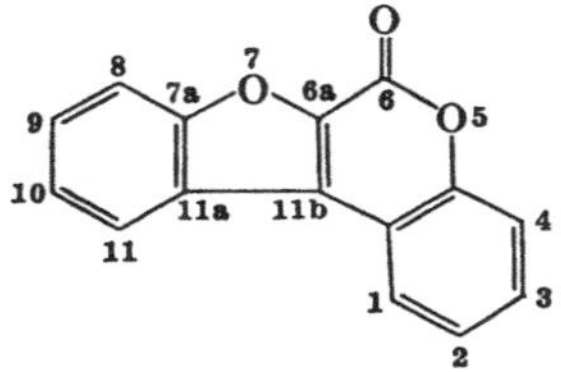				
2,10-Di-Me	274–274.5	A		150
2,10-Di-Me, 6a-OH, 6a, 11b-dihydro, 11b-COOMe	223	E	243(–), 284.5(–), 315i(–)	150
2,10-Di-Me, 6a-OMe, 6a,11b-dihydro,11b-COOH	256–260	A		150
2,10-Di-Me, 6a-OMe, 6a,11b-dihydro, 11b-COOMe	259–260	I		150
2,10-Di-Me, 6a-OAc, 6a,11b-dihydro, 11b-COOMe	177.5–178; 260(poly-morph.)	F, P		150
2,4,8,10-Tetra-Me, 6a-OH, 6a, 11b-dihydro, 11b-COOMe	224–226	E	242.5(–), 280.5(–), 287.5(–), 316i(–)	150
2,4,8,10-Tetra-Me, 6a-OMe, 6a,11b-dihydro, 11b-COOMe	286	E		150
2,3,9,10-Tetra-Me, 6a-OH, 6a,11b-dihydro, 11b-COOMe	223–225	E	242(–), 284.5(–), 320i(–)	150
2,3,9,10-Tetra-Me, 6a-OMe, 6a,11b-dihydro, 11b-COOMe	238	E		150
2,10-Di-OMe, 6a-OH, 6a,11b-dihydro, 11b-COOMe	207–209	E	248.5(–), 296.5(–), 320(–)	150
2,6a,10-Tri-OMe, 6a,11b-dihydro, 11b-COOMe	216.5–217	P		150

(Table continued)

TABLE 1 (*continued*)

Substituents	M.p. (°c)	Solvent for crystallization[a]; appearance	Remarks (u.v. spectrum, λ_{max} (mμ) (log ϵ))	References
(c) Furo(3′,2′–6,7)benzofuro(3″,2″–3,4)coumarin				
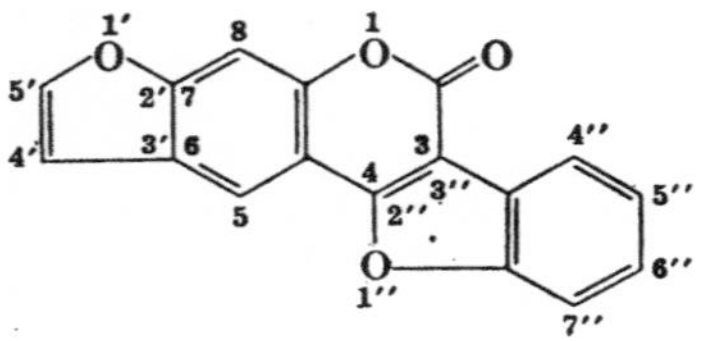				
5″,6″-Di-OH	300	—		124
5″,6″-Di-OMe	286–287	—	239(4.60), 280(3.98), 355(4.55)	124
5″,6″-Di-OMe, 4′,5′-dihydro	245	N; cream		108
5″, 6″-(:O_2CH_2) (erosnin)	330, 350	—	239(4.58), 285(3.95), 354(4.30), 366(4.26)	116, 124
5″,6″-(:O_2CH_2),4′,5′-dihydro (dihydroerosnin)	326–327	N; cream	245(4.17), 316(3.27), 357(3.91)	108
(d) Benzfuro(3,2-c)pyrano(3,2-g)–1–benzopyran-7-one				
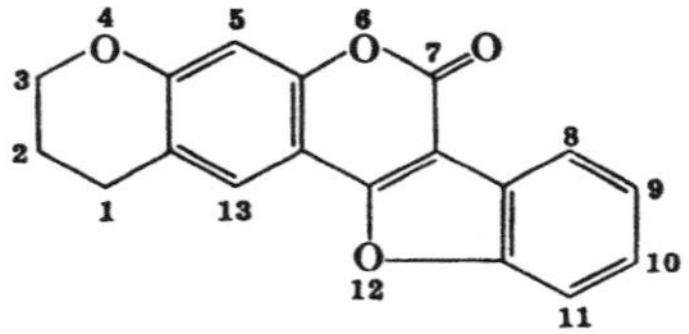				
3,3-Di-Me, 10-OH (isopsoralidin)	284–287	P	209(ϵ,44,700), 246(ϵ,22,400), 306(ϵ,8,200), 349(ϵ,29,100)	32, 114, 149
3,3-Di-Me, 10-OMe	172–174, 197–200, 198–205	C/P, E/O, N	211(ϵ,42,100), 244(ϵ,22,200), 306(ϵ,9,200), 346(ϵ,20,800)	32, 114, 120, 149
3,3-Di-Me, 10-OAc	220, 224	J		32, 114

Substituents	M.p. (°C)	Solvent for crystallization[a]; appearance	Remarks (u.v. spectrum, λ_{max} (mμ) (log ϵ))	References
(e) Furo(3,2-c: 4,5-g′)bis–1–benzopyran-6-one				
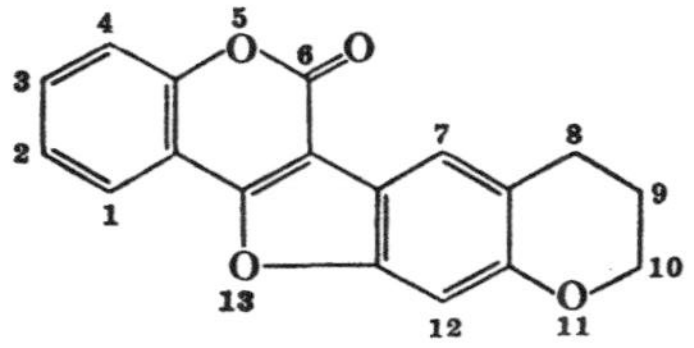				
10,10-Di-Me	187	A; pale cream		120
10,10-Di-Me, 3-OMe	220	A	250(4.02), 307(3.57), 346(4.08)	120
E. *Furo(3,2-c)–1–benzopyran-4-ones*				
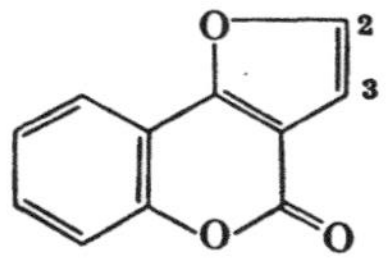				
Unsubstituted	90—91	—		155
2,3-Dihydro	143–144	—		155
8-Br	156–157	—		155
8-Br, 2,3-dihydro	162–163	—		155
8-NO_2	162–163	—		155
8-NO_2, 2,3-dihydro	186–187	—		155
2-Me, 3-Ph	199	—		156
2-$CONH_2$, 3-CN	350	H		157
2-HO(C_6H_4CO)-*o*	177–179	K	red-brown color with $FeCl_3$	133
2-AcO(C_6H_4C=NNHPh)-*o*	258–260	—		133

(Table continued)

TABLE 1 (*continued*)

Substituents	M.p. (°C)	Solvent for crystallization[a]; appearance	Remarks (u.v. spectrum, λ_{max} (mμ) (log ϵ))	References
2-AcO(C_6H_4CO-)-*o*	180–181	K		133
2-AcO(C_6H_4C=NNHPh)-*o*	190–192	B		133
2-AcO(C_6H_4C=NN(Me)Ph)-*o*	198–200(dec.)	K		133
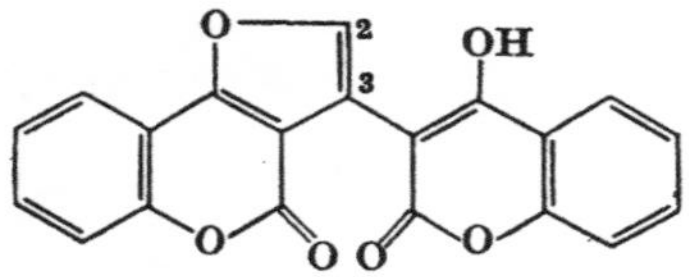				
2,3-Dihydro	256	—	Me ether (m.p. 247°)	126
2-Me	303	P_1	Me ether (m.p. 238°), $EtNH_2$ salt (m.p. 226–227°), EtCO deriv. (m.p. 217°), PrCO deriv. (m.p. 181°), Ac deriv. (m.p. 252°)	128, 129, 130, 158
2-Et	219–221	—	Ac deriv. (m.p. 228°), EtCO deriv. (m.p. 217–218°), PrCO deriv. (m.p. 202°)	158
2-Pr	279–280	—	Ac deriv. (m.p. 189°), EtCO deriv. (m.p. 210°), PrCO deriv. (m.p. 187–188°)	158
2-CH_2OH, 2,3-dihydro	278–279	R		132
2-CH_2OAc, 2,3-dihydro	195–196	A	Ac deriv. (m.p. 220°), Me ether (m.p. 219–220°)	132
2-OAc	—	—	Ac deriv. (m.p. 233°)	129

[a] A, acetic acid; B, acetic anhydride; C, acetone; D, acetonitrile; E, benzene; F, butanone; G, chloroform; H, N,N′-dimethylformamide; I, dioxan; J, ethanol; K, ethyl acetate; L, ether; M, ethoxyethanol; N, large amount of ethyl acetate; O, light-petroleum; P, methanol; P_1, pyridine; Q, skellysolve B; R, water.
i, inflexion; s, shoulder.

(+)-dolineone (**3**) already found in the plant[1]. Nepseudin can be dehydrogenated with manganese dioxide to give the dehydro compound (**4**), which on oxidation with alkaline hydrogen peroxide, gives two acids, 2,3,4-trimethoxybenzoic acid and the benzofuran acid (**5**). On hydrolysis **4** gave the deoxybenzoin (**6**).[2] As a result of these degradative experiments, nepseudin has been shown to be the furoisoflavanone (**1**) similar to neotenone, but with a unique 2′,3′,4′-trimethoxylated ring A replacing the normal 2′,4′,5′-oxygenation pattern.

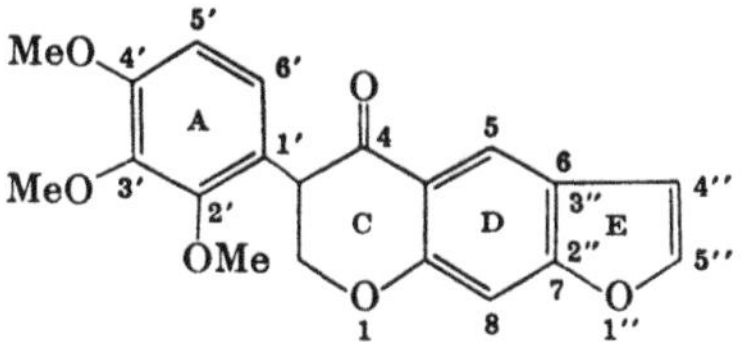

(**1**)

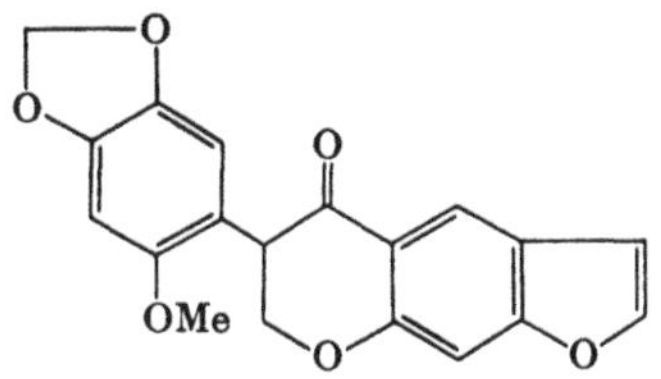

(**2**)

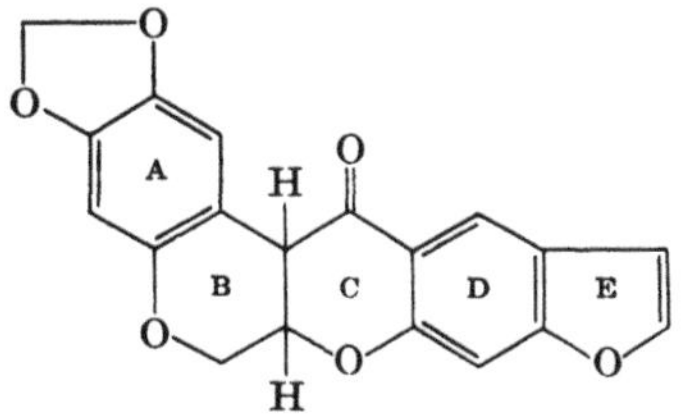

(**3**)

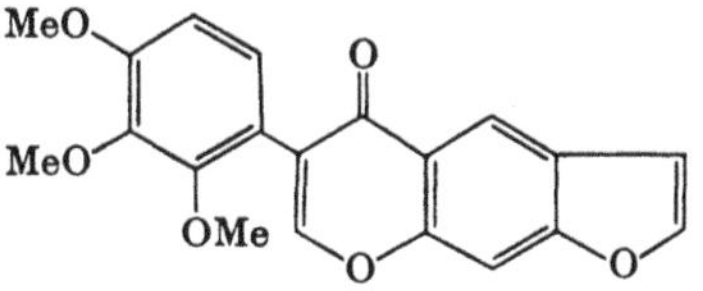

(**4**)

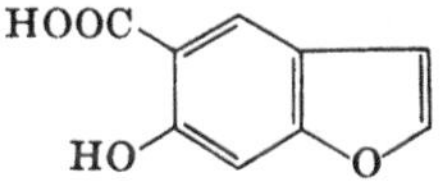

(**5**)

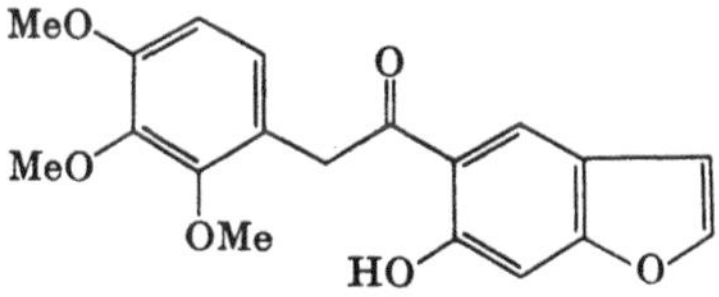

(**6**)

The nuclear magnetic resonance spectrum of nepseudin could be interpreted in terms of structure **1**. The proton integral showed three methoxyl groups and the presence of the furan ring was indicated by the characteristic spin–coupled protons. No methylenedioxy group was present and two of the four aromatic protons were adjacent and spin coupled.[3]

The total synthesis of **4**, confirming the assigned structure, has been achieved.[4] Hoesch condensation of 6-hydroxy-2,3-dihydrobenzofuran with 2,3,4-trimethoxybenzyl cyanide yielded 6-hydroxy-5-(2,3,4-trimethoxyphenylacetyl)-2,3-dihydrobenzofuran (**7**). Treatment of **7** with ethyl orthoformate–pyridine–piperidine gave the dihydro compound (**8**). On dehydrogenation with *N*-bromosuccinimide, **8** yielded dehydronepseudin (**4**). On alkaline hydrolysis **4** gave 6-hydroxy-5-(2,3,4-trimethoxyphenylacetyl)benzofuran (**6**).[1,2,5] The parent furoisoflavone (**4**) could be reformed by treating **6** with ethyl orthoformate–pyridine–piperidine (Eq. 1).

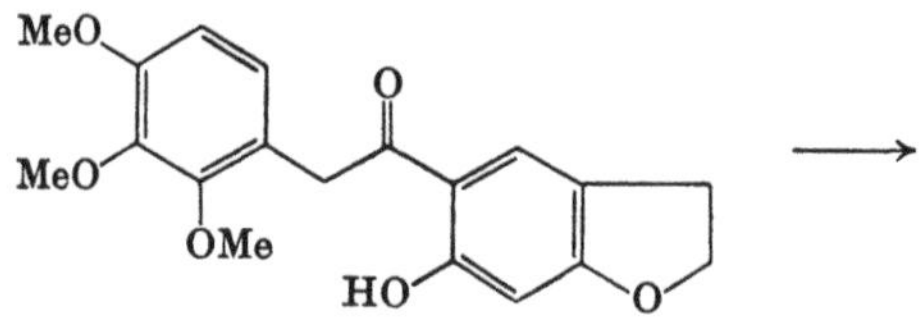

(**7**)

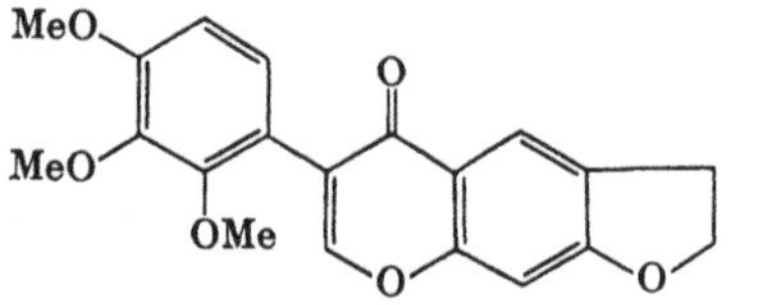

$\longrightarrow$ **4** $\rightleftharpoons$ **6** (1)

(**8**)

The furoisoflavone (**4**), when reduced with sodium borohydride, gave the furoisoflavanol (**9**); presumably a product of 1,4-reduction followed by further reduction of the resulting ketone. On oxidation with a chromic oxide–pyridine complex the alcohol (**9**) gave nepseudin.[6]

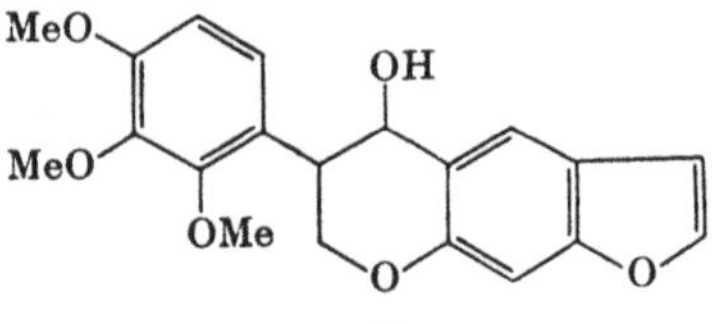

(**9**)

B. Neotenone

(±)-Neotenone, $C_{19}H_{14}O_6$, was isolated from the root of the leguminous plant *N. pseudopachyrrhiza*, *P. erosus*,[5a] *N. edulis*, and *N. amboensis*.[5b] It is optically inactive, contains one methoxyl group, and gives a positive Labat test for a methylenedioxy group. A purple coloration in the Durham test, and a green one in the Rogers and Calamari test indicated that the compound might be a rotenoid. The infrared and ultraviolet data are very similar to those of the rotenoid, elliptone. As with a rotenoid, the action of manganese dioxide in refluxing acetone, produced a dehydro compound with a shift of the carbonyl absorption to 1658 cm^{-1}. However, neotenone is not a rotenoid, but a new furoisoflavanone (**2**); the dehydro compound having structure (**10**).

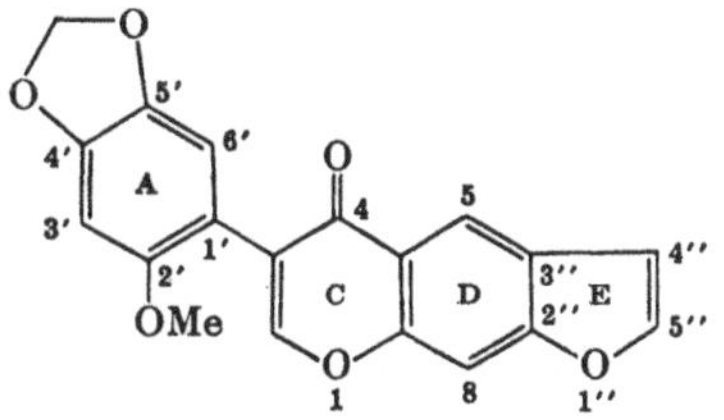

(**10**)

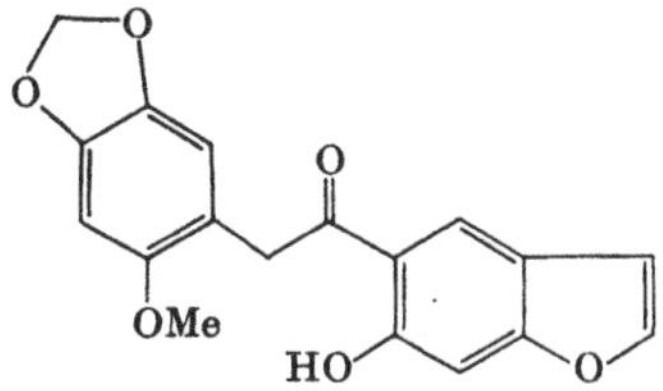

(**11**)

On alkaline hydrolysis, dehydroneotenone (**10**) yielded the deoxybenzoin (**11**) and formic acid, establishing the isoflavone character of **10**, and the parent furoisoflavanone (**2**) was resynthesized from the deoxybenzoin (**11**). The latter was treated with ethyl orthoformate–pyridine–piperidine[1,5a] to give the ketone (**10**), which when reduced with potassium borohydride, gave the isoflavanol (**12**), presumably a product of 1,4-reduction followed by further reduction of the resulting ketone. The alcohol (**12**) was identical with the borohydride reduction product of neotenone (**2**). On dehydration with phosphorus oxychloride and pyridine the fully conjugated compound (**13**) was obtained, and on oxidation with chromic oxide the alcohol was converted into neotenone.

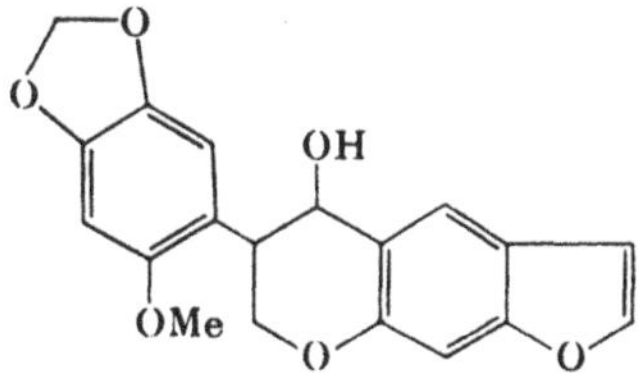

(**12**)

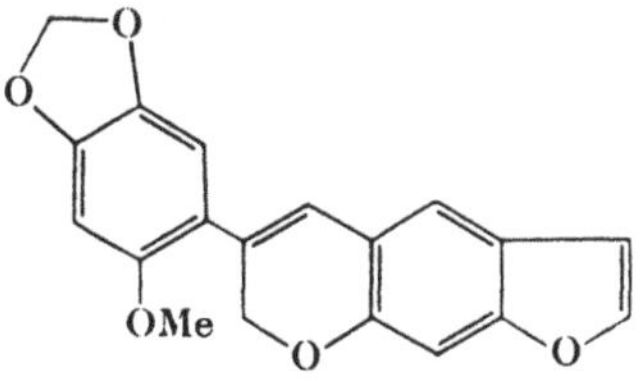

(**13**)

Oxidation of **10** with alkaline hydrogen peroxide produced a mixture of the methylenedioxy acid and an acid (**5**) was synthesized by a base–catalyzed cyclization of the Hoesch reaction product of resorcinol and chloroacetonitrile, namely, 6-hydroxycoumaranone (**14**). The latter compound was converted by a Haung–Minlon reduction into the phenol (**15**), and then carboxylated to give the salicylic acid (**16**). The ester of **16** was dehydrogenated over palladium, and hydrolysed to the acid (**5**), identical to that obtained by oxidative degradation.

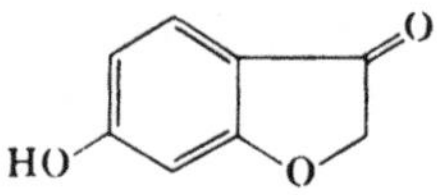

(**14**)

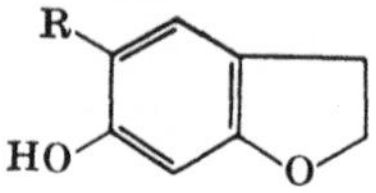

(**15**) R=H
(**16**) R=COOH

The results from the proton magnetic resonance spectrum are in full agreement with the assigned structure **2**.[3]

The synthesis of dehydroneotenone (**10**) has recently been achieved.[7] Hoesch condensation of 6-hydroxy-2,3-dihydrobenzofuran with 2′-methoxy-4′,5′-(methylenedioxy)phenylacetonitrile gave the deoxybenzoin, which upon treatment with ethyl orthoformate in the presence of a pyridine–piperidine mixture yields 2,3-dihydro-6-(2′-methoxy-4′,5′-(methylenedioxy)phenyl)-5*H*-furo(3,2-g)-1-benzopyran-5-one. The latter was dehydrogenated with *N*-bromosuccinimide to **10** (Eq. 2).

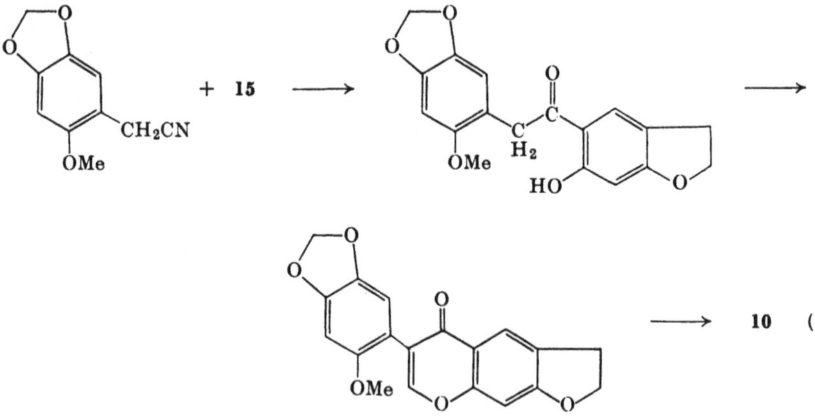

10 (2)

2. Synthetic Furoisoflavanones

A. Angular Furoisoflavones

(1) Introduction of a furan nucleus into an isoflavone skeleton (Tanaka's method). 5′-Carbethoxyfuro(2″,3″–7,8)-2-methylisoflavone (**18a**) was obtained from 8-formyl-7-hydroxy-2-methylisoflavone (**17a**) and ethyl bromomalonate[8] by Tanaka's procedure for benzofuran synthesis.[9] Hydrolysis of the reaction product (**18a**) gave the free acid (**18b**), which underwent decarboxylation to yield furo(2″,3″–7,8)-2-methylisoflavone (**19a**)[8] (Eq. 3). Row and Seshadri[10] obtained **18a** and **19a** by an alternate route; the melting points of these compounds differed markedly from those prepared by the Japanese authors[8] who have confirmed the proposed structures.

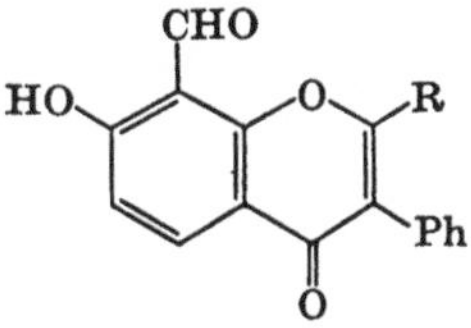

(**17a**) $R=CH_3$;
(**17b**) $R=H$

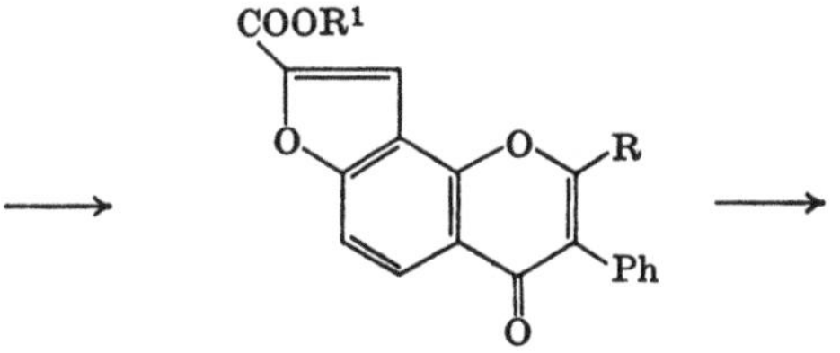

(**18a**) $R=CH_3$, $R^1=C_2H_5$;
(**18b**) $R=CH_3$, $R^1=H$;
(**18c**) $R=H$, $R^1=C_2H_5$;
(**18d**) $R=R^1=H$

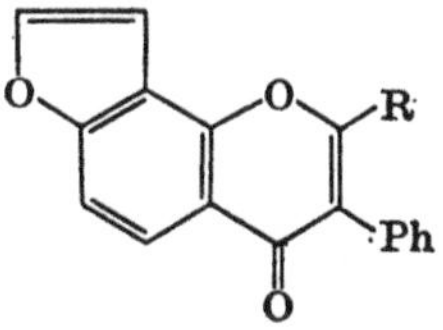

(3)

(**19a**) $R=CH_3$;
(**19b**) $R=H$

Adoption of the same procedure for the synthesis of furo(2″,3″–7,8)-isoflavone (**19b**) presented more difficulties than in the case of the 2-methyl derivative (**19a**).[11] Thus, treatment of the ester (**18c**) with ethanolic sodium hydroxide solution gave the free acid (**18d**) together

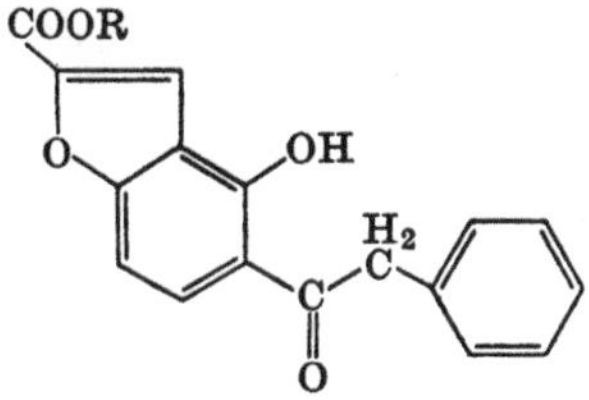

(**20a**) $R=C_2H_5$;
(**20b**) $R=H$

with a small quantity of benzyl 2-carbethoxy-4-hydroxy-5-benzofuranyl ketone (**20b**). On the other hand, aqueous sodium carbonate in acetone produced **18d** together with a small amount of **20a**.

Starting with the isoflavones (**21**), the furan ring was built up in the 7,8-position by a similar method[12] to give furo(2″,3″–7,8)-2′-methoxyisoflavone (**22**) and furo(2″,3″–7,8)-3′,4′-(methylenedioxy)isoflavone (**23**)[13] (Eq. 4). Demethylation of **22a**, with aluminum chloride in nitrobenzene, followed by bromination, gave the furoisoflavone (**24**).[14]

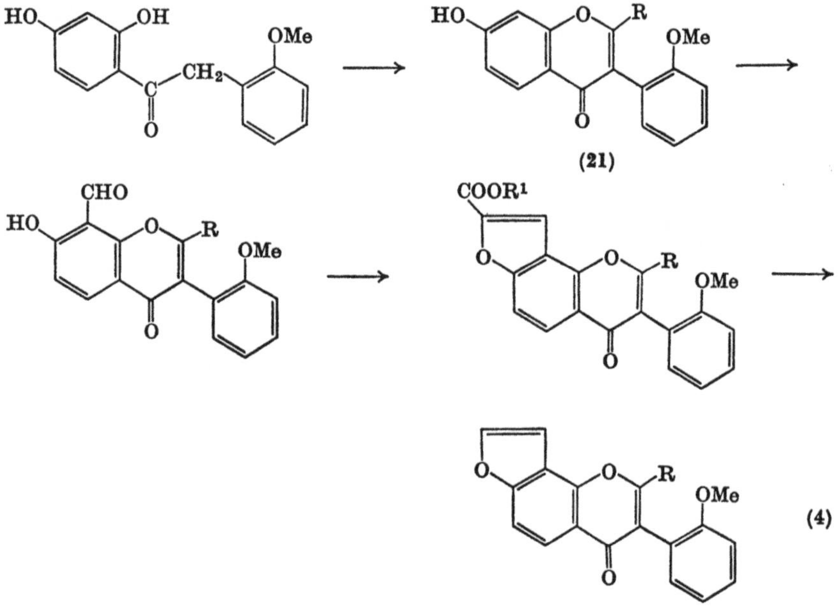

(**21**)

(4)

(**22a**) R=H;
(**22b**) R=CH_3;
(**22c**) R=$CH_2OC_2H_5$

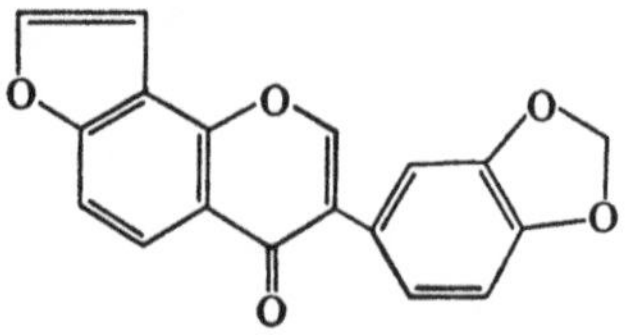

(**23**)

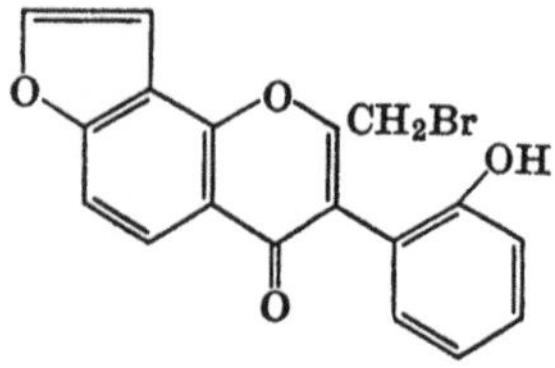

(**24**)

A modification of the above method is the use of osmium tetroxide and potassium periodate for the conversion of the allyl group in 8-allyl-7-

hydroxyisoflavones into the acetaldehyde derivatives, followed by ring closure of the latter by polyphosphoric acid to the corresponding furoisoflavones[15] (Eq. 5).

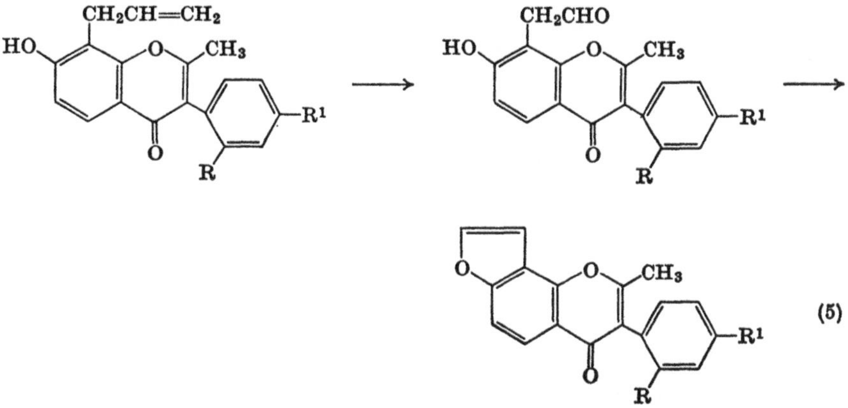

(5)

Introduction of the furan nucleus has also been effected by treatment of the hydroxyallylisoflavones with hydrobromic acid, followed by cyclization of the adducts to **25**,[16] **26**,[16] and **27**.[17] Dehydrogenation of **27** has been

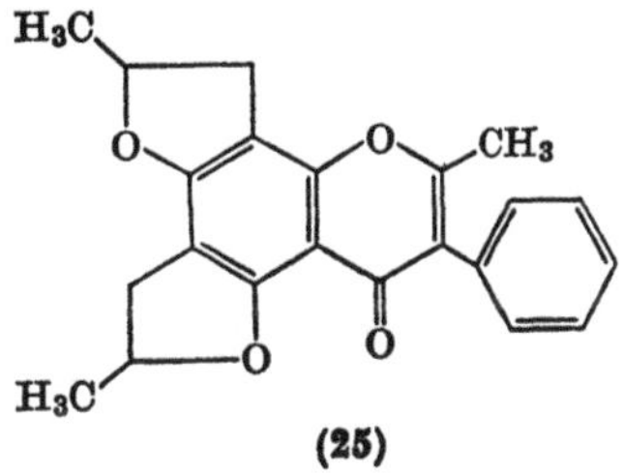

(25)

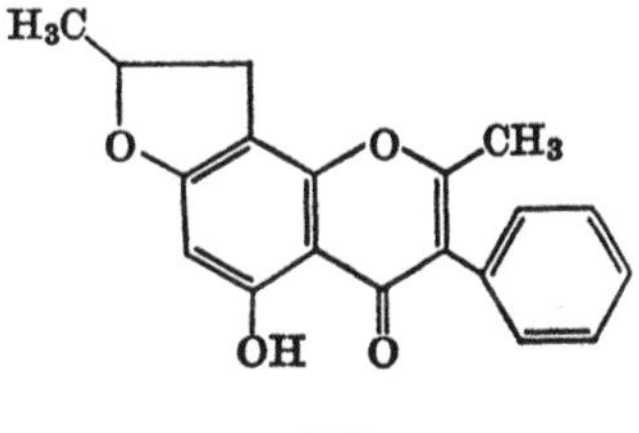

(26)

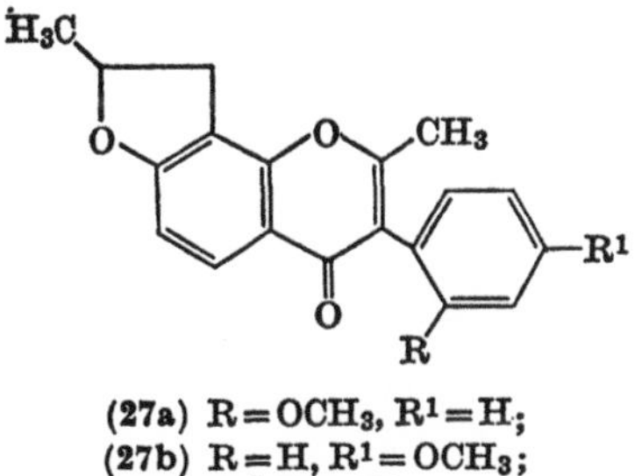

(27a) R = OCH_3, R^1 = H;
(27b) R = H, R^1 = OCH_3;
(27c) R = R^1 = H

brought about by the action of *N*-bromosuccinimide.[17]

(2) Ethyl orthoformate method (Venkataraman). The method is based on the completion of the isoflavone structure using a suitable coumarone derivative, which on condensation with ethyl orthoformate in pyridine containing a little piperidine[18] gives the furoisoflavones. The method has been applied successfully in the synthesis of furo(2″,3″–7,8)-isoflavone (**19b**) from benzyl karanjyl ketone (5-benzofuranyl ketone) (**28**).[11]

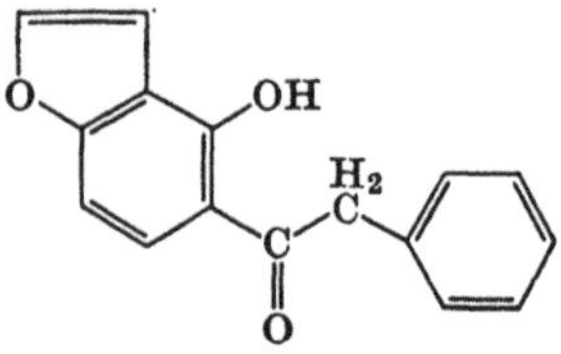

(28)

This synthesis has been recently applied to the confirmation[19] of the structure of the naturally occurring isoflavone, munetone, isolated from *Mundulea suberosa* Benth, and originally assigned structure **30**.[20] Hoesch condensation of dihydrotubanol with 2-methoxybenzyl cyanide gave **29**, which by reaction with ethyl orthoformate, followed by dehydrogenation, gave a product of the expected structure (**30**), which was found to be different from natural munetone[19] (Eq. 6). The nuclear magnetic resonance spectrum of munetone and its chemical characteristics confirm its structure as **31**, and not **30**.[19]

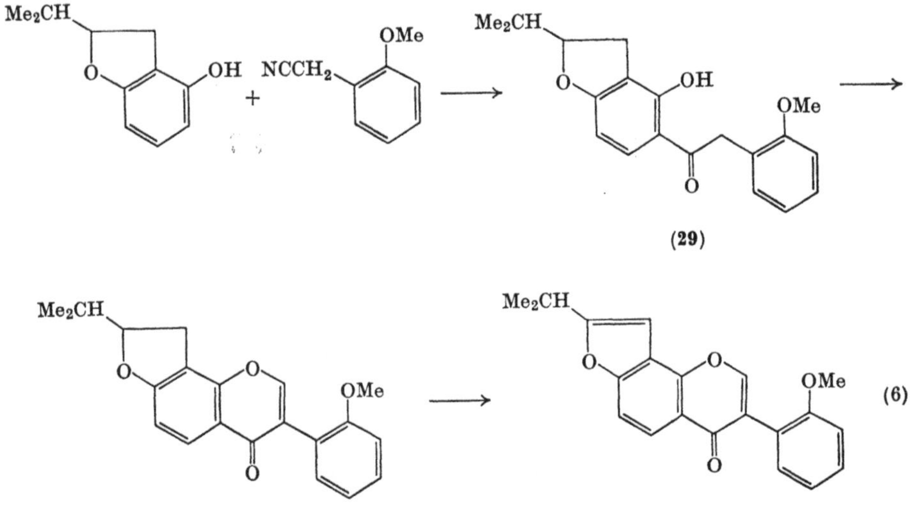

(29) (30) (6)

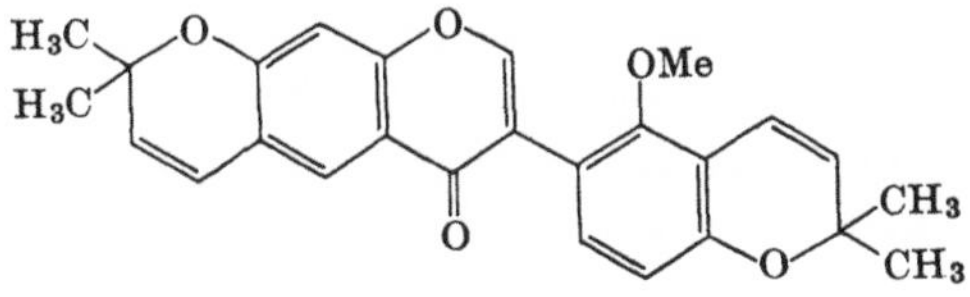

(31)

Elliptol isoflavone (furo(2″,3″–7,8)-2′,4′,5′-trimethoxyisoflavone) (**32**) was obtained from elliptone (**33**).[21,22] Isoelliptol isoflavone (furo-(2″,3″–6,7)-2′,4′,5′-trimethoxyisoflavone) (**34**) may be similarly obtained

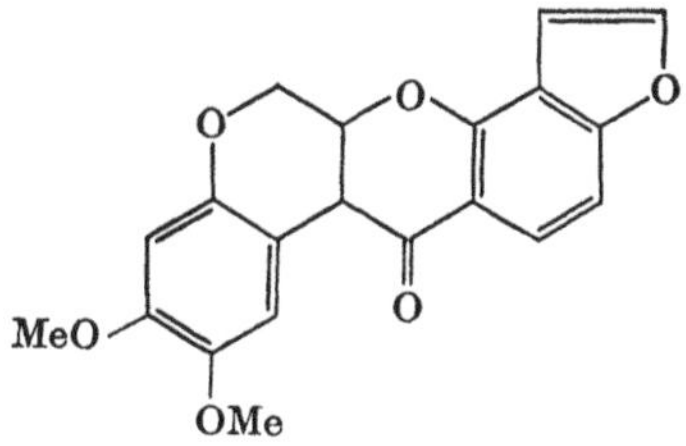

(33)

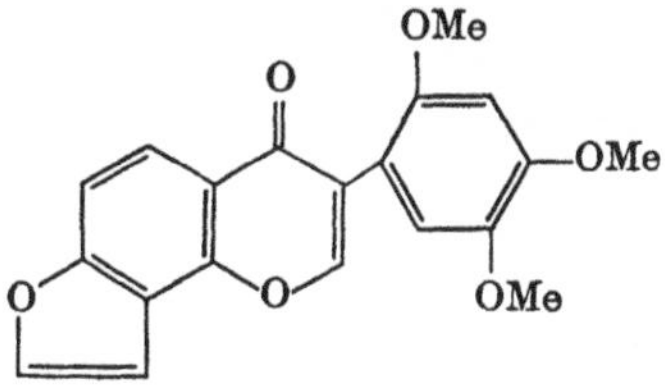

(32)

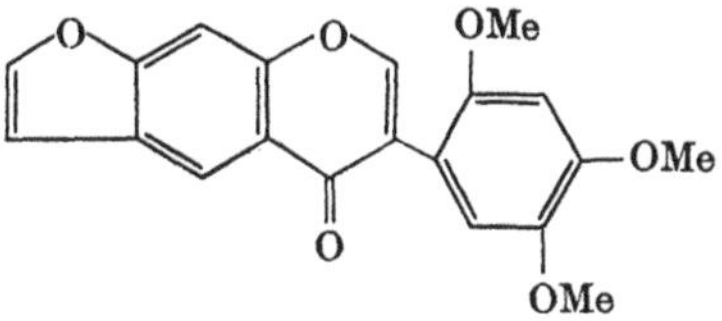

(34)

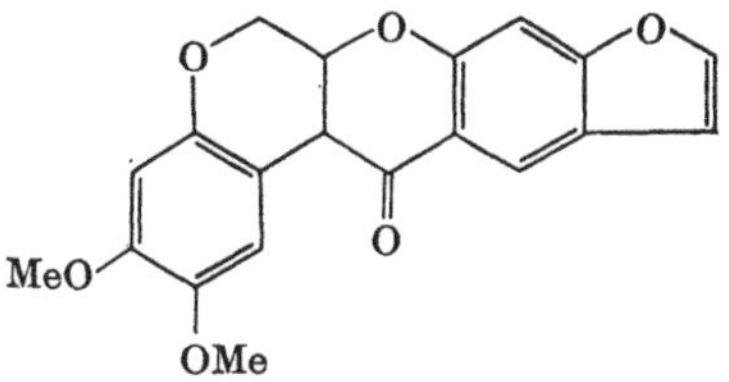

(35)

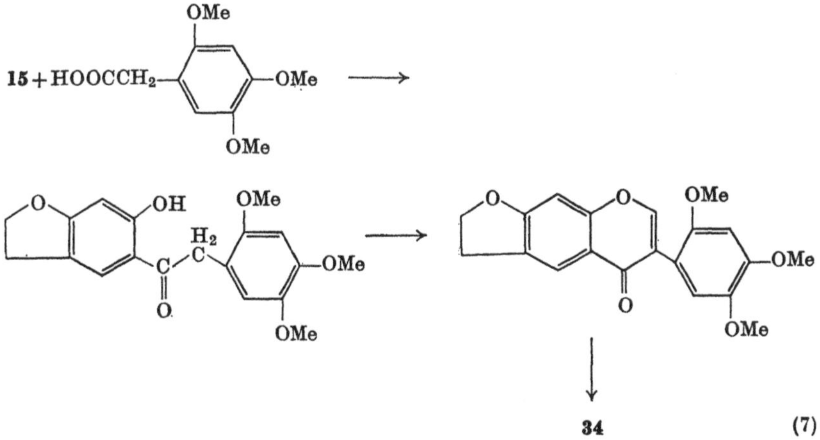

(7)

from isoelliptone (**35**) which has recently been isolated.[23] Fukui and coworkers[24] reported the synthesis of **32** from 7-hydroxy-2′,4′,5′-trimethoxyisoflavone.[25] **34** has recently been synthesized[23] by the method used for the synthesis of dehydronepseudin (**4**)[4] (Eq. 7).

B. Linear Furo(3″,2″-6,7)isoflavones

Synthesis was achieved by ring closure of **36**, obtained by Hoesch condensation of 6-hydroxycoumaran and phenylacetonitrile, by acetylation[26] and formylation,[27] followed by dehydrogenation of the dihydrofuroisoflavones (**37**), with a palladium–charcoal catalyst,[28, 29] to the corresponding furoisoflavones (**38**).

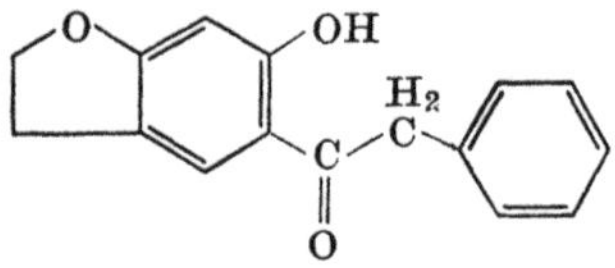

(**36**)

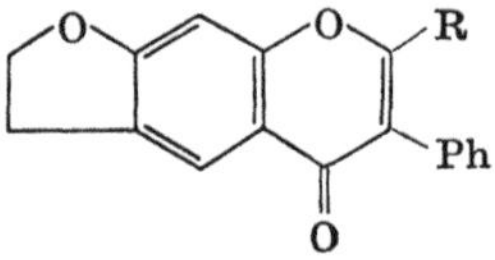

(**37**)

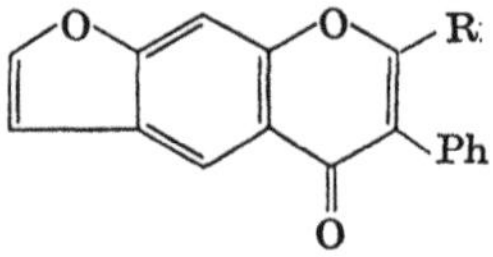

(**38**)

On the other hand, catalytic hydrogenation of **37** using Adam's catalyst is a convenient method for the synthesis of linear 2-methyl-4″,5″-dihydrofuro(3″,2″-6,7)isoflavanones (**39**).[30] The solutions of the latter compounds in sulfuric acid are invariably yellow in the cold, deepening on warming; while solutions of the parent furoisoflavones remain almost colorless even on warming. **39b** is highly toxic to fresh water fish.[31]

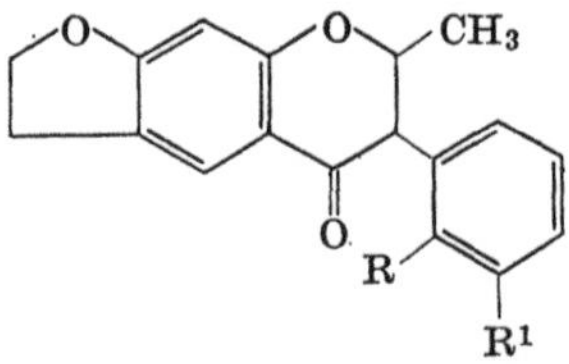

(**39a**) $R = R^1 = H$;
(**39b**) $R = R^1 = OCH_3$;
(**39c**) $R = R^1 == O_2CH_2$

III. Coumaranochromans

1. Introduction

Red sandalwood (*Pterocarpus santalinus*, Linn), camwood (variety of *Baphia nitida*), barwood (*Baphia nitida*, Lodd), and narra wood have been known under the group name, Insoluble Red Woods, and used as vegetable dyes from ancient times. They belong to the genus *Pterocarpus*, and the presence of colorless crystalline components has been established. Some of these belong to the isoflavonoid group; the most important are homopterocarpin (**40**) and pterocarpin (**41**), which was isolated in 1874 by Cazeneuve[39] and by several workers from most of the species of this genus (Table 2).

TABLE 2. Naturally Occurring Pterocarpin Group of Compounds

Pterocarpin group	Sources	References
Pterocarpin	*Pterocarpus santalinus* (Red sandal wood)	39
	Pterocarpus dalbergoides (Andaman padauk)	152
	Pterocarpus macrocarpus (Burma padauk)	152
	Pterocarpus osun	153
	Pterocarpus indicus (Narra wood)	154
	Baphia nitida (Camwood)	66
	Sophora subprostrata	51
Homopterocarpin	*Pterocarpus santalinus*	38
	Pterocarpus sojauxii	152
	Pterocarpus osun	153
	Pterocarpus indicus	154
l-Maackiain	*Maackia amurensis*	47
dl-Maackiain	*Sophora japonica*	51
Trifolirhizin	*Trifolium pratense*	37, 47, 50
l-Maackiain-β-D-glucoside	*Sophora subprostrata*; *Sophora flavescens*	51
Sophojaponicin	*Sophora japonica*	51
Pisatin	*Pisum sativum*	44, 49, 56

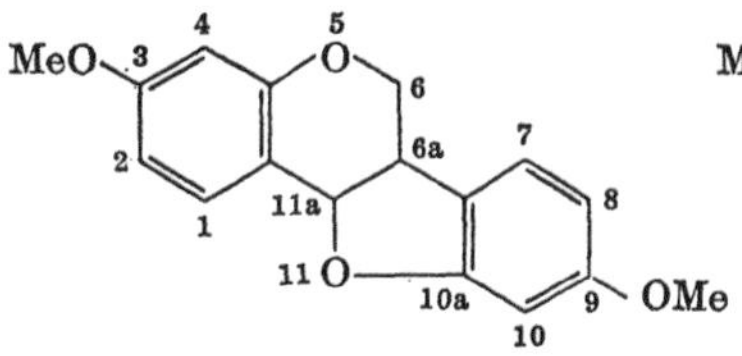

(**40**)

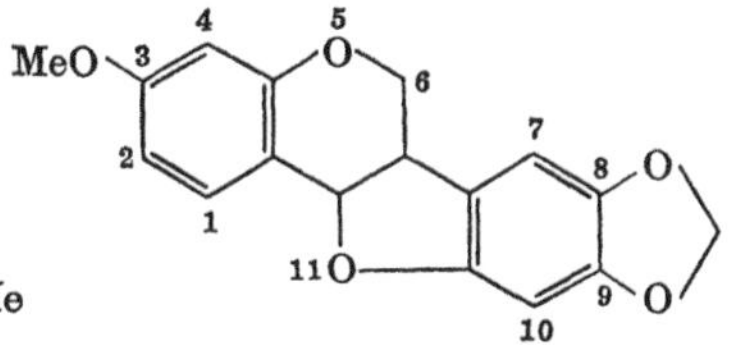

(**41**)

2. Naturally Occurring Coumaranochromans

A. Homopterocarpin

Homopterocarpin (6a,11a-dihydro-3,9-dimethoxy-6*H*-benzofuro-(3,2-c)-1-benzopyran) (**40**), $C_{17}H_{16}O_4$, contains two methoxyl groups; carbonyl and hydroxyl groups are absent. The structure of homopterocarpin was established by Robertson and coworkers[32] and independently by Späth and coworkers[33]. It shows high *laevo* rotation, and the structure allocated to homopterocarpin includes a benzylic oxygen; hence, undesirable changes are liable to occur in acid media.

Oxidation of homopterocarpin with potassium permanganate in acetone solution gives a mixture of 2-hydroxy-4-methoxybenzoic acid (**42**) and 2-carboxy-5-methoxyphenoxyacetic acid (**43**) by destruction of the oxygen rings. The production of **43** indicates the presence of at least one ether oxygen.

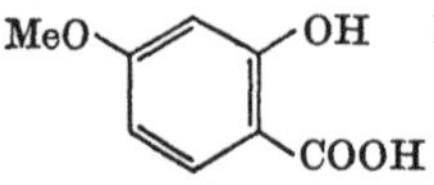

(**42**)

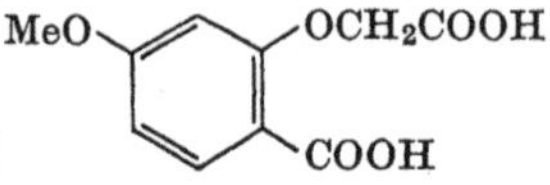

(**43**)

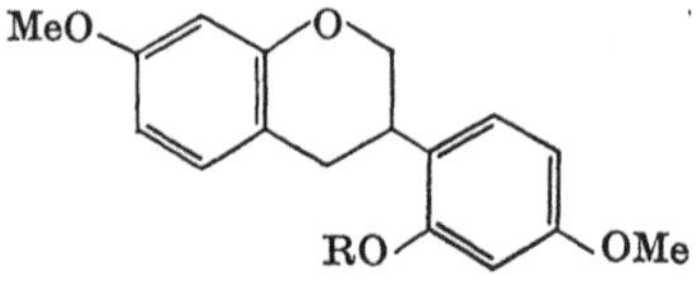

(**44**)

Reduction of homopterocarpin with palladium and charcoal proceeds slowly at room temperature and much more rapidly at 100°c to give the phenolic 1-dihydrohomopterocarpin (**44**, R = H) by the rupture of the ether linkage of the dihydrofuran ring; thus behaving similarly to rotenone, in which the tubanol residue is converted to tetrahydrotubanol (**45**). Application of the Clemmensen reduction to homopterocarpin produces **44** (R = H).

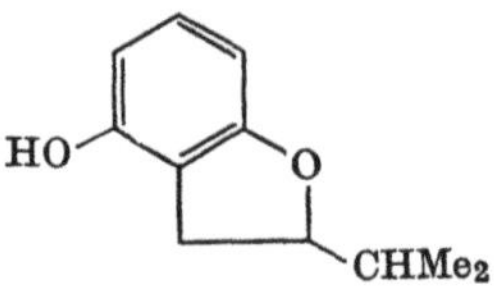

(**45**)

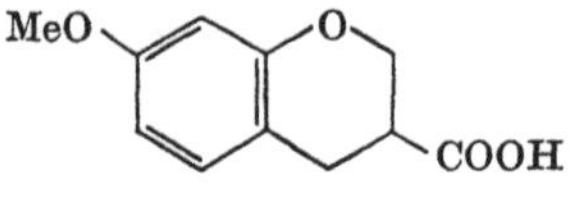

(**46**)

Oxidation of the dihydro compound (**44**, R = H) with potassium permanganate gives *dl*-7-methoxychroman-3-carboxylic acid (**46**); thus establishing the presence of the 7-methoxychroman nucleus in *l*-dihydrohomopterocarpin, and hence in the parent compound. The production of this acid taken in conjunction with the results of the hydrogenation establishes the presence of two oxygen atoms in cyclic systems. The *dl*-acid (**46**) was synthesized by cyclization of 2-formyl-5-methoxyphenoxypropionic acid (**47**) with sodium acetate and acetic anhydride to 7-methoxy-Δ^3-chromen-3-carboxylic acid (**48**) which was readily hydrogenated to **46** (Eq. 8). Oxidation of 1-dihydrohomopterocarpin with chromic acid led to a quinone (**49**), while under the same conditions, the methyl ether (**44**, $R = CH_3$) yielded the isoflavanone (**50**, R = H), which unlike homopterocarpin, is sensitive to alkali and readily affords the oxime and undergoes similar reactions.

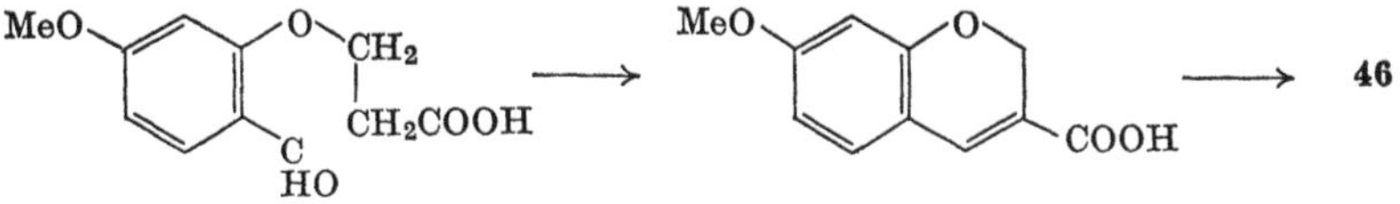

(8)

(47) (48)

Permanganate oxidation gives an alcohol (**50**, R = OH), easily dehydrated to the corresponding isoflavone.

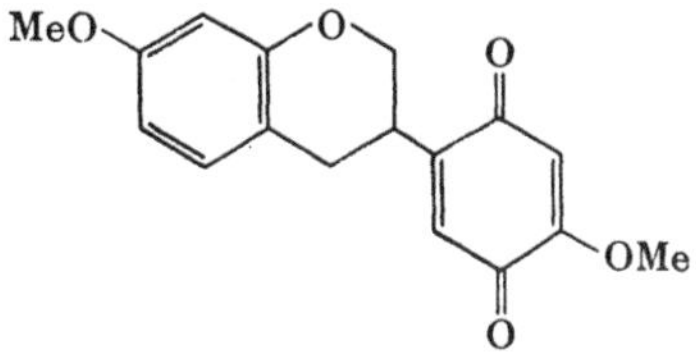

(49)

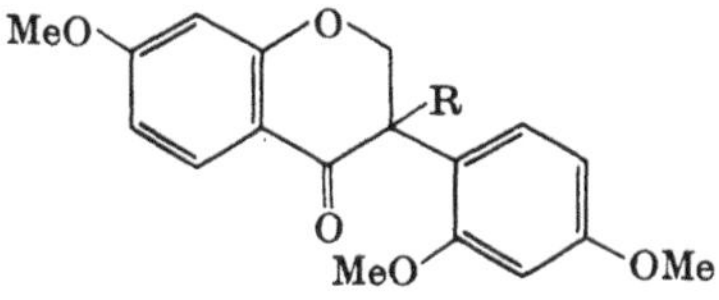

(50)

Dehydrogenation of homopterocarpin (**40**) yields dehydrohomopterocarpin (**51**), which on oxidation yields **52**, thus providing collateral evidence for the tetracyclic system of homopterocarpin.

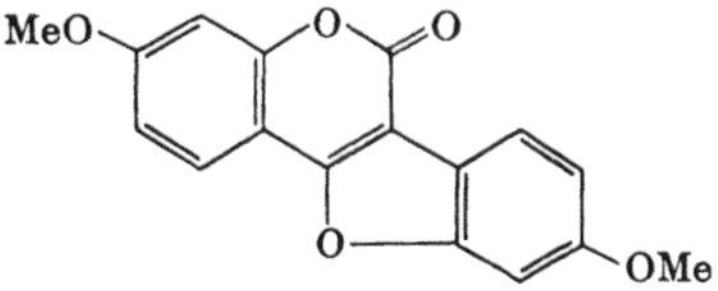

(52)

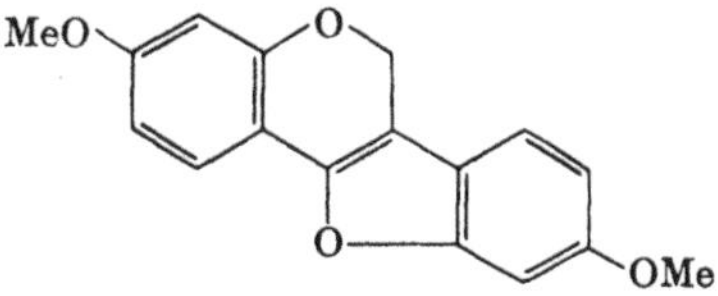

(51)

Homopterocarpin undergoes a remarkable reaction when heated with palladium and charcoal, since the product is the coumarin **53**.[34] This reaction locates the 4′-methoxyl group.

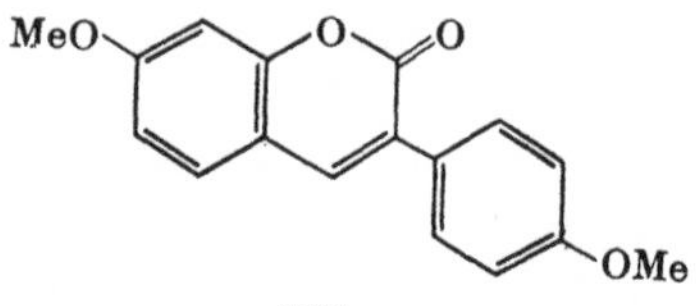

(**53**)

Demethylhomopterocarpin (**54**), a naturally occurring furoisoflavonoid extracted from the Western Indian hardwood, *Andira inermis* (Wright) H.B.K.,[35] has been synthesized. It gives homopterocarpin (**40**) on methylation[36] (Eq. 9).

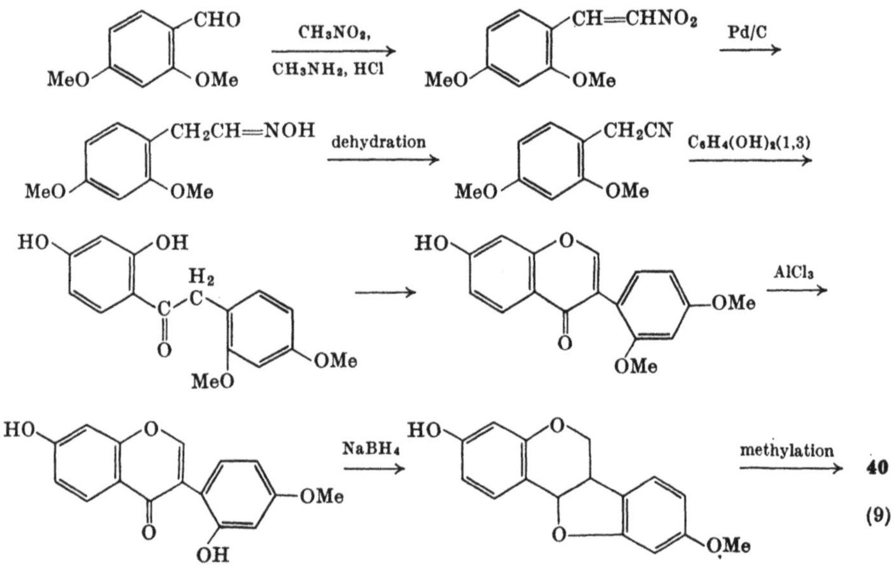

(9)

(**54**)

B. Pterocarpin

Pterocarpin (3 - methoxy - 6a,12a - dihydro - 6*H* - (1,3)dioxolo(5,6)-benzofuro(3,2-c)–1–benzopyran), $C_{17}H_{14}O_5$, behaves similarly to its congenor. Thus, reduction with palladium and charcoal gives the phenolic *l*-dihydropterocarpin, which upon oxidation with potassium permanganate yields *dl*-7-methoxychromen-3-carboxylic acid (**48**). Structure **55** for pterocarpin was regarded as correct since Robertson and coworkers[34] confirmed the structure of *l*-dihydropterocarpin methyl ether by comparison of its infrared spectrum with that of synthetic 2′,7-dimethoxy-3′,4′-methylenedioxyisoflavan (**56**). However, recently, Bredenberg and Shoolery[37] noticed that the rotenoids,[38, 39] erosnin,[40] sophorol,[41, 42] jamaicin,[43] and pachyrrhizin[44] occurring in Papilunaceous plants and biogenetically relating to pterocarpin, possess a methylenedioxy or dimethoxyl group in the 8,9-positions. The structure of pterocarpin (**55**) presented by Robertson and coworkers[34] would be doubtful from the biogenetical point of view, since it ought to possess a methylenedioxy grouping in the 9,10-positions. On the basis of the nuclear magnetic resonance spectral analysis, Bredenberg and Shoolery amended Robertson's formula for pterocarpin (**55**), and proposed formula **41**, which possesses the methylenedioxy grouping in the 8,9-positions. The H′ resonance spectrum,[37] not only confirms the presence of a methylenedioxy grouping (τ, 5.92), but also requires the presence of two uncoupled hydrogen atoms attached to a benzene ring, such that one is flanked by one oxygen (τ, 6.72), and the other by two (τ, 6.43). Structure **41** is the only one to satisfy these conditions, and structure **57** is the correct one for dihydropterocarpin methyl ethers.

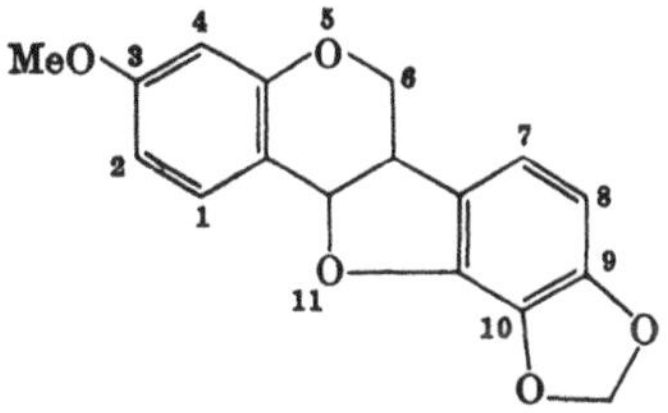

(**55**) Robertson and coworkers

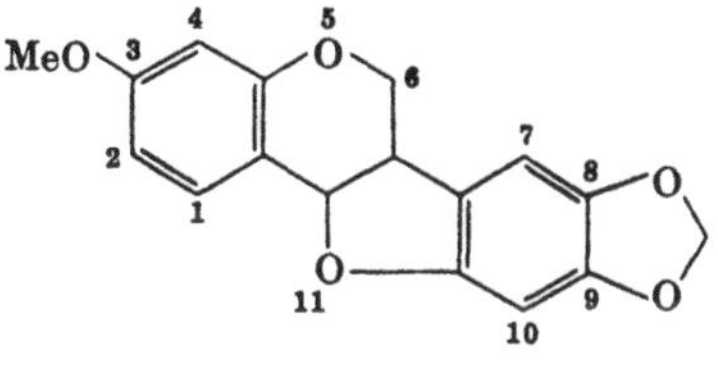

(**41**) Bredenberg and Shoolery

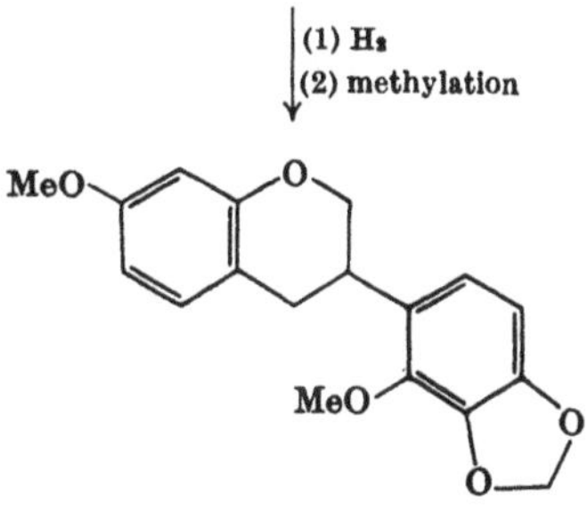

(**56**)

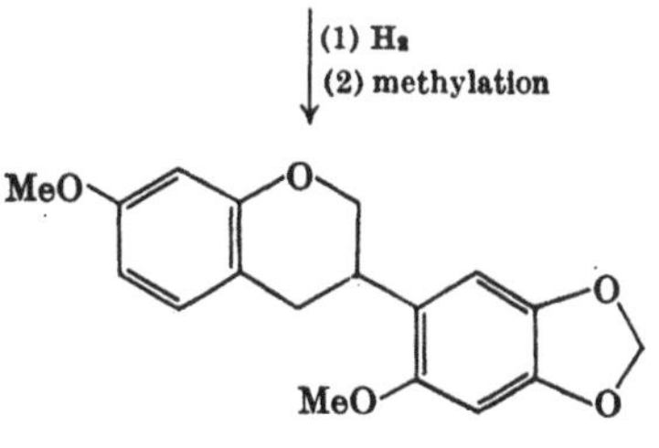

(**57**)

The fundamental nucleus of pterocarpin (**41**) and homopterocarpin (**40**) has been synthesized[45] via the isoflavone (**58**). Reduction of the latter with sodium borohydride or lithium aluminum hydride produces a mixture of stereoisomeric flavin-4-ols (**59**) which is cyclized by acids. Spontaneous ring closure in alkaline medium[46] has been reported in the case of 2′-hydroxy-7,4′,5′-trimethoxyisoflavone. Hydrogenation of the isoflavone (**58**) leads to the isoflavanone (**60**), which can also be cyclized by acids, but the product, a benzofuran derivative (**61**), resists[34] saturation to the desired material **62** (Eq. 10).

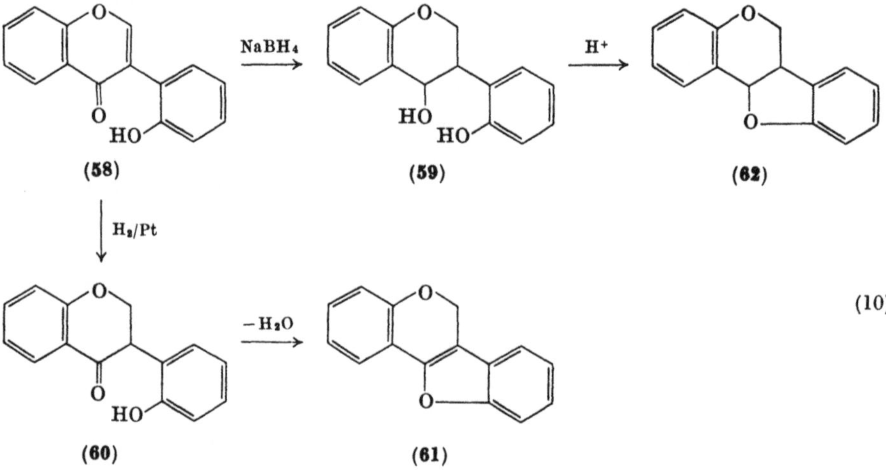

(10)

Homopterocarpin (**40**) has been synthesized[47] by dealkylation of 7,2′,4′-trimethoxyisoflavone (**63a**) with aluminum chloride to yield the trihydroxyisoflavone (**63b**), which on selective methylation with methyl iodide gives 2′,4′-dihydroxy-7-methoxyisoflavone (**63c**). Sodium borohydride reduction of the latter compound, followed by methylation, gave *dl*-homopterocarpin. Interpretation of the nuclear magnetic resonance spectrum suggests the assignment of the *cis* configuration to homopterocarpin. The synthesis of 3,4-dehydrohomopterocarpin and of (±)-dihydrohomopterocarpin has been achieved.[48]

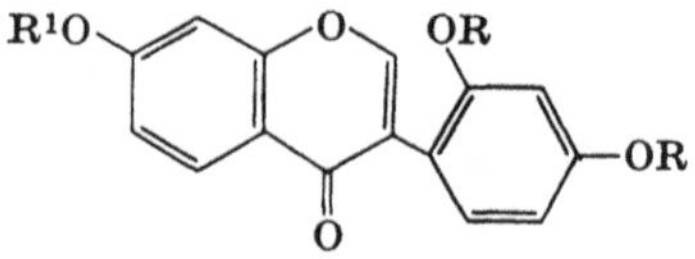

(**63a**) $R = R^1 = CH_3$;
(**63b**) $R = R^1 = H$;
(**63c**) $R = H, R^1 = CH_3$

C. Maackiain

Bredenberg[49] and Hietala[50] have isolated trifolirhizin (**64**) from *Trifolium pratense* L. (Red clover), and it was stated that it would be the D-glucoside of norpterocarpin. On the other hand, Suginome[47] isolated a compound, named maackiain (6a,12a-dihydro-6*H*-(1,3)dioxolo(5,6)-benzofuro(3,2-*c*)(1)benzopyran-3-ol) from the heartwood of *Maackia amurensis* Rupr. et Maxim. var. *Buergeri* (Maxim) C. K. Schneid. Maackiain has later been proved to be norpterocarpin.[51] The structure of maackiain (**65**) was proposed by Suginome[52] independently from the norpterocarpin formula of Bredenberg and Shoolery[37] on the basis of the ultraviolet spectral analysis and the biogenetical relationship with the coexisting sophorol (**66**).[41, 42] The identity of the methyl ether of maackiain and pterocarpin has later been reported.[51]

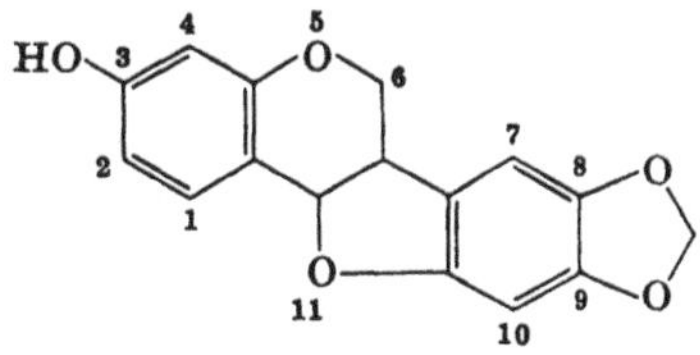

(**65**)

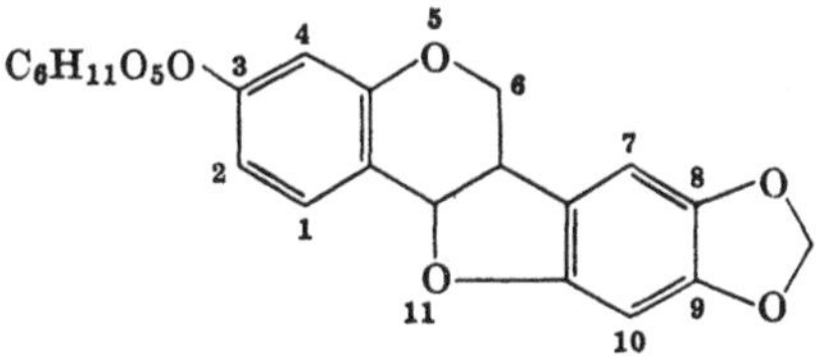

(**64**)

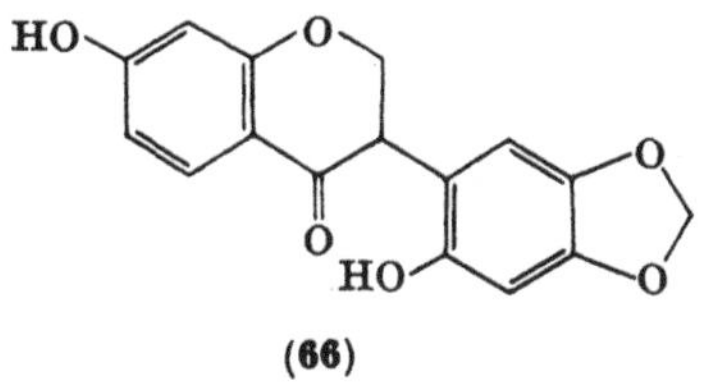

(**66**)

Shibata and coworkers[51, 53] have recently isolated *l*-maackiain mono-β-D-glucoside and *l*-pterocarpin from *Sophora subprostrata* and *d*-maackiain mono-β-D-glucoside (sophojaponicin) and *dl*-maackiain from *Sophora japonica*. The authors reported that *l*-maackiain mono-β-D-glucoside seems to be identical with trifolirhizin (**64**) and proposed the name sophojaponicin. Later, Suginome[52] has reported on *l*-maackiain, and almost at the same time Cocker and coworkers[54] published their study on inermin, $C_{16}H_{12}O_5 \cdot \frac{1}{2}H_2O$, which is identical with *l*-maackiain. Inermin affords pterocarpin on methylation, and is a constituent of *Andira inermis*.

Acid hydrolysis of the glucoside of the pterocarpin-type compound fails to give the crystalline aglycone.[49] The ultraviolet spectral curve of the ethereal solution of the resinous acid, produced by hydrolysis of sophojaponicin, is similar to that of anhydrosophorol (**67**) (Eq. 11).

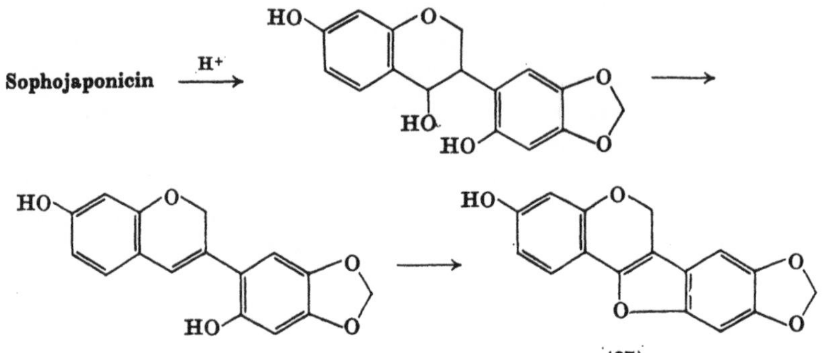

(11)

(**67**)

Although an unsuccessful result of the catalytic hydrogenation of trifolirhizin has been reported,[49, 53] sophojaponicin has been reduced. Dihydrosophojaponicin, and its pentaacetate gave ultraviolet spectra similar to those of *dl*-dihydromaackiain and its diacetate. On hydrolysis with dilute sulfuric acid, dihydrosophojaponicin affords *d*-dihydromaackiain.

D. Pisatin

Pisatin (3-methoxy-6a-hydroxy-8,9-methylenedioxy-6a,11a-dihydro-6*H*-benzofuro(3,2-*c*)–1–benzopyran) (**68**), an antifungal substance of molecular formula, $C_{17}H_{14}O_6$, has been isolated from fungal-infected pods of *Pisum sativum*.[55–57] Infrared, ultraviolet spectra, and chemical

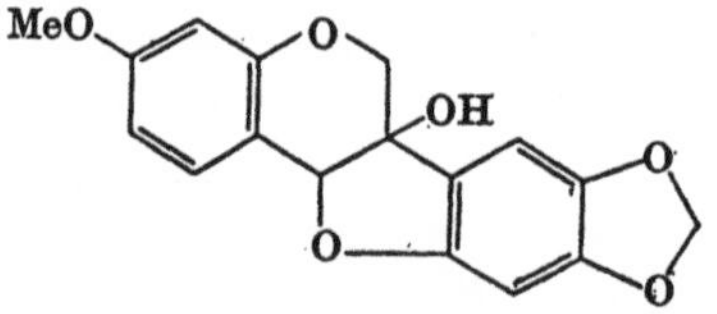

(**68**)

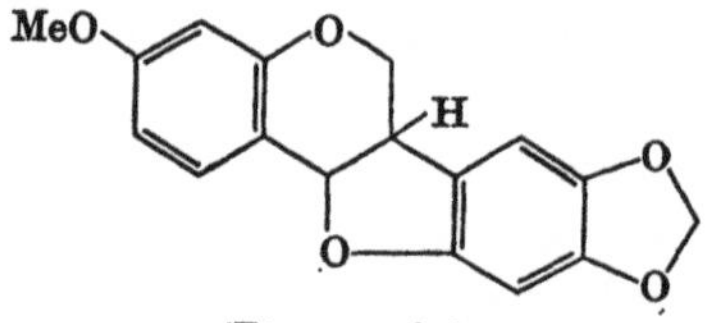

(**Pterocarpin**)

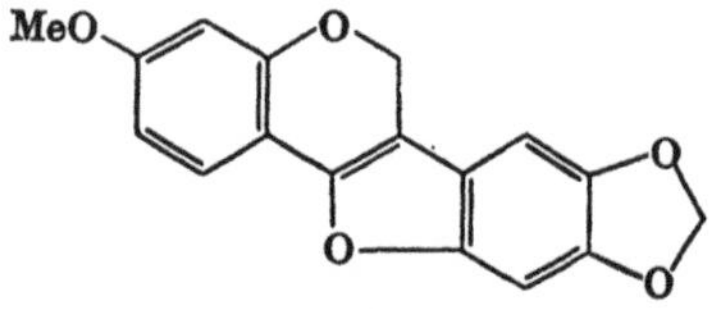

(**69**)

evidence, including a positive Labat test, showed pisatin to have one methoxyl group, a methylenedioxy group, and an alcoholic (nonphenolic) hydroxyl group,[58] but otherwise provided little information about the structure. From an analysis of its nuclear magnetic resonance spectrum in carbon tetrachloride, pisatin is assigned structure **68.** Examination of the aromatic protons by their n.m.r. spectra indicates that in pisatin and in pterocarpin (**41**) the methylenedioxy group is located across the 4′ and 5′ positions, and hence pisatin is 6a-hydroxypterocarpin.[37, 59]

The presence of the hydroxyl group confers some interesting chemical properties on pisatin. Thus, in the presence of acid, pisatin readily loses one molecule of water, and optical activity to give anhydropisatin,

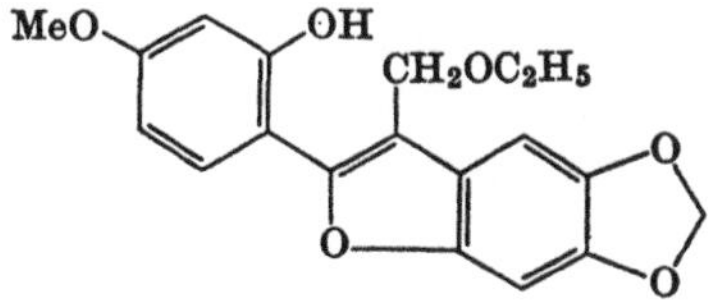

(70)

41

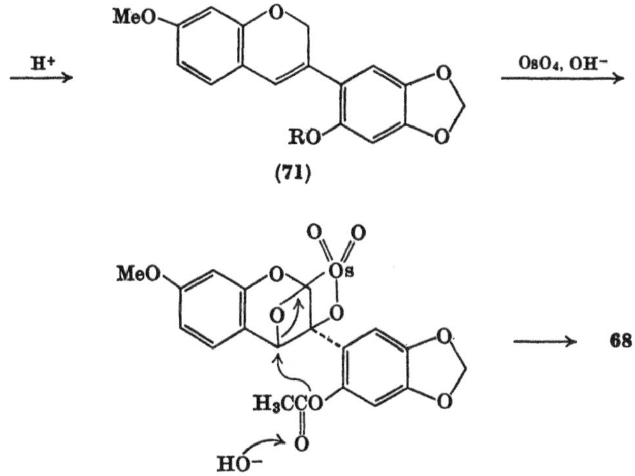

(71)

(73)

68 (12)

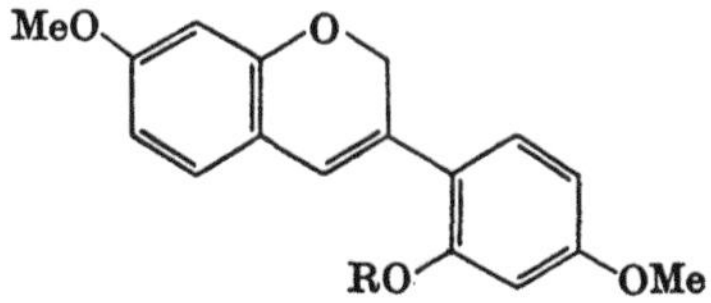

(72)

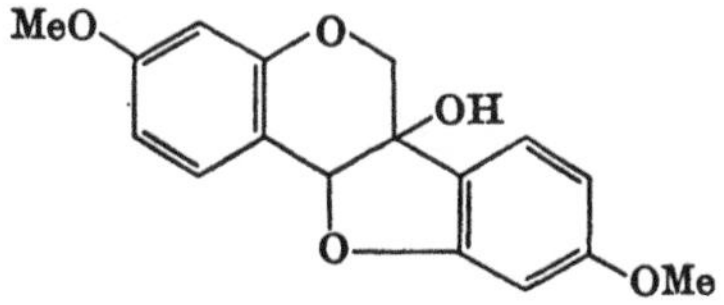

(74)

$C_{17}H_{12}O_5$;[58] its n.m.r. spectrum is consistent with the assigned structure **69.**[59] Direct comparison of anhydropisatin with *O*-methylanhydrosophorol, the methyl ether of anhydrosophorol (**67**),[41–42] established their identity and thus, confirmed structure **69** for anhydropisatin.

When alcoholic solutions of anhydropisatin are exposed to diffuse daylight, fission of the pyran ring and alcoholysis across the 1, 2-oxygen–carbon bond occurs to give the phenolic ether (**70**). The formation of a different product when the reaction is carried out in methanolic solution confirms this interpretation.

dl-Pisatin (**68**) and the analog (**74**) have recently been obtained via oxidation of the isoflavenes (**71, 72,** R = OAc respectively) with osmium trioxide (cf. **73**), followed by action of alkali[60] (Eq. 12). The isoflavenes (**71, 72,** R = H) have been obtained by controlled action of acid on pterocarpin (**41**) and homopterocarpin (**40**), respectively.

Biogenetic studies suggest that pisatin may also result from some modification of an isoflavonoid intermediate.[58]

E. Neodulin

The name edulin[61] has been replaced by neodulin.[61, 5a] It is one of the constituents of *Neorautanenia edulis*[5b, 62] and *N. amboensis*,[5b] related on the one hand to pterocarpin (**41**) and on the other hand to dolichone (**75**), a rotenoid from a hindered plant. The tuber of *N. edulis* is used as a fish poison by natives in the Northern Transvaal region of Southern Africa.[62, 63]

Neodulin (**76**), $C_{18}H_{12}O_5$, is optically active, gives a positive Labat test[64] for methylenedioxyphenyl, and does not contain methoxyl or *C*-methyl groups. Hydrogenation of **76** over palladized charcoal gave **77** and **78.** On the other hand, hydrogenation of **76** over palladium under pressure gave mainly **78** which was readily methylated to **79.** The latter could be obtained by hydrogenation of neotenone (**2**) over palladium.[65–66] That **78** is soluble in alkali, accounts for the opening of a dihydrobenzofuran structure (cf. **77**), and simulates the hydrogenolysis of pterocarpin (**41**) to the phenol.[32, 34]

Oxidation of neodulin with nitric acid (5%) gave styphnic acid; permanganate oxidation produced oxalic acid. Oxidation of **79** with potassium permanganate in acetone led to **82.** The latter compound was also obtained upon hydrogenation of **76** over Raney nickel. Manganese dioxide oxidation of **82** or reduction of dehydroneotenone (**10**) produced **83,** which can be dehydrogenated with *N*-bromosuccinimide to **10.** Treatment of **80**[1] with aluminum chloride in benzene effected demethyla-

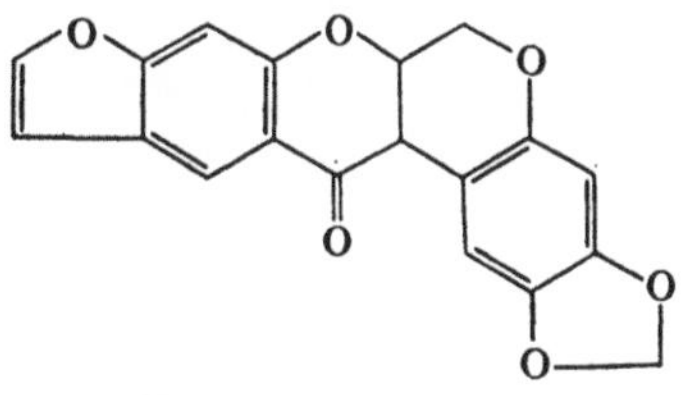

(75) **Dolichone**

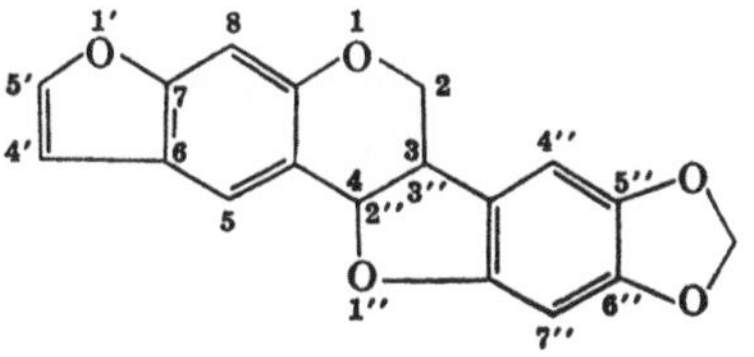

(76)

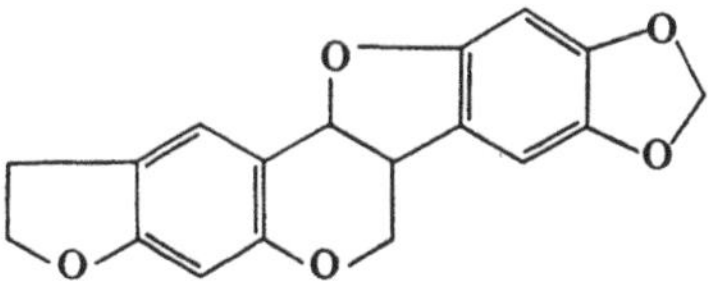

(77)

tion to **81**. The latter compound was readily converted to neodulin (**76**) on refluxing with acetic acid.[5b]

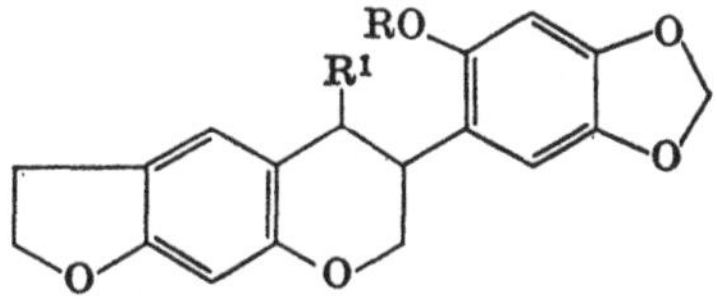

(78) R=R¹=H
(79) R=CH_3, R¹=H
(80) R=CH_3, R¹=OH
(81) R=H, R¹=OH

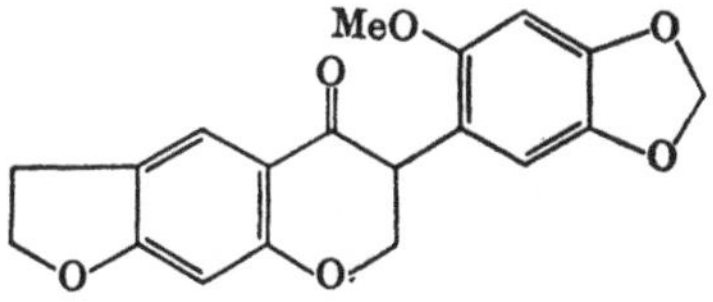

(82)

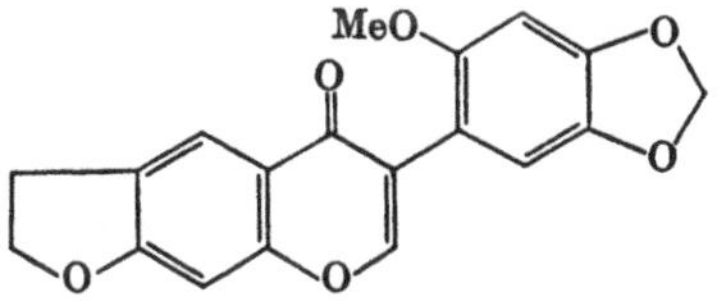

(83)

F. Phaseollin

Phaseollin (**84**), $C_{20}H_{18}O_4$, has recently been isolated from French beans (*Phaseolus vulgaris* L.).[67] It appears to be produced after fungal inoculation, and its biological activity indicates that it has the properties of phytoalexin. Phytoalexins have been defined as a class of antifungal compounds produced by host plants following fungal infections;[68–70] pisatin (**68**) is a substance of this type from the garden pea, *Pisum sativum*.[55]

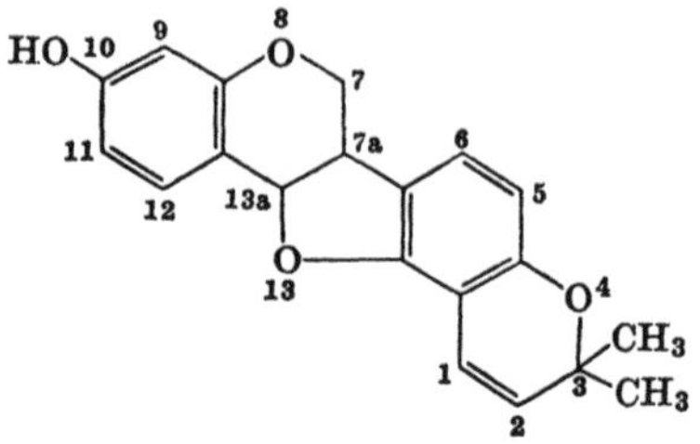

(**84**)

Phaseollin is a phenol, and contains no other hydroxylic groups. Methoxyl, carbonyl, and carboxyl groups are absent. The ultraviolet and infrared spectra of phaseollin are similar to those of homopterocarpin (**40**); this suggested that both substances contain the same coumarano-coumarin ring system which is present in pisatin (**68**)[58] and in several other substances from plants of Leguminosae, sub-family Papilionaceae. Detailed analysis of nuclear magnetic resonance spectra of phaseollin and homopterocarpin in $CDCl_3$ confirm that the former has structure **84**. Moveover, this analysis[71] indicates that phaseollin is closely related to the aglycone of the antifungal trifolirhizin (**64**).[49] This aglycone differs only in the position of the methylenedioxy grouping. The close agreements of p*K* and spectral values confirm that the phenolic groups in both substances are similarly located with respect to the other substituents in the molecule, and hence that the phenolic hydroxyl group in phaseollin is correctly located on C_{10}.

IV. Coumaronoflavan-4-ols

1. Naturally Occurring Coumaronoflavan-4-ols

This group is represented by naturally occurring cyanomaclurin.

A. Cyanomaclurin

Cyanomaclurin (1,3,8,10a - tetrahydroxy - 5a,10a - dihydro - 11*H* - benzofuro(3,2-b)-1-benzopyran), $C_{15}H_{12}O_6$, a colorless compound, derives its name from the deep blue solution it forms with alkali.

Perkin[72] isolated it from the heartwood of *Artocarpus integrifolia* (Jackwood). Structure **85** was proposed for cyanomaclurin by Appel and Robinson.[73] Using Perkin's method the heartwood was extracted with hot water. Freudenberg and Weinges[74] also adopted the same procedure, but separated out a small quantity of racemic cyanomaclurin by counter-current distribution between ether and water. From the light–petroleum and benzene extracts of heartwood, obtained from different places, Venkataraman and coworkers[75] isolated four related compounds: artocarpin, artocarpetin, artocarpanone, and isoartocarpin.

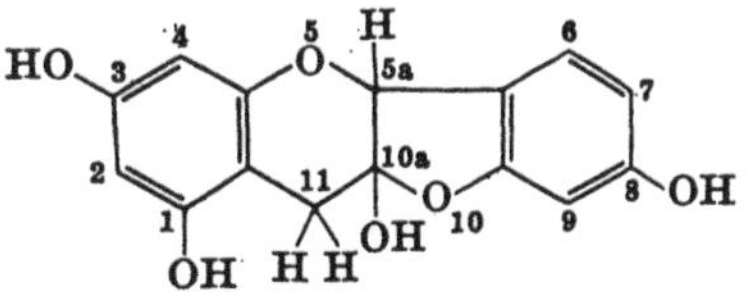

(85)

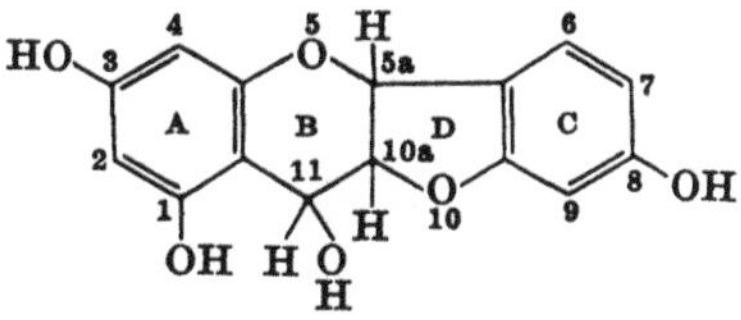

(86)

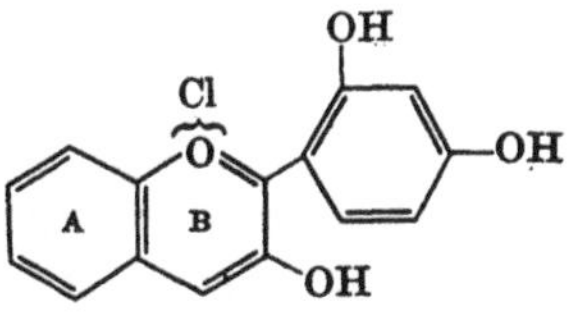

(87)

Cyanomaclurin yielded an acetate and trimethyl ether, conversion of cyanomaclurin into morindin chloride (**87**) also proceeded as recorded by Appel and Robinson[73] and by Chakravarty and Seshadri;[76] thus showing that the dihydrofuran ring of cyanomaclurin is unstable under the effect of sodium carbonate, followed by acidification.

There is no positive chemical proof which will conclusively support structure **85** or **86** for cyanomaclurin. Recently, Freudenberg[77] has favored structure **85** for cyanomaclurin. On the other hand, Chakravarty and Seshadri[76] have reported that cyanomaclurin acetate does not react with *N*-bromosuccinimide, a reaction characteristic of catechin acetates,[78] which have a reactive benzylic $\rangle CH_2$ group. Further, cyanomaclurin has high optical rotation, a property in which it differs from catechin, and resembles peltogynol (**88**)[79] which has the flavan-4-ol structure. However, oxidation of the $\rangle$CHOH group, which was successful in the case of peltogynol methyl ether, does not proceed with cyano-maclurin methyl ether.

A study of the nuclear magnetic resonance spectra of cyanomaclurin and its derivatives, for the indication of a >CH_2 group, helps to differentiate between structure **85** and **86**, since a >CH_2 group is present in

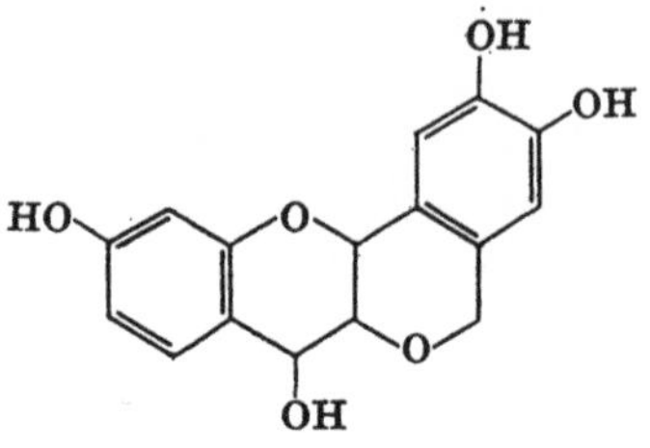

(88)

85 and absent in **86**. Whereas, (+)-catechin and (−)-epicatechin, closely related compounds which are used for comparison, showed the presence of a >CH_2 group, cyanomaclurin did not. Detailed studies indicate that the nuclear magnetic resonance spectral data agree with the requirements for **86**, and exclude structure **85**. Structure **86** is also supported by the spin–spin decoupling experiment.[76]

Based on structure **86** for cyanomaclurin, attempts were originally made to prepare the corresponding tetracyclic(3′,2′-furo)flavylium salts, as a stage in the synthesis of cyanomaclurin and its analogs. The flavylium salts have been synthesized according to Pratt and Robinson's[80] method. 6-Hydroxy- and 6-methoxycoumarones have been condensed with salicylaldehyde, resorcylaldehyde, its monomethyl ether, phloroglucinaldehyde, its dimethyl ether, and triacetate to give a number of 3′,2′-furoflavylium salts (**89**).

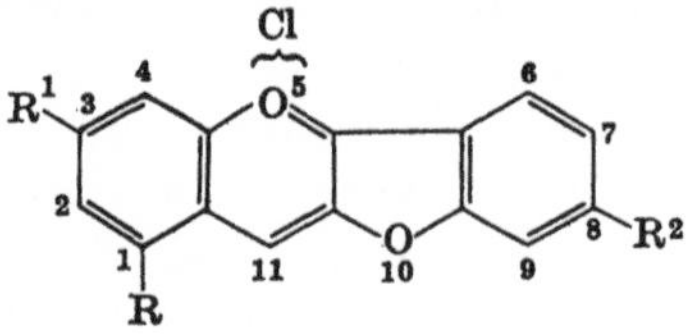

(89)

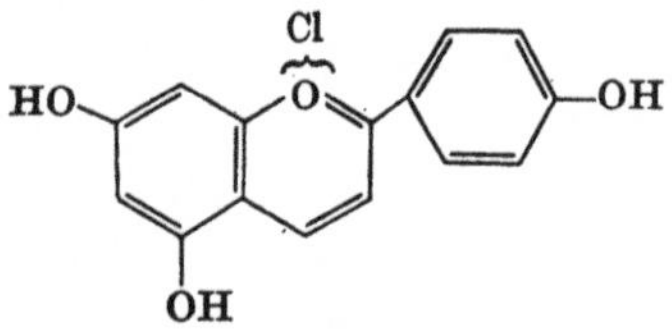

(90)

The furoflavylium salts resemble the flavylium salts of the geseneridin type (**90**), though differing in their spectral characteristics; introduction of the furan ring seems to have a hypsochromic effect.[81] On treatment with aqueous sodium acetate or pyridine, they change into color bases (**91**), when there is a hydroxyl group in a suitable position, e.g. 1,3, or

8 and a keto group preferably in the 7-position.[82] The partial methyl ether (**89**, $R = R^1 = OCH_3$, $R^2 = OH$) having only a 4′-hydroxyl group gave an anhydro base (**92**) with a keto group in the 4′-position.[83] On the other hand, when no free hydroxyl group was present, the flavylium salt underwent easy change to the corresponding chalcone (**93**). Hydroxyflavylium chlorides (**89**, $R = H$, $R^1 = R^2 = OH$), by treatment with pyridine and acetic anhydride, underwent conversion into the corresponding pseudoacetates (**94**).[84]

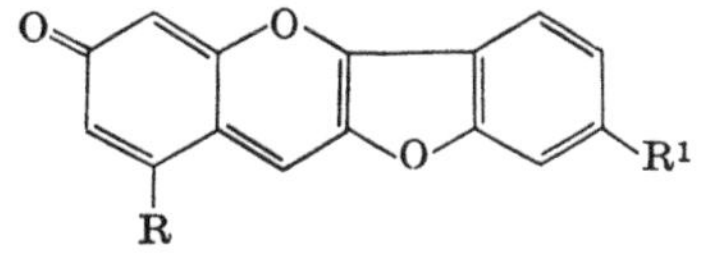

(**91**)

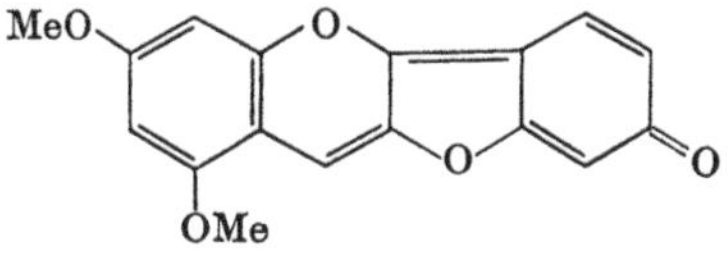

(**92**)

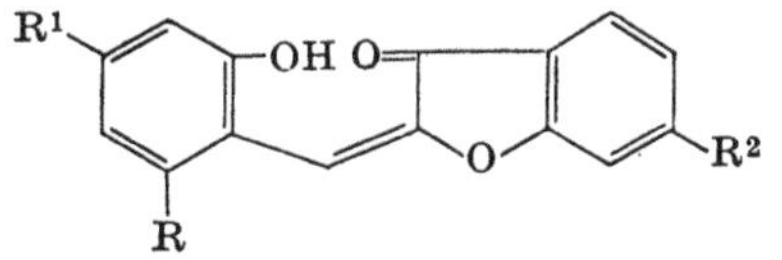

(**93**)

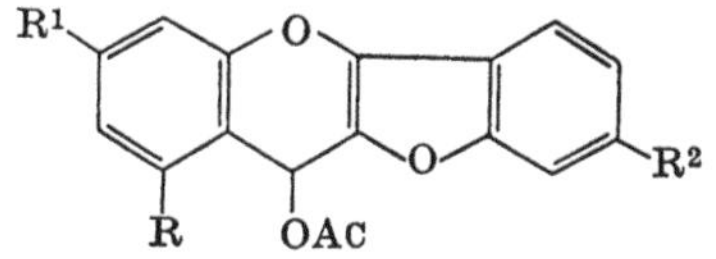

(**94**)

To explain the ready conversion of cyanomaclurin (**86**) under mild alkaline conditions into **87**,[76,81] it has been pointed out that 5,7,4′-trimethoxy-3′,2′-furoflavylium chloride is readily converted with hot aqueous sodium carbonate to the trimethoxy chalcone (**93**, $R = R^1 = R^2 = OCH_3$), which reproduces the furoflavylium salt; thus showing that the furan ring is unaffected under these conditions. The same is true of the corresponding trihydroxyflavylium chloride which does not form morindin chloride (**87**); thus excluding furan–ring fission during conversion of **86** to **87**. This observation may support the dipyran formula (**95a**), which has recently been suggested[81, 85] for cyanomaclurin, and may indicate that the second pyran in the 4,2′-position undergoes easy opening; its instability could be expected because it contains two axial linkages. The recent achievement of the synthesis of (±)-*O*-trimethyl cyanomaclurin (**95b**)[85a] has confirmed structure **95a** for cyanomaclurin.

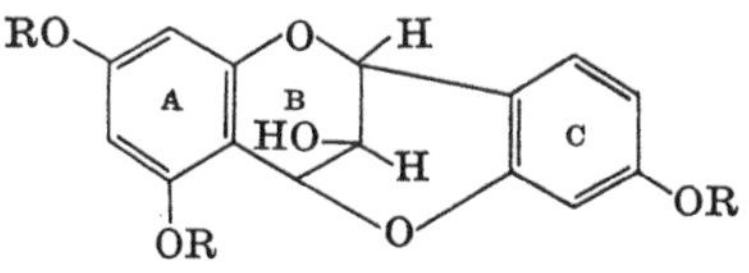

(**95a**) $R = H$; (**95b**) $R = CH_3$

2. Synthetic 11*H*-benzofuro(3,2-b)–1–benzopyran-11-ones

The synthesis of a number of 11*H*-benzofuro(3,2-b)–1–benzopyran-11-ones has recently been reported;[80] it involves the drastic demethylation of the appropriate 2′-methoxyflavonols. Treatment of 2′,6′-dimethoxyflavonol (**96**, R = H) and 2′,4′,6′-trimethoxyflavonol (**96**, R = OCH_3) with sulfuric acid (70 %) at the boiling point gave the corresponding chromonobenzofurans (**97** and **98**), respectively. The attempted cyclization of 2′-hydroxyflavonols, under similar conditions, failed.[86]

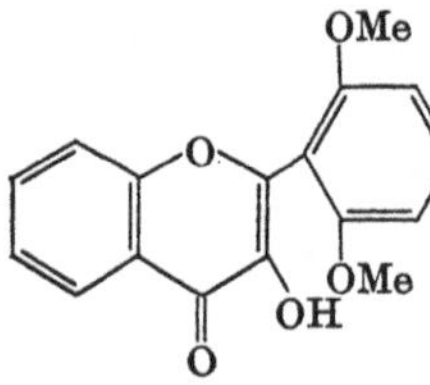

(96)

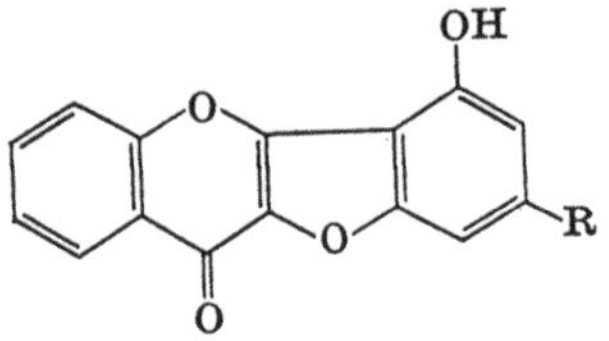

(97) R = H
(98) R = OH

A ready acid–catalyzed dehydration of 2,2′-hydroxybenzoylcoumaran-3-ones (**100**) to yield **101** has recently been achieved.[87] When 2,4-dihydroxy- (**99**, R = OH, R^1 = H) and 2-hydroxy-4,6-dimethoxy-ω-(2-methoxycarbonylphenoxy)acetophenone (**99**, R = R^1 = OCH_3) are treated with potassium carbonate in acetone,[82] the resulting 2,2′-hydroxybenzoylcoumaranones (**100**, R = OH, R^1 = H) and (**100**, R = R^1 = OCH_3) yield **101** (R = OH, R^1 = H) and **101** (R = R^1 = OCH_3), respectively by the action of methanolic hydrogen chloride. **101** (R = OCH_3, R^1 = H) is obtained upon methylation of **101** (R = OH, R^1 = H) (Eq. 13).

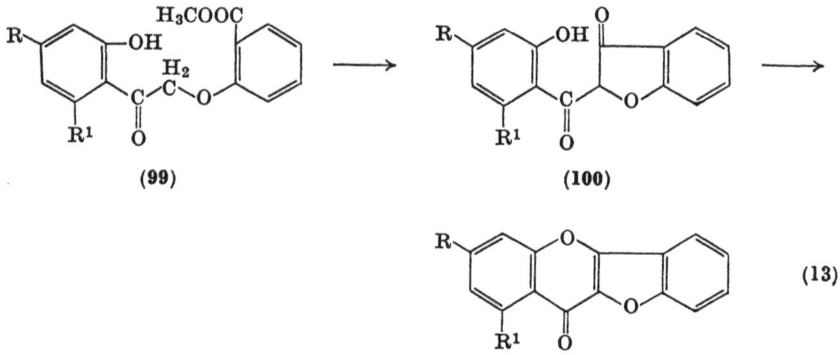

(13)

(101)

V. Coumaronocoumarins

This group is represented by naturally occurring coumestrol, wedelolactone, trifoliol, medicagol, and psoralidin. The coumaronocoumarins are of structural interest in that they are related to the coumaranochromans and the 3-arylcoumarins.

The names coumaronocoumarin, benzofurano-α-benzopyrone, and coumarinobenzofuran have been applied to the class of compounds of which coumestrol is a representative.[88] The trivial name, coumestan, has been proposed for the skeletal structure of the heterocyclic, four–ring system having the systematic name, 6*H*-benzofuro(3,2-c)–1–benzopyran-6-one.[89] The name is occasionally employed and the list of naturally–occurring coumestans is growing. However, the trivial names are used in this text.

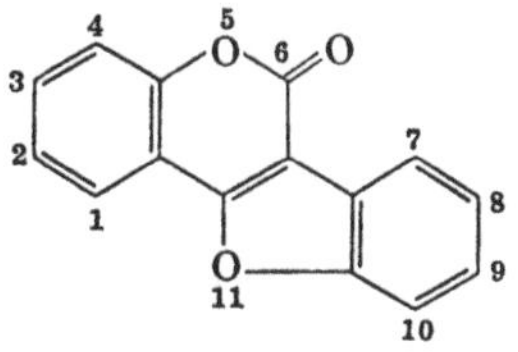

(Coumestan)

1. Naturally Occurring Coumaronocoumarins

A. Coumestrol

Coumestrol (3,9-dihydroxy-6*H*-benzofuro(3,2-c)–1–benzopyran-6-one or 3-benzofurancarboxylic acid, 2-(2,4-dihydroxyphenyl)-6-hydroxy-, δ-lactone) (**102**), $C_{15}H_8O_5$, is an oestrogenic substance found in ladino clover, in alfalfa,[90, 91] and in other legumes.[92] The compound **102** contains two free hydroxyl groups since it forms a diacetate and dimethyl ether.[90] The monomethyl derivative of **102** has been obtained by selective alkylation of coumestrol diacetate.[93] Fusion yielded only resorcinol and β-resorcylic acid, showing the number and position of the hydroxyl groups on the two rings. The idea that it might be a 3-phenylcoumarin arose from its oestrogenic properties, which are also found in isoflavones and in certain synthetic 3-phenylcoumarins. A coumarin–type structure was demonstrated by the fact that its blue fluorescence was not affected by ammonia, the bathochromic shift with alkali was much less than that expected from a flavone, and methylation under alkaline conditions brought about ring opening with the formation of a trimethoxy acid.

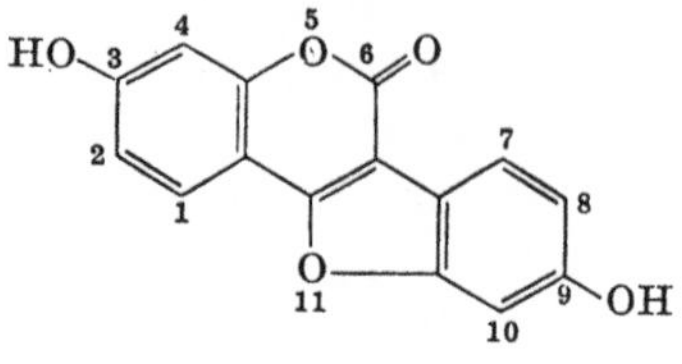

(102)

The furo type of structure **102** has been verified by a series of stepwise degradations, similar to those described for wedelolactone.[94, 95] Confirmation of structure **102** was accomplished by a series of progressive degradations involving: (a) alkaline methylation of 2-(2′,4′-dimethoxyphenyl)-6-methoxybenzofuran-3-carboxylic acid (**103**); (b) decarboxylation of the acid (**103**) by pyrolysis to yield 2-(2′,4′-dimethoxyphenyl)-6-methoxybenzofuran (**104**); (c) cleavage of the double bond by ozonolysis to form **105**; and (d) hydrolysis of the ester to 2,4-dimethoxy- (**106**), and 2-hydroxy-4-methoxybenzoic acid (**42**) (Eq. 14).

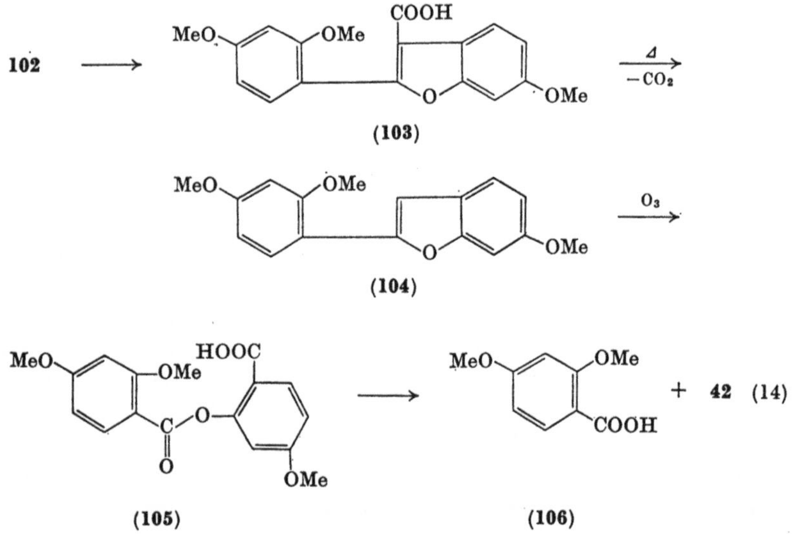

Synthesis of coumestrol (**102**) has been achieved, 2,4-dimethoxyphenylacetonitrile (**107**) is condensed with resorcinol to give α-(2,4-dimethoxyphenyl)-2,4-dihydoxyacetophenone (**108**). Treatment of the latter with methyl chloroformate yields 3-(2,4-dimethoxyphenyl)-4,7-dihydroxycoumarin (**109**), which on heating with aniline hydrochloride, yields coumestrol[96] (Eq. 15).

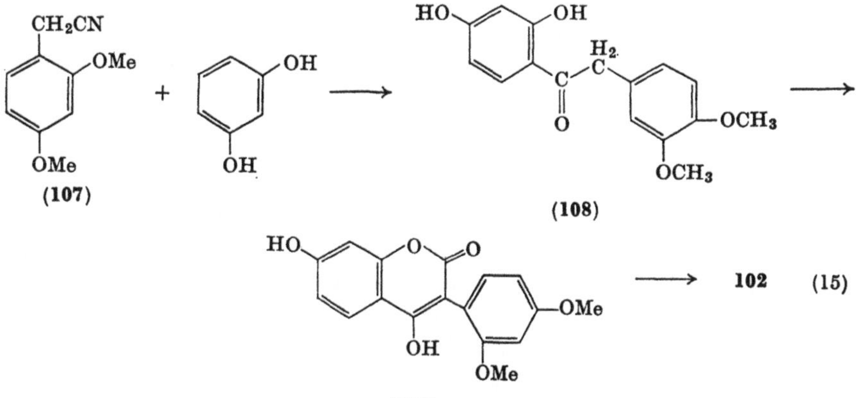

(107) (108) (109) (15)

Thermal condensation of *o*-methoxyphenylmalonate (**110**) with phenol, followed by treatment with pyridine hydrochloride produces coumestan (**111**)[97] (Eq. 16).

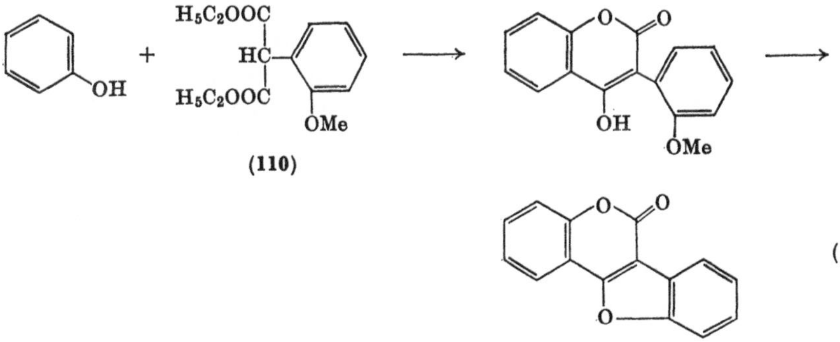

(110) (111) (16)

Synthesis of **111** has also been effected by treatment of the keto nitrile (**114**, $R = R^1 = H$), obtained by condensation of *o*-methoxyphenylacetonitrile (**112**) with ethyl *o*-methoxybenzoate (**113**, R = H) in the presence of hydrobromic acid[98, 99] (Eq. 17).

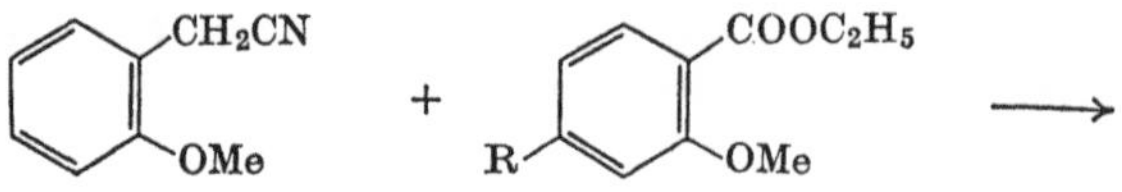

(112) (113)

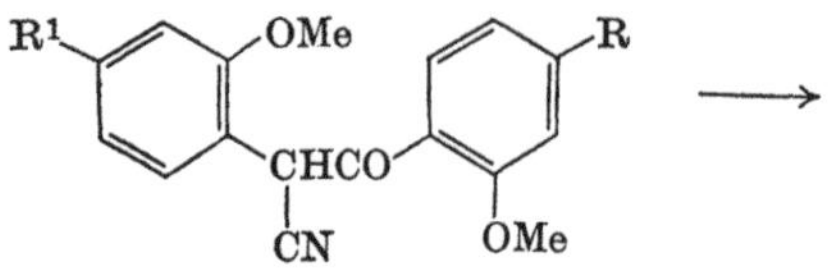

111 (17)

(114)

Treatment of 2,4-dimethoxybenzoyl-2,4-dimethoxyphenylacetonitrile (**114**, $R = R^1 = OCH_3$), obtained by condensation of **107** and **113** ($R = OCH_3$), with pyridine hydrochloride or hydriodic acid,[100] and/or hydrobromic acid in acetic acid[99] brought about demethylation, hydrolysis, and cyclization to yield coumestrol (**102**). This method has been used successfully for the synthesis of compounds analogous to coumestrol to ascertain structural features necessary for oestrogenic activity.[88,101]

Coumestrol and related compounds have been synthesized by hydrogen peroxide oxidation of appropriately substituted 2′-hydroxy-3-methoxyflavylium salts.[102] 2′,4′,7-Trihydroxy-3-methoxyflavylium chloride (**115a**), oxidized with hydrogen peroxide in aqueous methanol, gives the 3-carbomethoxybenzofuran (**116**), which rapidly lactonizes on acidification to **117**. 9-*O*-Methylcoumestrol (**118**, $R = OCH_3$, $R^1 = OH$) and 3-*O*-methylcoumestrol (**118**, $R = OH$, $R^1 = OCH_3$) have also been obtained by similar oxidation of 2′,4′-dihydroxy-3,7-dimethoxyflavylium chloride (**115b**) and 2′,7-dihydroxy-3,4′-dimethoxyflavylium chloride (**115c**).

The coumarinobenzofurans (**118**, $R = OCH_3$, $R^1 = H$) and (**118**, $R = R^1 = H$) were similarly prepared by oxidation, and subsequent debenzylation of the flavylium chlorides (**115d** and **115e**), obtained by acid condensation of the appropriate *o*-hydroxybenzaldehydes with ω-methoxy-2-benzyloxyacetophenone, respectively.[102] There are discrepancies between the properties of **118** ($R = R^1 = H$) and the compound recently described by Govindachari and coworkers.[88]

B. Wedelolactone

Wedelolactone (1,8,9-trihydroxy-3-methoxy-6*H*-benzofuro(3,2-c)–1–benzopyran-6-one) (**119**), $C_{16}H_{10}O_7$, is a greenish–yellow lactone, isolated from the leaves of *Wedelia calendulacea* (*Compositae*).[94,103,104] It is an

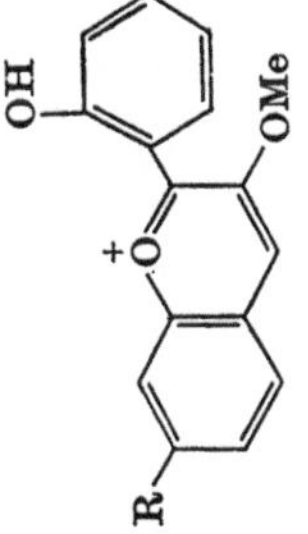

(115a) R = R[1] = OH;
(115b) R = OMe, R[1] = OH;
(115c) R = OH, R[1] = OMe;
(115d) R = OMe, R[1] = $OCH_2C_6H_5$;
(115e) R = H, R[1] = $OCH_2C_6H_5$

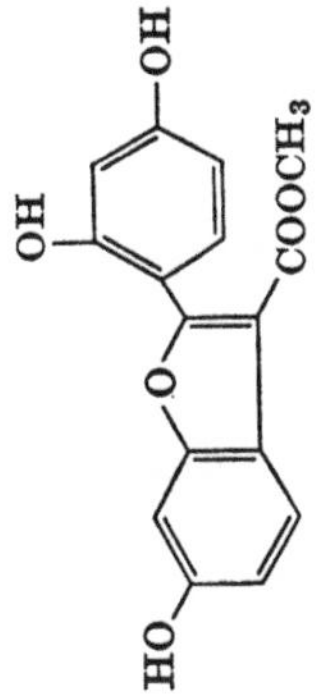

(116)

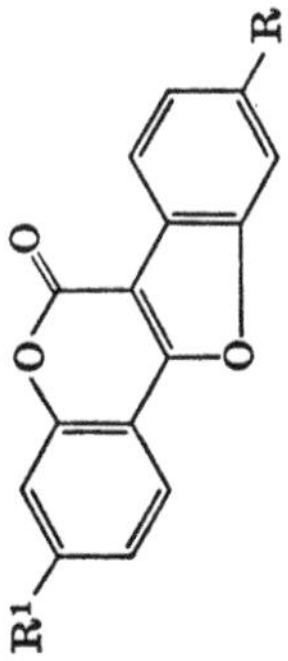

(118)

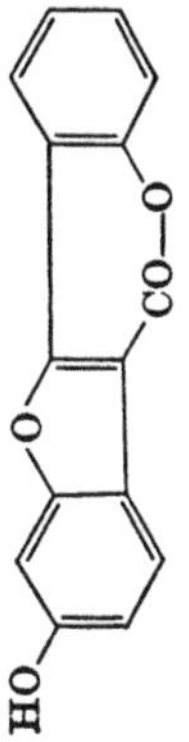

(117)

example of the angular coumaronocoumarins, which have been obtained by cyclodehydration of 3-substituted coumarins.[103–106] It contains one methoxyl group, forms a triacetate, and has an absorption band at 1707 cm^{-1}. It has the characteristics of an unsaturated δ-lactone, and one inert oxygen atom. Complete methylation with methyl sulfate gives tri-*O*-methylwedelolactone (**120**). The lactone properties are seen in the alkaline hydrolysis of the trimethyl ether to a phenolic acid, tri-*O*-methylwedelic acid (**121a**), $C_{19}H_{18}O_8$, which relactonizes when heated in acid conditions, but yields methyl tetra-*O*-methylwedelate when treated with diazomethane. Hydrolysis of this ester yields tetra-*O*-methylwedelic acid (**121b**), $C_{20}H_{20}O_8$, which is readily decarboxylated to give the benzofuran (**122**), whose structure is verified by ozonolysis to the aldehyde ester (**123**), $C_{19}H_{20}O_8$, which is readily hydrolyzed to give 6-hydroxyveratraldehyde and 2,4,6-trimethoxybenzoic acid (Eq. 18).

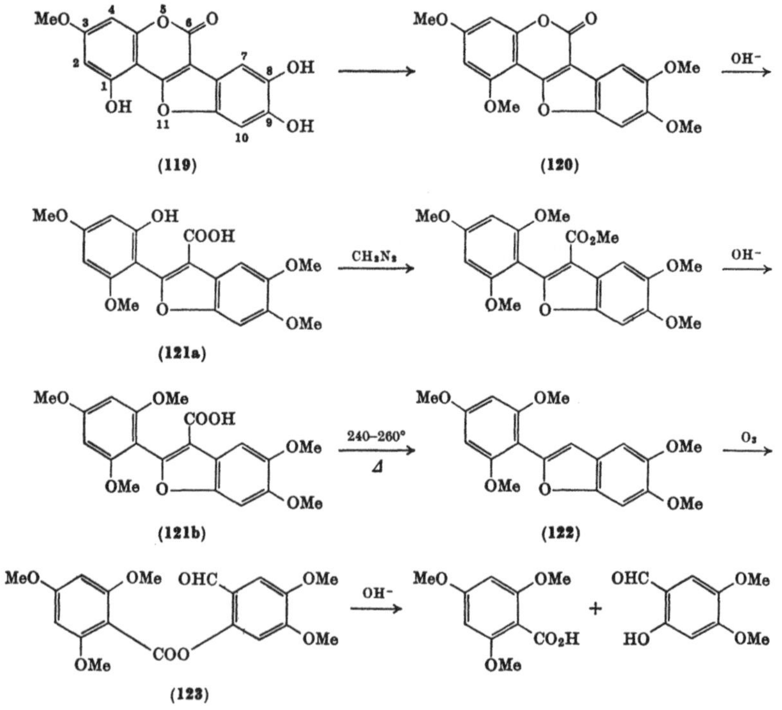

(18)

The methoxyl groups of wedelolactone were located by similar studies, after ethylation.[103, 104] The lactone ring was opened, the phenolic hydroxyl group ethylated, and the carboxyl group removed to give the benzofuran. As expected, ozonolysis led to 4,5-diethoxy-2-hydroxy-benzaldehyde, but the other product appeared to suffer spontaneous decarboxylation, so the result has no value for orientation purposes. A useful result was obtained upon hydrolysis of the ozonide, followed by reduction with lithium aluminum hydride to yield the benzyl alcohol; thus proving that wedelolactone is a 7-methoxycoumarin derivative (Eq. 19).

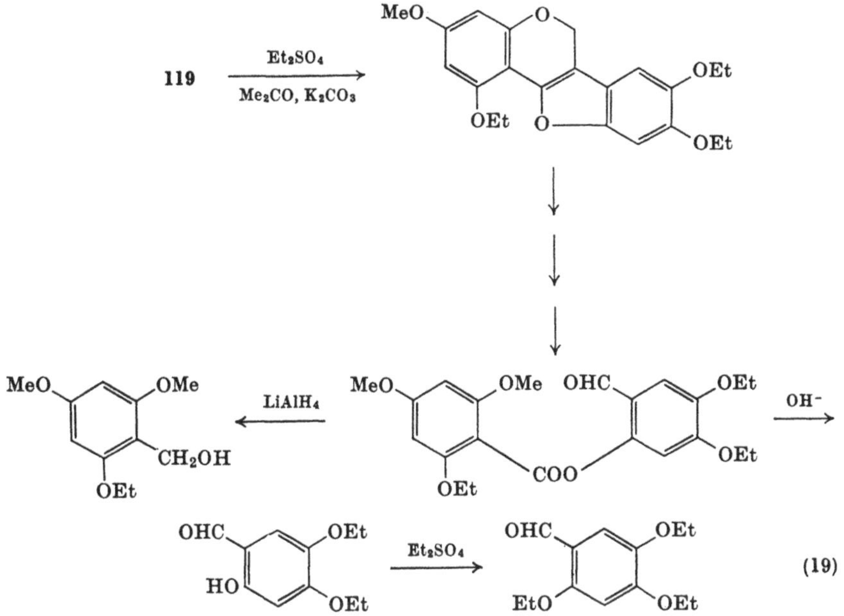

(19)

The deoxybenzoin (**124**) prepared by means of the appropriate Hoesch reaction, can be converted into the 4-hydroxycoumarin (**125**) by the ethyl carbonate–sodium technique, and when heated with aniline hydrochloride, is converted into the coumaronocoumarin, tri-*O*-methylwedelolactone (**120**).[104, 105] Since the 7-methoxyl group in a benzopyrone is the least easily attacked by acids, controlled treatment of the trimethyl ether (**120**) with hydrogen iodide produces **119**, identical with the natural product[107] (Eq. 20).

Dehydrogenation of catechol in presence of 4,5-dihydroxy-7-methoxy-coumarin, obtained by partial methylation of 4,5,7-trihydroxycoumarin with methyl sulfate and sodium carbonate, and potassium ferricyanide, yielded wedelolactone (**119**)[106] (Eq. 21). Similarly, the angular coumar-onocoumarin (**126**) was obtained by dehydrogenation of catechol in

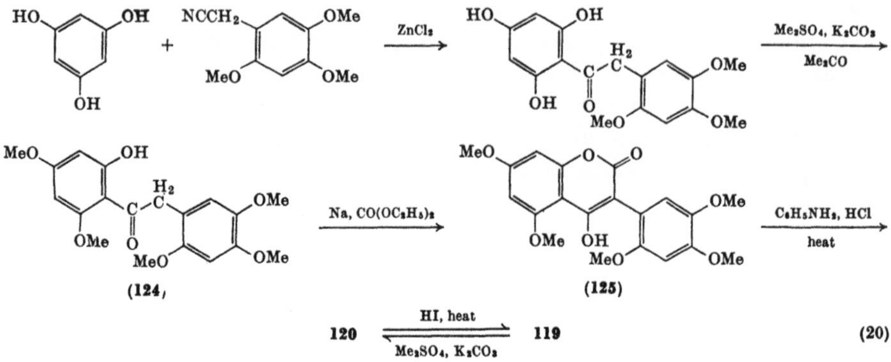

$$120 \underset{Me_2SO_4,\ K_2CO_3}{\overset{HI,\ heat}{\rightleftharpoons}} 119 \qquad (20)$$

presence of 4-hydroxycoumarin and a mixture of sodium acetate and potassium iodate[106] (Eq. 21).

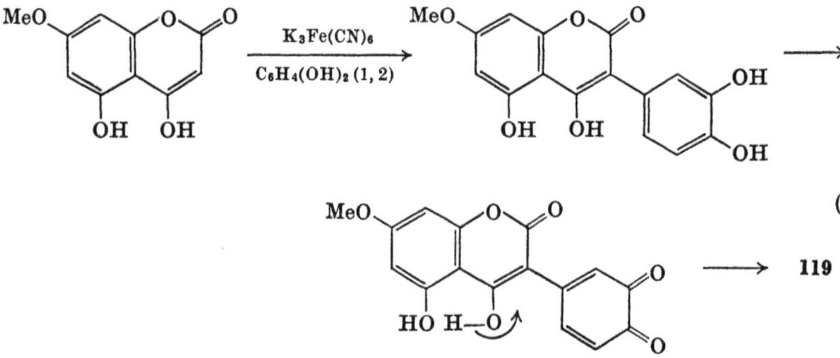

$$(21)$$

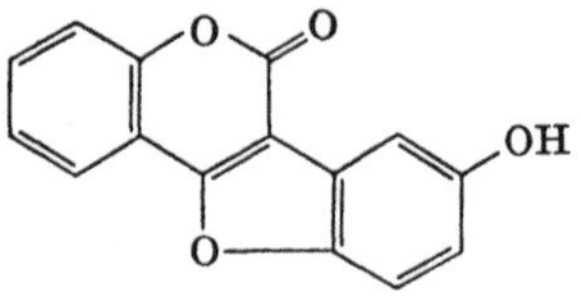

(126)

Synthesis of tri-O-methylwedelolactone (**120**) has recently been achieved.[108] The keto nitrile (**127**), obtained by condensation of 2,4,5-trimethoxybenzyl cyanide and ethyl 2,4,6-trimethoxybenzoate in the presence of sodium hydride, was treated with pyridine hydrochloride to yield the phenol (**128**), which was readily methylated to give **120** (Eq. 22).

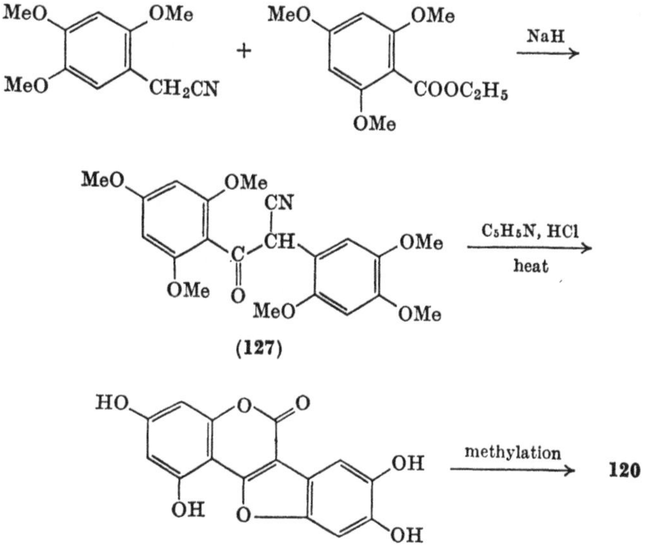

(127)

(128)

(22)

C. *Trifoliol*

Trifoliol (3,7-dihydroxy-9-methoxy-6*H*-benzofuro(3,2-c)–1–benzopyran-6-one) (**129**), $C_{16}H_{10}O_6$, is obtained from ladino clover. Briefly, isolation involved suspending the acetone solubles from ladino clover meal in chloroform, extracting with alkali, and reacidifying the aqueous extract to precipitate a sediment. Countercurrent distribution of this sediment yielded a crude crystalline material from which trifoliol was obtained.[109]

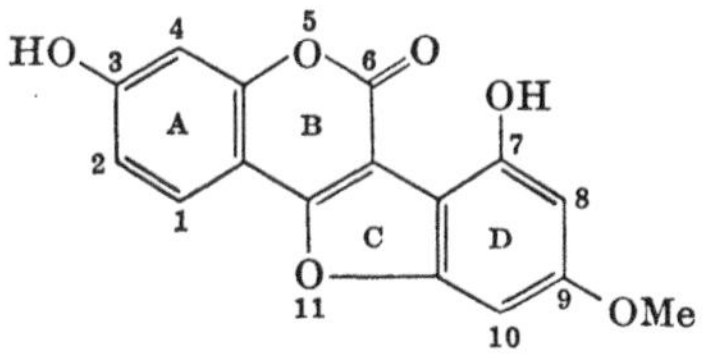

(129)

The compound contains one methoxyl group, and the formation of diacetyl and dimethyl derivatives indicates two phenolic hydroxyl groups. The ultraviolet spectrum shows that trifoliol and coumestrol (**102**) are structurally related, and furthermore, that the methoxyl group of trifoliol is probably located at the 9-position. Nuclear magnetic resonance spectrum shows that the only positions for the oxygen substituents in the D ring are C_7 and C_9. Confirmation of structure **129** was accomplished by a series of progressive degradations: (a) alkaline methylation to a tetramethyl ether (**130**, R = CH_3), which was hydrolyzed to the acid (**130**, R = H); (b) decarboxylation by pyrolysis of the acid (**130**, R = H) to yield 2-(2′,4′-dimethoxyphenyl)-6,8-dimethoxybenzofuran (**131**); (c) cleavage of the double bond of **131** by ozonolysis, and reduction of the product with triethylphosphite to an aldehyde, **132**; and (d) hydrolysis to 2-hydroxy-4,6-dimethoxybenzaldehyde and 2,4-dimethoxybenzoic acid(Eq. 23).

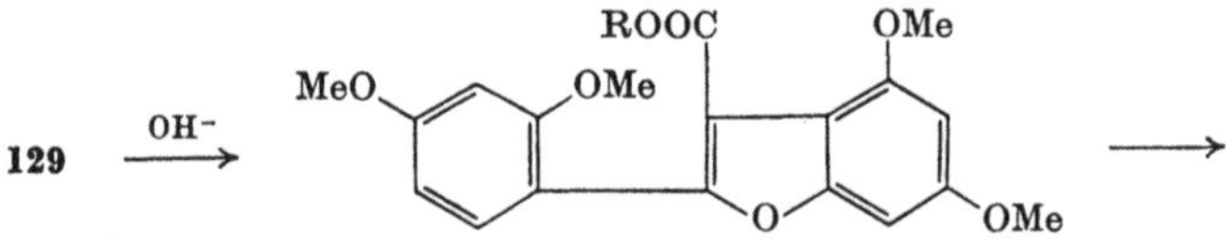

(**130**)

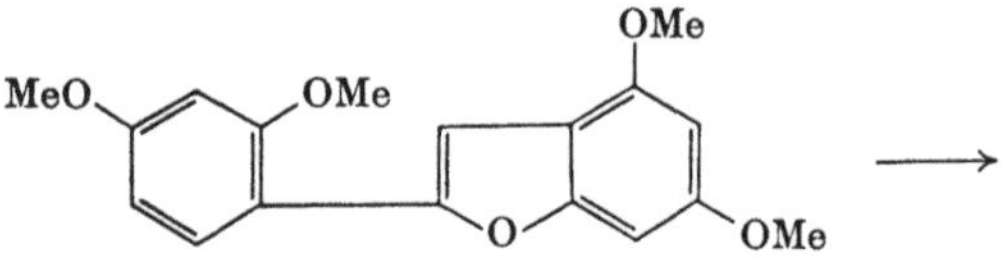

(**131**)

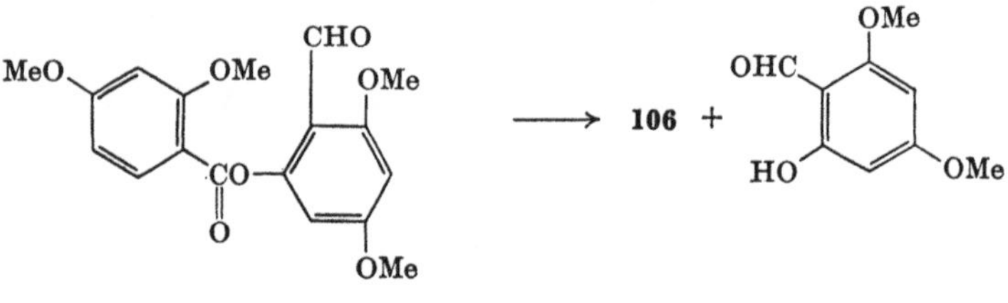

(**132**) (23)

Structure **129** was confirmed unequivocally by its synthesis, starting with 5-benzoyloxy-7-hydroxy-3-methoxy-2′,4′-dibenzyloxyflavylium chloride (**133**).[110] Peroxide oxidation of the flavylium salt gave the intermediate **134**, which was methylated to **135**, and then debenzylated to yield 7-benzoyloxy-3-hydroxy-9-methoxycoumestan (**136**). Alkaline hydrolysis of the latter compound furnished 3,7-dihydroxy-9-methoxycoumestan (trifoliol) (Eq. 24).

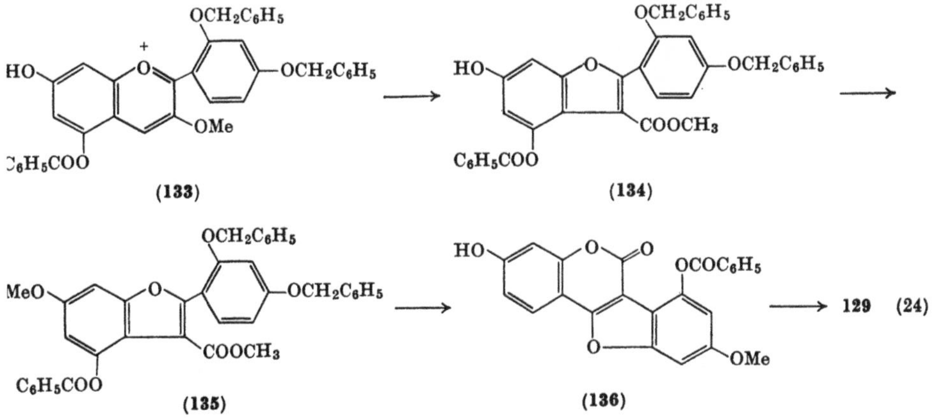

Trifoliol (**129**) is the first reported natural product containing a substituted phloroglucinol-like structure in ring D. This is of particular interest because it was thought that only the A ring of flavonoids arises from phloroglucinol.[111] Trifoliol has no oestrogenic activity, in contrast to the parent phenol **137** which is active as coumestrol (**102**).[109]

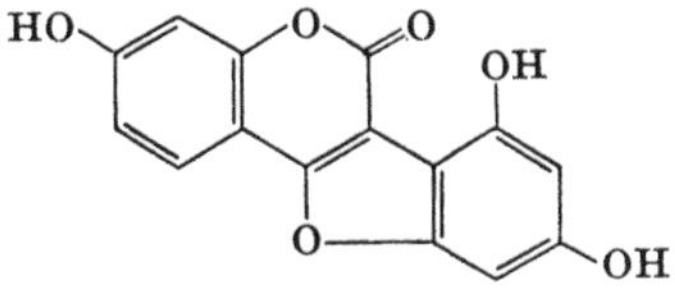

(137)

D. Medicagol

Medicagol (3-hydroxy-8,9-dioxymethylene-6*H*-benzofuro(3,2-*c*)-1-benzopyran-6-one) (**138**), $C_{16}H_8O_6$, is obtained as a mixture with pisatin from alfalfa meal. The name medicagol is proposed from the genus name for alfalfa, Medicago.[111a] The basic ring pattern of medicagol is very similar to that of the antifungal agent, pisatin (**68**) or its anhydro derivative (**69**).[57, 60] The similarity of the ultraviolet spectrum of this inseparable mixture to that of coumestrol (**102**) suggested their relation to coumestans. Since coumestans can be degraded systematically by methylative ring opening (**139**, $R = COOCH_3$), hydrolysis (**139**, $R = COOH$), and decarboxylation to their more soluble benzofuran derivatives (**139**, $R = H$), this offered a means of purification. The benzofuran could then be used in elucidating the structure of the parent phenolic from which it was derived. Thus, using this procedure with the medicagol mixture, the methoxybenzofuran derivative of medicagol (**139**, R=H) was purified by fractional crystallization of the ether.

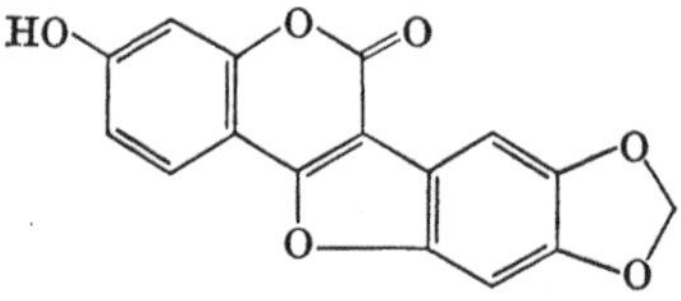

(**138**)

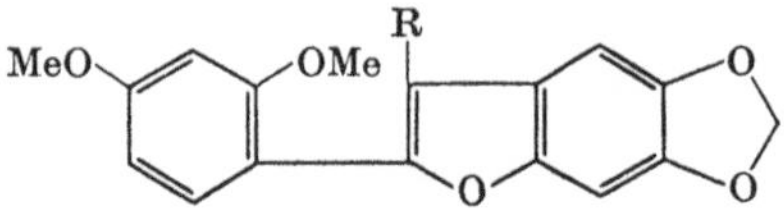

(**139**)

Medicagol has a hydroxyl group in the 3-position. The methoxybenzofuran of medicagol, $C_{17}H_{14}O_5$, could be derived from a coumestan having a methylenedioxy-, and a hydroxyl group. The presence of a methylenedioxy group was further indicated by a positive Hansen test,[112] and by the nuclear magnetic resonance and infrared spectra.

Strong acid hydrolysis of medicagol acetate mixture gave 3,8,9-trihydroxycoumestan (**141**), thus further indicating that the medicagol structure must be 7-hydroxy-11,12-methylenedioxycoumestan. Confirmation of this structure was obtained by synthesis from 4,7-dihydroxycoumarin (**140**) which was oxidatively coupled with catechol.[106] Methylenation of **141** with diiodomethane gave **138**. Recently, medicagol has been synthesised via hydrogen peroxide oxidation of 3-methoxy-6,7-methylenedioxy-2′,4′-dihydroxyflavylium chloride.[113]

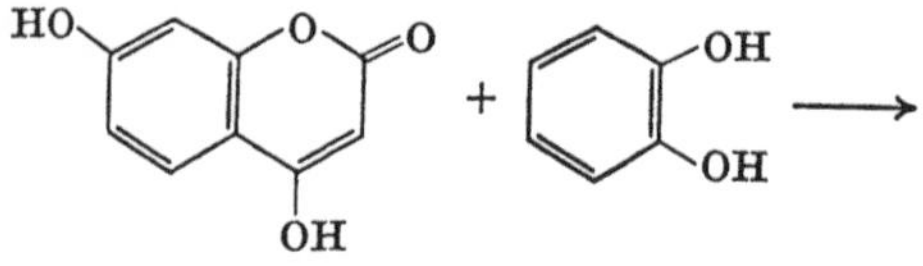

(**140**)

138

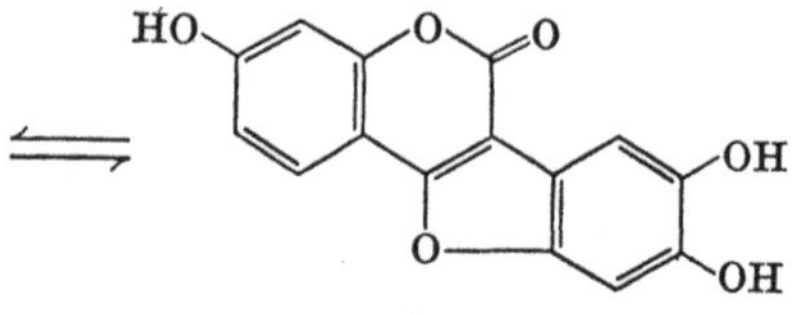

(25)

(**141**)

E. Psoralidin

Psoralidin (2-isopentenyl-3,9-dihydroxy-6*H*-benzofuro(3,2-*c*)–1–benzopyran-6-one) (**142**, R=H), $C_{20}H_{16}O_5$,[114] is a phenolic coumarin, first isolated from the pericarp of the seeds of *P. corylifolia* L. by Chakravarti and coworkers.[115] This group of workers assigned the molecular formula $C_{16}H_{14}O_4$, leading to untenable structure, and suggested the presence of a phenolic hydroxyl group, a lactone group, and an isopentenyl group, as well as the formation of acetone on chromic acid oxidation. Recently, it has been shown[114] that psoralidin is without doubt the isopentenyl derivative of coumestrol;[102] nevertheless, it has a quite different source, being isolated from the alcoholic extract of the seed kernel of *Psoralea corylifolia* L, whence several other coumarins have been obtained. The compound contains two hydroxyl groups, one conjugated δ-lactone, one isopentenyl group, and an intramolecular ether linkage.

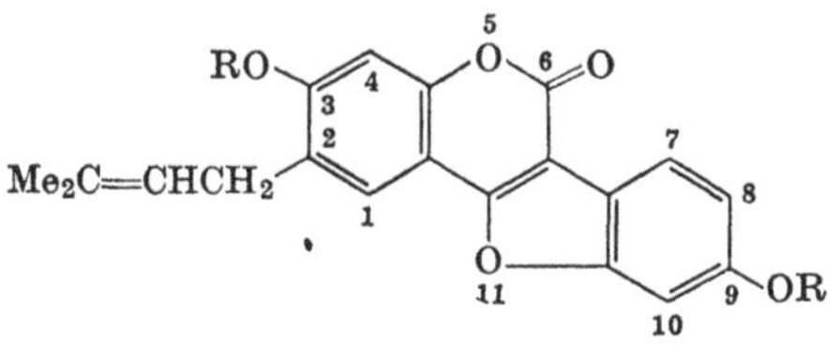

(142)

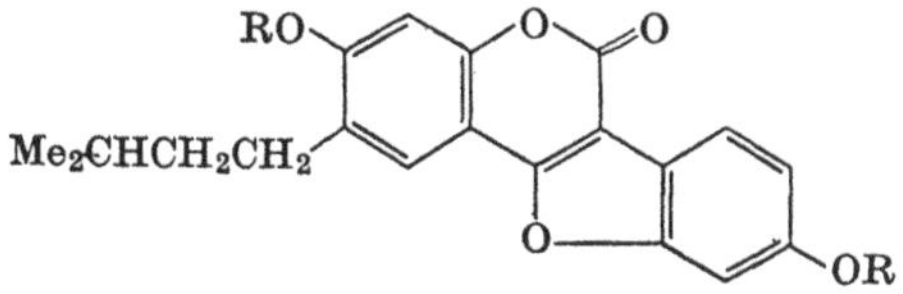

(143)

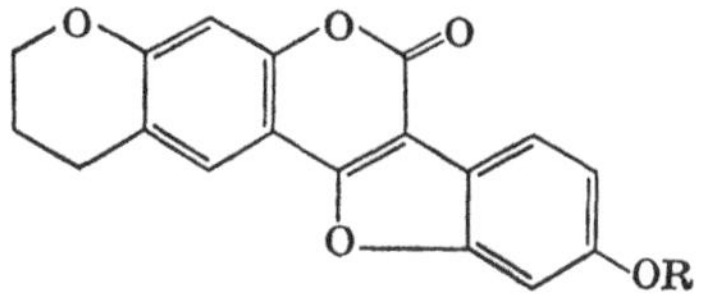

(144)

Psoralidin (**142**, R = H) and its dimethyl ether (**142**, $R = CH_3$) smoothly absorbed one molecule of hydrogen to yield dihydropsoralidin (**143**, R = H) and dihydropsoralidin dimethyl ether (**143**, $R = CH_3$) respectively. The ultraviolet absorption spectra of the dihydro compounds are similar to those of psoralidin and its dimethyl ether, thus establishing the presence of a non-conjugated linkage in the molecule of psoralidin. The formation of acetone, on chromic acid oxidation of psoralidin dimethyl ether,[114–115] establishes the presence of the $Me_2C{=}C$— group in psoralidin. Further, dihydropsoralidin on treatment with fuming nitric acid gives isocaproic acid; thus confirming the presence of the $Me_2C{=}CHCH_2$— group directly attached to the aromatic nucleus.[114–115] Upon treatment with methanolic hydrochloric acid, psoralidin undergoes isomerization to the monohydroxychroman (**144**, R = H) having identical ultraviolet absorption spectra with those of psoralidin.[116] The formation of the chroman (**144**, R = H), named isopsoralidin[117] establishes that the isopentenyl group is situated at the *ortho* position to one of the two phenolic hydroxyl groups in psoralidin. On fusion with potassium hydroxide, psoralidin gives,[120] as expected, resorcinol, and β-resorcylic acid. The presence of a coumarin-type structure is demonstrated by the fact that psoralidin dimethyl ether (**142**, R = CH_3) on treatment with methanolic alkali and methyl sulfate yields psoralidin trimethyl ether (**145**). After saturation of the side chain, dihydropsoralidin dimethyl ether (**143**, R = CH_3) on similar treatment with alkali and methyl sulfate yields methyl dihydropsoralidinate trimethyl ether (**146**, R = CH_3), which on saponification affords dihydropsoralidinic acid trimethyl ether (**146**, R = H). The latter compound can be decarboxylated to the benzofuran (**147**).

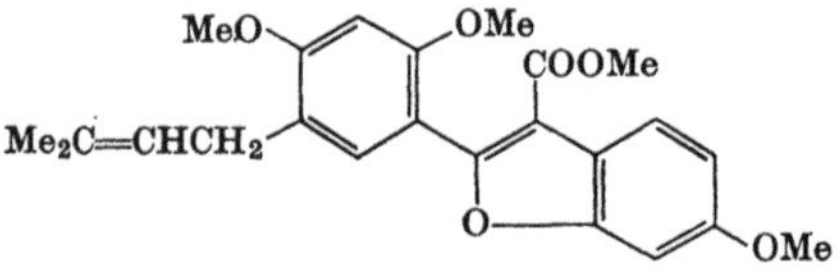

(145)

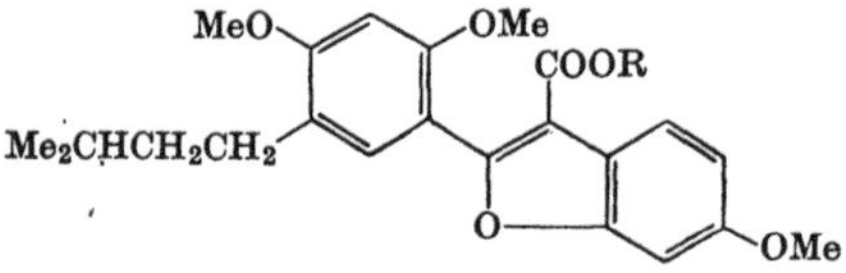

(146)

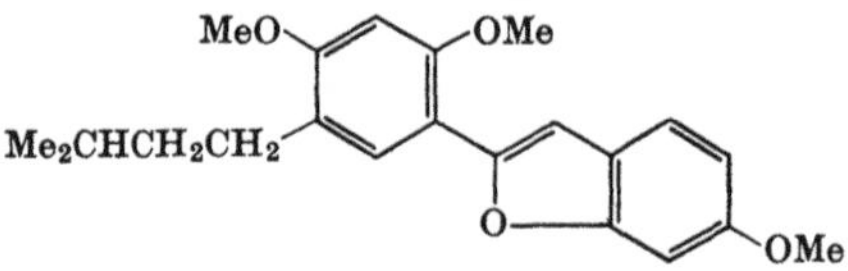

(147)

Osmium tetroxide hydroxylates the furan ring in **147** to the diol **148**, and the diol tautomeric form **149** upon oxidation with sodium periodate yields 2-hydroxy-4-methoxybenzaldehyde and the isopentylbenzoic acid (**151a**) which has been synthesized by an unambiguous method.[114]

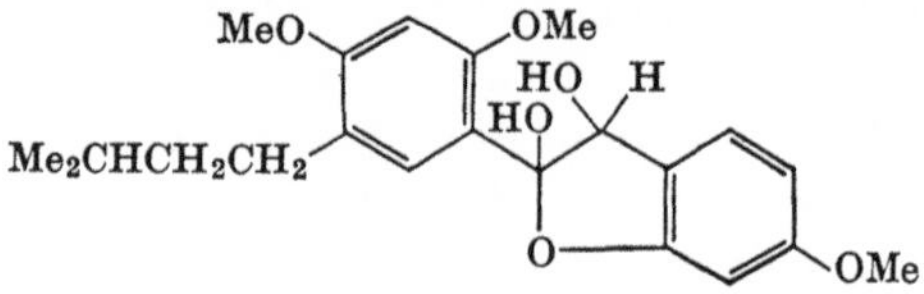

(148)

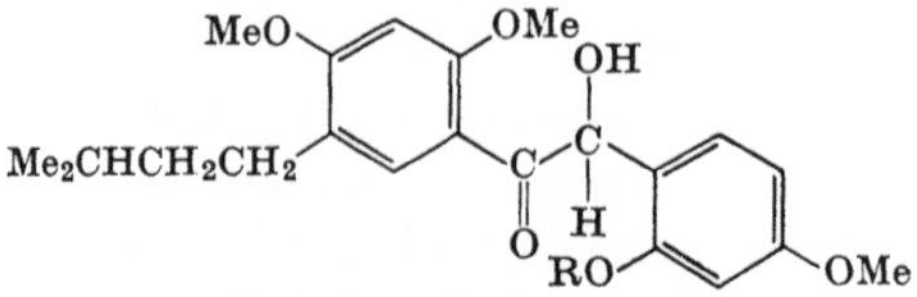

(149) R = H (**150**) R = CH_3

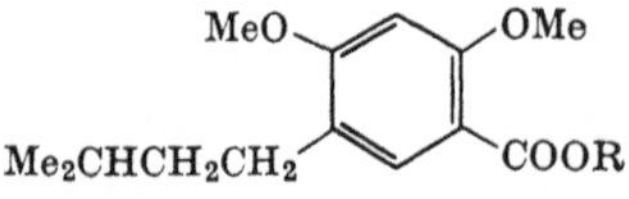

(151a) R = H; (**151b**) R = CH_3

Thus, psoralidin (**152**, R = H) is an addition to the family of coumestans. Dihydropsoralidin (**143**, R = H) but not psoralidin, which is labile toward acids, has been synthesized by standard methods.[118, 119, 121] Treatment of 2,4-dihydroxy-5-isopentyl-2,4-dimethoxybenzyl ketone **152**, obtained by Hoesch reaction of 2,4-dimethoxybenzyl cyanide and 4-isopentylresorcinol with ethyl chloroformate, followed by treatment with alkali, and then with acid gave 4,7-dihydroxy-6-isopentyl-3-(2′,4′-dimethoxyphenyl)coumarin (**153**). Demethylation of the latter compound with aniline hydrochloride resulted in the formation of dihydropsoralidin (**143**, R = H); its dimethyl ether (**143**, R = CH_3) is identical with that obtained from dihydropsoralidin[118, 119, 121] (Eq. 26).

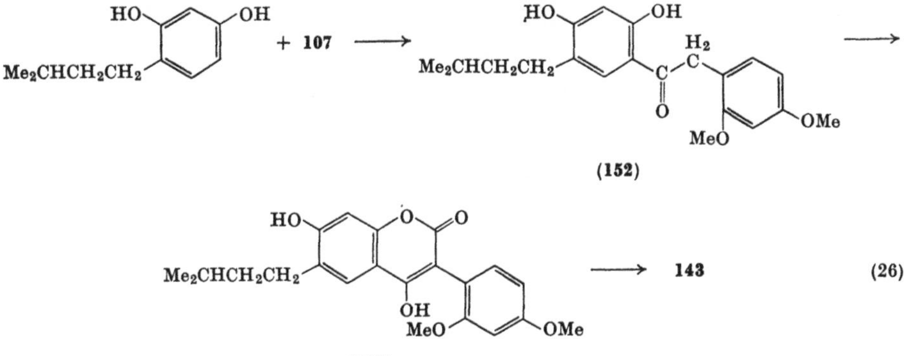

(26)

Treatment of the keto nitrile, α-(2,4-dimethoxy-5-isopentylbenzoyl)-2,4-dimethoxybenzyl cyanide (**150**), obtained by condensation of methyl 2,4-dimethoxy-5-isopentylbenzoate (**151b**) and 2,4-dimethoxybenzyl cyanide (**107**), in the presence of pyridine hydrochloride, led to the formation of **143** (R = H)[122] (Eq. 27).

$$\textbf{151b} + \textbf{107} \xrightarrow{\text{NaH}} \textbf{150} \longrightarrow \textbf{143} \qquad (27)$$

Synthesis of isopsoralidin (**144**, R = H) has been similarly achieved either from 2,4-dimethoxybenzyl-7-hydroxy-2,2-dimethylchroman-6-yl (**154**),[118] or from the keto nitrile (**155**).[122]

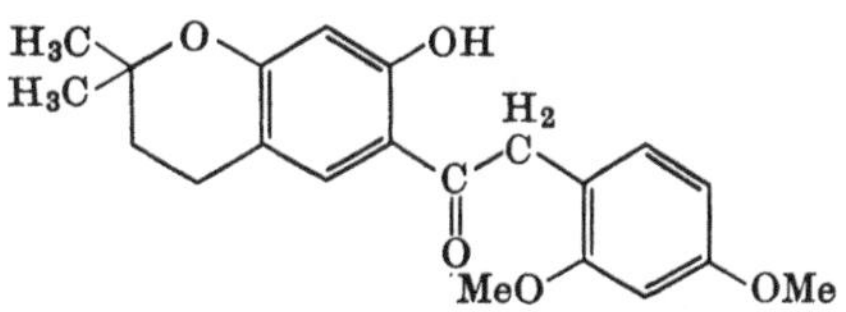

(154)

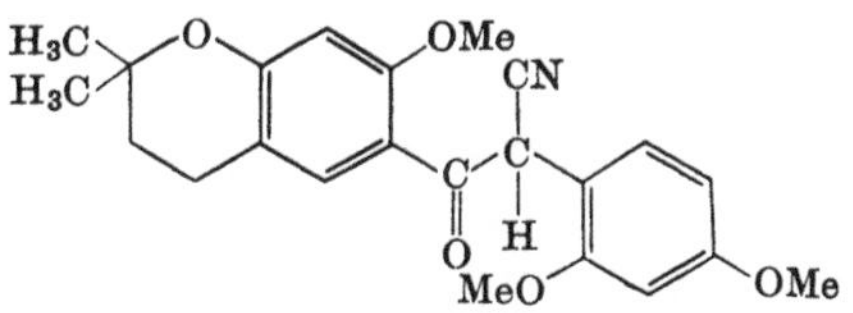

(155)

VI. 3-Arylfurocoumarins

The sole representative of this group which is known to occur naturally is pachyrrhizin, $C_{19}H_{12}O_6$. It was isolated from the seeds of yam beans, *Pachyrrhizus erosus* as green–yellow needles. It is a 3-phenylcoumarin, an unusual arrangement similar to that in isoflavones and found elsewhere only in a few complex compounds, e.g. wedelolactone.[119] Infrared and ultraviolet spectra suggested a lactone ring; a fact which has been confirmed by methylative ring opening.[44] The usual tests showed that one methoxyl group and a methylenedioxy grouping were present. Oxidation of pachyrrhizin with sodium periodate–potassium permanganate reagent gave 6-methoxypiperonylic acid (**157**). On the other hand, treatment of pachyrrhizin with hydrogen peroxide yielded the anhydride of 2-methoxy-4,5-methylenedioxyphenylmaleic acid (**158**). Oxidation of dihydropachyrrhizin (**159**), together with ozonization, gave 6-hydroxycoumaran-5-aldehyde (**160**) (Eq. 28).

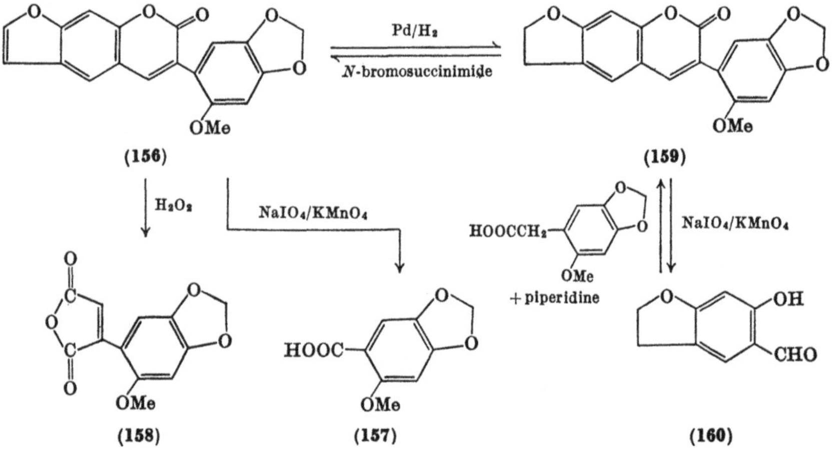

(28)

Structure **156** assigned to pachyrrhizin[44] was confirmed by synthesis.[123] The methyl ester of 2-methoxy-4,5-methylenedioxyphenylacetic acid was condensed with **160** in the presence of piperidine, followed by dehydrogenation of the reaction product **159**, with *N*-bromosuccinimide in dimethylaniline, to yield **156**.

Recently, synthesis of the methyl ether of 4-hydroxy-3-phenyl-4′,5′-dihydrofuro(3′,2′–6,7)coumarin (**161**) has been achieved from 6-hydroxy-5-(ω-phenylacetyl)coumaran (**162**).[29] 4-Hydroxy-3-phenylfuro(3′,2′–6,7)coumarin (**163**) was formed by dehydrogenation of **161** with palladium and charcoal. It was also obtained from **162** via the intermediates, dihydrofuroisoflavone (**164**), furoisoflavone (**165**), and 6-hydroxy-5-(ω-phenylacetyl)benzofuran (**166**) (Eq. 29).

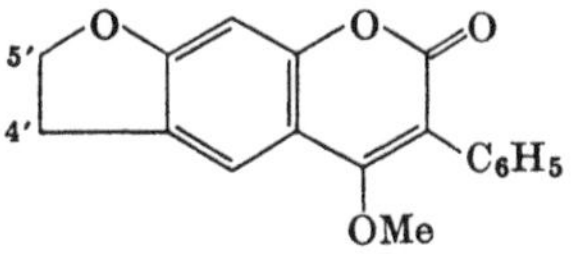

(**161**)

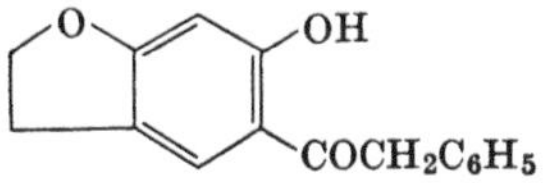

(**162**)

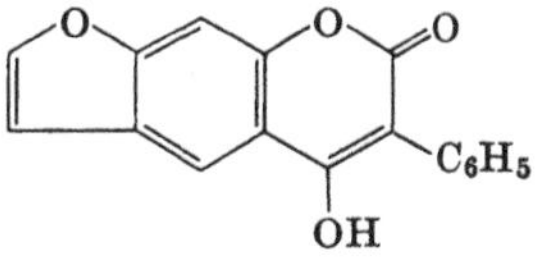

(**163**)

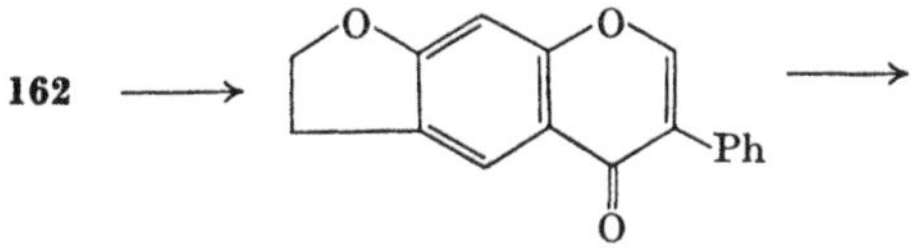

(**164**)

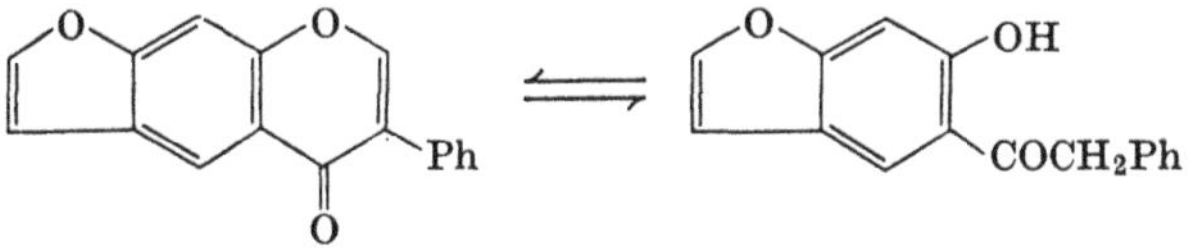

(**165**) (**166**) (29)

VII. Coumaronofurocoumarins

This group is represented by the natural occurrence of erosnin (**167**), $C_{18}H_8O_6$, isolated from the seeds of *phacyrrhizus erosus* (yam beans). It decomposes at high temperature without melting, readily sublimes at 230–50°/0.02 mm, exhibits blue fluorescence in methoxyethanol, and is a very insoluble substance. Eisenbeiss and Schmid[40] relating these characteristics to a pachyrrhizin–like (3-phenylcoumarin) derivative[44] have succeeded in preparing a methyl ether methyl ester, although erosnin contains no hydroxyl groups. Positive Eegrieve test[43] illustrates the presence of a methylenedioxy grouping. Alkaline methylation of erosnin gave the acid (**168**), which was hydrogenated with the help of a rhodium catalyst to yield the dihydromethoxy acid (**169b**). The latter was decarboxylated to the benzofuran (**169a**) which was further decomposed by ozonolysis to produce 2-formyl-4,5-methylenedioxyphenyl-6-methoxycoumaran-5-carboxylate (**170**), which was identified by synthesis (Eq. 30).

The total synthesis of erosnin has recently been achieved.[124] The dehydrogenative condensation of 5-hydroxypsoralen (**171**) and catechol

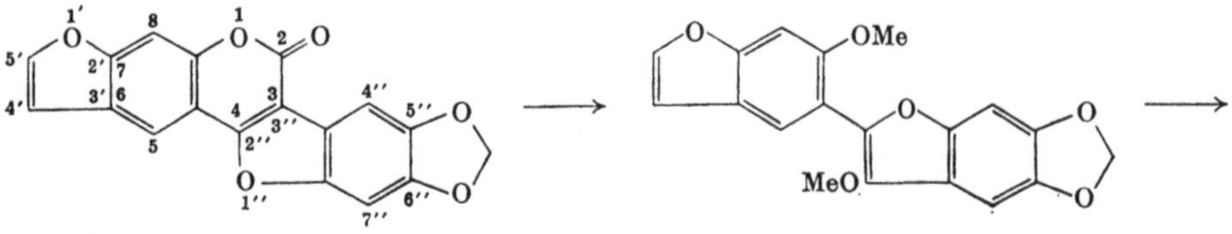

(167) (168)

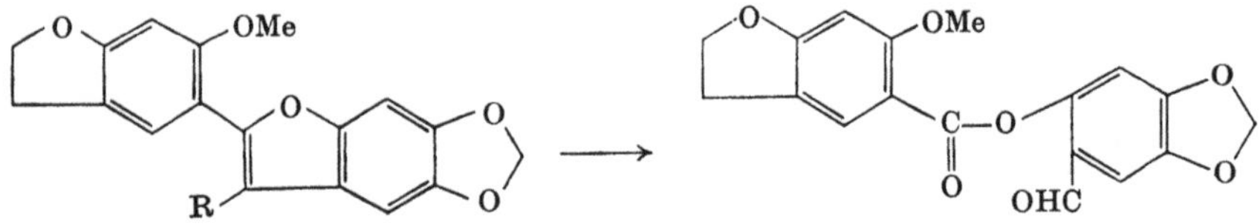

(169a) R=H; (169b) R=COOH (170)

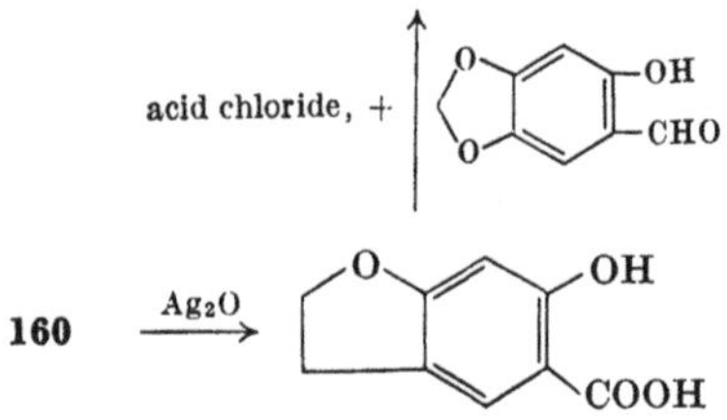

(30)

in the presence of potassium ferricyanide according to the method of Wanzlick and coworkers[106], gave furo(3′,2′–6,7)-5″,6″-dihydroxybenzofuro(3″,2″–3,4)coumarin (**172**), which on condensation with methylene iodide yielded erosnin (**167**). Methylation of **172** gave the corresponding dimethyl ether (**173**).

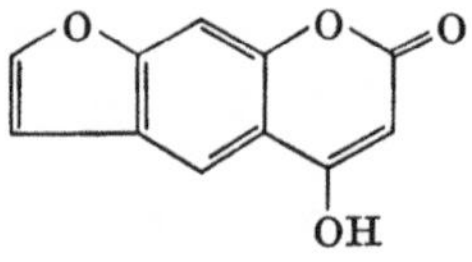

(171)

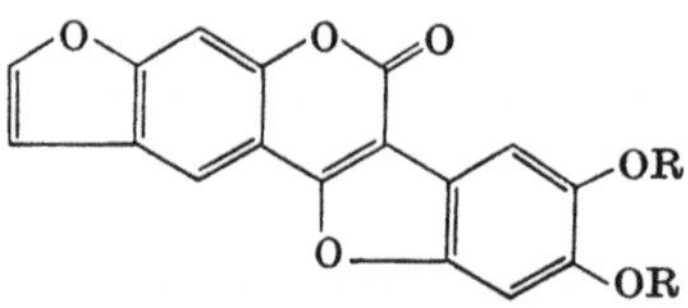

(172) R=H (173) R=CH_3

The appropriate keto nitrile, prepared by means of the Hoesch reaction, can be converted to dihydroerosnin (**174**).[98, 100, 108] Attempts to bring about dehydrogenation of **174** to erosnin (**167**) by action of *N*-bromosuccinimide were unsuccessful[108] (Eq. 31).

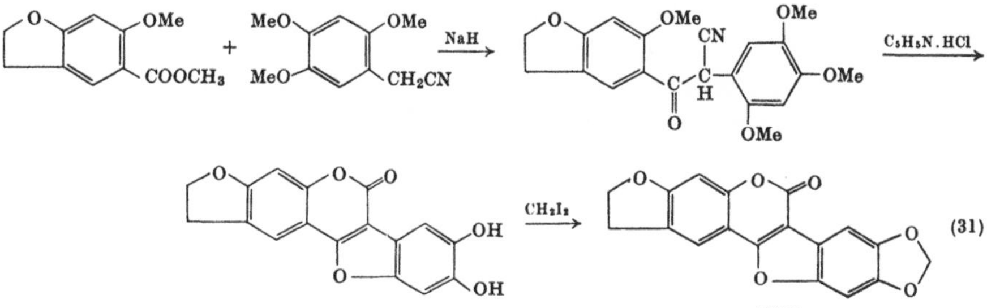

(174) (31)

VIII. Furo(3,2-c)-1-benzopyran-4-ones

Recognition of the coumarin moiety as an anticoagulant species, dates back to the report of Link and coworkers[125] concerning the toxic agent of spoiled sweet clover hay. The toxic, often fatal material causing serious hemorrhagic conditions in cattle was found to be 3,3′-methylene-bis(4-hydroxycoumarin) (dicoumarol) (**175**, R = H). Following this discovery, the problem of structure–activity relationships was examined by numerous investigators. Thus, during the study of substances with feeble toxicity, rapidly eliminated from the organism, and without cumulative effect, **176** ($R = R^1 = H$) was obtained by condensation of chloroacetaldehyde acetal with 4-hydroxycoumarin[126] (Eq. 32).

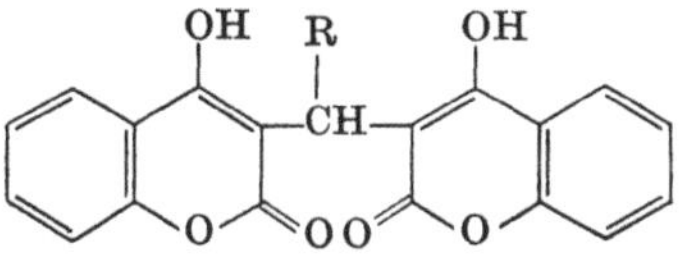

(175)

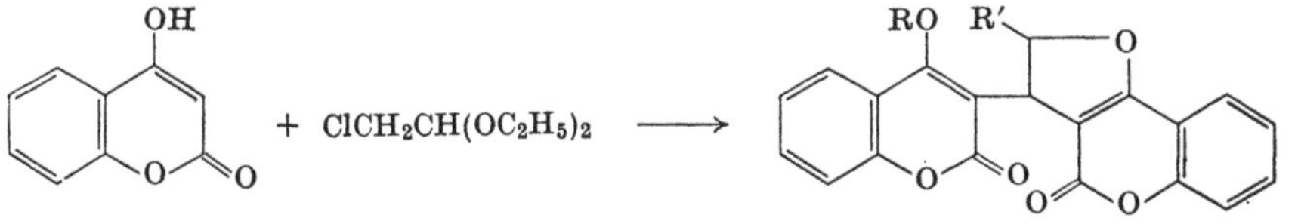

(176) (32)

Treatment of the powerful anticoagulant, 1,1-bis(4-hydroxy coumarinyl)-2-propanone (**175**, $R = COCH_3$), obtained from 4-hydroxy-coumarin and $CH_3COCH{=}NOH$,[127] with a sulfuric–acetic acid mixture gave **177** ($R = CH_3$).[128] The latter can readily be obtained upon boiling pelentanic acid (**175**, R = COOH) with acetic anhydride, followed by deacylation.[129, 130] Dehydration of **175** (R = COOH) with phosphorus oxychloride or thionyl chloride yielded a chloride, which on hydrogen chloride elimination, produced **177** (R = OH).[131]

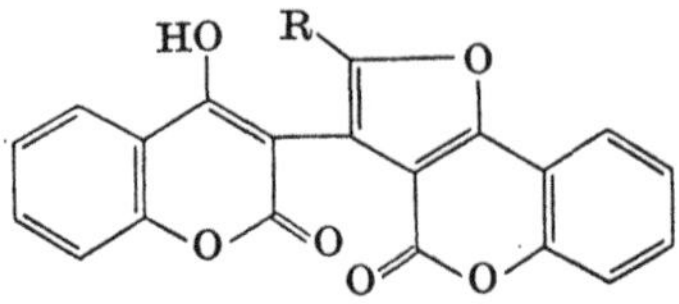

(177)

Condensation of 4-hydroxycoumarin and glyceraldehyde in a water medium produced **176** (R = H, $R^1 = CH_2OH$).[132] Treatment of 4-hydroxycoumarin with tetraiodohexamethylenetetramine or treatment of the latter compound with **175** gave 2-(2-hydroxybenzoyl)-4-furo-(3,2-c)–1–benzopyran-4-one (**178**), which presented a long lasting anticoagulant activity without having hemorrhagic properties.[133]

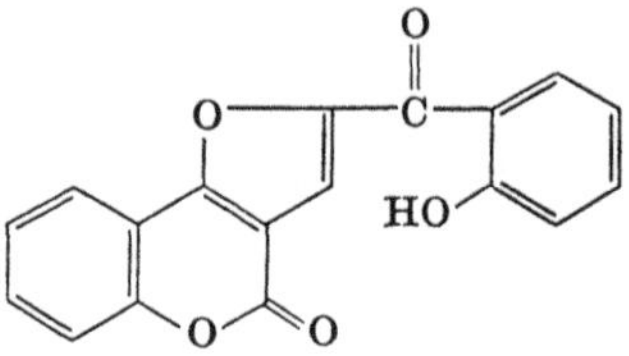

(178)

IX. Physiological Activity

Furoisoflavonoids have shown oestrogenic, insecticidal, piscicidal, and antifungal activity, and recently there has been considerable interest in the presence of oestrogens in the food consumed by grazing animals. Certain furoisoflavones show a higher fish posion activity than their flavonoid isomers.

The activity associated with the coumarin–type molecule is currently[134] considered to be a vitamin K antagonism in which the anticoagulant competes with vitamin K in its blood clotting role. Prothrombin production is believed to arise from a combination of vitamin K with an apoenzyme (AE) to form the active enzyme (AEK). The coumarins (B) can compete successfully with the K vitamin, which acts as the prosthetic group of the enzyme, for the apoenzyme, due to their structural similarity, and thus they exert antivitamin K activity. The equation, $AEK + B \rightleftharpoons AEB + K$ expresses this relationship as a reversible system; thus vitamin K can reverse the coumarin activity if present in sufficient concentration. Reduction of prothrombin production ensues if the form AEB predominates. In keeping with this concept, Chimielewska and Cieslak[135] have suggested that coumarin–type antivitamin K compounds (**179** and **180**), as well as **181** and **182** (3-substituted derivatives of 2-hydroxy-1,4-naphthoquinone) which are known hemorrhagic agents, lack the active center but contain the hemiketal linkage, and being similar in shape, successfully compete with vitamin K for the apoenzyme.

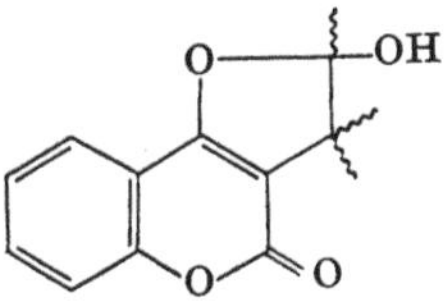

(179)

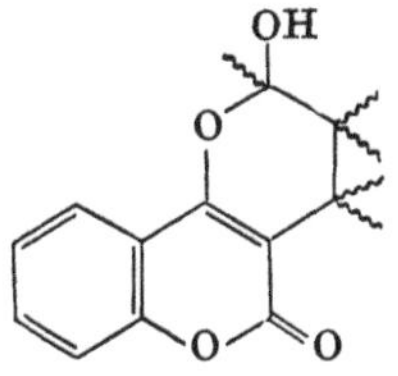

(180)

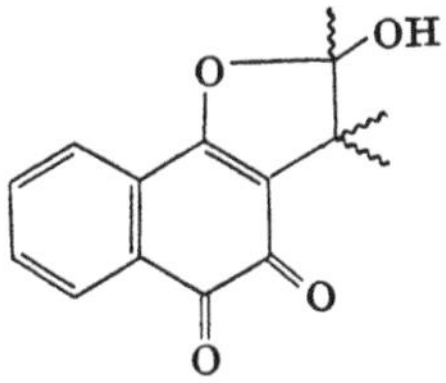

(181)

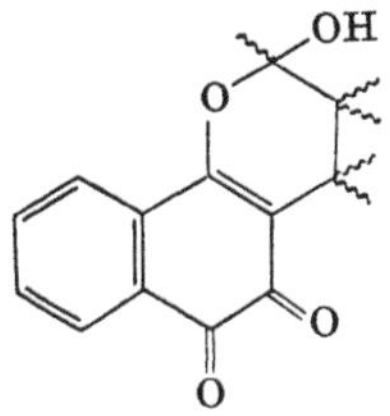

(182)

A discussion of oestrogenic activity in coumarin–type compounds may include a brief consideration of the closely related isoflavones, which were the first of the plant phenolics to show such activity. The isoflavones may well be the source of coumarin oestrogens, e.g. coumestrol (**102**) in the plant. Recognition of the role of the isoflavones as oestrogens probably stems from the observations of Bennetts and coworkers.[136] This finding was confirmed by the isolation of genestein (**183**), an isoflavone, from the plant *Trifolium subterraneum* L., which proved to be the principal oestrogenic substance.[137–139] The activities of isoflavones are well represented by Micheli and coworkers.[140]

The matter of oestrogenic activity of plant phenolics was revived when Bickoff and coworkers[91, 95] isolated the oestrogenic coumarino-coumarone, coumestrol, which was approximately 30 times as potent

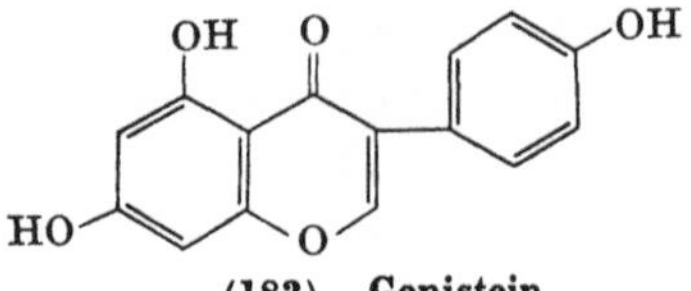

(183) Genistein

as **183**, although still far short of diethylstilbestrol (**185**). It is worth mentioning here, that as early as 1946, Mentzer and coworkers[141] had anticipated the oestrogenic activity of coumarins on purely theoretical grounds.

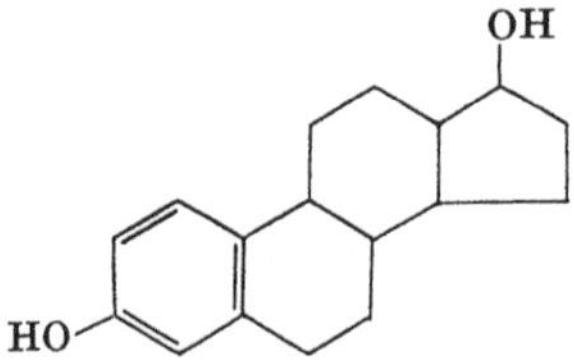

(184) Estradiol

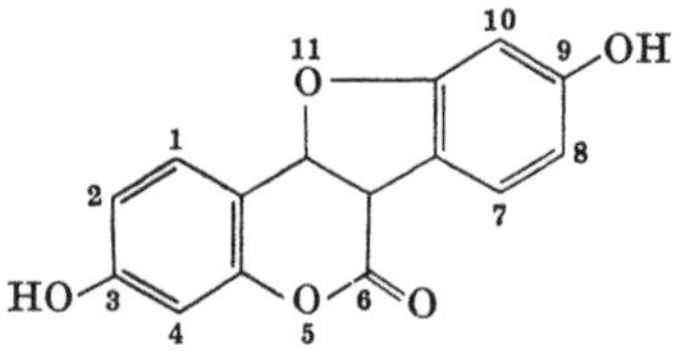

(102) Coumestrol

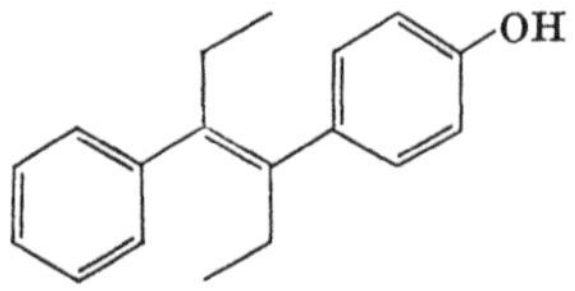

(185) Diethylstilbestrol

The relationship of coumestrol structure **102** to its biological activity has been investigated.[119, 140] A number of correlations of structure with activity was observed. The phenolic hydroxyl groups in positions 3 and 9 were quite important since their absence led to inactivity. Some return of activity was noted with the presence of a hydroxyl group at position 3. Etherification of the phenolic groups results in reduced activity and etherification of the hydroxyl group in 9-position reduces activity more than similar treatment of the hydroxyl group in position 3. Other nuclear substitution, for example in the 1- and 10-position almost eliminates activity. Acetylation of the hydroxyl groups of coumestrol results in a very little loss in activity, possibly due to the facile hydrolysis of these groups *in vivo* with regeneration of coumestrol.

The ether link bridging the 10a- and 11a-positions appears to contribute substantially to activity since its absence produces a virtually inactive compound. 7,4′-Dimethoxy-3-phenylcoumarin, closely related to coumestrol, is devoid of oestrogenic activity. It has been suggested that the greater activity of coumestrol, as compared with the isoflavones, may be due to the ketonic character of the isoflavone at position 4. Reduction of the double bond at the 6a,11a-positions brings about loss of activity even though the ether linkage is still present.

The oestrogenic activity is said to be attributed to its stilbene-like structure analogous to that of diethylstilbestrol (**185**), and to the natural estradiol (**184**).[142]

The ether bridge in coumestrol (**102**) stabilizes the double bond in the 6a,11a-positions to maintain the stilbene-like structure. When this ring is opened, this double bond is free to resonate and keto-enol tautomerism exists.[142]

The close relationship between the naturally occurring oestrogenic isoflavones, e.g. genistein (**183**), and the 3-phenyl-4-hydroxycoumarin, is further emphasized by considering the addition of water across the double bond of an isoflavone to give 2-hydroxyisoflavanone, followed by enolization to 2,4-dihydroxyisoflav-3-en.[143] Bate-Smith, quoted by Biggers,[142] suggests the possibility that coumestrol may be related to the isoflavones by postulating the conversion of an isoflavanol to coumestrol (Eq. 33) as in the isoflavanol corresponding to daidzein. That the above postulation had a solid foundation is born out by the findings of Grisbach and Barz[144]. The reader's attention should be directed to the recent comprehensive study of Micheli and coworkers,[140] in which the authors examine the question of oestrogenic structure-activity relationships of coumestrol, plant phenolics, and synthetic oestrogens.

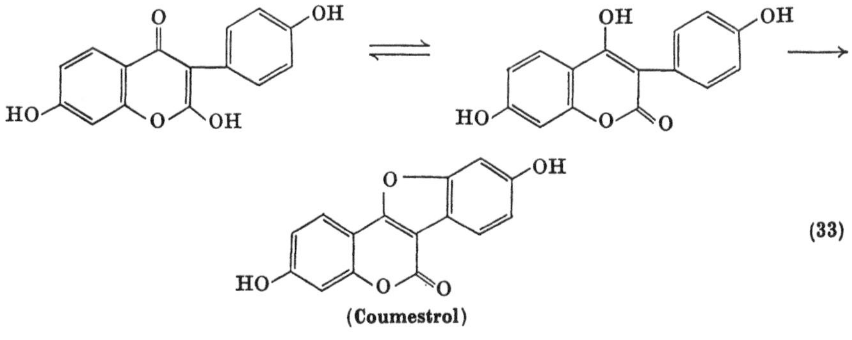

(Coumestrol)

(33)

The α-pyrone ring structure appears to be as important as the furan structure for maintenance of the oestrogenic activity of coumestrol. Opening of the lactone ring to form the potassium salt of the resulting *o*-hydroxycinnamic acid did not decrease activity, undoubtedly due to the ready conversion of the salt back to coumestrol by the acidity in the stomach of the animal. However, the corresponding *o*-methoxycinnamic acid is very greatly reduced in activity.

The production of antibiotic substances by plants in response to injury or infection has recently received attention. In some cases, the accumulation of these compounds and the concomitant shifts in metabolism are believed to be mechanisms for disease resistance in the plant. One such compound is the antifungal agent, pisatin (**68**), which was isolated from the pods of garden peas inoculated with fungal spore.[57] Coumestrol levels in clover are much higher in virus–infected plants than in the comparable virus–free specimens.[145]

X. References

1. L. Crombie and D. A. Whiting, *Tetrahedron Letters*, (18) 801 (1962).
2. L. Crombie and D. A. Whiting, *Chem. Ind.* (*London*), 1946 (1962).
3. L. Crombie and J. W. Lown, *J. Chem. Soc.*, 775 (1962).
4. K. Fukui and M. Nakayama, *Experientia*, **19**, 621 (1962).
5. (a) L. Crombie and D. A. Whiting, *J. Chem. Soc.*, 1569 (1963); (b) C. van der M. Brink, J. J. Dekker, E. C. Hánekom, D. H. Meirring and G. J. H. Rall, *J.S. African Chem. Inst.*, **18**(1), 21 (1965); *Chem. Abstr.*, **63**, 14843 (1965).
6. F. Fukui and M. Nakayama, *Chem. Ind.* (*London*), 1843 (1963).
7. F. Fukui and M. Nakayama, *Experientia*, **20**, 668 (1964); *Chem. Abstr.*, **62**, 4017 (1965).
8. T. Matsumoto, Y. Kawase and M. Nanbu, *Bull. Chem. Soc. Japan*, **31**, 688 (1958).
9. S. Tanaka, *J. Am. Chem. Soc.*, **73**, 872 (1951).
10. L. R. Row and T. R. Seshadri, *Proc. Indian Acad. Sci.*, *Ser. A*, **34**, 187 (1951).
11. K. Fukui and Y. Kawase, *Bull. Chem. Soc.*, *Japan*, **32**, 693 (1959).
12. Y. Kawase, K. Ogawa, S. Miyoshi and K. Fukui, *Bull. Chem. Soc. Japan*, **33**, 1240 (1960).
13. K. Fukui, M. Nakayama, and K. Okazaki, *Nippon Kagaku Zasshi*, **85**,441 (1964).
14. Y. Kawase and Ch. Numata, *Bull. Chem. Soc. Japan*, **35**, 1366 (1962).
15. K. S. Raizada, P. S. Sarin and T. R. Seshadri, *J. Sci. Ind. Res.* (*India*), *Ser. B*, **19**, 76 (1960).
16. P. S. Sarin, P. S. Sehgal and T. R. Seshadri, *J. Sci. Ind. Res.* (*India*), *Ser. B*, **16**, 206 (1957).
17. P. S. Sarin, P. S. Sehgal and T. R. Seshadri, *J. Sci. Ind. Res.* (*India*), *Ser. B*, **16**, 61 (1957).
18. V. R. Sathe and K. Venkataraman, *Curr. Sci.* (*India*), **18**, 378 (1949); *Chem. Abstr.*, **44**, 8916 (1950).
19. S. F. Dyke, *Proc. Chem. Soc.*, 179 (1963).

20. N. L. Dutta, *J. Indian Chem. Soc.*, **33**, 716 (1956); **35**, 165 (1959); **39**, 475 (1962).
21. S. H. Harper, *J. Chem. Soc.*, 595 (1942).
22. S. H. Harper, *J. Chem. Soc.*, 1099, 1424 (1939).
23. K. Fukui, M. Nakayami, A. Tanaka and S. Sasatani, *Bull. Chem. Soc., Japan*, **38**, 845 (1965).
24. K. Fukui, M. Nakayama and M. Hatanaka, *Bull. Chem. Soc. Japan.*, **36**, 872 (1963).
25. K. Fukui, M. Nakayama, M. Okamoto and Y. Kawase, *Bull. Chem. Soc. Japan*, **36**, 397 (1963).
26. W. Barker and R. Robinson, *J. Chem. Soc.*, 1981 (1925).
27. G. G. Goshi and K. Venkataraman, *J. Chem. Soc.*, 513 (1934).
28. L. R. Row and T. R. Seshadri, *J. Sci. Ind. Res.* (*India*), *Ser. B*, **15**, 495 (1956).
29. K. Fukui and M. Nakayama, *Nippon Kagaku Zasshi*, **85**, 444 (1964).
30. A. S. R. Anjaneyulu, M. V. R. K. Rao, G. P. Sastry, L. R. Row and G. S. R. S. Rao, *J. Sci. Ind. Res.* (*India*), *Ser. B*, **20**, 69 (1961).
31. S. K. Pavaram and L. R. Row, *J. Sci. Ind. Res.* (*India*), *Ser. B*, **17**, 272 (1958).
32. A. McGookin, A. Robertson and W. B. Whalley, *J. Chem. Soc.*, 787 (1940).
33. E. Späth and J. Schlager, *Chem. Ber.*, **73**, 1 (1940).
34. A. Robertson and W. B. Whalley, *J. Chem. Soc.*, 1440 (1954).
35. W. Cocker, T. Dahl, C. Dempsey and T. B. H. McMurry, *J. Chem. Soc.*, 4906 (1962).
36. W. Cocker, T. B. H. McMurry and P. A. Staniland, *J. Chem. Soc.*, 1034 (1965).
37. J. B. Sen Bredenberg and J. N. Shoolery, *Tetrahedron Letters*, (9), 285 (1961).
38. N. Finch and W. D. Ollis, *Proc. Chem. Soc.*, 176 (1960).
39. P. Cazeneuve, *Chem. Ber.*, **7**, 1798 (1874).
40. J. Eisenbeiss and H. Schmid, *Helv. Chim. Acta.*, **42**, 61 (1959).
41. H. Suginome, *Tetrahedron Letters*, (19), 16 (1960).
42. H. Suginome, *J. Org. Chem.*, **24**, 1655 (1959).
43. O. A. Stamm, H. Schmid and J. Büchi, *Helv. Chim. Acta*, **41**, 2006 (1958).
44. E. Simonitsch, N. Frei and H. Schmid, *Monatsh. Chem.*, **88**, 541 (1957).
45. H. Suginome and T. Iwadare, *Bull. Chem. Soc. Japan*, **33**, 567 (1960).
46. K. Aghermurthy, A. S. Kukla and T. R. Seshadri, *Curr. Sci.*, **30**, 218 (1961).
47. H. Suginome and T. Iwadare, *Experientia*, **18**, 163 (1962).
48. W. J. Bowyer, J. N. Chatterjea, S. P. Dhoubhadel, B. O. Handford, and W. B. Whalley, *J. Chem. Soc.*, 4212 (1964).
49. J. B.-Sen Bredenberg and P. K. Hietala, *Acta Chem. Scand.*, **15**, 696, 936 (1961).
50. P. K. Hietala, *Ann. Acad. Sci. Finnicae Ser. A. II Chimica* **100**, 61 (1960).
51. S. Shibata and Y. Nishikawa, *Chem. Pharm. Bull.*, **2**, 167 (1963).
52. H. Suginome, *Experientia*, **18**, 161 (1963).
53. S. Shibata and Y. Nishikawa, *Yakugaku Zasshi*, **81**, 1635 (1961).
54. W. Cocker, T. Dahl, C. Dempsey and T. B. H. McMurry, *Chem. Ind.* (*London*), 216 (1962).

55. I. A. M. Cruickshank and D. R. Perrin, *Nature*, **187**, 99 (1960).
56. I. A. M. Cruickshank and D. R. Perrin, *Australian J. Biol. Sci.*, **14**, 336 (1961).
57. D. R. Perrin and W. Bottomley, *Nature*, **191**, 76 (1961).
58. D. R. Perrin and W. Bottomley, *J. Am. Chem. Soc.*, **84**, 1919 (1962).
59. D. D. Perrin and D. R. Perrin, *J. Am. Chem. Soc.*, **84**, 1922 (1962).
60. A. J. Birch, B. Moore and S. K. Mukerjee, *Tetrahedron Letters*, (15), 673 (1962); C. W. L. Bevan, A. J. Birch and S. K. Mukerjee, *J. Chem. Soc.*, 5991 (1965).
61. B. L. Van Duuren and P. E. Groenewoud, *J.S. African Chem. Inst.*, **3**, 28 (1950).
62. J. M. Watt and G. Breyer-Brandwijk, *Medical and Poisonous Plants of Southern Africa*, (Ed. A. L. Livingston), Edinburgh, 1932, p. 77.
63. B. L. Van Duuren and P. W. G. Groenewoud, *J.S. African Chem. Inst.*, **3**, 35 (1950).
64. J. A. Labat, *Bull. Soc. Chim. Biol.*, **15**, 1344 (1933).
65. B. L. Van Duuren, *J. Org. Chem.*, **26**, 5013 (1961).
66. F. B. LaForge and L. F. Smith, *J. Am. Chem. Soc.*, **51**, 2574 (1929).
67. I. A. M. Cruickshank and D. R. Perrin, *Life Sciences*, (9), 680 (1963).
68. K. O. Müller and H. Dörger, *Arb. Biol. Reichsanstalt. Land-u. Fortstwirtschr., Berlin*, **23**, 189 (1940).
69. K. O. Müller, *Aust. J. Biol. Sci.*, **11**, 275 (1958).
70. K. O. Müller, *Phytopathol. Z.*, **27**, 237 (1956).
71. D. R. Perrin, *Tetrahedron Letters*, No. 1, 29 (1964).
72. A. G. Perkin, *J. Chem. Soc.*, 715 (1905).
73. H. Appel and R. Robinson, *J. Chem. Soc.*, 752 (1935).
74. K. Freudenberg and K. Weinges, *Ann. Chem.*, **613**, 61 (1958).
75. K. G. Dave, S. A. Telang and K. Venkataraman, *Tetrahedron Letters*, (1), 9 (1962).
76. G. Chakravarty and T. R. Seshadri, *Tetrahedron Letters*, (18), 787 (1962).
77. K. Freudenberg, *Experientia*, **16**, 101 (1960).
78. A. K. Ganguly, T. R. Seshadri and P. Subramanian, *Proc. Indian Acad. Sci., Ser. A*, **46**, 25 (1957).
79. W. R. Chan, W. C. C. Forsyth and C. H. Hassel, *J. Chem. Soc.*, 3174 (1958).
80. D. D. Pratt and R. Robinson, *J. Chem. Soc.*, 1577 (1922).
81. G. Chakravarty and T. R. Seshadri, *Indian J. Chem.*, **2**, 319 (1964).
82. S. K. Arora, A. C. Jain and T. R. Seshadri, *J. Indian Chem. Soc.*, **37**, 285 (1962).
83. F. M. Irvine and R. Robinson, *J. Chem. Soc.*, 127, 2086 (1957).
84. K. Freudenberg, Karimullah and G. Steinbrunn, *Ann. Chem.*, **518**, 37 (1935).
85. P. N. Nair and K. Venkataraman, *Tetrahedron Letters*, (5), 317 (1963).
85a. G. D. Bhatia, S. K. Mukerjee and T. R. Seshadri, *Tetrahedron Letters*, (18), 1717 (1966).
86. E. M. Philbin, T. S. Wheeler and F. O'Cinnéide, *Chem. Ind.* (*London*), 715 (1961).
87. R. Bryant, *J. Chem. Soc.*, 5140 (1965).
88. T. R. Govindachari, K. Nagarajan and P. C. Parthasarathy, *Tetrahedron*, **15**, 129 (1961).
89. C. Mentzer, *La Theorie Biogenetique et Son Application au Classement des Substances d'Origine Végetalé*, Editions du Museum, Paris, 1960, p. 36.

90. E. M. Bickoff, A. N. Booth, R. L. Lyman, A. L. Livingston, C. R. Thompson and G. O. Kohler, *J. Agr. Food Chem.*, **6**, 536 (1958).
91. E. M. Bickoff, A. N. Booth, R. L. Lyman, A. L. Livingston, C. R. Thompson and F. DeEds, *Science*, **126**, 969 (1957).
92. R. L. Lyman, E. M. Bickoff, A. N. Booth and A. L. Livingston, *Arch. Biochem. Biophys.*, **80**, 61 (1959).
93. L. Jurd, *J. Org. Chem.*, **24**, 1786 (1959).
94. T. R. Govindachri, K. Nagarajan and B. R. Pai, *J. Chem. Soc.*, 629 (1956).
95. E. M. Bickoff, R. L. Lyman, A. L. Livingston and A. N. Booth, *J. Am. Chem. Soc.*, **80**, 3969 (1958).
96. O. H. Emerson and E. M. Bickoff, *J. Am. Chem. Soc.*, **80**, 4381 (1958).
97. C. Deschampo-Vallet and C. Mentzer, *Compt. Rend.*, **251**, 736 (1960).
98. J. N. Chatterjea and S. K. Roy, *J. Indian Chem. Soc.*, **34**, 98 (1957).
99. J. N. Chatterjea, *J. Indian Chem. Soc.*, **36**, 254 (1959).
100. Y. Kawase, *Bull. Chem. Soc. Japan*, **32**, 690 (1959); *Chem. Abstr.*, **56**, 1437 (1962).
101. Y. Kawase, *Bull. Chem. Soc., Japan*, **35**, 573 (1962).
102. L. Jurd, *J. Org. Chem.*, **29**, 3036 (1964).
103. T. R. Govindachari, K. Nagarajan and B. R. Pai, *J. Chem. Soc.*, 545 (1957).
104. T. R. Govindachari, K. Nagarajan and B. R. Pai, *J. Chem. Soc.*, 548 (1957).
105. W. J. Bowyer, A. Robertson and W. B. Whalley, *J. Chem. Soc.*, 542 (1957).
106. H. Wanzlick, R. Gritzky and H. Heildeprein, *Chem. Ber.*, **96**, 305 (1963).
107. N. R. Krishnaswamy and T. R. Seshadri, *J. Sci. Ind. Res. (India), Ser. B*, **16**, 268 (1957).
108. J. N. Chatterjea and N. Prasad, *Chem. Ber.*, **97**, 1252 (1964).
109. A. L. Livingston, E. M. Bickoff, R. E. Lundin and L. Jurd, *Tetrahedron*, **20**, 1963 (1964).
110. L. Jurd, *Tetrahedron Letters*, (18), 1151 (1963).
111. T. A. Geissman, *Modern Methods of Plant Analysis*, (Ed. K. Paech and M. V. Tracey), Vol. III, Springer-Verlag, 1955, p. 450.
111a. A. L. Livingston, S. C. Witt, R. E. Lundin and E. M. Bickoff, *J. Org. Chem.*, **30**, 2353 (1965).
112. O. R. Hansen, *Acta Chem. Scand.*, **7**, 1125 (1953).
113. L. Jurd, *J. Pharm. Sci.*, **54**, 1221 (1965); *Chem. Abstr.*, **63**, 13227 (1965).
114. H. N. Khastgir, P. C. Duttagupta and P. Sengupta, *Tetrahedron*, **14**, 275 (1961).
115. K. K. Chakravarti, A. K. Bose and S. Siddiqui, *J. Sci. Ind. Res. (India), Ser. B*, **7**, 24 (1948).
116. T. Nakabayashi, T. Tokorayama, H. Miyazaki and S. Isono, *J. Pharm. Soc. Japan*, **73**, 669 (1953); *Chem. Abstr.*, **47**, 10348 (1953).
117. P. Yates and G. H. Stout, *J. Am. Chem. Soc.*, **80**, 1694 (1958).
118. D. Nasipuri and G. Pyne, *J. Chem. Soc.*, 3105 (1962).
119. D. Nasipuri and G. Pyne, *J. Sci. Ind. Res. (India), Ser. B*, **21**, 51, 148 (1962).
120. P. C. Duttagupta, H. N. Khastgir and P. Sengupta, *Chem. Ind. (London)*, 48 (1960).
121. E. M. Bickoff, R. L. Lyman, A. L. Livingston and A. N. Booth, *Arch. Biochem. Biophys.*, **88**, 262 (1960).
122. J. N. Chatterjea, K. D. Banerji and N. Prasad, *Chem. Ber.*, **96**, 2356 (1963).

123. P. Rajagopalan and A. I. Kosak, *Tetrahedron Letters*, (21), 5 (1959).
124. K. Fukui and N. Nakayama, *Tetrahedron Letters*, (30), 2559 (1965).
125. M. A. Stahmann, C. F. Huebner and K. P. Link, *J. Biol. Chem.*, **138**, 513 (1941).
126. K. Fučik, Z. Procházka, L. Lábler, F. Kanhuser and F. Jancik, *Bull. Soc. Chim. France*, 626 (1949); *Chem. Abstr.*, **43**, 1493 (1949).
127. K. Fučik, Z. Procházka, L. Lábler and J. Štrof, *Nature*, **166**, 830 (1950).
128. K. Fučik and L. Lábler, *Chem. Listy*, **45**, 496 (1951); *Chem. Abstr.*, **46**, 7568 (1952).
129. K. Fučik, Z. Procházka and J. Štrof, *Chem. Listy*, **45**, 484 (1951); *Chem. Abstr.*, **47**, 8739 (1953); *Chem. Listy*, **45**, 488 (1951); *Chem. Abstr.*, **47**, 8739 (1953).
130. K. Fučik, *Austrian Pat.* 171,693 (1952); *Chem. Abstr.*, **47**, 11258 (1953).
131. K. Fučik and Z. Procházka, *Czech. Pat.*, 85,251 (1955); *Chem. Abstr.*, **50**, 9450 (1956).
132. M. Eckstein and H. Pazdro, *Roczniki Chem.*, **38**, 1115 (1964); *Chem. Abstr.*, **61**, 16043 (1964).
133. S. Chechi, *Gazz. Chim. Ital.*, **90**, 295 (1960); *Chem. Abstr.*, **55**, 11401 (1961).
134. J. B. Lippincott, *New and Nonofficial Drugs*, Philadelphia, Pa., 1963, p. 618.
135. I. Chimielewska and J. Cieslak, *Tetrahedron*, **4**, 135 (1958).
136. H. W. Bennetts, E. J. Underwood and F. L. Shier, *Australian Vet. J.*, **22**, 2 (1946).
137. D. H. Curnow, T. J. Robinson and E. J. Underwood, *Australian J. Exptl. Biol. Med. Sci.*, **26**, 171 (1948).
138. R. B. Bradbery and D. E. White, *J. Chem. Soc.*, 3447 (1951).
139. J. D. Biggers and D. H. Curnow, *Biochem. J.*, **58**, 278 (1954).
140. R. A. Micheli, A. N. Booth, A. L. Livingston and E. M. Bickoff, *J. Med. Pharm. Chem.*, **5**, 321 (1962).
141. C. Mentzer, P. Gley, D. Mohle and D. Billet, *Bull. Soc. Chim. France*, 271 (1946).
142. J. D. Biggers, *Symposium on Pharmacology of Plant Phenolics*, Academic Press, London, 1958, p. 69.
143. R. B. Bradbery, *Australian J. Chem.*, **6**, 447 (1953).
144. H. Grisebach and W. Barz, *Chem. Ind.* (*London*), 690 (1963).
145. E. M. Bickoff, A. L. Livingston, A. N. Booth, C. R. Thompson, E. A. Hollwell and E. G. Beinhart, *J. Animal Sci.*, **19**, 4 (1960).
146. K. Fukui, M. Nakayama and H. Sesita, *Bull. Chem. Soc. Japan*, **37**, 1887 (1964).
147. S. Shibata, Y. Nishikawa, *Yakugaku Zasshi*, **82**, 777 (1962).
148. R. B. Desai and N. J. Ray, *J. Indian Chem. Soc.*, **35**, 83 (1958).
149. G. Pyne and D. Nasipuri, *J. Sci. Ind. Res.*, (*India*), *Ser. B*, **21**, 148 (1962).
150. E. Cerati, 'Contribution á l'etude des produits de condensation du dioxosuccinate de methyle avec le paracresol et quelques autres phenols parasubstitués', *Thèsis*, Annales Scientifique de L'Université de Besançon, 3° Serie-Chemie, 1963, fasc. 1.
151. J. N. Chatterjea and P. P. Dhoubhadel, *J. Indian Chem. Soc.*, **38**, 669 (1961).
152. F. E. King, C. B. Cotterill, D. H. Godson, L. Curd and T. J. King, *J. Chem. Soc.*, 3695 (1953).
153. A. Akisanya, C. W. Bevan and J. Hirst, *J. Chem. Soc.*, 2679 (1959).

154. B. T. Brooks, *Chem. Zentr.*, **II**, 649 (1911).
155. G. Kobayashi, Y. Kuwayana and T. Tsuchida, *Yakugaku Zasshi*, **85**, 310 (1965); *Chem. Abstr.*, **63**, 6983 (1965).
156. J. Reisch, *Angew. Chem.* **74**, 783 (1962); *Chem. Abstr.*, **59**, 2815 (1963).
157. H. Junck, *Monatsh. Chem.*, **96**, 1421 (1965); *Chem. Abstr.*, **64**, 3508 (1966).
158. K. Fučik, Z. Procházka and J. Štrof, *Collection Czech. Chem. Communs.*, **16**, 304 (1951); *Chem. Abstr.*, **47**, 8740 (1953).

CHAPTER VII

Chromanochromanones (The Rotenoids)

I. Introduction

The family Fabaceae or Papilionaceae (Leguminosae) includes a series of related genera, namely *Derris*, *Lonchocarpus*, *Tephrosia*, and *Mundulea*, whose members contain in their roots a number of chromanochromanones having considerable insecticidal activity. Preparations of such plants as an aid to catching fish, have been used by the native people of many tropical and subtropical countries as early as 1848.[1] They are familiar as garden sprays and dusts (derris), and they have relatively little effect when ingested by animals or man.[2, 3]

The rotenone–bearing plants have been shown to be toxic to many widely different species of insects; they are of special value for the control of plant feeding pests, especially where toxic resins are not desired, and for the control of cattle grubs.[4, 5]

The following account deals with members of the rotenone series in which a five membered oxygen ring is present, either as a dihydrofuran residue (dihydrofurochromanochromanones), e.g. rotenone and sumatrol, or as a furan system (furochromanochromanones), e.g. elliptone, pachyrrhizone and malaccol. The chemistry of the rotenone series containing the 2,2-dimethyl-Δ^3-pyran residue (pyranochromanochromanones), e.g. deguelin, tephrosin, toxicarol, is outside the scope of this text.

II. Nomenclature

Because of the cumbersome nature of the systematic names for the derivatives and degradation products of the rotenone group, nomenclature based on widely accepted trivial names has developed, but this can cause confusion. Thus rotenol refers to the phenol (**A**);[6, 7, 8] it has been applied to a different compound,[9] and recently has been used for yet a third.[10] The latter terminology is used in this section and accords with current practice.

TABLE 1. The Rotenoids

Compound (formula)	M.p. (°C)	Solvent for crystallization[a]; appearance	(α)D	Remarks (λ_{max} (mμ) (log ε))	References
Rotenone (**2**)	185, (185–186; dimorph.) 163, (181–183)	N	−228(C_6H_6); −226(C_6H_6)	217(4.47), 236(4.18), 244(4.11), 295(4.23)	11, 19, 48, 88
Rotenone α-oxime (cf. **51**)	237	I		235(4.36), 282(4.24), 300(4.06), 312(3.90)	48, 79
Rotenone β-oxime (cf. **52**)	249	I			48, 79
Rotenone iso-oxime (cf. **52**)	210–211, 208, 230	B/I		287(4.17)	6, 48
Rotenone enol acetate (acetylrotenone)	159–160, 164–165 (136–137 dimorph.)	I I	−156(C_6H_6), −172(C_6H_6)	220(4.40), 255(4.15), 315(4.04), 354(4.54), 372(4.46), 250(4.24), 354(4.47), 373(4.44)	11, 88
Mutarotenone(6aβ,12aβ,5′β- and 6aα, 12aα,5′β-rotenone mixture)	138–142, 145–146	I/O	−80 to −86, −99.7		48, 88
d-Epirotenone(6aα,12aα,5′β-rotenone)	86–89(solvate), 89–91(solvate)	L	+25.6, +30, +72.5		48, 88, 131
Rotenone,hydroxy compound (**70**)	amorphous, 94(solvated)	L	−208	220(4.35), 286(3.79)	10, 11
(−)-6′,7′-Dihydrorotenone	213–214, 212	C/I	−215(C_6H_6), −222.5(C_6H_6)	237(4.23), 295(4.28)	11, 130
(−)-Dihydrorotenone enol acetate	174–175	I	−164 (C_6H_6)		11
Dihydrorotenone enol acetate	208.5–210.5	I	−161 (C_6H_6)		11
Dihydrorotenone oxime	259	E/D		235(4.27), 242i(4.13), 283(4.18). 302(3.97), 315(3.83)	48
d,l-Nordihydrorotenone (**83**)	150–152	I			10
6′,7′-Dihydrorotenone, hydroxy compound (**74**)	amorphous				10
6a,12a-Dehydrorotenone (**3**)	216, 218, 226–228	E, I, E/L; yellow			29, 38, 48
12,12a-Dehydrorotenone	160–161, 164–165	C/L	−272.5(C_6H_6)	255(4.15), 265(4.12), 360(4.47), 375(4.41)	10, 11
12,12a-Dehydro-6′,7′-dihydrorotenone (**89**, R=H)	210–211	I			10
6a,12a-Dehydro-6′,7′-dihydrorotenone (**72**)	225, 226–227, 228–230	H, I; pale yellow			10, 11, 49
Dehydrotetrahydrorotenone (Dehydrodihydrorotenonic acid) (**6**)	218, 221, 223(dec.)	I, I/O, yellow		241(4.40), 276(4.39), 304(4.26), $\lambda_{max}^{EtOH-NaOH}$ 204i(4.42), 270(4.46), 282i(4.42), 316(4.36)	11, 49, 58
Rotenone (**7**)	295–297, 298	E/I, A; yellow		265(4.40), 298(4.33), 345(3.95)	29, 38, 48
Dehydrodihydrorotenone	273, 275(dec.)	A, I			49
(±)-Rotenolone	139–140	L			38
6aα,12aα,5′β-Rotenolone (cf. **96**)	powder	(O/B)	+26($CHCl_3$)		38
6aα,12aβ,5′β-Rotenolone (cf. **98**)	211–212	I	−78($CHCl_3$)		38

(Table continued)

TABLE 1 (*continued*)

Compound (formula)	M.p. (°C)	Solvent for crystallization[a]; appearance	$(\alpha)_D$	Remarks (λ_{max} (mμ) (log ε))	References
6aβ,12aα,5′β-Rotenolone (cf. **97**)	249–251(dec.)	E/I	+420($CHCl_3$)		38
6aβ,12aβ,5′β-Rotenolone (cf. **95**)	88(solvate)	L	−189($CHCl_3$)		38
6′,7′-Dihydrorotenolone (**100**)	187	K	−79($CHCl_3$)		38
6aα,12aα,12β,5′β-Rotenolol (cf. **106**)	185(solvate)	L	+120($CHCl_3$)		38
6aβ,12aβ,12α,5′β-Rotenolol (cf. **105**)	125–126(solvate)	D, L	−154($CHCl_3$)		38, 131
6aα,12aβ,12β,5′β-Rotenolol (cf. **100**)	259	L	−15($CHCl_3$)		38
6aα,12aβ,12α,5′β-Rotenolol (cf. **102**)	157	E/F	−50($CHCl_3$)		38
6aα,12aα,12α,5′β-Rotenolol (cf. **99**)	257(dec.)	E, L/E	+196($CHCl_3$)		38
6aβ,12aα,12β,5′β-Rotenolol (cf. **101**)	180	E, L/E	+228($CHCl_3$)		38
(±)-Isorotenone (**43**)	170–171, 172	B	±0		8, 11, 48, 88, 81a
(−)-Isorotenone	176(184 dimorph.), 185		−80(C_6H_6) +8($CHCl_3$)	243(4.69), 248(4.64), 261(4.18), 279(4.05) 330(3.65)	11, 20, 93b
(+)-6aα,12aα-Isorotenone	182, 183–184	I	+82, +75(C_6H_6)		11, 88
(±)-Isorotenone oxime	218(dec.), 230			242(4.70), 247(4.67), 278(3.93), 310(3.45) 322(3.40)	17, 48
(±)-Isorotenone iso-oxime	191–192, 192				8, 48
Isorotenone enol acetate	134			255(4.49), 265(4.38), 274(4.30), 312(3.85), 333i(4.11), 352(4.31), 370(4.24)	38
(±)-Isorotenone-12a-hydroxy compound (**91**)	193	E/L, I		251(4.38), 258(4.40), 280i(4.03), 290(4.11), 297i(3.95)	11, 38
6a,12a-Dehydroisorotenone (**4**)	192, 196(dec.)	I			11
Isorotenone (**57**)	251	E/L		295(4.36), 274(4.40), 261(4.38), 403(3.88)	38
(±)-Isorotenolone A (**95, 96**)	208.9	E		246(4.76), 252(4.76), 264(4.32), 285(4.25), 297(3.98). 325(3.72), $\lambda_{max}^{EtOH-NaOH}$ 243(4.84), 247(4.82), 260(4.39), 279(4.21), 332(3.83)	38
(±)-Isorotenolone A methyl ether	152	E			38
(±)-Isorotenolone B (**97, 98**)	242–243	E		246(4.69), 248(4.69), 262(4.25), 283(4.16), 335(3.70). $\lambda_{max}^{EtOH-NaOH}$ 242(4.69), 260(4.18), 278(4.03), 330(3.54)	38
(±)-Isorotenolone B methyl ether	166	E			38
6aβ,12aα-Isorotenolone A (**95**)	96–97(solvate)	L	−83.5(C_6H_6), −52($CHCl_3$)	246(−), 252(−), 264(−), 285(−), 298(−), 336(−)	38
6aβ,12aα-Isorotenolone B (**97**)	220(dec.)	E/O	+366($CHCl_3$)		38
(±)-Isorotenolone A acetate	165(179 dimorph.)	L			11, 38

Compound (formula)	M.p. (°C)	Solvent for crystallization[a]; appearance	$(\alpha)_D$	Remarks (λ_{max} (mμ) (log ε))	References
(±)-Isorotenolone C (**108a**)	211.2	E/L	±0	241(4.76), 247i(4.69), 264i(4.07), 281(4.11), 292i(3.98), 335(3.74). $\lambda_{max}^{EtOH-NaOH}$ 239(4.72), 245(4.64) 259i(4.02), 281(4.02), 290i(3.95), 333(3.70)	38, 104
(±)-Isorotenolone C methyl ether	157, 160			240(4.69), 246(4.62), 278(4.06), 282(4.07), 333 (3.73)	38, 104
(±)-Isorotenolone C monoacetate	142.3, 145				38, 104
(±)-Isorotenolone C toluene-*p*-sulfonate	187	C/K			38
(±)-Isorotenolol A	216(dec.)	E, J/K			11, 38
(±)-Isorotenolol B_1 (*cis*) (**99–100**)	263–264(dec.)	J/I			38
(±)-Isorotenolol B_2 (*trans*) (**101–102**)	213	I			38
(±)-Isorotenolol B methyl ether	262(dec.)	I			38
(±)-Isorotenolol B_1 dimethyl ether	170	M			38
(±)-Isorotenolol B_2 dimethyl ether	89 (dec.)				38
(−)-6aβ,12aβ,12α-Isorotenolol (**105**)	175–176	E/K	−157($CHCl_3$)		38
(+)-6aα,12aα,12β-Isorotenolol (**106**)	175–176	E	+160($CHCl_3$)		38
Isorotenolol C	197–198	J/K		252(4.29), 259(4.27), 289(4.06), 295(4.09)	38
Sumatrol (natural) (**112**)	192–194, 195, 196–198	E/I	−182(C_6H_6), −184(C_6H_6)	brown tinge with $FeCl_3$	108, 109, 110
Dihydrosumatrol (**114**)	184–185	E/L	−32 ($CHCl_3$)	brown color with $FeCl_3$	108, 109
Tetrahydrosumatrol (**115**)	222–223	E/L	+122($CHCl_3$)	olive green with $FeCl_3$	108, 109
Dehydrodihydrosumatrol (**116**)	235	E/I	−63 ($CHCl_3$	deep-green color with $FeCl_3$	108, 109
Dehydrotetrahydrosumatrol (**117**)	218	I; pale yellow		deep-green color with $FeCl_3$, diacetate from (B) (m.p. 193–194°)	108, 109
(±)-Malaccol (natural) (**122**)	225, 244, 249–250	E/I; pale yellow	±0($CHCl_3$)	oxime from (L) (m.p. 240°(dec.))	27, 110, 111
(±)-β-Malaccol (**126**)			±0 (C_6H_6), +96(acetone)	248i(ε,30,300), 253(ε,36,100), 288(ε,15,100), 357(ε,1800)	110
(±)-α-Malaccol (**122**)				246(ε,29,200), 253.5(ε,34,100), 264i(ε,13,600), 287(ε,12,000), 362(ε,2500)	110
Tetrahydromalaccol	222	I		brownish green with $FeCl_3$, triacetyl deriv. from (I) (m.p. 195°)	27
Dehydromalaccol	257	E/I; bright yellow			27
6a,12a-Dehydro-α-malaccol (**122a**)	246–250, 258 (sublim.)	E/I		233(ε,28,300), 240i(ε,27,900), 257(ε,27,000), 315(ε,14,700)	27, 110
6a,12a-Dehydro-β-malaccol (**126a**)	257			235i(ε,22,000), 260(ε,32,000), 275(ε,32,000), 312(ε,16,800)	27, 110

(Table continued)

TABLE 1 (*continued*)

Compound (formula)	M.p. (°C)	Solvent for crystallization[a]; appearance	$(\alpha)_D$	Remarks (λ_{max} (mμ) (log ε))	References
(±)-Elliptone (**128**)	173–175, 176–177	I	±0°(C_6H_6)	α-oxime from (L) (m.p. 259°), β-oxime from (L) (m.p. 261°)	115
(−)-Elliptone	160	I	−18 (C_6H_6), +55(acetone)	α-oxime from (L) (m.p. 222°), β-oxime from (L/O) (m.p. 236°), monoacetate from (I) (m.p. 200°)	27, 115
(±)-Tetrahydroelliptone (**131**)	205	I			115
(−)-Tetrahydroelliptone	217	I	+61(acetone)	diacetate from (I/O) (m.p. 140–142°)	115
Dehydroelliptone (**129**)	264(purple melt)	E/I	±0 ($CHCl_3$)		115
Amorphin	151–152		−67.9		122, 123, 124
Amorphigenin (**157**)	191–192(solvate)	C,L,B/O	−200.9	acetate (m.p. 152–154°), oxime (m.p. 225–227°)	122, 123, 124
Dehydroamorphigenin (**158**)	224–225		−50		123
(±)-Dolineone (**161**)	233–235	L	+135 ($CHCl_3$)	237(4.56), 275(3.84), 305(3.71), 335(3.50)	126
6a,12a-Dehydrodolineone (**162**)	270(dec.)	E/L			126
Pachyrrhizone (**164**)	272	B, D, A	+96($CHCl_3$)	oxime from (I/J) (m.p. 267–268°(dec.)), phenylhydrazone from (I/J) (m.p. 267–268°(dec.)), acetate from (L) (m.p. 147–148°)	127
(±)-Pachyrrhizone	250	B	0.0±5°($CHCl_3$)		53
Dehydropachyrrhizone (**165**)	255–260(dec.), 259–261	P; pale yellow			53, 126
Pachyrrhizonone (**166**)	349, 343	G; orange			53, 126

[a] A, acetic acid; B, acetone; C, benzene; D, carbon tetrachloride; E, chloroform; F, cyclohexane; G, cyclopentane; H, dioxan; I, ethanol; J, ethyl acetate; K, light-petroleum; L, methanol; M, pentane; N, trichlorobenzene; O, water; P, cyclopentanone.

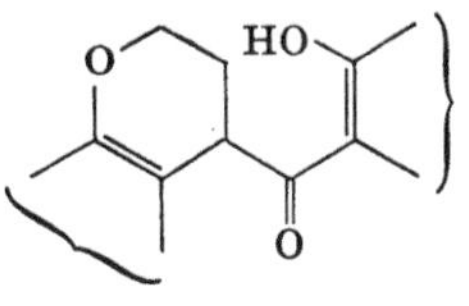

(A)

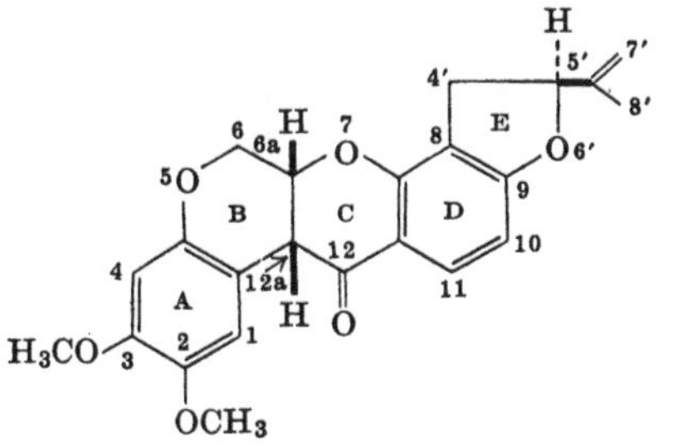

(1)

Similarly, the numbering of the rotenone system is performed in various ways, but that used in formula **1** is widely accepted[11] and is adhered to in this work. The individual rings in rotenone and similar molecules are distinguished by the letters A, B, C, D and E.

A knowledge of the structure and chemistry of rotenone accordingly leads to an understanding of the whole series of the rotenone–bearing plants. Partly for this reason and partly because of the intrinsic chemical interest of rotenone this key substance will be discussed in some detail.

III. Rotenone

1. Isolation

Rotenone (**2**), one of the most powerful vegetable insecticides, was first isolated from *Robinia nicou*,[12] and was named nicouline. The same plant, now called *Lonchocarpus nicou*, has been shown by Clark[13] to contain rotenone; identical with Geoffroy's nicouline.[13] Later it was isolated from *Milletia taiwaniana* Nagai, native to Formosa,[14] and was found to be identical with derrin,[15] and with tubotoxin,[16] obtained from *Derris elliptica* which is widely cultivated. The principal economic species are the various kinds of derris, *Derris elliptica* and *D. malaccensis* from Malaya and East India, and cube or timbo, *Lonchocarpus utilis* and *L. urucu* from South America.

2. Physical Properties

Rotenone, $C_{23}H_{22}O_6$,[17,18] is a colorless dimorphous substance; m.p. 163°,[19,20] 165–166°,[11] 180°,[21] 183°,[19] and 185–186°;[11] $(\alpha)_D^{20}$ $-105°$ (acetone)[22] and $-230°$ (benzene).[23] The solubility of rotenone in a number of organic solvents has been reported.[22] Under ordinary conditions, rotenone forms (1 : 1) crystalline solvates with benzene, carbon tetrachloride, and chloroform and with acetic acid (2 : 1).[24]

The ultraviolet, infrared[11, 23] and proton magnetic resonance spectra of rotenone and related compounds[25] have been explored. Specific rotations, molecular–rotation differences and rotatory–dispersion studies have been also reported.[26, 27]

Isolation of rotenone and related compounds from *Derris elliptica*,[28] and from commerical cube–timbo resin[11] has been reported in detail.

3. Structure

Rotenone, itself, although one of the most complex members, was the chief object of earlier structural investigations. The accepted formula **2** was advanced in 1932 by four independent schools.[29–32] For details of the early history of the subject, reference should be made to the review literature.[30, 33–37] Recent rotenoid chemistry[11, 38, 39] has been concerned, in the main, with establishing the relative and absolute stereochemistry, with clarification of many interesting reactions which were not given much attention during the earlier structural work, and synthetic studies.[40–47]

Rotenone, optically active, contains two methoxyl groups, a reactive ketone system capable of enolization, three oxygen atoms in different heterocyclic rings, and a reactive double bond. A feature of rotenone and other chromanochromanones, is the ease with which they undergo dehydrogenation to the chromenochromone system. Rotenone is converted to dehydrorotenone (**3**) by the action of mild oxidizing agents, e.g. potassium permanganate, potassium ferricyanide,[6] and preferably, manganese dioxide.[48] This change results in the loss of the asymmetric centers at C_{6a} and C_{12a}, but the asymmetric atom $C_{5'}$, remains and the compound is optically active.

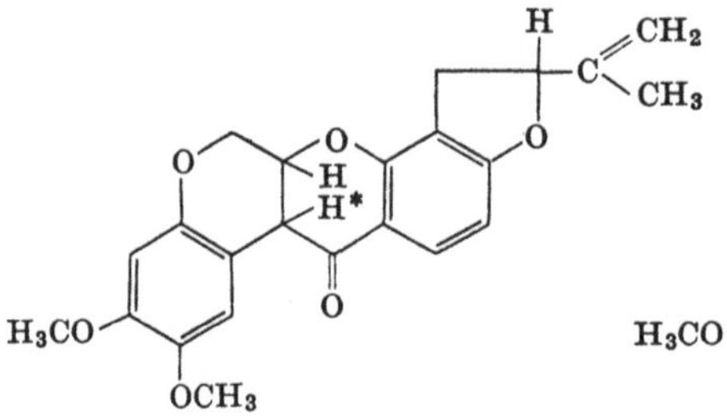

(2)

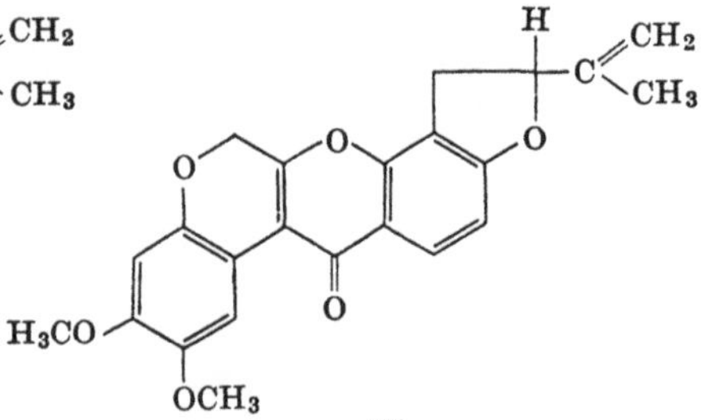

(3)

However, when dehydrorotenone (**3**), is isomerized to dehydroisorotenone (**4**) all asymmetry is lost and the substance is optically unresolvable.

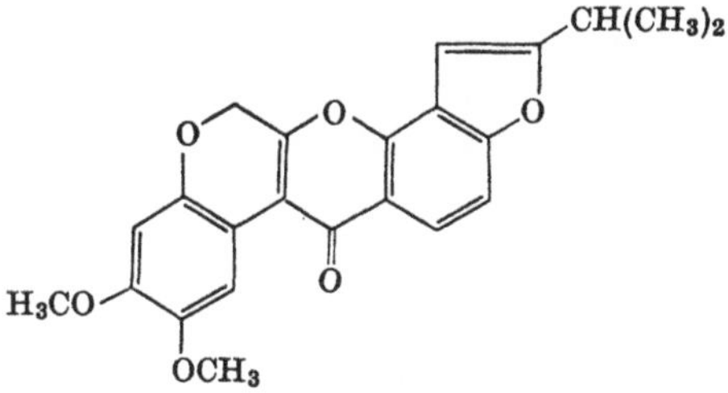

(**4**)

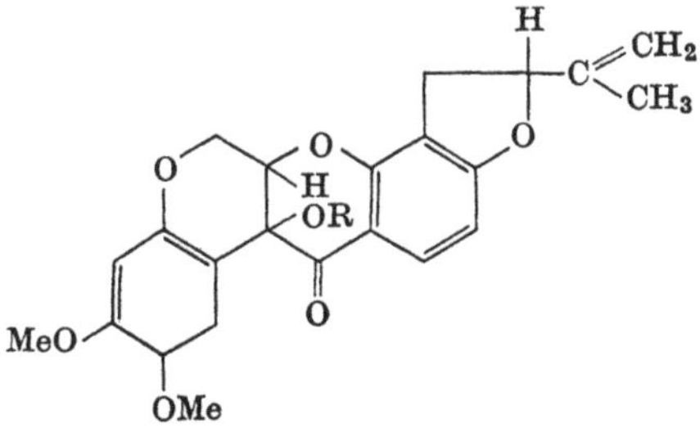

(**5**)

Various acetylrotenolones (**5**, R = $COCH_3$) are produced from the rotenones, most probably by acetoxylation of the *C* atom with an asterix (cf. **2**) with sodium acetate and iodine.[6, 49] Dehydrorotenone (**3**) is recovered upon hydrolysis of rotenolone acetate to the rotenolone (**5**, R = H), followed by dehydration.[49]

Unlike rotenone, dehydrorotenone (**3**) does not yield ketonic derivatives and the carbonyl group exhibits a behavior similar to that in the chromone system. Thus, it is no longer very sensitive to bases, and it no longer possesses enolizable hydrogen; in contrast to the carbonyl group of **2** which enolizes and can then be methylated and/or acetylated.[50, 51]

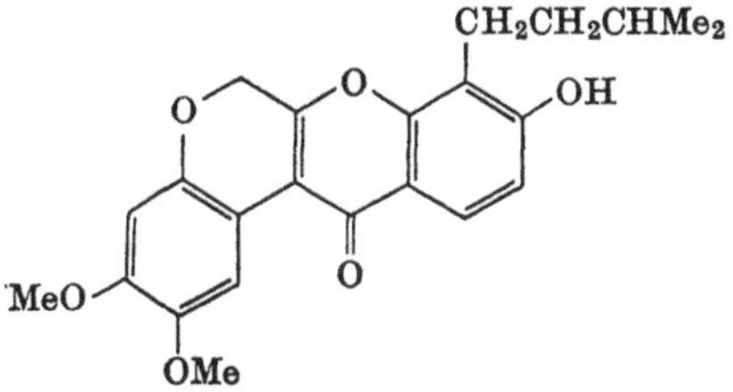

(**6**)

Hydrogenation of dehydrorotenone (**3**) with a platinum catalyst is accompanied by fission of the oxide linkage to give the optically inactive dehydrodihydrorotenonic acid (tetrahydrodehydrorotenone) (**6**).[49, 52]

Chromic acid oxidizes **3** to rotenonone (**7**)[29] as do nitric acid,[53] manganese dioxide,[48] and plumbic acetate;[48] the last being the best reagent. With ethanolic potassium hydroxide solution or zinc and alkali, both condensed oxygen rings of dehydrorotenone (**3**) undergo hydrolytic fission to give derrisic acid (**8**).[6, 20, 31]

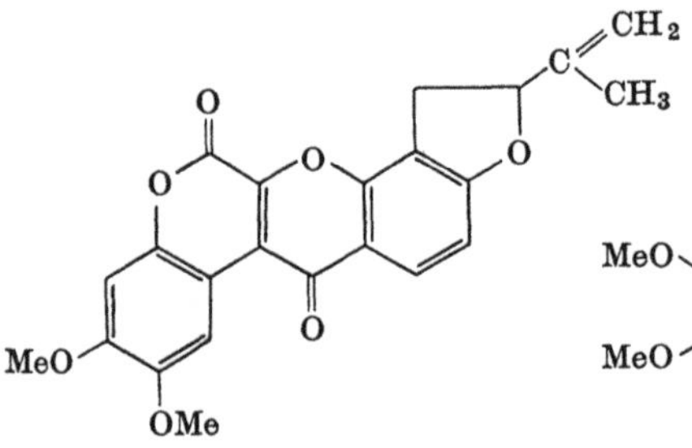

(**7**)

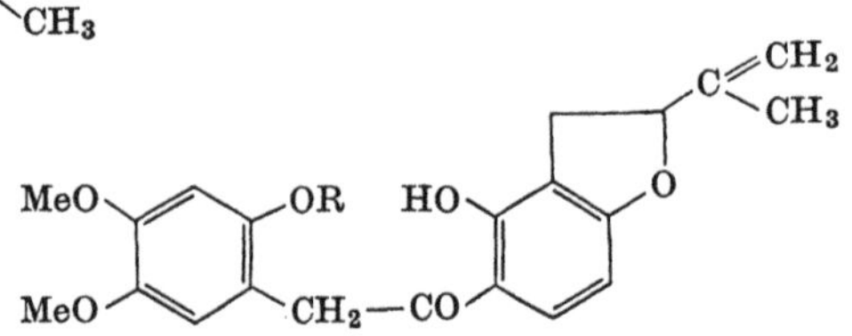

(**8**) R=CH_2COOH
(**9**) R=H

The nature of derrisic acid (**8**) is evident, since it has been shown to contain a free phenol group, develops a color reaction with ferric chloride, and may be acetylated.[6] **8** is closely related to derritol (**9**), which readily yields **8** on reaction with bromoacetate.[29, 54, 55] Iso-, and dihydroderrisic acids are obtained by the same reaction from iso-, and dihydroderritol. The acid (**8**) is oxidatively degraded by alkaline hydrogen peroxide to derric acid (**10**),[49] which is oxidized by potassium permanganate to rissic acid (**12a**).[56] The latter melts with a loss of a carboxy group, to form monobasic decarborissic acid (decarboxyrissic acid) (**12b**).[8, 30, 57]

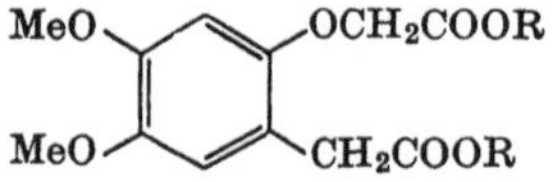

(**10**) R=H
(**11**) R=Et

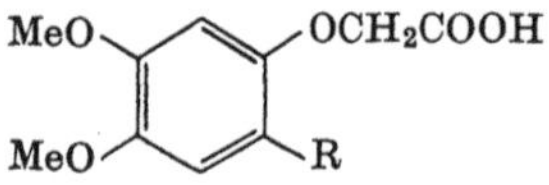

(**12a**) R=COOH;
(**12b**) R=H

The assigned structure **12b** has been established by synthesis from 3,4-dimethoxyphenol and ethyl iodoacetate,[57] and those of **10** and **12a** were also confirmed by synthesis[32] (Eq. 1).

The constitution of rotenone (**2**) is most readily understood by

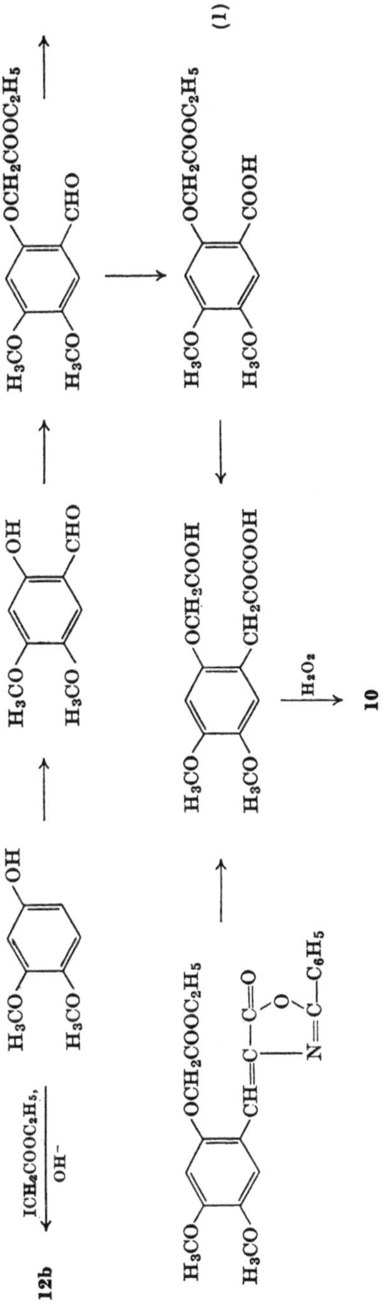
12b
ICH2COOC2H5,
OH⁻
H2O2
10
(1)

considering the reactions of tetrahydrodehydrorotenone (**6**) which is hydrolysed to tetrahydroderrisic acid (**13**, R = H).[31, 58] Similarly, alkaline hydrolysis of dehydrorotenone (**3**) gives the parent acid, derrisic acid (**8**).[59] **13** is a monobasic ketonic acid, which gives an intense ferric chloride reaction in alcohol, in agreement with the presence of a carbonyl group in the *ortho* position to the hydroxyl group. It is methylated to the methyl ester (**13**, R = CH_3).[31]

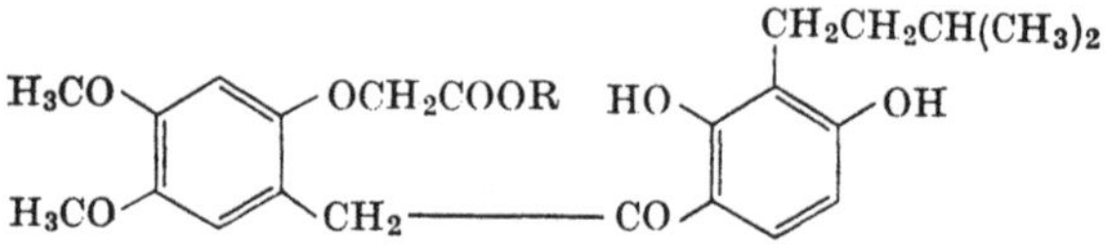

(**13**)

Regeneration of **6** is effected by refluxing with acetic anhydride and sodium acetate followed by the removal of the acetyl group from the acetate of **6**, formed intermediately.[60] Similarly, derrisic acid (**8**) is converted into dehydrorotenone (**3**).[31] Thus, these recyclization reactions leave no doubt as to the constitution of all these compounds and, indeed, of rotenone itself.

The structure of tetrahydroderrisic acid (**13**, R = H) is evident, since it behaves as an *o*-hydroxy ketone, and is smoothly cleaved by alkali into tetrahydrotubanol (**14**) and derric acid (**10**). Moreover, it has been synthesized:[60] 2-hydroxy-4,5-dimethoxybenzaldehyde can be transformed by standard methods into methyl 2-cyanomethyl-4,5-dimethoxyphenoxyacetate (**15**), which under Hoesch conditions condenses with **14** producing the desired product. The structure of **14** has been established by synthesis[61, 62] (see later).

With the structure of the two halves of **13** (R = H) established, it is possible to reconstitute the formula of the latter compound. Moreover, with the structure of derrisic acid (**8**) and its relation to dehydrorotenone (**3**) known, the nature of the groupings which compose the skeleton of the rotenone molecule (**2**) could be deduced.

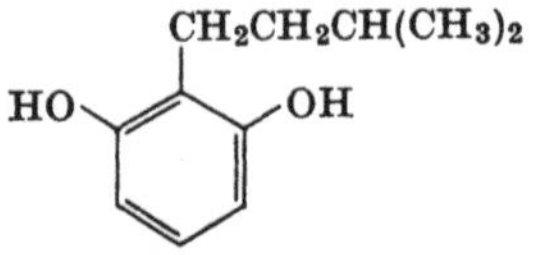

(**14**)

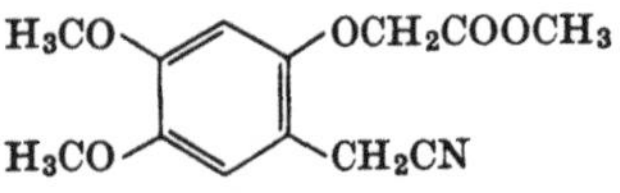

(**15**)

The above characteristic reactions may be summarized as follows:

rotenone (**2**) ⟶ tetrahydrorotenone (**56**)

rotenone (**2**) ↓ dehydrorotenone (**3**)

tetrahydrorotenone (**56**) ↓ tetrahydrodehydrorotenone (**6**)

dehydrorotenone (**3**) ⇅ ($+2H_2O$ ↓, $-2H_2O$ ↑) derrisic acid (**8**)

tetrahydrodehydrorotenone (**6**) ⇅ ($+2H_2O$ ↓, $-2H_2O$ ↑) derric acid (**10**) + tetrahydrotubanol (**14**)

derrisic acid (**8**) ↓ H_2O_2 derric acid (**10**)

derric acid (**10**) ↓ $KMnO_4$ rissic acid (**12a**)

rissic acid (**12a**) ↓ Δ decarborissic acid (decarboxyrissic acid) (**12b**)

A. Dehydrorotenol

Rotenone, being a chromanone or isoflavanone, is readily affected by alkali which removes a proton from the 12a-position and thus leads to a 3-chromen, known as 6a,12a-dehydrorotenol (**16**).

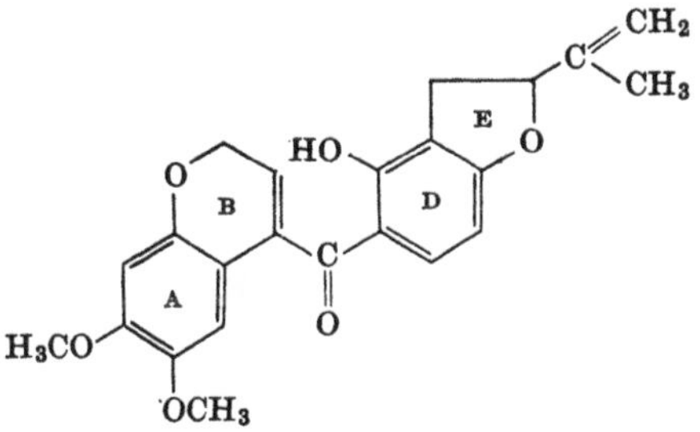

(**16**)

The reaction is essentially one in which flavanones are converted into chalkones; however, recyclization is very difficult to avoid, and the compound has only, recently, been isolated and studied.[48] As expected, even a weak base such as sodium acetate promotes rapid regeneration of the chromanochromanone ring system.

In more vigorous conditions, sodium hydroxide eventually converts rotenone into an isomer, 6,6a-dehydrorotenol (**17**), which can be recyclized to the chromanochromanone only by bases powerful enough to reverse the prototropic shift. Sodium acetate is not sufficiently strong but the base need not be as strong as sodium hydroxide and in practice, wet potassium carbonate is found to be adequate.[48]

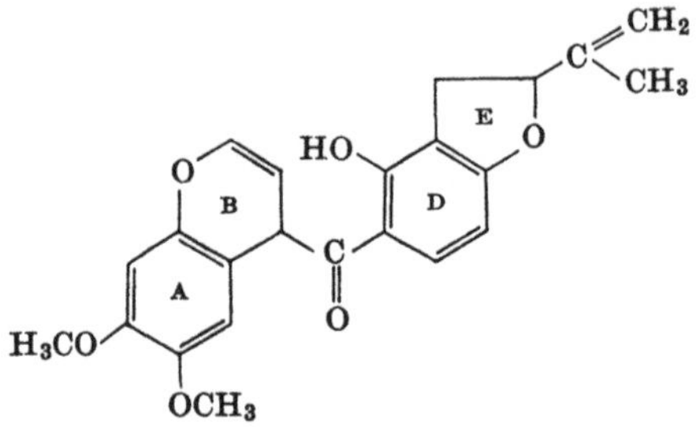

(17)

The existence of dehydrorotenol in alkaline solution accounts for many details of rotenone chemistry.

Controlled alkaline degradation (alcoholic potassium hydroxide solution) of rotenone produces optically active tubaic acid **(18)** ($C_{12}H_{12}O_4$) (m.p. 128°).[9, 63–65] This compound is phenolic, monobasic, contains a reactive double bond and can produce an optically active dihydrotubaic acid (**19**) (m.p. 168°).[9] It is isomerized on fusion with alkali or by treatment of an acetic acid solution with a little sulfuric acid, with migration of the ethylenic linkage, to give optically inactive isotubaic acid (rotenic acid) (**20**) (m.p. 183°, rapid heating).

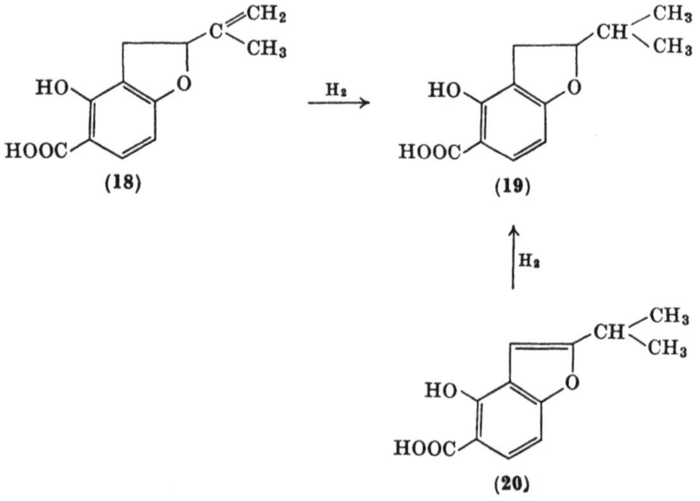

This is a characteristic property of rotenone and its derivatives, e.g. isorotenone and in this case is accompanied by loss of an asymmetric carbon atom. Isotubaic acid (**20**) can be hydrogenated with difficulty to give dihydroisotubaic acid (**21**), a racemic mixture of *d*- and *l*-dihydrotubaic acids, which can be resolved.[9,20] Tubaic acid, dihydrotubaic, and isotubaic acids produce intense coloration with alcoholic ferric chloride, in agreement with their formulae.

The hydrogenation of **18** parallels the behavior of **2** under similar conditions. By energetic hydrogenation **18** gives a mixture of optically active dihydrotubaic acid and optically inactive tetrahydrotubaic acid (**22**); the latter being formed by the opening of the isopropyldihydrofuran ring, saturation of the resulting side chain and consequent loss of asymmetric centre. If the hydrogenation is carried out in alkaline medium, the furan ring is opened by catalytic hydrogenolysis without saturation of the double bond to give an intermediate iso- or β-dihydrotubaic acid (**21**)[66] which is isomeric with dihydrotubaic, and dihydroisotubaic acids. Further hydrogenation of β-dihydrotubaic acid in a neutral medium, and with a platinum oxide (PtO_2) catalyst, gives **22**.[65]

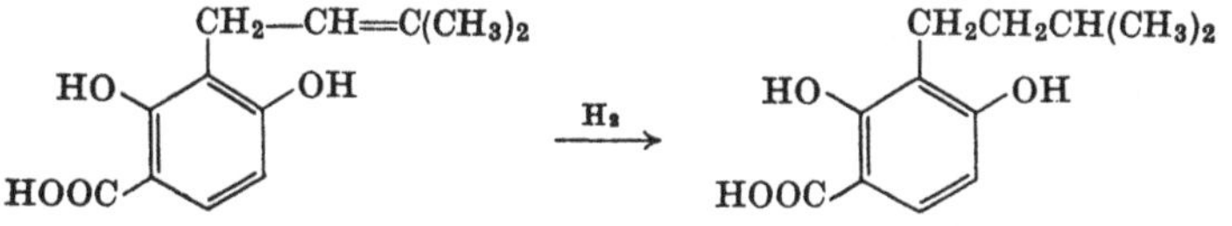

(**21**) (**22**)

The behavior of **18** on hydrogenation bears a striking similarity to the reactions of a number of codeine derivatives, which contain the system

$$\overset{1}{O}\overset{2}{C}H\overset{3}{C}=\overset{4}{C}H$$

also present in **18**. According to Schöpf,[67] the grouping is comparable to a conjugated system, and hydrogenation first takes place by a 1,4-addition of hydrogen followed by the formation of a double bond at the 2,3-position.

Tubaic acid (**18**) gives the optically active phenol, tubanol (**23**), and similarly, the various derivatives of **18** give the corresponding tubanol derivatives, including isotubanol and tetrahydrotubanol (**14**).[9,20] The latter is identical with the product obtained from **6** by alkaline degradation; thus establishing the similarity between these portions of the molecule.

The observations described so far suggest that tubaic acid is a coumaran

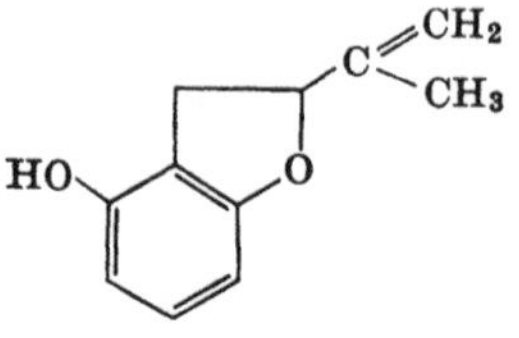

(23)

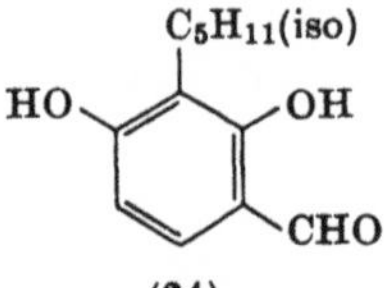

(24)

derivative. The only other obvious possibility would be a 2,2-dimethylchromone formula which is ruled out by its lack of an asymmetric carbon atom. Further points in favor of the accepted structure are that ozonolysis gave formaldehyde, and that **22** and **20** have been synthesized. Thus the carboxylation of **14**[64] by Kolbe's method gave **22**, identical with the acid obtained by the oxidation of the diacetate of the aldehyde (**24**), which is produced by a Gattermann reaction on **14** and subsequent deacetylation of the product.[68]

The synthesis of isotubaic acid (**20**) has been achieved by Reichstein and Hirt.[69] Their route began with furan and is one of the rare examples in which a benzene ring is built onto a furan ring, instead of vice versa. Application of the Perkin reaction, with sodium succinate and succinic anhydride, to 2-formyl-5-isopropylfuran (**25**) gave the ester (**26**), which after hydrolysis and decarboxylation gave isotubanol (**28**), a reaction which undoubtedly proceeds by way of **26** and **27**. Carboxylation of **28** gave **20** (Eq. 2).

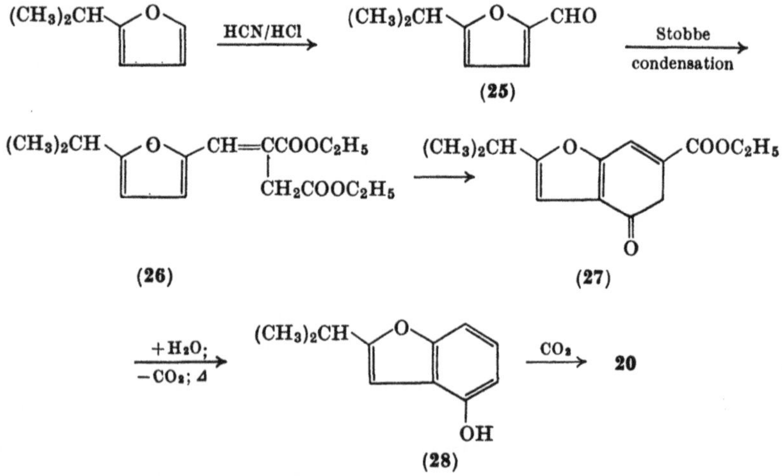

(2)

An alternative synthesis of isotubanol (**28**) through the stages **29** and **30** has been established[70] (Eq. 3).

The constitution of rotenone is largely solved by considerations based only on the chemistry of tubaic acid. Historically, different considera-

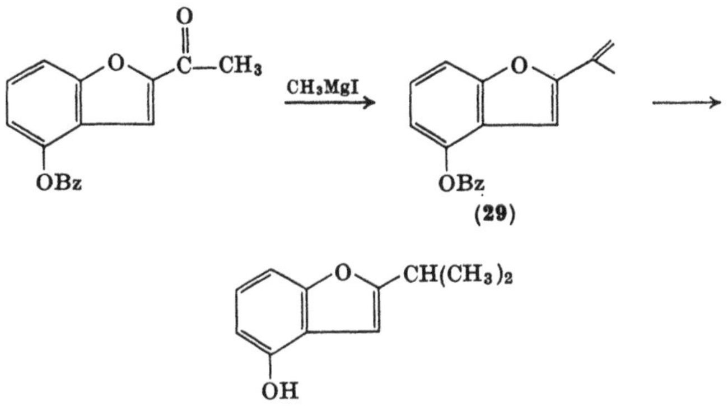

(29)

(30)

(3)

tions were more important. A degradation, important in determining structure, is effected by cleaving rotenone (**2**) with alkali in the presence of a reducing agent, e.g. zinc. The main products are derritol (**9**) and rotenol (**31**), which are readily separated by dilute alkali in which the monohydric *o*-hydroxy ketone, rotenol, is almost insoluble.[6, 71]

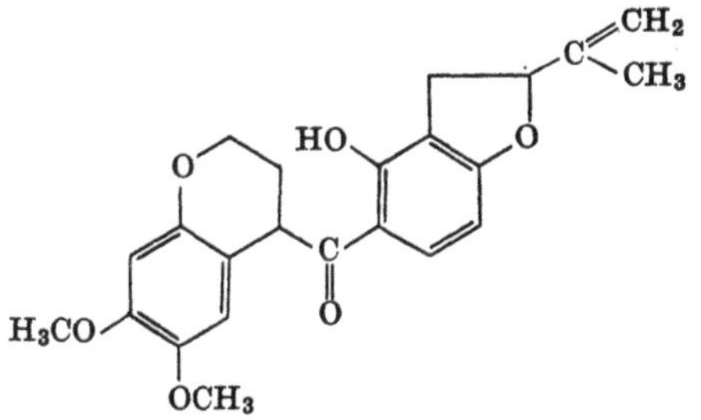

(**31**)

Derritol (**9**), $C_{21}H_{22}O_6$ (m.p. 161°), an intensely yellow substance, is a diphenol. Treatment of an alkaline solution of rotenone,[71] i.e.of 6a, 12a-dihydrorotenol, with zinc, eliminates two carbon atoms. This reaction which has been interpreted by Crombie and coworkers[48] as equivalent to the well–known α-ketol reduction of the same reagent, is here followed by a retroaldol condensation, as indicated below. At the same time, a straightforward saturation (not without precedent) of the olefinic intermediate leads to rotenol (Eq. 3a).

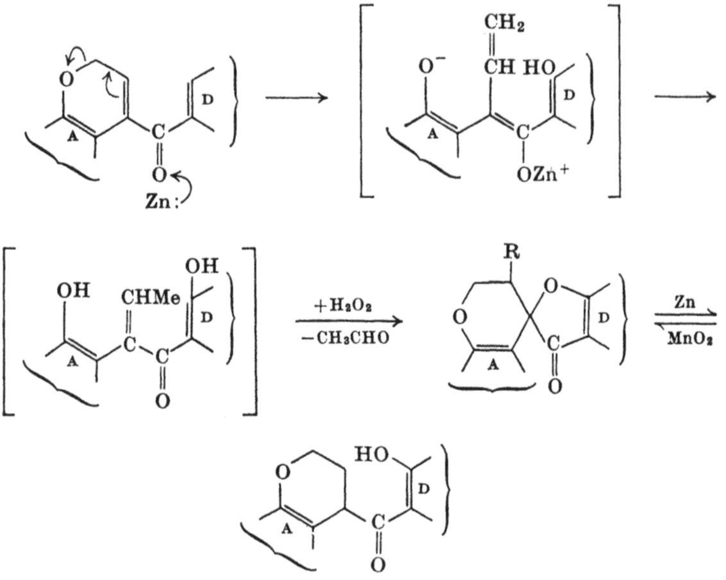

(3a)

(**32**)

Alkaline fusion of **9** yields isotubaic acid (**20**), but tubaic acid (**18**) is formed by permanganate oxidation (tubaic acid is isomerized by alkali to isotubaic acid). The permanganate oxidation of **9** leads to the formation of 2-hydroxy-4,5-dimethoxybenzoic acid (**33a**), and oxidation of derritol methyl ether to **18** and 2,4,5-trimethoxybenzoic acid (asaronic acid, m.p. 143°) (**33b**).[50] The methyl ether of **9**, with alkaline hydrogen peroxide, yields 2,4,5-trimethoxyphenylacetic acid (**34**) (homoasaronic acid, m.p. 87°).[66] Steric hindrance prevents the methylation of the second phenolic group of **9**; the same is true in the reaction of **9** with bromoacetate to yield derrisic acid (**8**).[54, 55] The condensation of **9** with ethyl oxalate[29] in the presence of sodium acetate regenerates rotenonone (**7**), and ethyl oxalyl chloride[55] has also been used for the same purpose, thus foreshadowing one of the best methods for preparing isoflavones.

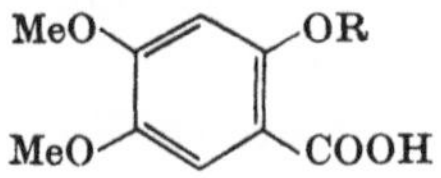

(**33a**) R = H;
(**33b**) $R = CH_3$

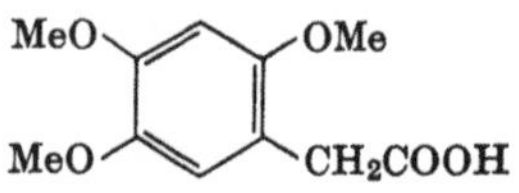

(**34**)

O-Methyltetrahydroderritol has been synthesized[29] by condensation of tetrahydrotubanol (**14**) and **34** in the presence of zinc chloride. Only traces of the requisite product were obtained from the attempted condensation of **14** and 2,4,5-trimethoxyphenylacetonitrile by the Hoesch method.[68]

B. Rotenol

(**31**), $C_{23}H_{24}O_6$ (m.p. 119°), is a colorless phenolic ketone, which is obtained from rotenone (**2**) by cleaving with alkali and zinc. It forms an oxime (m.p. 187°). When fused with potassium hydroxide, it gives a high yield of isotubaic acid (**20**).[7] Hydrogen peroxide oxidation of **31**, or of **14**, gives nerotic acid (**35**), $C_{12}H_{14}O_5$, which occurs in two interconvertible forms (m.p. 87° and 131°).[30] Manganese dioxide[48] oxidizes the acetate of rotenol to that of 6a, 12a-dehydrorotenol (**16**), and rotenol itself to the spiran (**36**), which is presumably formed in a one–electron transfer via a mono- or a diradical (**38**).[48, 72] The spiran, being an α-ketol derivative, is readily converted into rotenol by reduction with zinc. Another reaction involving **16** is the oxidation of rotenone to a related spiran (**37**, R = OH). As the reagent is alkaline hydrogen peroxide, the result can be attributed to epoxidation of the double bond, followed by cyclization, resembling certain aspects of the Algar–Flynn–Oyamada flavanol synthesis.[48]

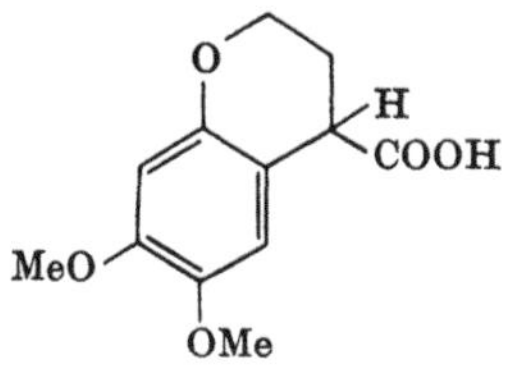

(**35**)

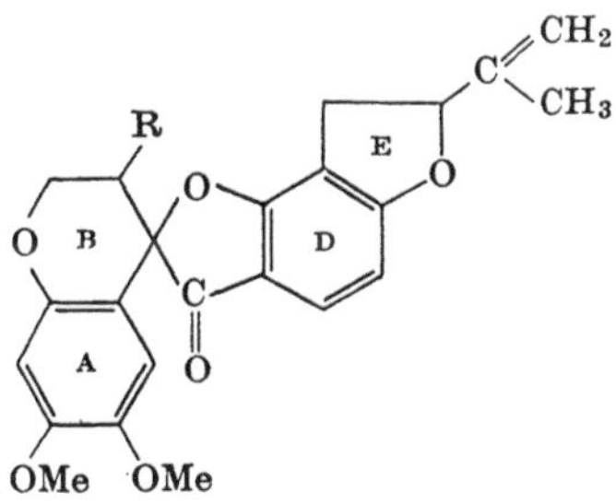

(**36**) R = H

(**37**) R = OH

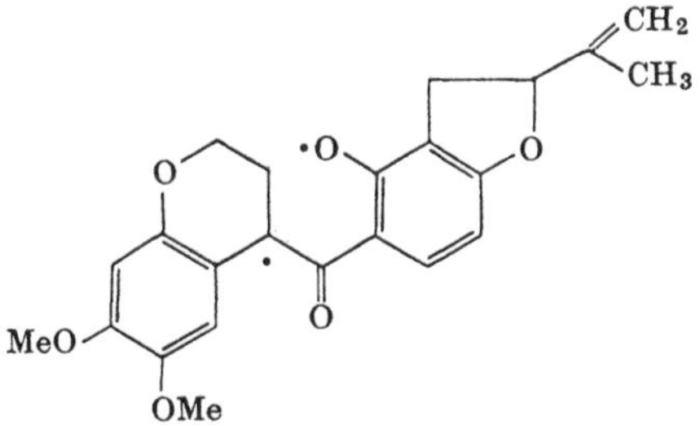

(**38**)

The structure of rotenol has been confirmed by synthesis.[73] The cyclization of ethyl derrate (**11**) with sodium gave the chromen ester (**39**), which was hydrogenated over a platinum catalyst to **40**. Subsequent dehydration and hydrolysis of **40** yielded toxicaric acid (**41**), which is obtained from rotenone (**2**) by alkaline oxidation[74] (Eq. 4). Reduction of **40** gave the ester (**42**) of nerotic acid.

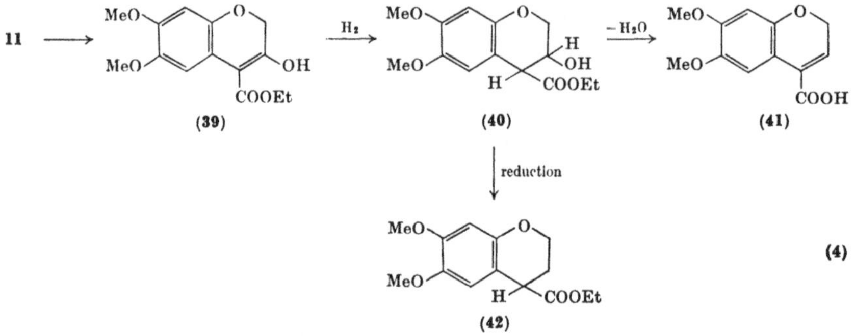

(39) (40) (41) (42) (4)

4. Isomerization of Rotenones (The Isorotenones)

Rotenone, all its degradation products, and its analogs, which contain the isopropyl side chain, are isomerized by sulfuric acid in acetic acid. The 6′,7′-olefinic linkage in rotenone (**2**) is readily shifted by acid treatment via carbonium ion intermediates to give isorotenone (**43**); centers 6a and 12a being in this case unaffected.[30, 48, 75, 76] The nature of the isomerization has been established by dehydrogenation of **43** to isotubaic acid (**20**)[77]. Dehydroisorotenone (**4**), and isoderrisic acid (**44**) are formed by orthodox reactions as well as by isomerization of normal isopropenyl derivatives.[20]

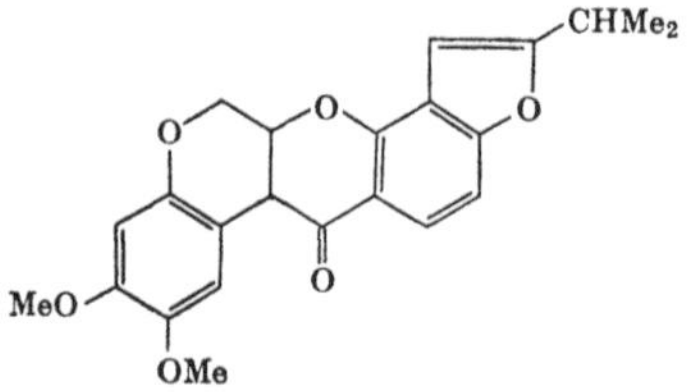

(43)

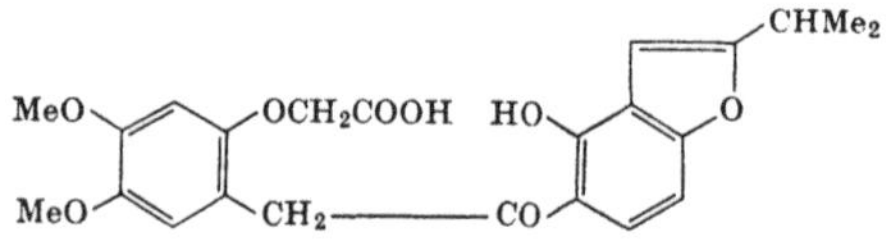

(44)

β-Dihydrorotenone (**45**) is converted into a saturated isomeride, allorotenone (**46**) upon treatment with sulfuric acid. The latter compound, frequently called *β*-dihydrorotenone, no longer shows phenolic properties.

Isorotenone is converted into a spiro compound analogous to **37**. The nature of this spiro compound is confirmed since it gives isorotenol

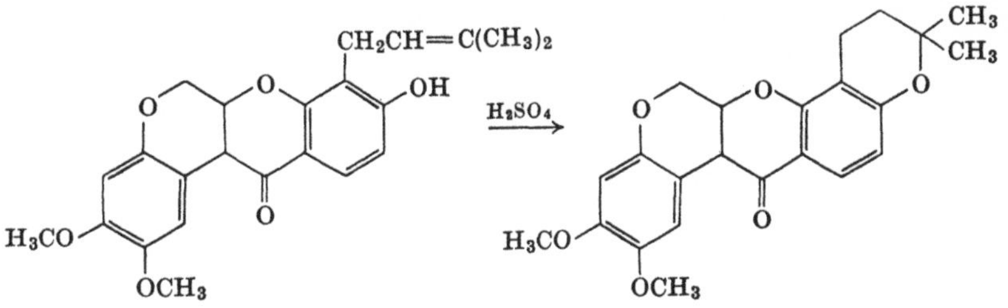

(**45**) (**46**)

(**48**), without the formation of isoderritol (**49**), when heated with zinc and alkali.[78] The spiro compound (**47**) is presumably also formed, by one–electron transfer reactions via a mono-, or a diradical (**50**)[48, 72] as in the case of rotenol. Treatment of **48** with lead tetraacetate gives the spiro compound (**47**).[48]

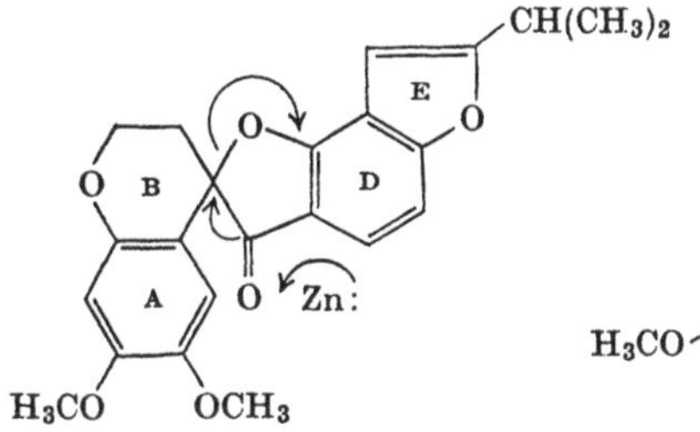

(**47**)

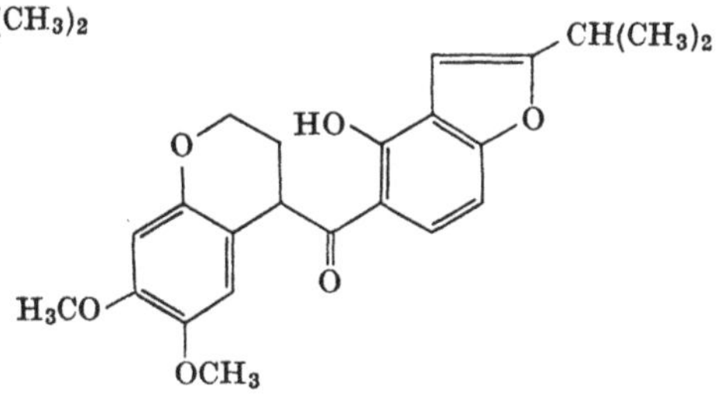

(**48**)

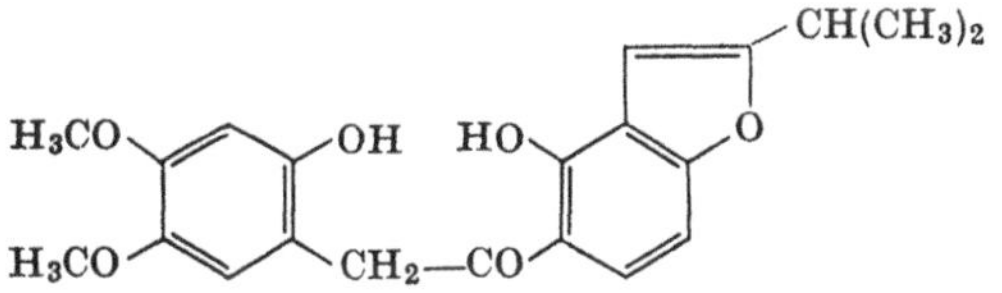

(**49**)

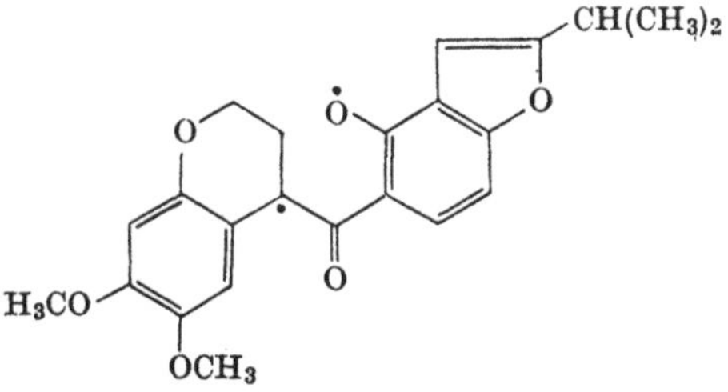

(**50**)

A. Isomeric Carbonyl Derivatives of Rotenone

Rotenone reacts with hydroxylamine in ethanol to form the α-oxime (m.p. 237°), and in pyridine to form the β-oxime (m.p. 249°).[79] These are dimorphous forms of which the higher melting isomer is the more stable. The hydrazone (m.p. 258°), and phenylhydrazone (m.p. 255°) are also known. Further, rotenone gives a different compound when oximated with hydroxylamine hydrochloride in the presence of sodium hydroxide, from that obtained by oximation with the same reagent in the presence of sodium acetate.[6] The latter is considered to be the normal oxime (**51**), and the former isoxazoline (isooxime, m.p. 210–211°) (**52**),[6, 30, 31, 48, 74] which would arise from attack on the ion as in **53**.[48] Spectral information supports the constitution of **52**. **51** has a free hydroxyl band near 3592 cm^{-1}, whilst the isooxime (**52**) has a strongly chelated hydroxyl group; there are two chelated bands near 3170 and 3080 cm^{-1}. The presence of the two bands suggests that the two tautomers **52** and **54** may be present. Similarly, oximes and isooximes have been obtained in the case of isorotenone (**43**) and dehydrorotenone (**3**).[48]

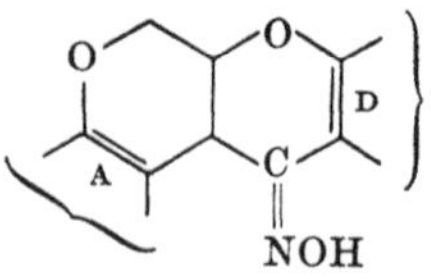

(**51**)

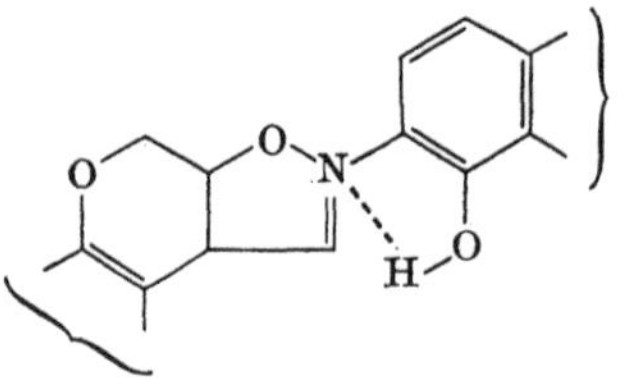

(**52**)

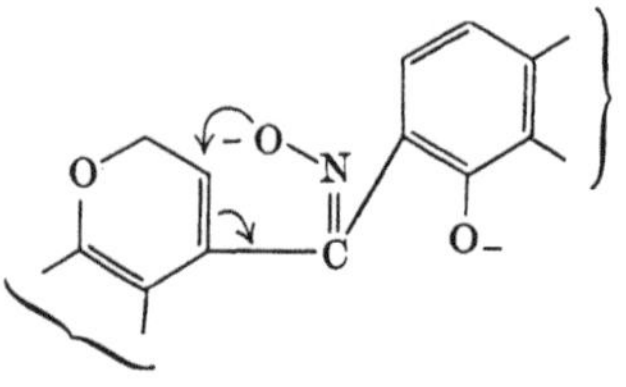

(**53**)

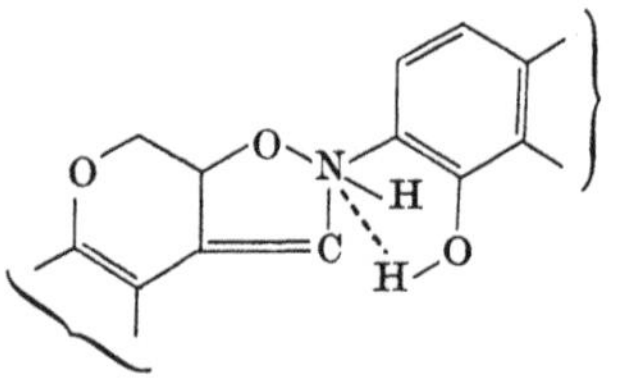

(**54**)

B. Hydrogenation

Catalytic hydrogenation of rotenone (**2**) involves addition of two hydrogen atoms to the isopropenyl group to produce dihydrorotenone.[65, 77, 80] It is dimorphic (m.p. 216° and 164°), the lower melting form being the less stable.[81] Hydrogenation may be accompanied by fission of the hydrofuran ring as in the case of tubaic acid (**18**) with the resultant loss of an asymmetric atom and formation of phenolic derivatives.[52, 71] The reaction takes the latter course, especially in alkaline solution. In neutral solution, hydrogenation gives optically active α-dihydrorotenone (**55**), isomeric with dihydroisorotenone, obtained by hydrogenation of isorotenone (**43**) together with β-dihydrorotenone (rotenonic acid) (**45**).[81a] **45** is an unsaturated phenol corresponding to β-dihydrotubaic acid (**21**), and on further hydrogenation gives tetrahydrorotenone (dihydrorotenoic acid) (**56**).[77, 82]

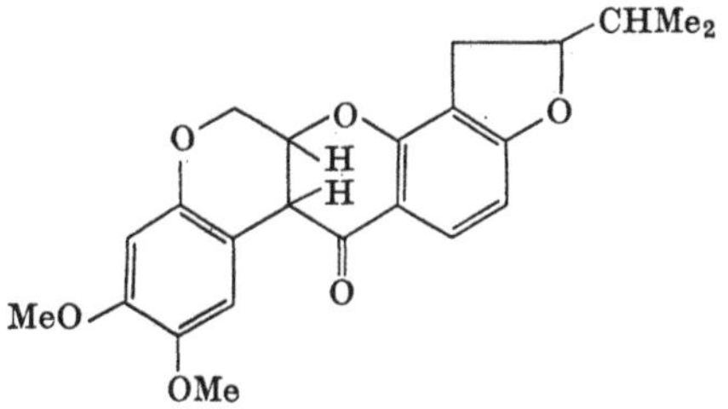

(**55**)

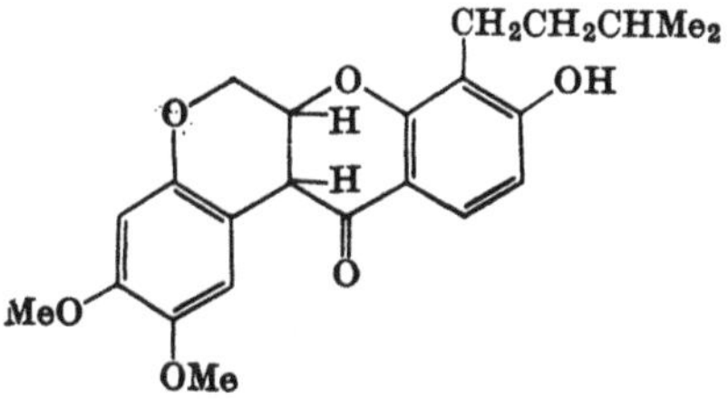

(**56**)

55 gives an enol acetate which is catalytically reduced to dihydrodesoxyrotenone.[51] Similarly, acetic anhydride and sodium acetate produce a diacetyl compound with **56**. Tetrahydrorotenone derivatives have been obtained upon hydrogenation.[31, 49, 65, 83]

C. Oxidation

Chromic acid oxidizes rotenone to the yellow substance, rotenonone (**7**) (m.p. 298°),[29] and so do nitrous acid,[53] manganese dioxide, and plumbic acetate; the last being the best reagent.[48] Analogous compounds are obtained similarly from isorotenone (**43**)[29] α-dihydrorotenone (**55**),[49] β-dihydrorotenone (**45**),[82] and tetrahydrorotenone (**56**),[49] namely, isorotenonone (**57**), α-dihydrorotenonone (**58**), β-dihydrorotenonone (**59**), and tetrahydrorotenonone (**60**), respectively. It is evident that the first step of this oxidation is the formation of the dehydro derivatives, as **7** and its analogs are also obtained from the corresponding dehydro compounds.[82]

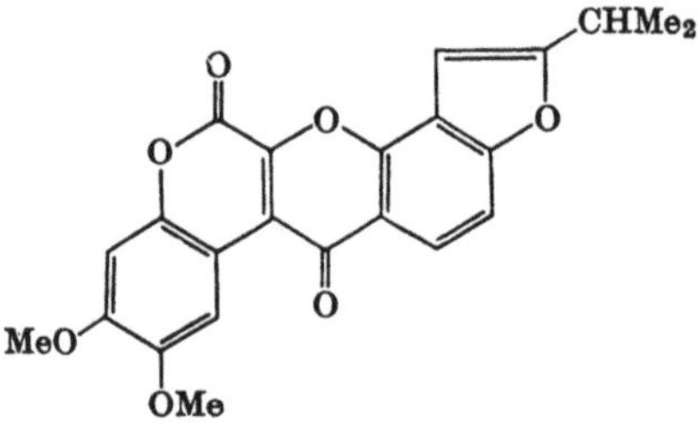

(57)

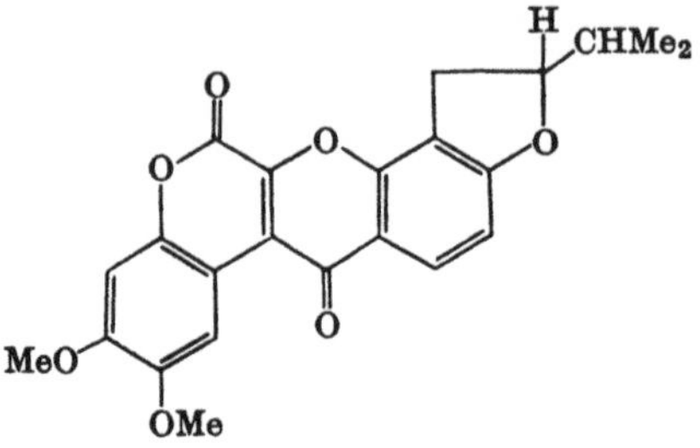

(58)

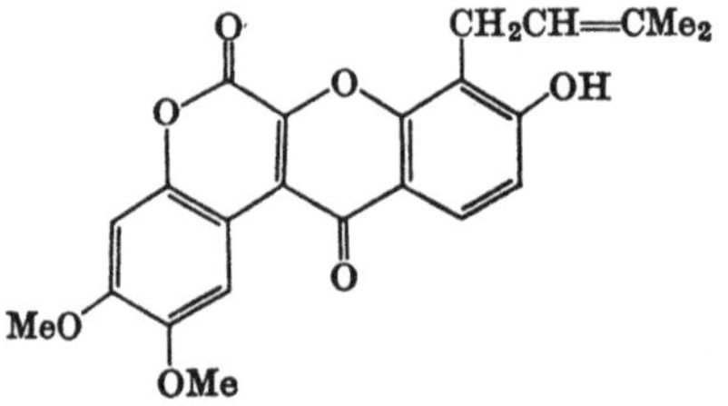

(59)

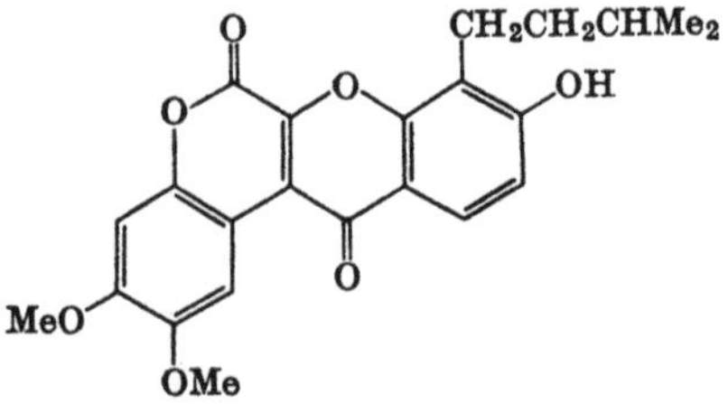

(60)

Rotenonone (**7**) is converted by vigorous alkaline degradation (20% ethanolic potassium hydroxide) to derritol (**9**) and oxalic acid, probably through the keto form (**62**) of the intermediate (**61**) (Eq. 5a)[29, 55] while **9** has been converted into **7** by interaction with diethyl oxalate in the presence of sodium acetate or the acid chloride of monoethyl oxalate.[55] By the action of mild alkali, **7** is converted into rotenon acid (**63**) (m.p. 250°), probably via the intermediate (**61**).[49] **63** is lactonized in acid solution to β-rotenonone (**64**) (m.p. 275°),[29] which rearranges in ethanolic potassium hydroxide to **7** (Eq. 5b).

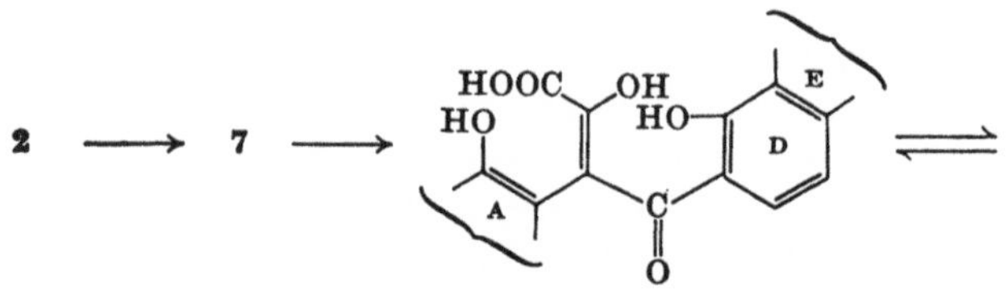

(61)

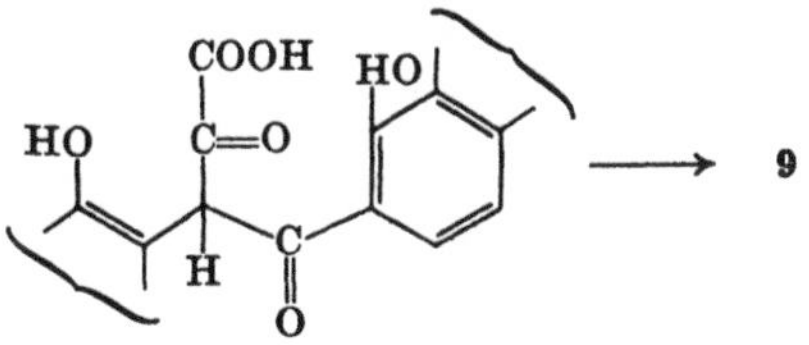

(62)

(5a)

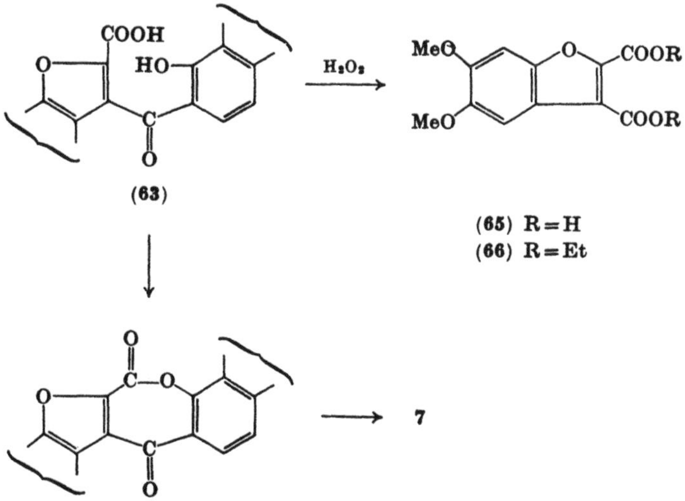

(63)

(65) R=H
(66) R=Et

(64)

(5b)

Alkaline hydrogen peroxide oxidation of **63** gives tubaic acid (**18**) and a dibasic acid, abutic acid (**65**),[30] which has been synthesized by two methods. In the first process,[84] the 3-chlorocoumarin (**67**), obtained from 4-hydroxyveratrol and ethyl-β-carbethoxy-β-chloropyruvate by von Pechmann condensation, was converted by treatment with alkali into **65**; a process which closely parallels the mechanism postulated by LaForge for the rotenonone–rotenonic acid (**7**–**63**) change,[49] and provides support for the structure of **63**. In the second method,[85] 2-hydroxy-4,5-dimethoxyglyoxylic ester (**68**) was condensed with ethyl bromoacetate to yield the phenoxy ester (**69**), which was cyclized with sodium ethoxide to the benzofuran diester (**66**) and subsequently hydrolyzed to give a product (**65**), identical with that prepared by the first route and with abutic acid (arbutic acid) (Eq. 6).

5. Synthesis of Rotenone

The partial synthesis of rotenone (**2**) has been described by different authors. Condensation of derritol (**9**) with ethyl bromoacetate in the

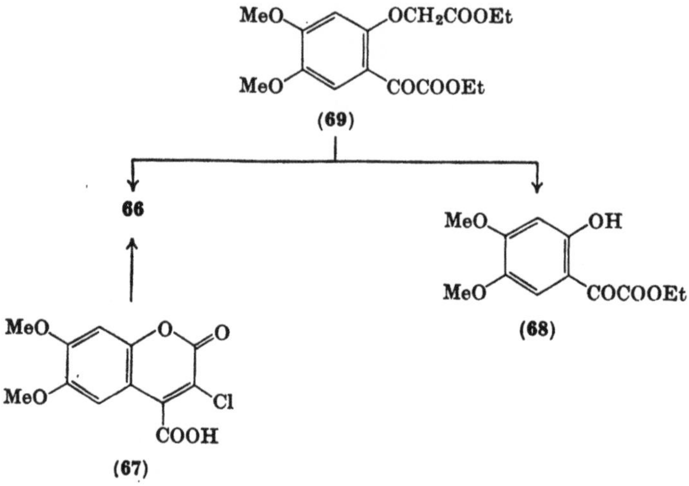

(6)

presence of sodium ethylate readily gave dehydrorotenone (**3**) and derrisic acid ethyl ester.[86] Hydrolysis of the latter ester to the acid (**8**), followed by treatment with acetic anhydride and sodium acetate yielded **3**.[29] Treatment of **3** with sodium borohydride gave the hydroxy compound (**70**), which upon oxidation with acetone under the influence of aluminum isopropoxide gave mutarotenone;[87a] the latter was readily converted to *l*-rotenone[88] (Eq. 7).

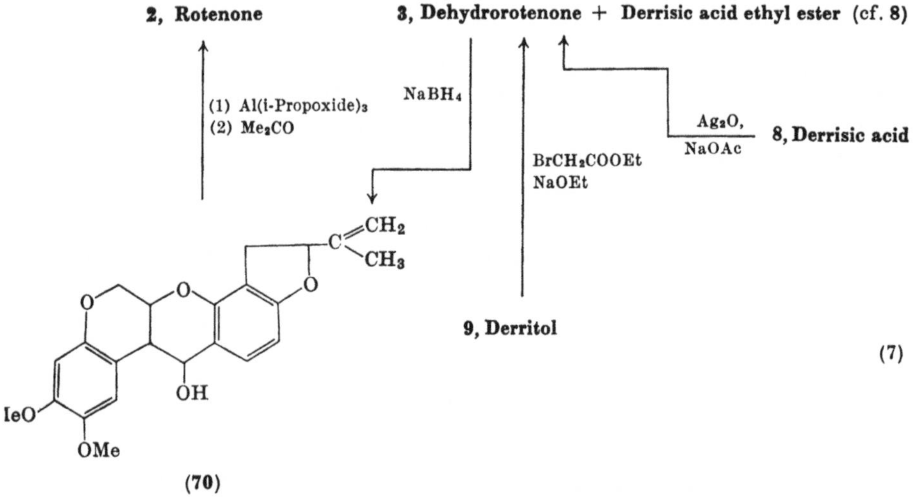

(7)

Miyano and Matsui[10] have partially synthesized rotenone and dihydrorotenone by Oppenauer oxidation of the hydroxy compounds (70) and
(71), respectively. The latter compounds can be obtained from derritol
(9) and dihydroderritol, respectively, via dehydrorotenone (3) and
dihydrodehydrorotenone (72), respectively, followed by treatment with
sodium borohydride as above.

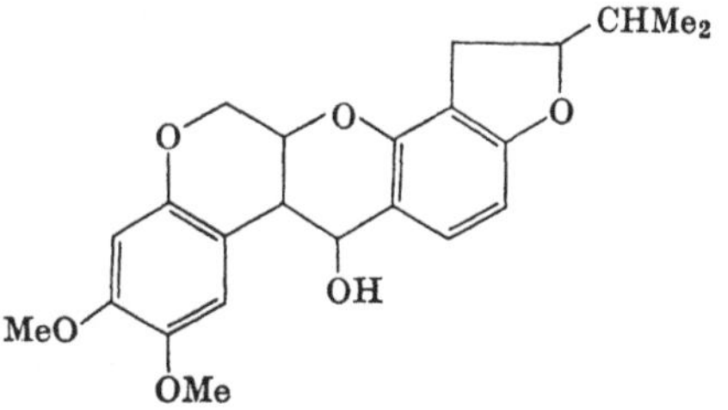

(71)　　　　　　　　　　　　　　　　　　(72)

An absolute synthesis of dihydrorotenone (cf. 72) has been achieved by
Miyano and Matsui.[40, 89] The necessary l-dihydrotubaic (19) was prepared by resolving the synthetic racemate by means of brucine. The
product 19 was decarboxylated to dihydrotubanol (73) and a Hoesch
reaction with the cyanide (15) was effected, to yield the optically active
dihydro derivative of derrisic acid (8), which was cyclized by the standard
method to dehydrodihydrorotenone (72); identical with material
prepared from the natural product. 72 was reduced with sodium borohydride to the saturated alcohol (71), which is identical with the substance made by acting on dihydrorotenone with borohydride, and can be
oxidized to dihydrorotenone, either by the Oppenauer method, or by the
use of manganese dioxide under conditions minimizing dehydrogenation
(Eq. 8).

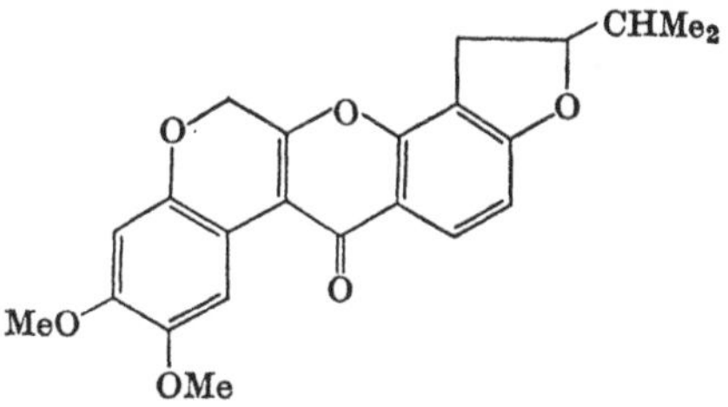

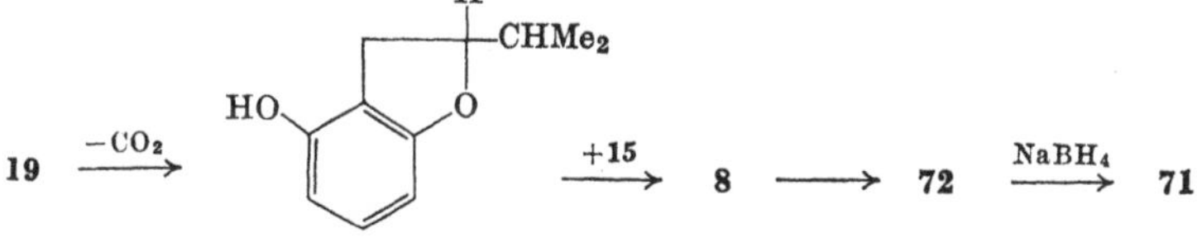

$$ 19 \xrightarrow{-CO_2} \qquad \xrightarrow{+15} 8 \longrightarrow 72 \xrightarrow{NaBH_4} 71 \qquad (8) $$

In this connection 3 has recently been obtained by condensation of
pyrrolidine enamine (74a) with tubaic acid (18) and/or tubaic acid
chloride. The latter compound, reacts similarly with morpholine
enamine (74b) to yield 3.[87b] Derrisic acid (8) is converted into 3 by dicyclo-
hexylcarbodiimide in the presence of a tertiary base, followed by mild
basic treatment of the isolabile intermediate (74c).[87c]

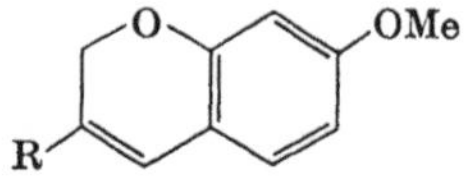

(**74a**) R = pyrrolidin;
(**74b**) R = morpholin

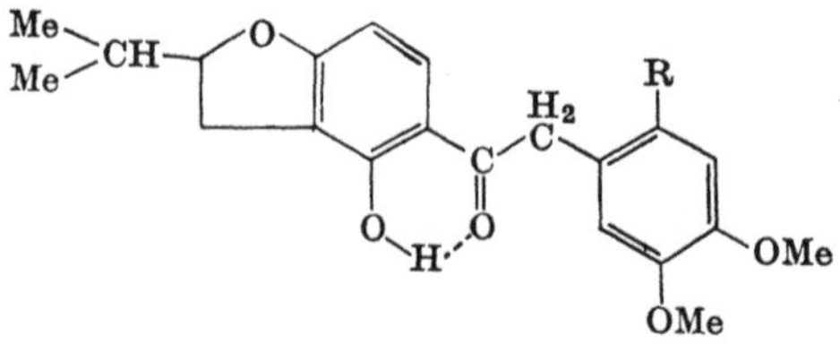

(**74c**) $R = OCH_2CO_2C(=NC_6H_{11})NHC_6H_{11}$

The absolute synthesis of *l*-rotenone has been achieved.[41] Starting from 2-acetyl-4-hydroxycoumarone (**75**), which had been synthesized in six steps from resorcinol by other workers,[90, 91] 4-hydroxy-2-(α-hydroxyisopropyl)coumarone (*dl*-hydroxydihydrotubanol) (**78**) was obtained by treatment of the benzyl ester (**76**)[40] with methylmagnesium iodide to yield 4-benzyloxy-2-(α-hydroxyisopropyl)coumarone (**77**). The latter was hydrogenated over Raney nickel.

Hoesch reaction of **78** with the cyanide (**15**), followed by saponification, resulted in a mixture of *dl*-hydroxydihydroderrisic acid (**79**, X = OH) (main product) and *dl*-chlorodihydroderrisic acid (**79**, X = Cl). The

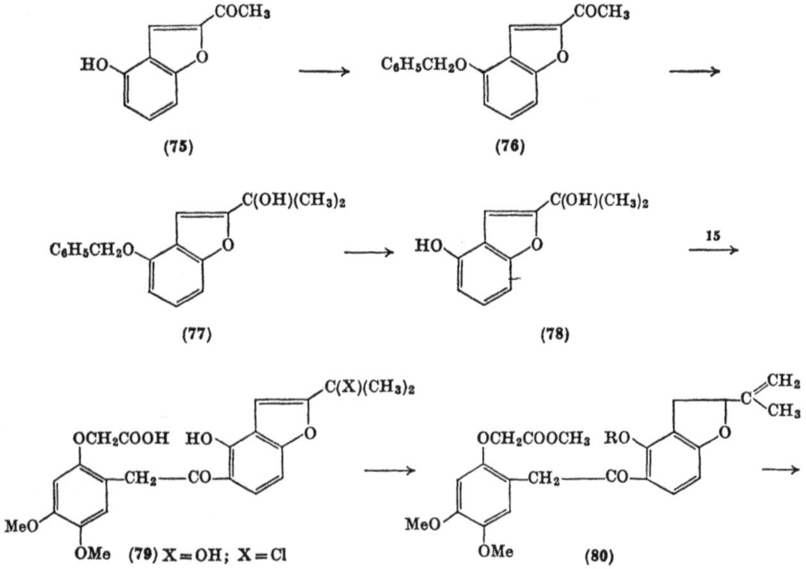

(75) (76) (77) (78) (79) X = OH; X = Cl (80)

8 ⟶ 1-α-Phenyl ethylamine salt ⟶ *l*-form of 8 $\xrightarrow[Ac_2O]{NaOAc}$ 3 ⟶

70 ⟶ Mutarotenone ⟶ *l*-Rotenone (9)

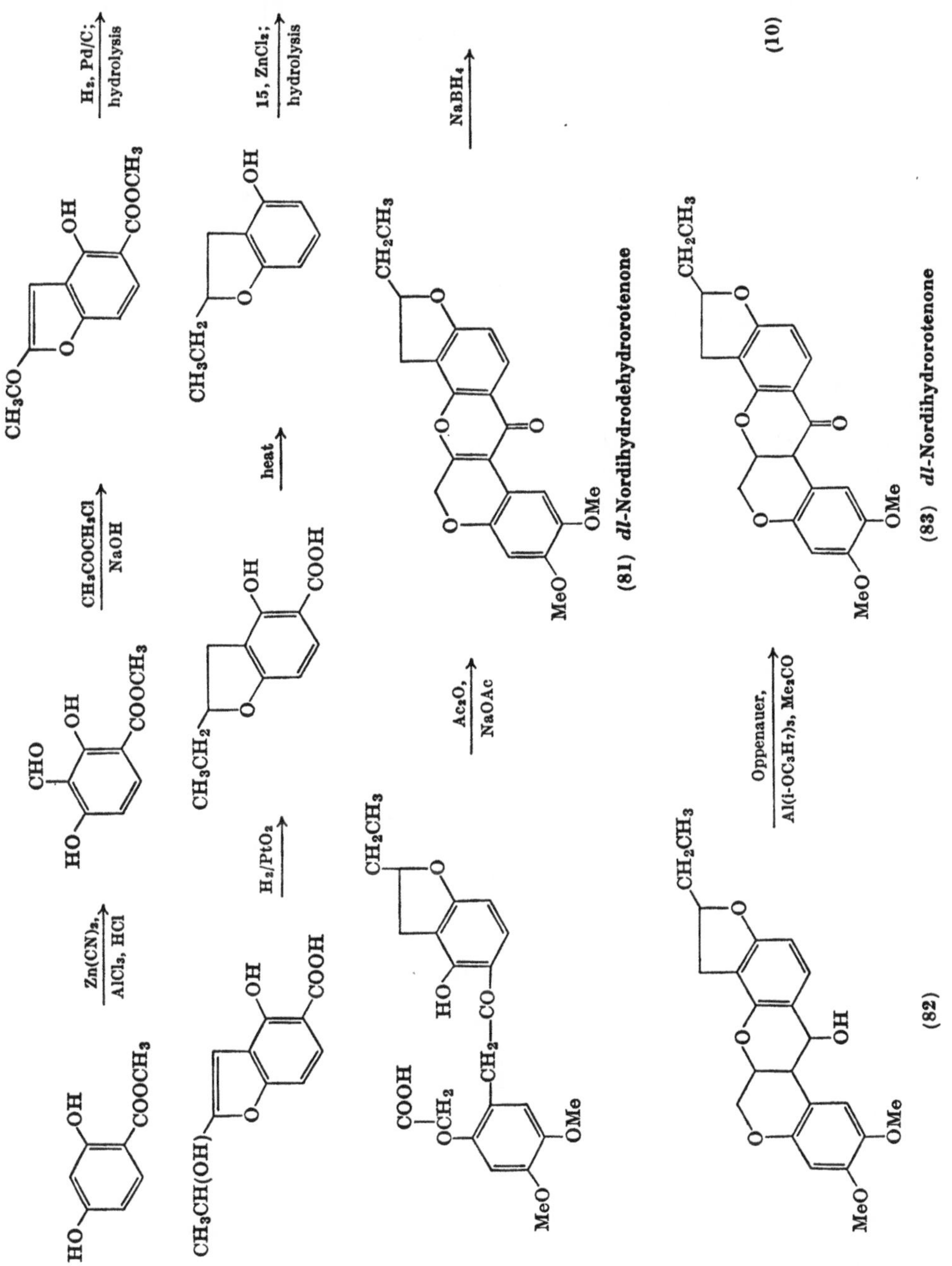

(10)

(81) *dl*-Nordihydrodehydrorotenone

(82)

(83) *dl*-Nordihydrorotenone

methyl ester of **79** (X = OH), upon dehydration with phosphorus bromide in cold pyridine, gave **80** (R = $P(OH)_2$), which was converted into the free *dl*-derrisic acid (**8**). *l*-Derrisic acid, obtained by resolving the racemate with 1-α-phenylethylamine, was cyclized to dehydrorotenone (**3**) by boiling with acetic anhydride in the presence of sodium acetate,[29] which was converted into the hydroxy compound (**70**) by sodium borohydride.[10, 87] On Oppenauer oxidation, the latter gave mutarotenone, a stereoisomer of the natural rotenone, which converted into the natural modification by thermal isomerization. The total synthesis consists of a sequence of 11 steps from the known compound **75** or 17 steps from the commerically available resorcinol (Eq. 9).

Radioactive rotenone has recently been biosynthesized by applying solutions of $CH_3^{14}COONa$ and $^{14}CH_3COONa$ for 10 to 20 days to the leaves of *Lonchocarpus nicou*, and drying and extracting the roots 55 hours later with chloroform. Both acetates yielded radioactive rotenone.[92]

The synthesis of *dl*-nordihydrorotenone (**83**) has been effected[93] via the intermediates **81** and **82** according to the following sequence of reactions (Eq. 10).

6. Stereochemistry of Rotenone

A. Optical Activity

Rotenone is optically active[33] due to the presence of three asymmetric carbon atoms (6a, 12a, 5′) (cf. **1**). Rotenone is *laevo* rotatory in benzene, and on oxidation gives *l*-dehydrorotenone (**3**), destroying the asymmetry of C_{6a} and C_{12a}; thus $C_{5'}$ must be the *laevo* rotatory center[93b]. Rotenone, when treated with a mild base gives mutarotenone,[88] which is an equilibrium mixture of rotenone and a diasteroisomer; these substances have the same stereochemistry at $C_{5'}$ but are epimeric at C_{6a} and C_{12a}.[88] Thus, the bases can invert the configuration at C_{6a} and C_{12a} simultaneously. The base–catalyzed equilibria involve the anions **84** and **85**.[48, 93a]

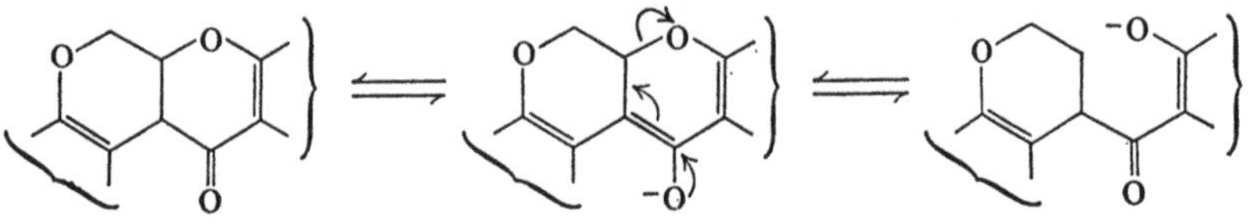

(**84**) (**85**)

All natural rotenoids with C_{6a} and C_{12a} as the only asymmetric centers show in optical activity rotatory dispersion measurements a rather similar, positive Cotton effect; thus making it very likely that all have the same configuration.[26] Rotenone and sumatrol (**112**) do not fall precisely into line because of the additional center at position 5′, but the iso compounds, e.g. isorotenone (**43**) conform to the rule.

Substance	Asymmetric centers	References
Rotenone (**1**)	C_{6a} C_{12a} $C_{5'}$	33
Sumatrol acetate (cf. acetyl deriv of **112**)	C_{6a} C_{12a} $C_{5'}$	88, 93a
Dihydrorotenone (**54**)	C_{6a} C_{12a} $C_{5'}$	33
Dehydrorotenone (**3**)	$C_{5'}$	33
Dehydrodihydrorotenone.	$C_{5'}$	
Derrisic acid (**8**)	$C_{5'}$	33
Elliptone (**128**)	C_{6a} C_{12a}	53
Pachyrrhizone (**164**)	C_{6a} C_{12a}	
Rotenonic acid	C_{6a} C_{12a}	33
Isorotenone (**43**)	C_{6a} C_{12a}	33

B. Configuration

The stereochemical problem could be solved by three determinations of absolute configuration, two determinations of absolute and one of relative configuration, or one determination of absolute and two of relative configurations. Büchi, Crombie, and coworkers[11] have chosen the second course.

Rotenone is cleaved by alkali to give tubaic acid which is readily hydrogenated to (−)-dihydrotubaic acid (**19**),[64] retaining the rotenone configuration **1** at position 5′. On exhaustive ozonolysis,[94, 95] followed by oxidation with hydrogen peroxide, **19** gave (+)-3-hydroxy-4-methylpentanoic acid (**86**). The absolute configuration of **86** was established as follows: synthesis and resolution gave the (−)isomer of the latter acid **87**, and this was degraded to (−)-2-methylpentan-3-ol (**88**), which can be related to L-glyceraldehyde by correlations in the literature.[96–98] (+)-3-Hydroxy-4-methylpentanoic acid from (−)-dihydrotubaic acid is thus related to D-glyceraldehyde and hence rotenone has the (R) configuration[99] at position 5′.

The absolute configuration at position 6a was determined. When rotenone was vigorously acetylated by acetic anhydride and sodium acetate, it gave the enol acetate (**89**, R = OAc)[50, 51] in which the double bond is adequately located by means of the ultraviolet spectrum, indicative of a greatly extended chromophore.[41] At the same time, the center

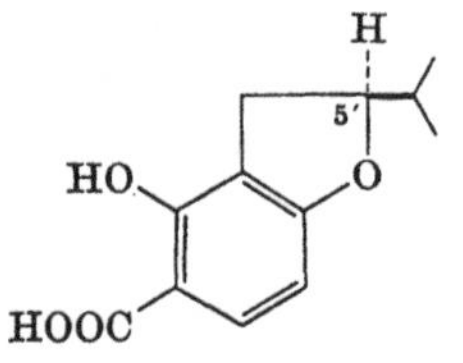

(19)

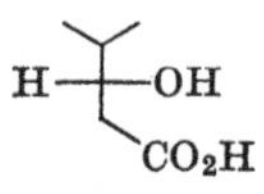

(86)

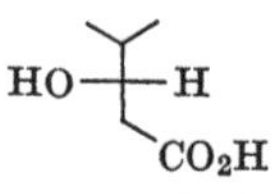

(87)

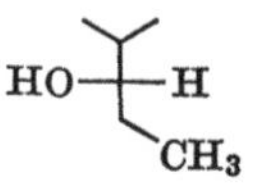

(88)

at position 6a epimerizes, and the diastereoisomers have to be separated. The enol acetate was hydrogenated to acetyl 6′,7′-dihydrorotenone[51] which can be made directly from 6′,7′-dihydrorotenone by the isopropyl acetate–acid technique; like acetylrotenone[88, 100] it exists in dimorphic form. Exhaustive ozonoloysis of acetyl-6′,7′-dihydrorotenone gave a mixture of acids which was purified by ion-exchange chromatography and shown to be (+)-3-hydroxy-4-methylpentanoic acid (**86**) together with D(−)-glyceric acid (**90**), which contains atom 6a of structure **89** (R = OAc). The configuration at position 12a has not been determined directly, but it must be (S) because it can be proved that the B/C ring fusion in rotenone is *cis*. Hence, natural rotenone has the (6aS, 12aS, 5′R) configuration.[11]

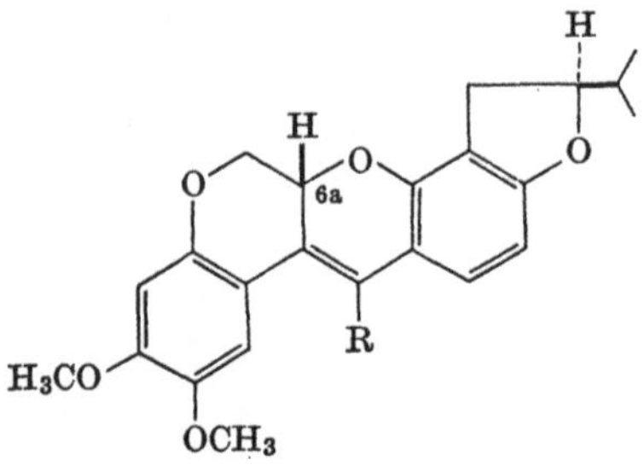

(89)

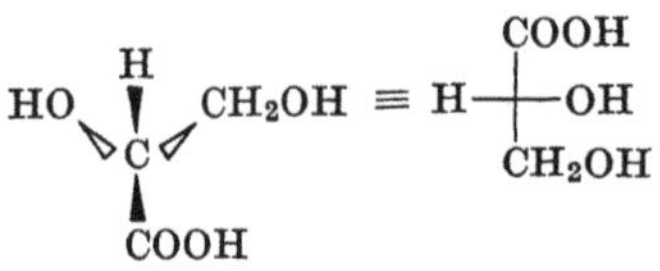

(90)

All natural rotenoids with atoms 6a and 12a as the only asymmetric centers show in their optical rotatory dispersion measurements, similar, positive Cotton–effect curves; thus making it very likely that all have the same configuration.[26] Rotenone and sumatrol (**112**) do not fall precisely into line because of an additional center at position 5′, but their isocompounds (as isorotenone, **45**) conform to the rule. Rotenone and sumatrol acetate exhibit positive Cotton–effect curves. Minor variations in the curves are the result of differences in the substitution patterns and are not of great significance. Hence all natural rotenoids are 6aS, 12aS.

Cahn, and coworkers[88] have established that natural rotenone has the stable form of B/C fusion which they thought might be *trans*, although they reserved judgment. Miyano and Matsui[10] claim to have determined that rotenone has the *trans* fusion form. 6a,12a-Dehydrorotenone (**3**) was treated with sodium borohydride to give an amorphous alcohol (**69**), which on Oppenauer oxidation, yielded a mixture of two diastereoisomers, mutarotenone,[88] one of which was natural rotenone. Such a sequence should lead to a product with a thermodynamically stable fusion; which confirms Cahn's view.[88] The proof that the junction is *cis* is owed to Büchi, Crombie, and coworkers.[10, 38, 48] Direct evidence in favor of a *cis* B/C fusion in rotenone itself was obtained[48] by reduction of (±)-isorotenone (**43**) with potassium borohydride to give a crystalline hydroxy compound (**91**) in more than 80% yield. The latter was also obtained by hydrogenation of **43** over Adams platinum, or by treating 6a, 12a-dehydroisorotenone (**4**) with borohydride. The hydroxy compound (**91**) is dehydrated by phosphorus oxychloride in pyridine, giving **92**, (R=H). The ultraviolet absorption of the latter is similar to that of isorotenone enol acetate (**92**, R=OAc).

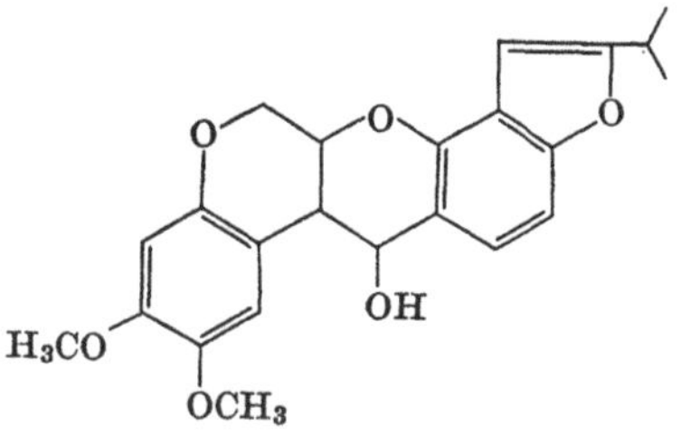

(**91**)

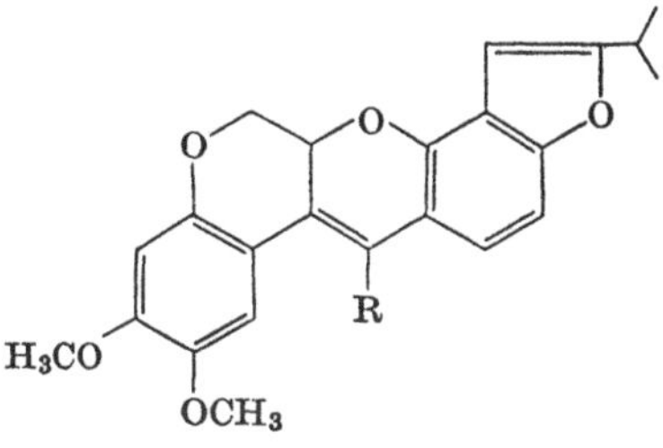

(**92**)

Only one alcohol is formed when two are possible, and this, in theory, is an indication that one side of the molecule is more approachable than the other. Since *trans* fusion gives a more or less flat molecule, whereas *cis* fusion results in a folded molecule with only one open face, the fact clearly favors the latter arrangement for isorotenone.

Stronger evidence has been found by consideration of the hydrogen stretching frequencies of the hydroxy compound (**71**) which occur at 3613 and 3566 cm^{-1}. The former frequency could correspond to a weakly bonded hydroxyl group, but the latter must be related to a more strongly bonded one, and evidently there is at least one acceptor site available for such bonding. The former frequency is found in compounds such as 1-hydroxytetralin (3614 cm^{-1}) and chroman-4-ol (3615 cm^{-1}) and might be attributed to hydrogen bonding with π electrons, i.e. with the aromatic nuclei. The latter frequency is low enough to suggest that it represents bonding to an appropriately placed oxygen atom, such as occurs in 2,4,5-trimethoxyphenylethanol (**93**, R = OCH_3) which has a band at 3547 cm^{-1} absent from the spectrum of 4,5-dimethoxyphenylethanol (**93**, R = H).

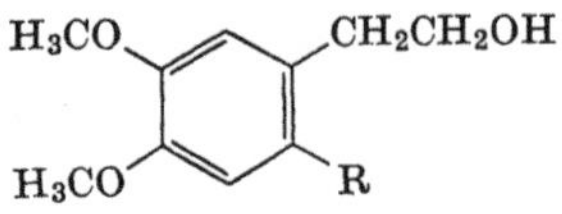

(**93**)

Models show that if the rings are *cis* fused and the hydroxyl group is α-oriented then it is only possible for bonding to occur with the oxygen atom at the 5-position (cf. diagram **94**). The conclusion that the junction is *cis* fused not only serves to explain the production of a single alcohol of type **91**, but also accounts for the resulting α-orientation, since the reducing agent would have been forced to approach from the B face.

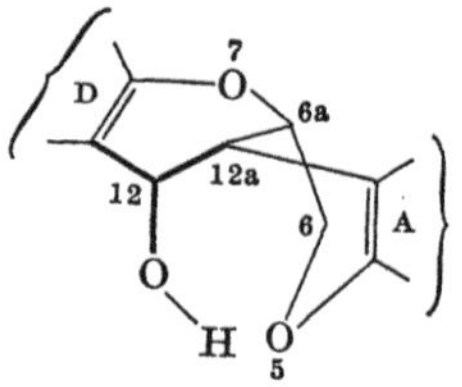

(**94**)

Study of the 1-proton resonances in rotenoids throws valuable light on the geometry of the B/C fusion.[25] In 6a, 12a-dehydroisorotenone (**4**), the 1-hydrogen atom is approximately coplanar with the 12-carbonyl group and shows the expected negative shieldings. The possible shielding of 1-hydrogen by aryl ring D has been estimated by Johnson and Bovey's method[101] and does not influence the interpretation. According to this nuclear magnetic resonance study, natural optically active rotenone has the *cis* B/C fusion. This applies also to isorotenone and various other rotenoids with the thermodynamically favored B/C fusion. By using the earlier establishment of the *cis* B/C fusion and complete absolute configuration of rotenone,[11] similar conclusions can be reached for natural rotenoids by comparing optical rotatory dispersion curves.[26] The latter method compares absolute configurations, but the present one is applicable to either optically active or (±)-6a, 12a compounds. The nuclear magnetic resonance method is also absolute in type, rather than dependent on correlation with a previously worked–out standard.

7. Rotenolones, Rotenolols, Isorotenolones, and Isorotenolols

Oxygenation of rotenone (**2**) or isorotenone (**43**) in the presence of alkali gives mixtures of ketols called rotenolones[102] and isorotenolones,[103] respectively. Their structures were uncertain and their stereochemistry unknown until recently[40a] partial assignments have been given to some of the rotenolones of the diols (rotenolols) derived by reduction of the 12-keto group.

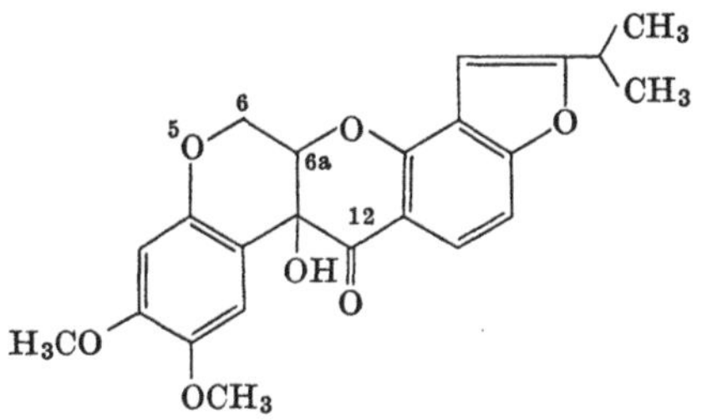

(Isorotenolone A and B)

The structure and stereochemistry of the rotenolones, rotenolols and the iso isomers have recently been studied by Crombie and Godin.[38, 104] From (–)-isorotenone two ketols, (±)-A, m.p. 205°, and (±)-B, m.p. 243°, were obtained. That, the hydroxyl group in each racemate is at position 12a, follows from their resistance to iodination, to enol acetylation, to dehydrogenation with manganese dioxide, and from their ready dehydration by acid to give 6a, 12a-dehydroisorotenone (**4**). Further evidence is the hydrogenolysis of their derived isorotenolols to keto-aldehydes (see below). The formation of isorotenolol or isoderritolol on treatment of either ketol with zinc and alkali is readily accommodated. In (±)-isorotenolone A the hydroxyl is strongly intramolecularly bonded to the keto group but this is not so for (±)-isorotenolone B; this has been supported by the carbonyl stretching vibrations measurements. (±)-Isorotenolone A is therefore assigned a *cis* B/C junction (**95–96**), where in suitable conformation the hydroxyl group is closer to the 12-ketonic–oxygen atom and much better aligned with its lone pair orbits[105] than is possible in the *trans* B/C structure (**97–98**) assigned to (±)-isorotenolone B.

Because of the alkaline preparative conditions, these isorotenolones are racemic at positions 6a and 12a but, by oxidizing (–)-isorotenolone

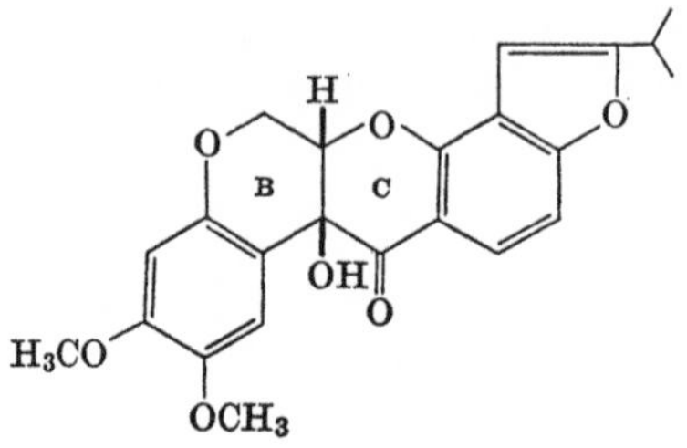

(95)

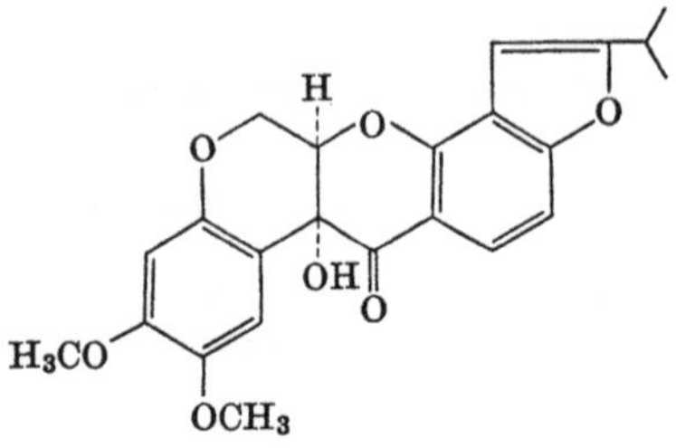

(96)

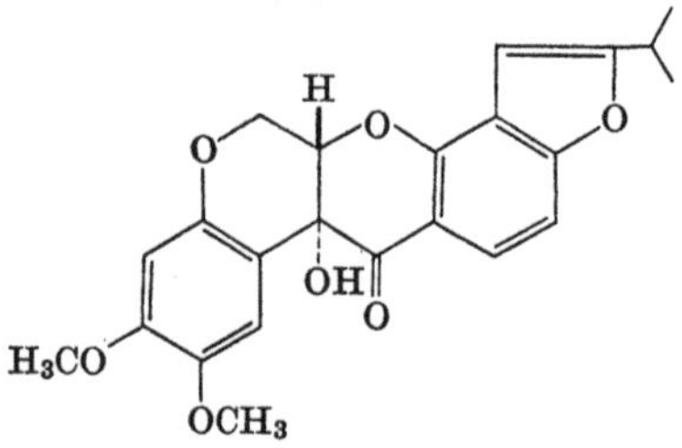

(97)

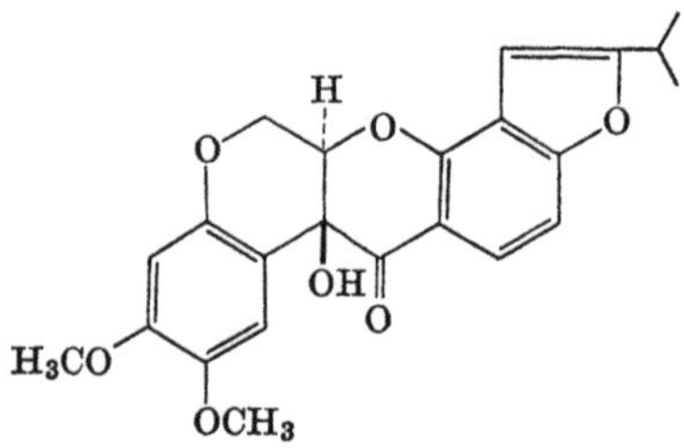

(98)

with chromic acid,[106] which does not racemize C_{6a}, the optically active stereoisomers **95** and in small yield **97** were obtained.

Reduction of (±)-isorotenolone B with sodium borohydride gave a

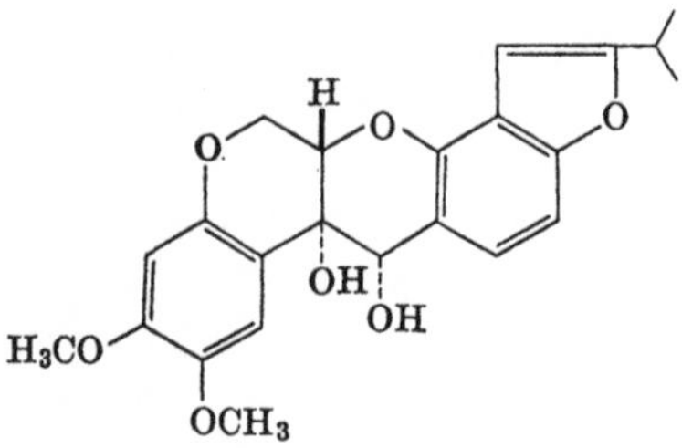

(99)

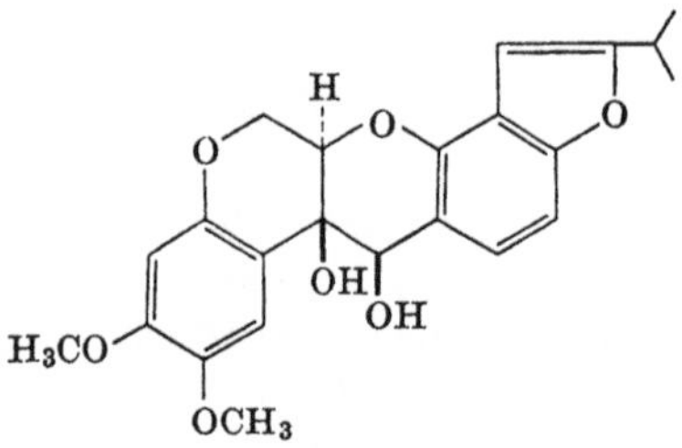

(100)

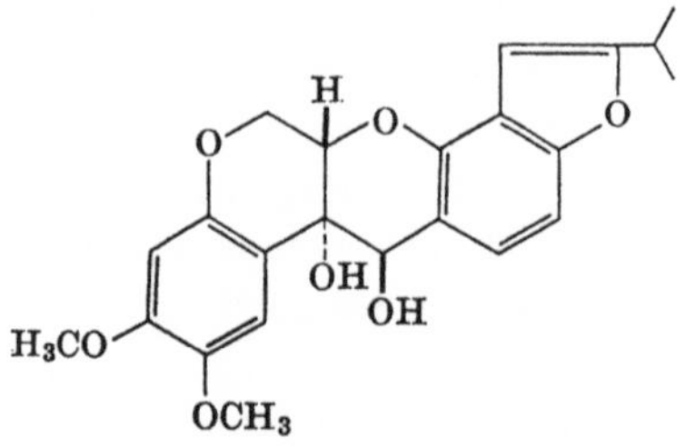

(101)

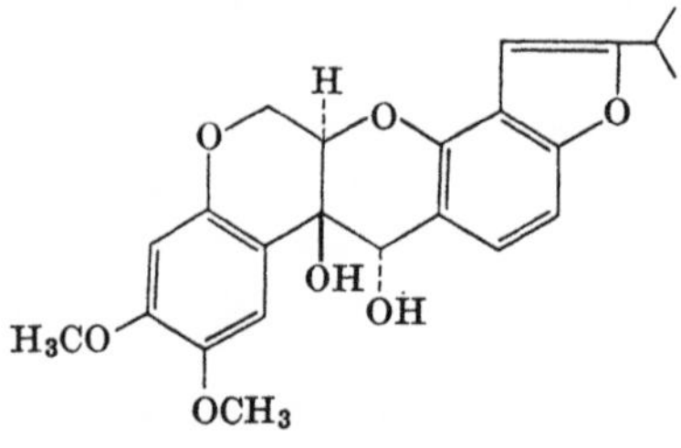

(102)

(±)-diol B_1, m.p. 264°, and a (±)-diol B_2, m.p. 213°. The first is the racemate (**99–100**), since it results when the 12, 12a-dehydro compound (**103–104**) is hydroxylated with osmium tetroxide, and because it is cleaved much more readily by lead tetraacetate than the (±)-diol B_2 (**101–102**). When shaken with manganese dioxide in acetone each diol reformed (±)-isorotenolone B.

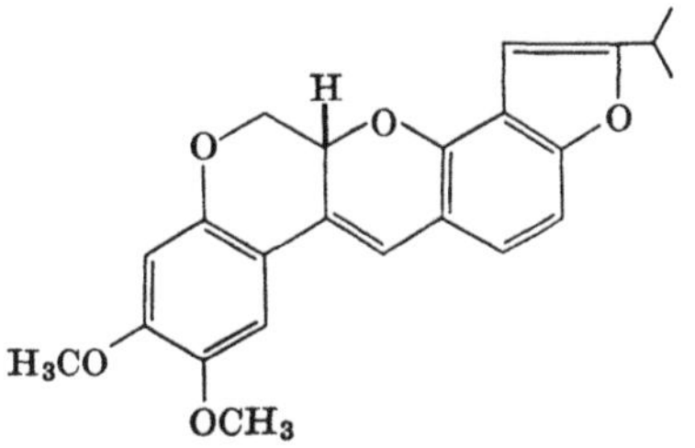

(**103**)

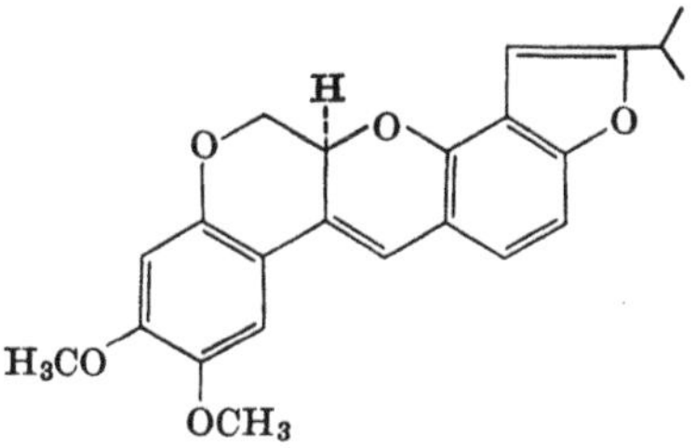

(**104**)

Reduction of (±)-isorotenolone A with borohydride gave only one (±)-diol A, m.p. 218°. Its rate of reaction with lead tetraacetate favors the racemate structure (**105–106**), which is to be expected if attack is from the less hindered face of the *cis* B/C ketol. When treated with periodic acid the (±)-diols A_1, B_1, and B_2 gave the racemic keto aldehyde (**107–108**).

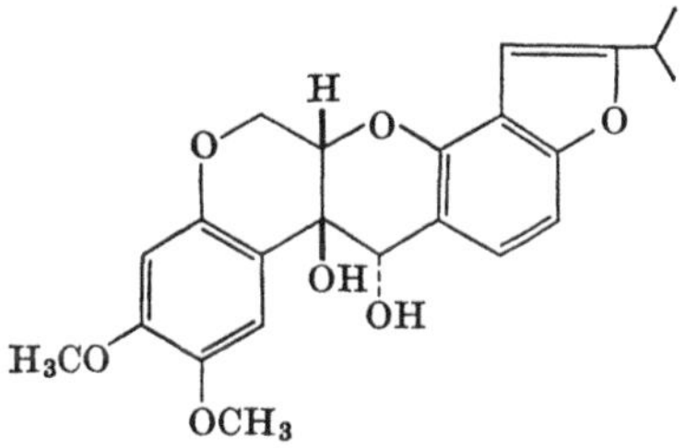

(**105**)

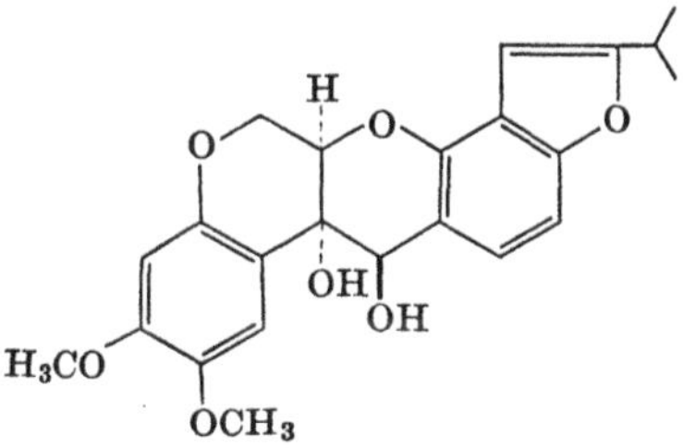

(**106**)

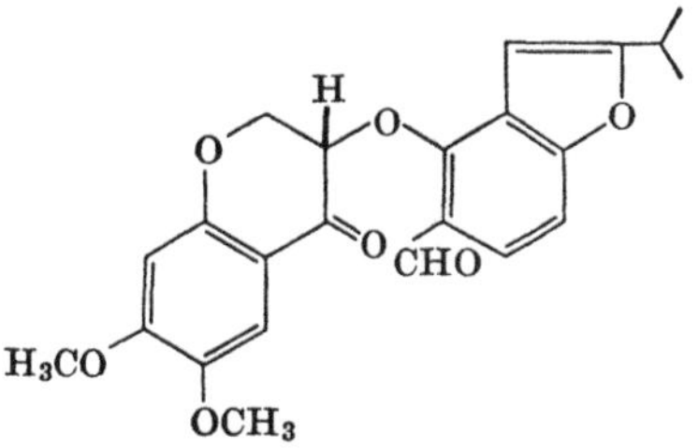

(**107**)

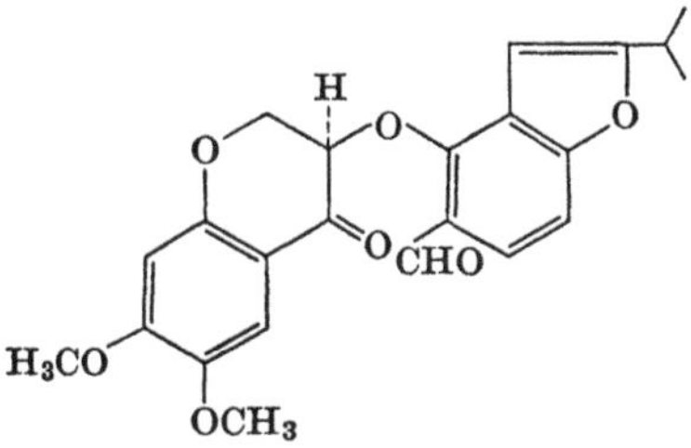

(**108**)

(±)-Isorotenolone C is made by heating (–)-isorotenone in methanolic potassium hydroxide (5%) with hydrogen peroxide. Analogous compounds can be obtained from dihydrorotenone. Isorotenolone C is isomeric with A and B, and contains one hydroxyl group which forms a methyl ether, and a toluene-*p*-sulfonate, and by treatment with ketene or refluxing sodium acetate–acetic anhydride, a monoacetate is obtained. Dehydration is difficult compared with that of isorotenone and a relationship with the spiro compound (**47** and **108a**) has been suggested, which is supported by close similarity of ultraviolet spectra.[107]

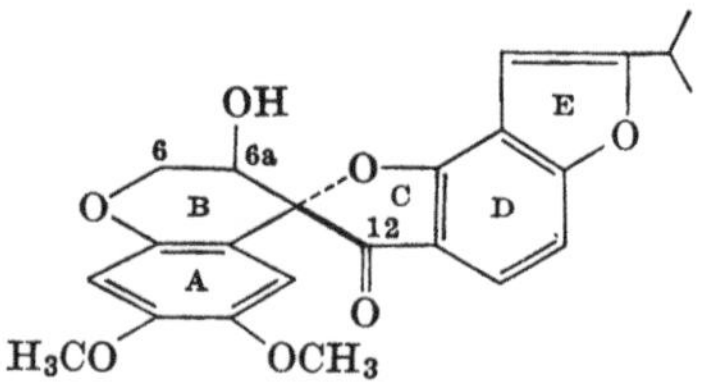

(108a) Isorotenolone C

108a gives isoderritol (**49**) and isorotenol (**48**) when treated with zinc and alkali and it is reduced to a diol by borohydride; oxidation of the latter with manganese dioxide gives back the original ketol.

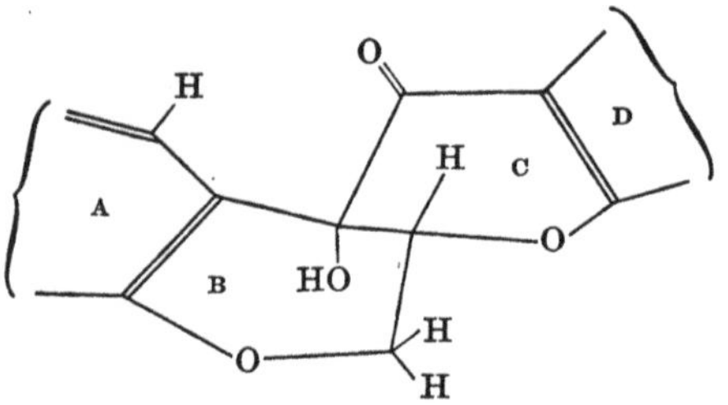

(109)

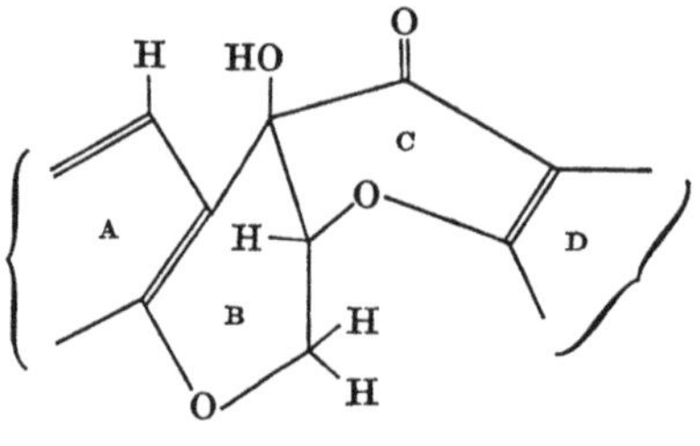

(110)

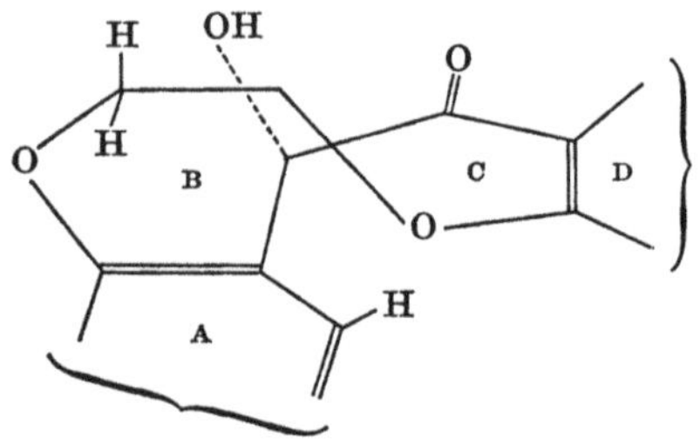

(111)

Alkaline oxygenation of natural (−)-rotenone is more complex, as four diastereoisomeric ketols instead of two pairs of enantiomorphs, are formed. The two *trans* B/C compounds (cf. **97**) and (cf. **98**) were separated, but the two *cis* B/C diastereoisomers did not separate clearly.[88] Only one diol was isolated from each of the latter *cis* B/C rotenolone diastereoisomers on reduction with borohydride.[38] Structural and stereochemical assignments rest on evidence similar to that for the isorotenolones.

Study of the 1-proton resonance in rotenolones throws valuable light on the geometry of the B/C fusion.[25] Rotenolone B and isorotenolone (**97–98**) and their methyl ethers[38] have τ values of about 2.0 for their 1-hydrogen atoms and this is consistent with the rigid and fairly flat *trans* B/C fusion (**109**), for in this the 1-hydrogen must be subject to negative carbonyl long-range shielding. On the other hand, the corresponding A compounds (**105–106**) and their methyl ethers, have the 1-hydrogen line near 3.3; indicating that it lies in a region of low positive or negative shielding and agrees with the *cis* B/C conformation (**110–111**).

IV. Sumatrol

Sumatrol (**112**), $C_{23}H_{22}O_7$, the phloroglucinol analog of rotenone, was first isolated from a Sumatra-type derris resin of unspecified species.[100, 108] The formation of an oxime establishes the presence of an active carbonyl group, and this fact together with the strong ferric chloride reaction, indicates that there is a phenolic group in the *o*-position to the carbonyl group. Apart from the phenolic hydroxyl group, **112**, like all other members of the rotenone series, possesses two methoxyl groups, and three different oxygen atoms. Conversion of the isopropenyl-dihydrofuran system into an isopropenyl furan residue, as rotenone was converted to the corresponding iso compound, has been unsuccessful.[108, 109]

Sumatrol cannot be dehydrated with a mixture of sulfuric and acetic acids. Dehydrosumatrol (**113**) and its derivatives are readily formed by the sodium acetate–iodine process; other oxidizing agents are less satisfactory.[108] Hydrogenation of sumatrol, with the aid of a platinum catalyst, similarly to rotenone, gives rise to two optically active products, dihydro- (**114**), and tetrahydrosumatrol (**115**). Dehydrogenation of the latter compounds by the iodine method yielded dehydrodihydro-(**116**), and dehydrotetrahydrosumatrol (**117**).

The dehydrosumatrols are readily hydrolyzed to acids of the derrisic acid type, for example **113** gives sumatrolic acid (**118**) and **115** gives tetrahydrosumatrolic acid (**119**).

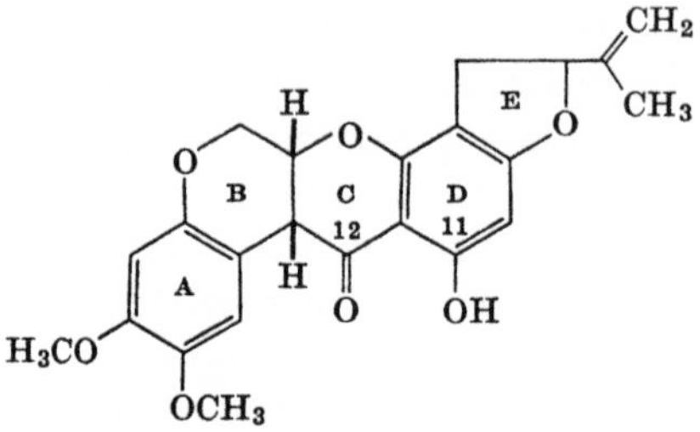

(**112**)

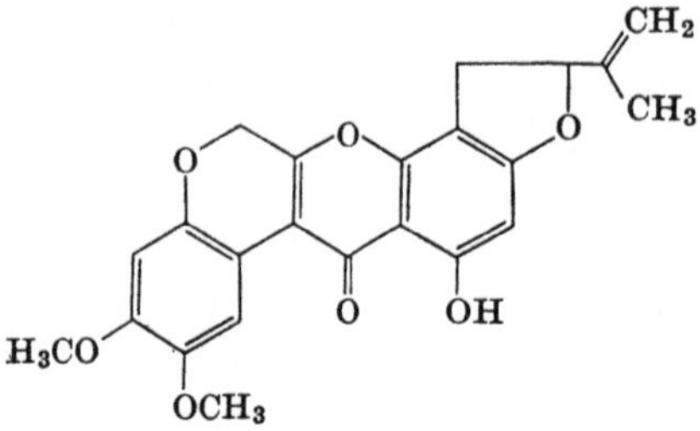

(**113**)

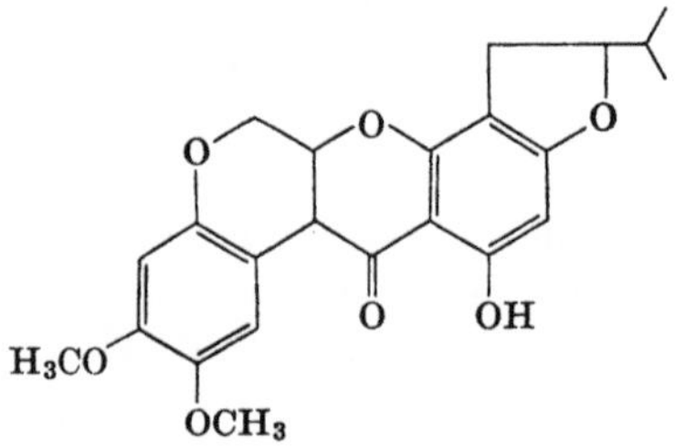

(**114**)

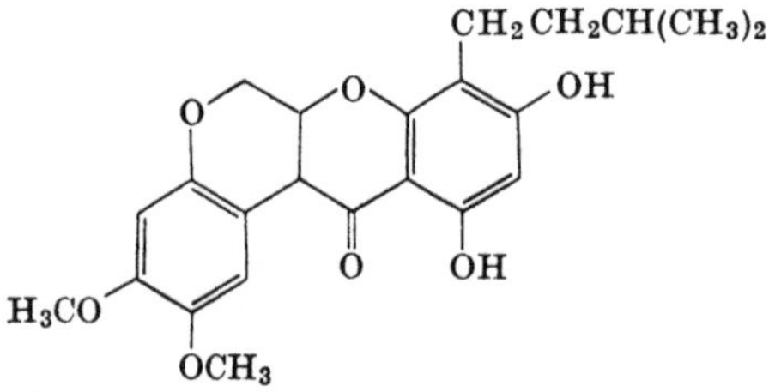

(**115**)

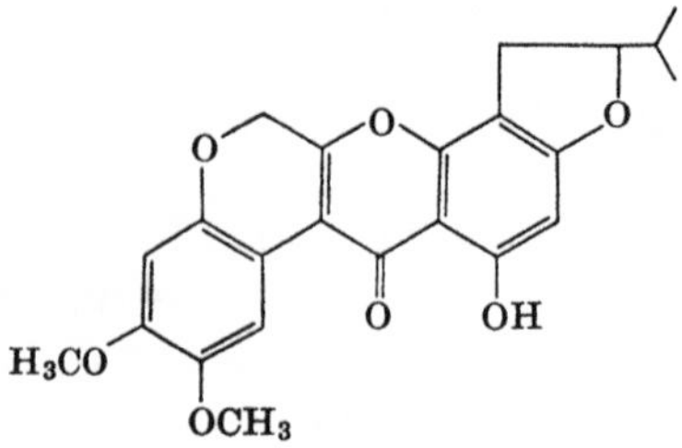

(**116**)

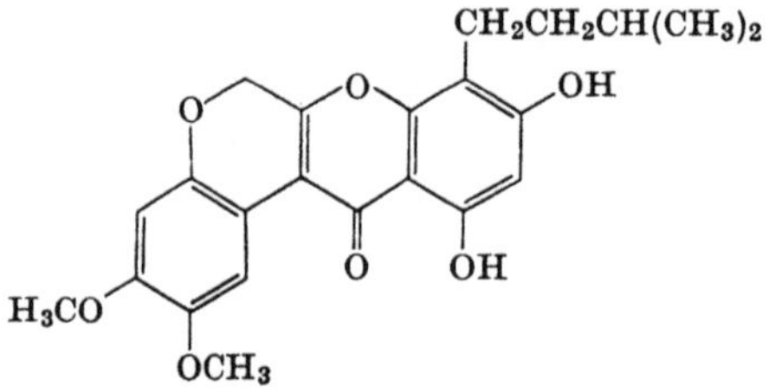

(**117**)

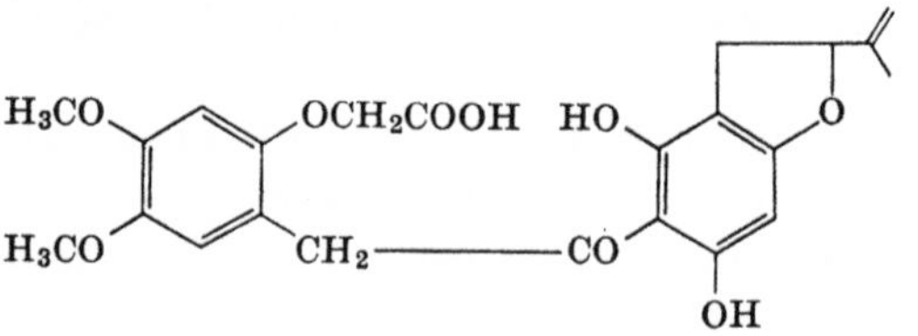

(**118**)

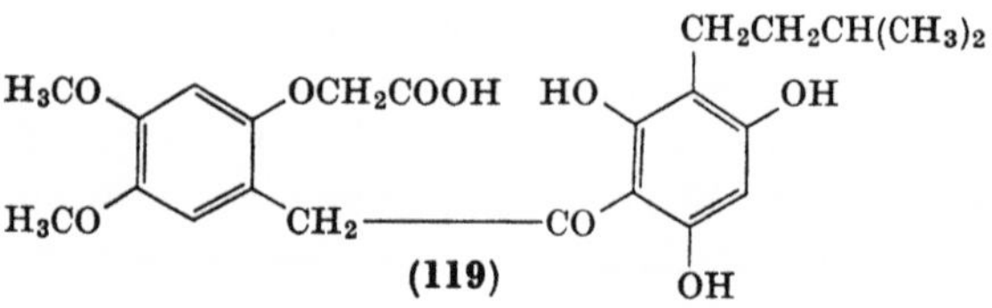

(**119**)

Degradative study of the natural (−)-sumatrol[108, 109] has led to the alternative linear orientation of the D/E rings (**120**). The angularly fused form (**112**) was preferred by earlier workers mainly because of analogy with the known natural rotenoids. Consequently, the synthesized dehydrotetrahydrosumatrol (**117**) has the alternative structure **121**. Synthesis was achieved[109] by condensation of methyl-2-cyanomethyl-5,6-dimethoxyphenoxyacetate (**15**) and isoamylglucinol to yield the acid (**119**), followed by cyclization of the latter acid with acetic anhydride in the presence of sodium acetate, and hydrolysis of the diacetate derivative of dehydrotetrahydrosumatrol so formed, to **118**. Recently, the angular orientation of the D/E rings of natural sumatrol has been established.[110]

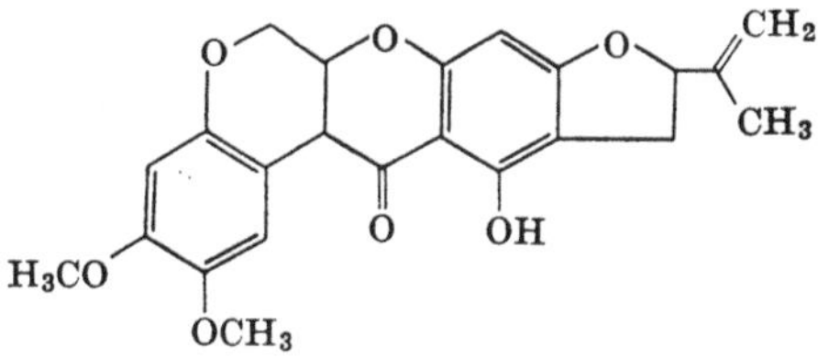

(**120**)

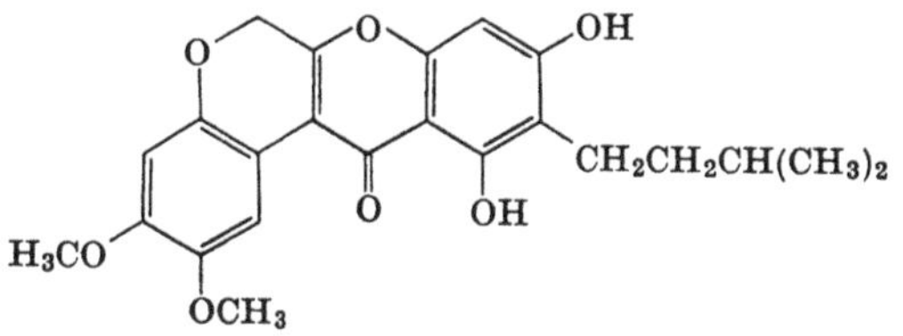

(**121**)

Comparison of the optical rotatory dispersion of **118** with derrisic acid (**8**), derived from natural rotenone, shows that both have the same type of negative plain curve; indicating that the 5′ center of sumatrol has the same absolute configuration as in rotenone.[110] 11-Acetylsumatrol and sumatrol have a positive Cotton effect superimposed on the above negative plain curve,[26] and rotenone also shows a positive Cotton effect, so that the stereochemistry of sumatrol appears to be the same as that of rotenone, i.e. 6aβ, 12aβ, 5′β (**112**).[11] Moreover, according to the nuclear magnetic resonance spectrum, natural sumatrol has a τ value of about 3.23 for its 1-hydrogen atom, indicating that it lies in a region of low positive or negative shielding, which agrees with the *cis* B/C fusion.[25]

Sumatrol, like rotenone, also contains asymmetric carbon atoms at positions 6a, 12a, and 5′ and exhibits parallel properties in its analogous derivatives. Thus, **113** is optically active by virtue of the asymmetry of $C_{5'}$, and **115**, in which the activity of this center has been destroyed, is still optically active because C_{6a} and C_{12a} are not affected by this hydrogenation process. As expected, **117** is optically inactive.[108]

V. Malaccol

Malaccol (**122**), $C_{20}H_{16}O_7$, was isolated in yellow prisms from *Derris malaccensis*. It is optically active, and appears to be 11-hydroxyelliptone, i.e. the phloroglucinol analog of elliptone, and undergoes the typical rotenone transformations.[27, 111–114] In agreement with the proposed structure (**122**), malaccol gives a Durham test, has two methoxyl groups, and an active carbonyl group as shown by the formation of an oxime. The strong ferric chloride color indicates the presence of a phenolic hydroxyl group in an *ortho* position to the carbonyl group.

Tetrahydromalaccolic acid (**123**) has been synthesized and cyclized to tetrahydrodehydromalaccol (**124** or **125**), which appears to have the requisite properties but has not been compared directly with the natural product.

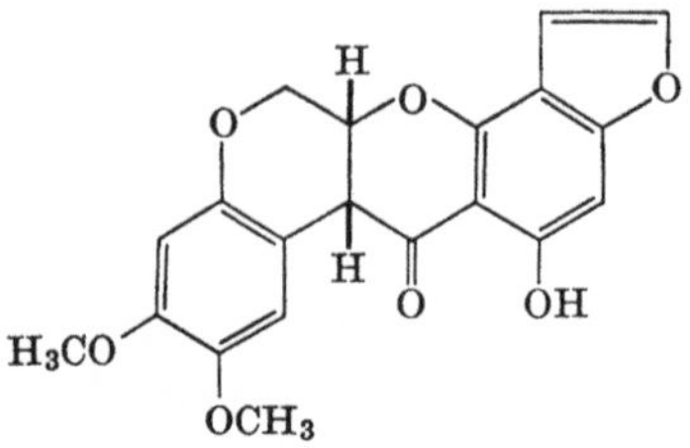

(**122**)

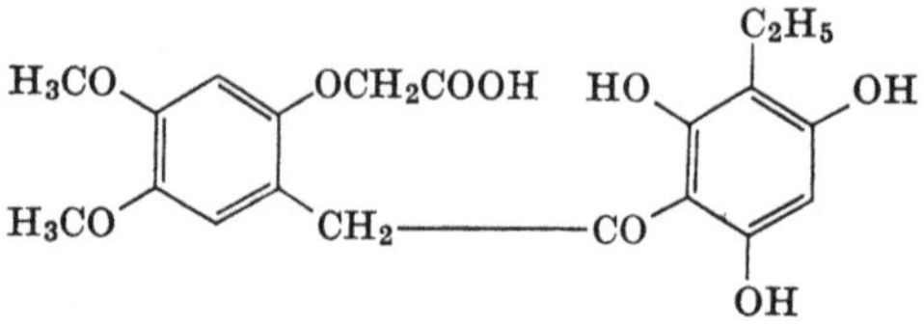

(**123**)

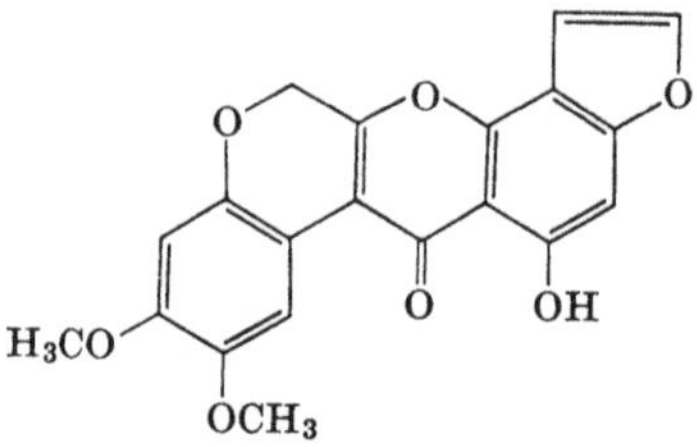

(**122a**)

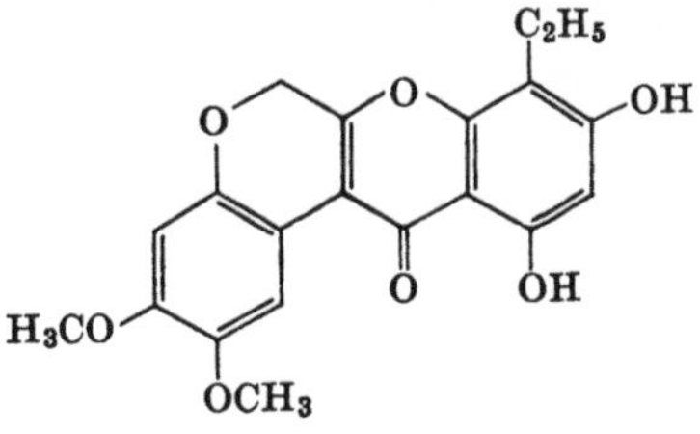

(124)

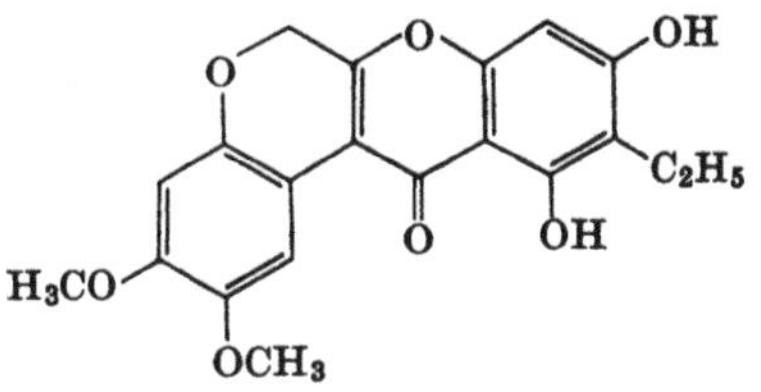

(125)

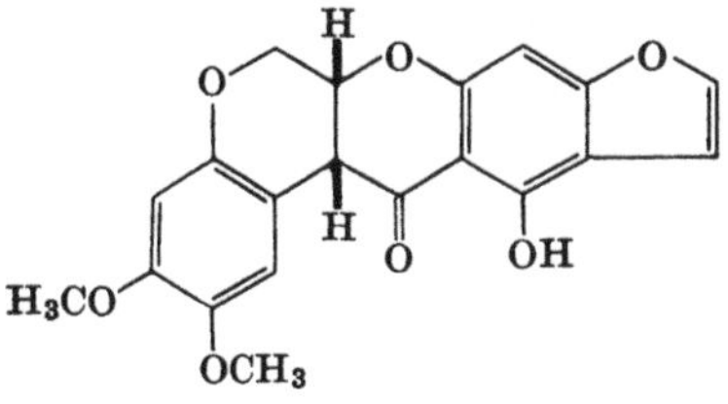

(126)

Natural (+)-malaccol, when boiled with ethanolic sodium acetate is racemized, and the (±)-product has been accepted as having an unaltered skeleton.[27] Crombie and Peace[110] have found that the infrared spectrum of its solution differs from that of the (−) form, and one compound must have the linear **126** and the other the angular fusion **122**; the D/E arrangement being altered because of the two possible ways of reclosing the ion from **127**, which results from treatment with base. Natural (+)-malaccol (α-malaccol) is the angular form **122**, whereas (±)-malaccol (referred to as (±)-β-malaccol)[110] is the linear form **126**.

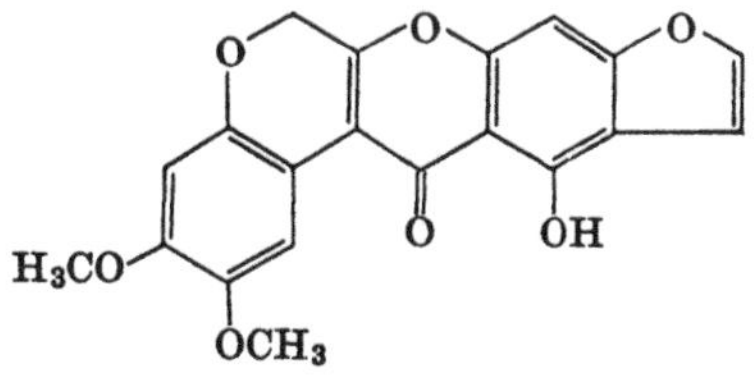

(126a)

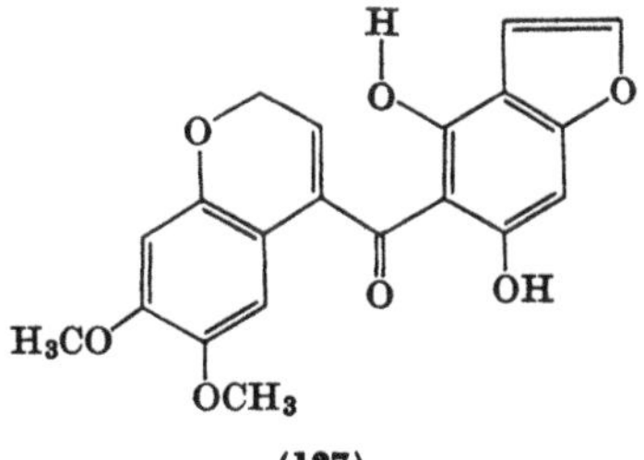

(127)

By treatment of natural (+)-malaccol with alcoholic sodium acetate, then with iodine, and finally zinc, a dehydromalaccol is obtained; (±)-β-malaccol gives a similar product. The dehydromalaccol is the linearly D/E–fused 6a, 12a-dehydro compound (cf. **126a**). However, if

(+)-malaccol is treated with sodium acetate and iodine rapidly added, the corresponding angular α-dehydro compound (**122a**) is produced. Natural (−)-sumatrol belongs to the angularly fused D/E class. Malaccol has a positive Cotton effect,[110] and the optical rotatory dispersion correlations with natural rotenone[26] indicate that it must therefore be 6aβ, 12aβ.[110]

VI. Elliptone

(±)-Elliptone (**128**), $C_{20}H_{11}O_6$, was first isolated from an ethereal solution of *Derris elliptica* that had been repeatedly treated with potassium hydroxide solution (5%).[79,115] The naturally occurring compound, (−)-elliptone[79,112,115] resembles[26,79,115] isorotenone (**43**). The structure of elliptone has been reviewed.[33,36] Treatment with sodium acetate causes racemization, and the ready formation of dehydroelliptone (**129**), by standard methods, demonstrated the presence of a chromanochromanone system. Similarly to isorotenone, an inactive

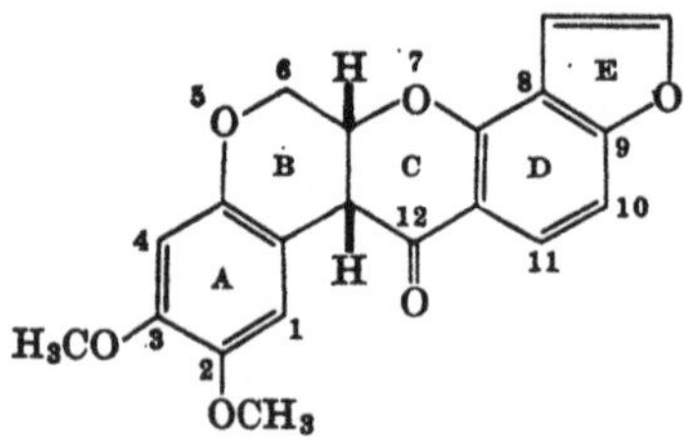

(128)

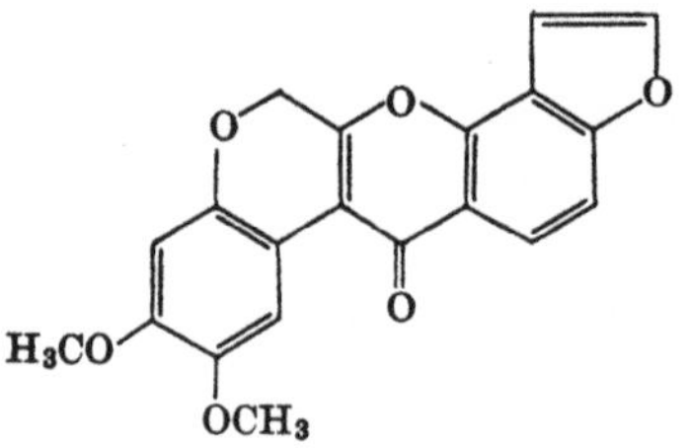

(129)

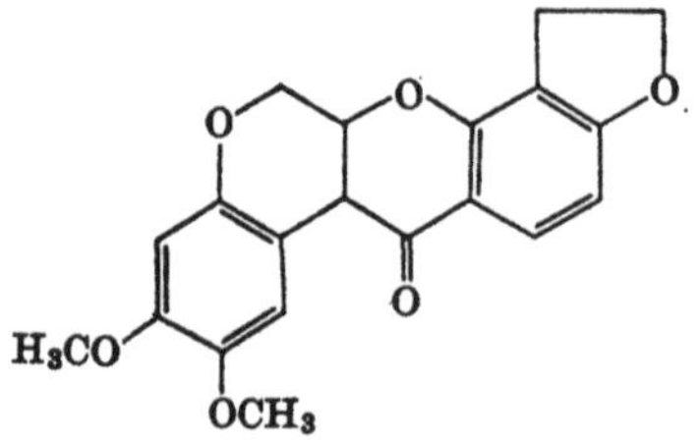

(130)

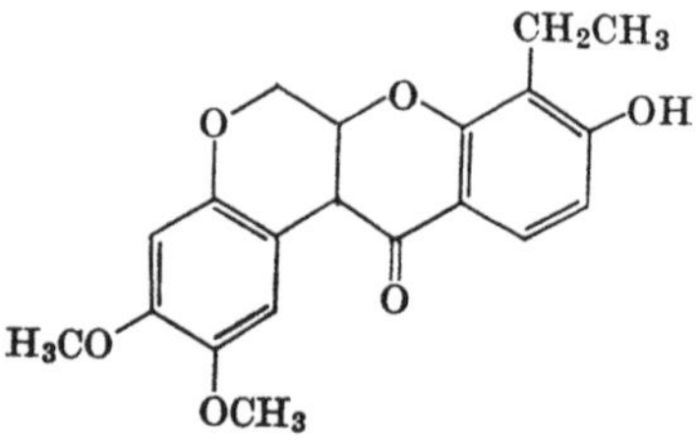

(131)

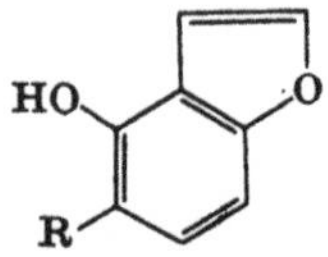

(132)

monoacetate is obtained upon acetylation of (–)-elliptone with acetic anhydride, probably an enol acetate.

On hydrogenation over Adam's catalyst, elliptone gave dihydro- (**130**), and tetrahydroelliptone (**131**). Alkali degradation of **128** gave derric acid (**10**) and karanjic acid (**132**, R = COOH), analogous to tubaic acid (**18**), and yielding 4-hydroxycoumarone (**132**, R = H) on decarboxylation. The degradation of dihydrodehydroelliptone (**133**) with zinc and alkali[79] gave dihydroelliptic acid (**134**), together with dihydroelliptol (**135**), analogous to derritol (**9**) (Eq. 11).

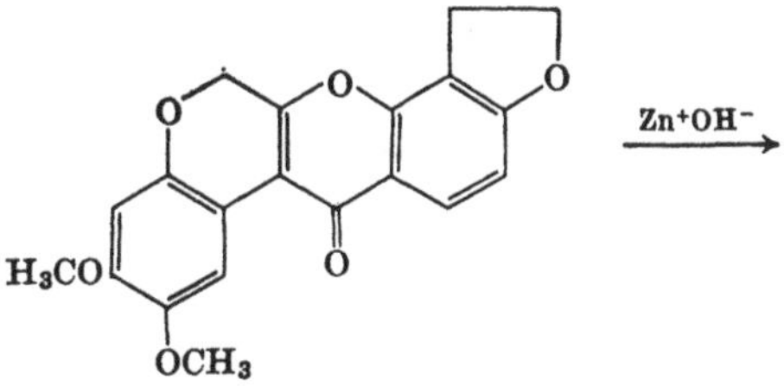

(**133**)

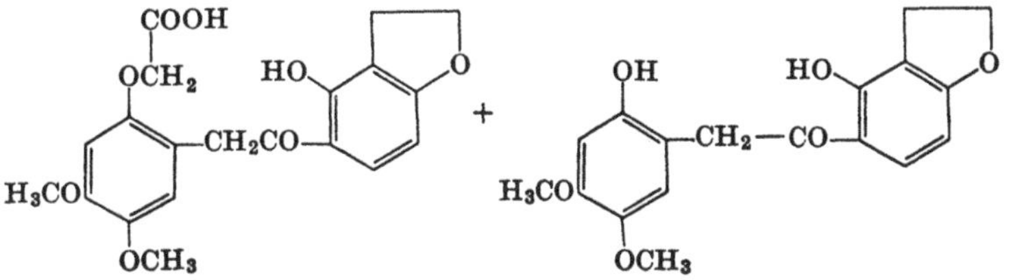

(**134**) (**135**) (11)

Oxidation of elliptol (**136**) with nitrous acid or chromic oxide produced elliptonone (**137**), which has been synthesized by heating **136** with a mixture of methyl oxalate and sodium acetate.

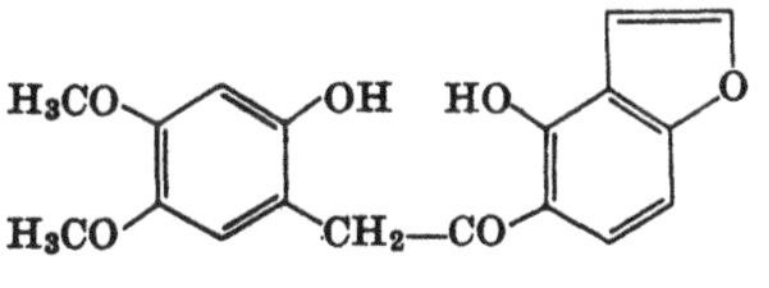

(**136**)

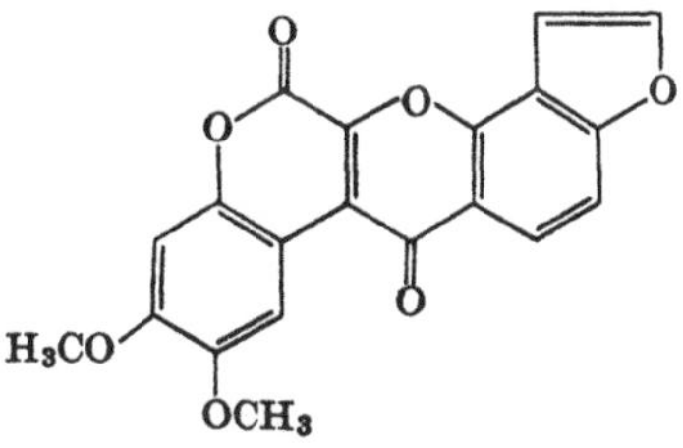

(**137**)

Tetrahydrodehydroelliptone (**140**) has been synthesized by cyclization of tetrahydroelliptic acid (**139**) which was obtained by the standard procedure from 2-ethylresorcinol (**138**) and methyl 2-cyanomethyl-4,5-dimethoxyphenoxyacetate (**15**)[113] (Eq. 12).

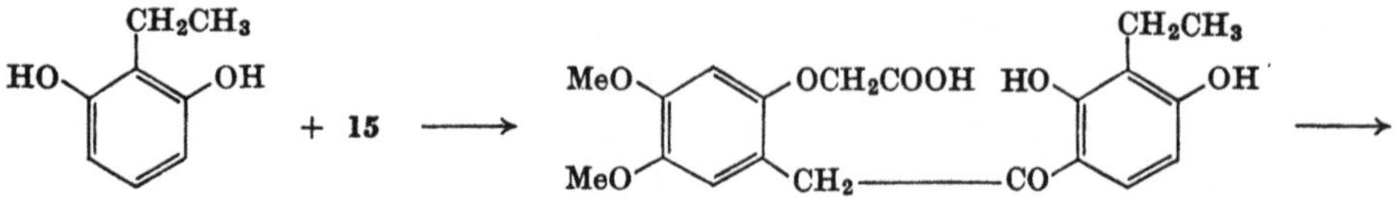

(**138**) (**139**)

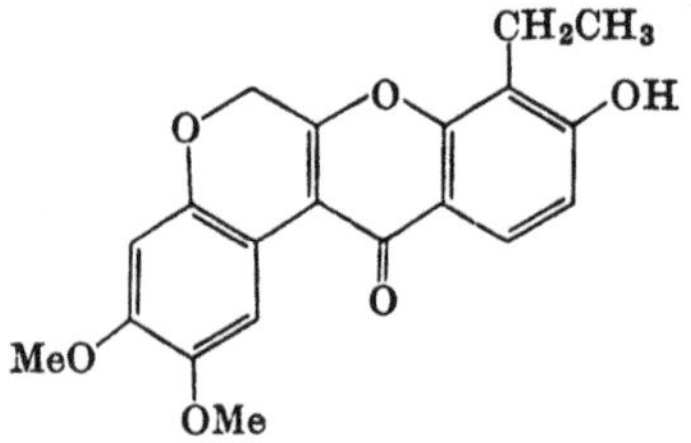

(12)

(**140**)

Elliptone exhibits a positive Cotton effect curve. It lacks the asymmetric center at $C_{5'}$ and hence is not subject to the initial negative rotational drift in the visible range, as is the case with rotenone (**2**).[26] According to the magnetic resonance criterion, natural optically active elliptone has τ value of about 3.21 for its 1-hydrogen atom; this is consistent with the *cis* B/C fusion.[25]

The synthesis of elliptone (**128**) has not yet been achieved. The parent nucleus of **128**, namely dehydrodesmethoxyelliptone (**143**) has been obtained by three routes: (a) 7-hydroxychromenochromone (**141**) is converted to 8-allyl-7-hydroxychromenochromone (**142**)[116] and the latter is treated with osmium tetroxide and potassium periodate to give an intermediate, which undergoes ring closure to yield **143**;[117]

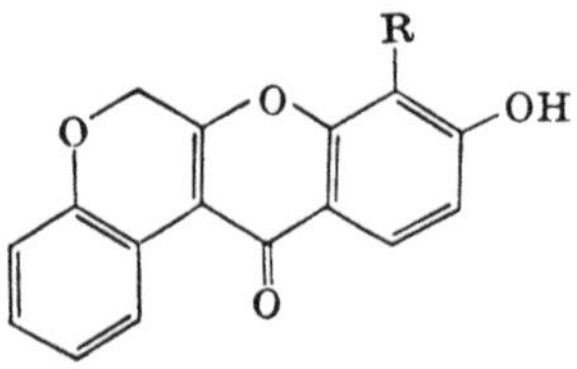

(**141**) R=H
(**142**) $R=C_3H_5$

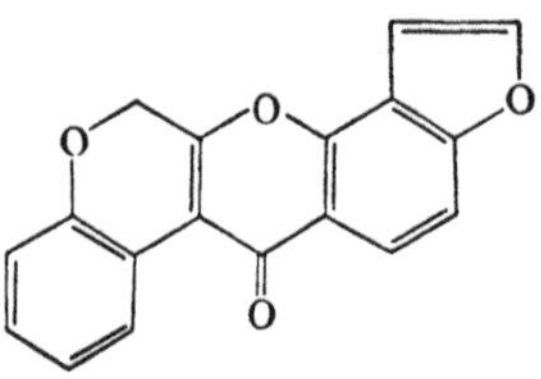

(**143**)

(b) by condensation according to Seshadri's method, i.e. chromen-ring formation of the 2-methyl derivative of 2′-methoxyflavone,[118] 2-ethoxy-

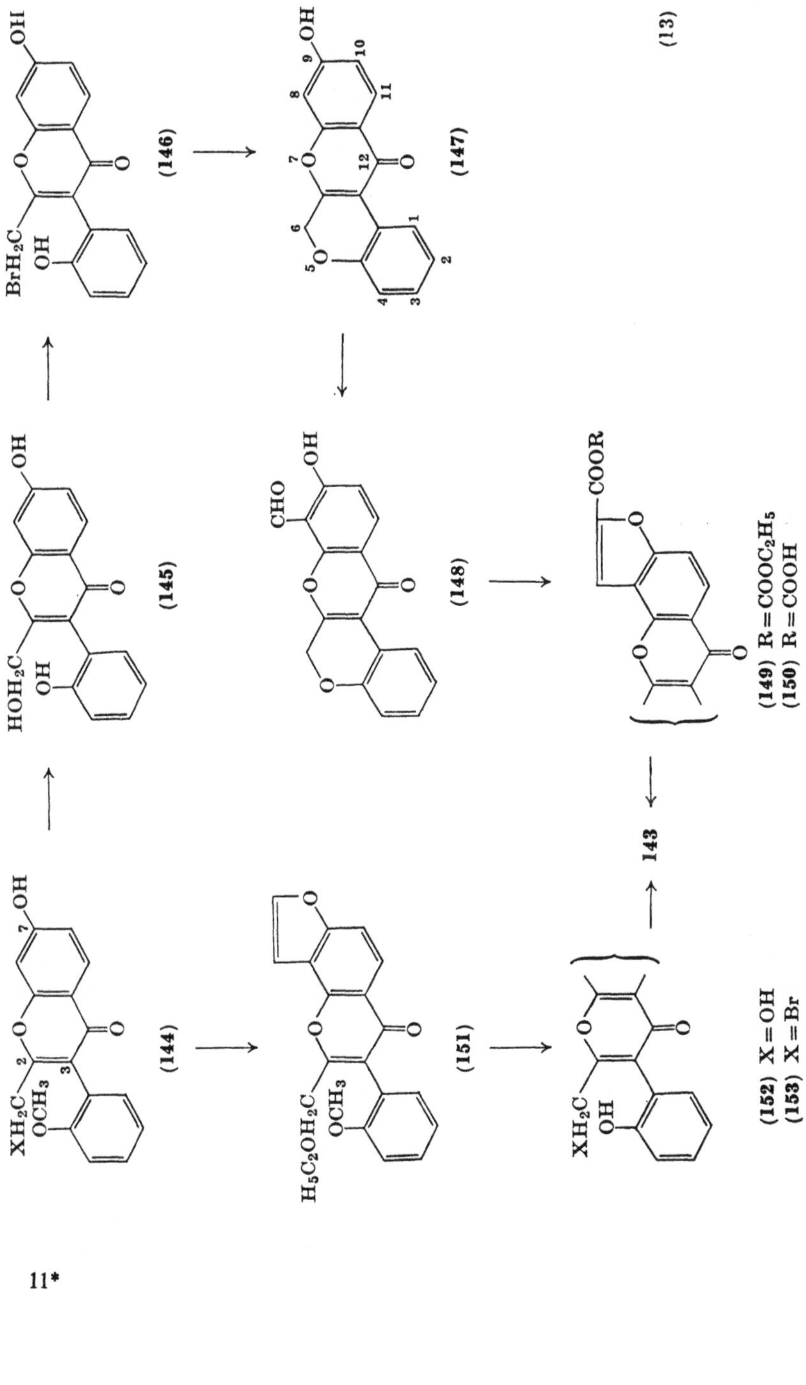
(144)
(145)
(146)
(147)
(148)
(151)
(149) R = COOC2H5
(150) R = COOH
(152) X = OH
(153) X = Br
143
(13)

methyl-7-hydroxy-2′-methoxyisoflavone (**144**, R = OEt) is readily converted by the action of aluminum chloride into the trihydroxy compound (**145**), which is made to react with hydrogen bromide in acetic acid to give the bromomethyl compound (**146**) which is readily converted into the chromenochromone (**147**).[10, 87] The latter, upon formylation with hexamine, gives the 8-formyl derivative (**148**), which is converted into furochromenochromone-2-carboxylic acid (**150**) by the action of ethyl bromomalonate and potassium carbonate, followed by hydrolysis of the formed ester (**149**). Decarboxylation of the acid (**150**) produces **143**; (c) Chromen-ring closure of 2-ethoxymethyl-2′-methoxyfuro(2,3-f) isoflavone (**151**), obtained from **144**.[119–121] Treatment of **151** with aluminum chloride in nitrobenzene yields the dihydroxy compound (**152**), which is converted into the bromo compound (**153**), and then finally into **143** (Eq. 13).

142, R = C_3H_5

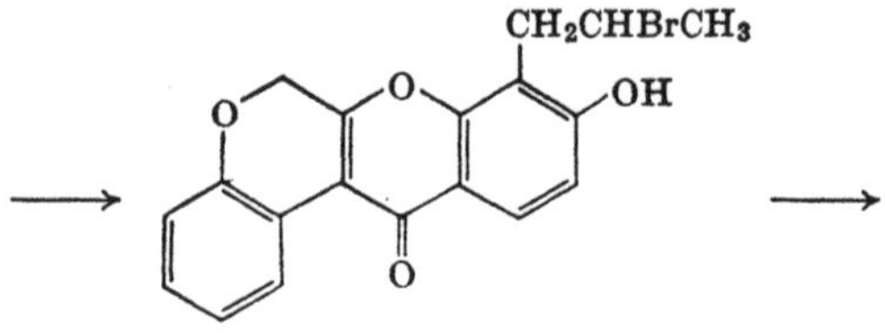

(**154**)

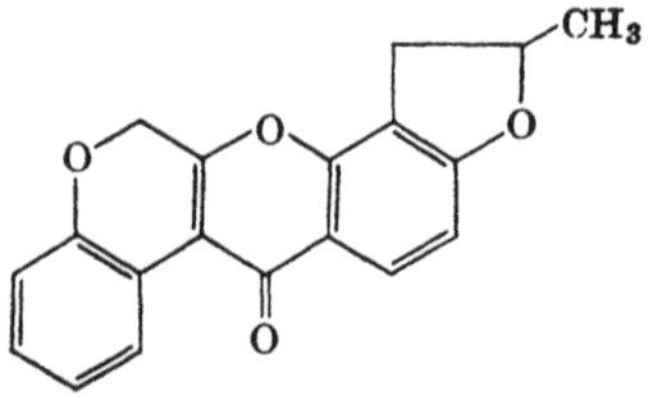

(**155**)

(14)

(d) Treatment of **142** with aqueous hydrobromic acid gives 8-(β-bromopropyl)-7-hydroxychromenochromone (**154**), which undergoes furan ring closure with pyridine to yield **155**[116] (Eq. 14).

VII. Amorphigenin

Amorphin, a glycoside which occurs in the insecticidal seeds of *Amorpha fruticosa*,[122] gives glucose, arabinose, and amorphigenin on acid hydrolysis. The genin is also formed by hydrolysis with β-glucosidase and gives a positive Durham test, like rotenone. Kondratenko and

Abubakirov[123] have presented further evidence of its rotenoid character and suggest that it has molecular formula $C_{22}H_{20}O_7$ (Acree, and coworkers[122] proposed $C_{22}H_{22}O_7$), and possesses a D/E system as in **156**, with the hydroxyl group unassigned and the rest of molecule as in rotenone. Recent results of Crombie and Peace[124] have led to structure **157** for amorphigenin, $C_{23}H_{22}O_7$; amorphin, with the sugar attachment at 8′, is thus the first rotenoid glycoside.

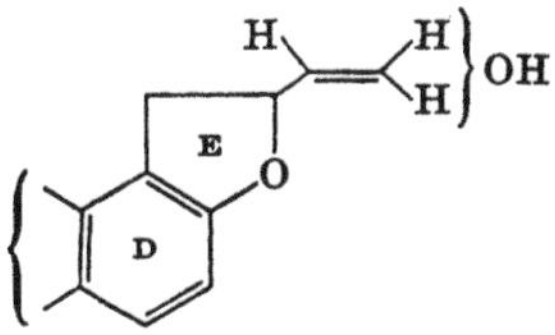

(156)

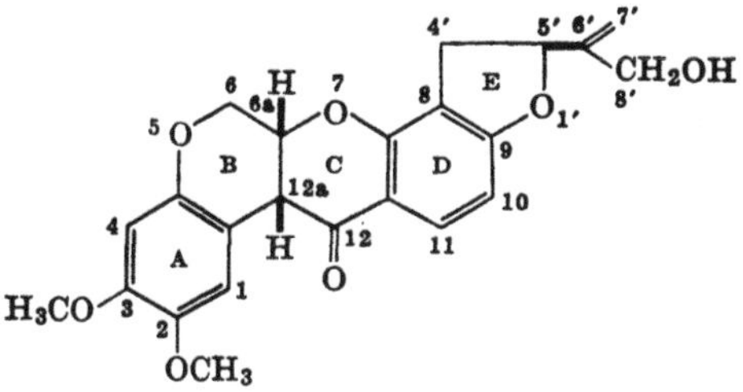

(157)

Amorphigenin (**157**) forms tenaceously solvated crystals (n.m.r. evidence) from methanol, benzene, and aqueous acetone, and the molecular formula was established from analysis of derivatives, the mass–spectral molecular weight of 410, and nuclear magnetic resonance proton counts on **157** and its derivatives. Infrared and ultraviolet spectra are closely similar to those of rotenone except for hydroxyl absorption (ν_{max} 3498 cm^{-1} in CCl_4; intramolecular bonding to 1′-O), and the n.m.r. spectrum shows two methoxyl groups with an aromatic proton pattern as in rotenone. The 1- and 4-hydrogen atoms give peaks at τ 3.19 and 3.52, respectively, and the 10, 11-hydrogen atoms give rise to the characteristic quartet, τ 3.47 and 2.12 (J 9 cps.), leaving no doubt that the E ring is angularly fused.

Chemical description of the B/C rings is given by 6a, 12a dehydrogenation which occurs with characteristic shift of the carbonyl frequency and τ value of the 1-hydrogen atom (1.90). Treating the dehydro compound (**158**) with nitrous acid gives the keto lactone (**159**), verifying the presence of the 5, 6, 6a, 12a, 12 system as in **157**. **158** is hydrolyzed to the derrisic acid analog (**160**), which is recyclized by hot acetic anhydride to the 8′-acetate of **158**.

The primary nature of the hydroxyl group is shown by two pieces of information. Firstly, the 8′-methylene band at τ 5.73 in amorphigenin and its derivatives, moves to 5.83 on acetylation (the unsplit band also demands no proton on the carbon bearing the CH_2OH group). Secondly,

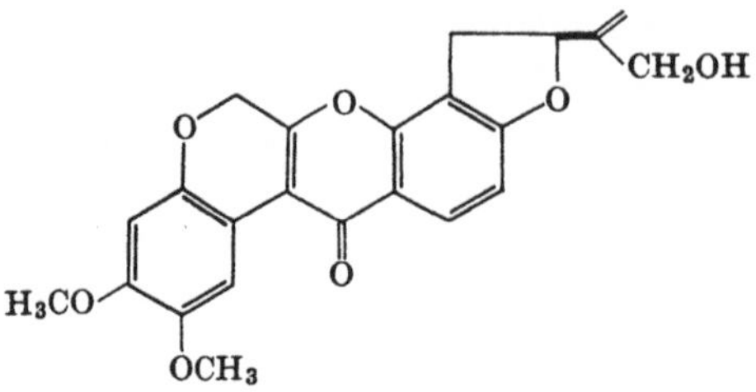

(158)

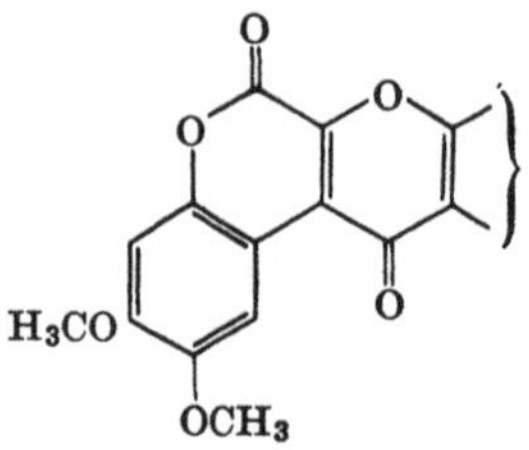

(159)

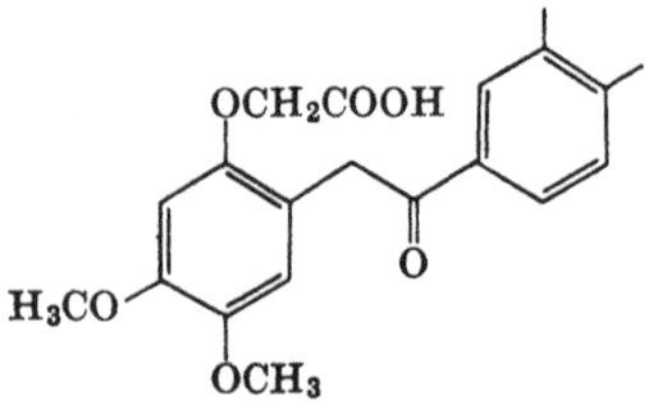

(160)

on careful oxidation with manganese dioxide, amorphigenin is oxidized to a 6′-formyl-12-ketone, with an aldehydic hydrogen peak at τ 0.46. The corresponding hydroxyrotenonone (cf. **159**) gives an analogous aldehyde when oxidized. The oxidation conditions, and increased absorption at 225 mμ, support an allylic primary alcohol system, and amorphigenin absorbs one molecule of hydrogen over a palladium catalyst. In addition amorphigenin has a band at τ 4.73 assignable to the two vinyl protons (cf. rotenone). With this in mind, the two unassigned carbon atoms must form a dihydrofuranoid ring E. The whole 5-carbon addendum to ring D is isoprenoid, since attachment of the acyclic residue is at 5′ and not 4′.

The τ value of the 1-hydrogen atom in amorphigenin indicates a *cis* B/C fusion[25] and the positive Cotton effect supports this, and indicates that the absolute configuration at positions 6a, 12a is as for rotenone.[26] As both derrisic acid and the corresponding compound from amorphigenin (**160**) (both of which possess asymmetry at $C_{5'}$) have similar negative plain curves, the rotenone configuration is assigned at this center.

The assigned structure **157** has been confirmed by hydrogenolysis of the 8′-hydroxyl group in amorphigenin and hydrogenation of the 6′,7′-olefinic link. The product is identical with (−)-6′,7′-dihydrorotenone from natural rotenone. (−)-12-Desoxy-6′,7′-dihydrorotenone is also formed in the hydrogenolysis by elision of the keto group.

VIII. Dolineone

Dolineone (dolichone) (**161**), $C_{19}H_{12}O_6$ (m.p. 235°, $(\alpha)_D^{20} + 135°$ ($CHCl_3$))[125, 126] is isolated from the root of the leguminous plant *Neorautanenia pseudopachyrrhiza* Harms. (*Dolichos pseudopachyrrhizus*), and from the seeds of *Pachyrrhizus erosus*. The mass–spectral molecular weight is in agreement with the molecular formula. The ultraviolet spectrum closely resembled that of elliptone (**128**), and suggested a generally similar structure, with a benzofuran system. Infrared data support the resemblance. Dolineone gives a green–blue Rogers and Calamari test but no Durham reaction; a methylenedioxy group is present according to the blue Labat test. The proton magnetic resonance spectrum is in agreement with structure **161**.

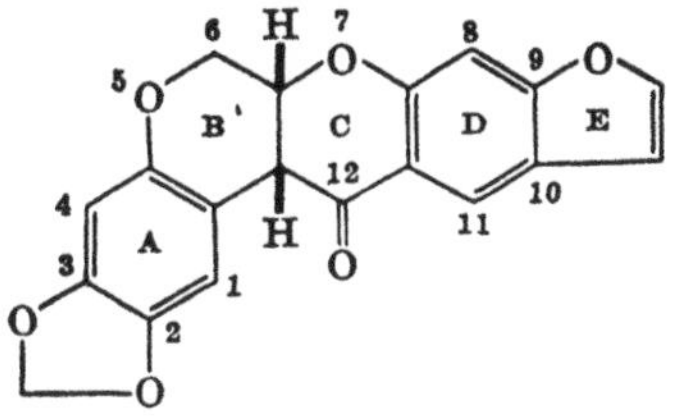

(**161**)

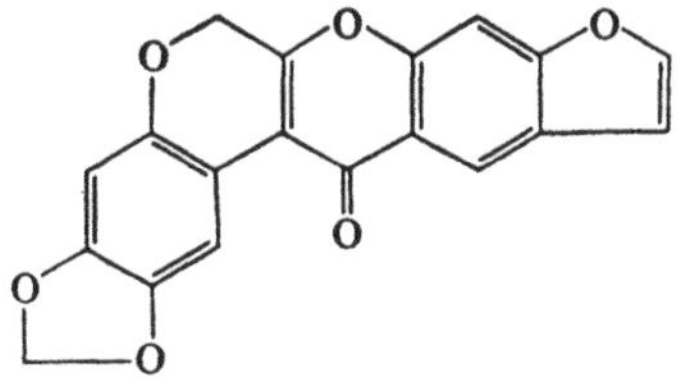

(**162**)

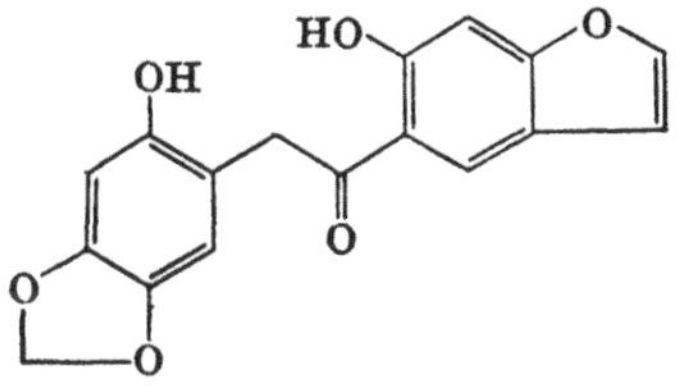

(**163**)

Dolineone is dehydrogenated with manganese dioxide to give 6a, 12a-dehydrodolineone (**142**) with a characteristic fall of the carbonyl frequency. The latter compound is further oxidized at C_6 with lead tetraacetate, and the infrared spectrum accounts for the presence of atoms 12, 12a, 6a, 6, and 5 as in **142**. Treatment of **141** with zinc dust and alkali yields the deoxybenzoin (**143**). The unchelated hydroxyl group can be easily methylated.

The nuclear magnetic resonance spectrum indicates that the B/C fusion in **141** is *cis* as the 1-hydrogen atom is located at τ 3.44 and is therefore not being strongly negatively shielded by the carbonyl group. Dolineone, like rotenone, has a positive Cotton effect.[26] It is therefore likely that the *cis* B/C fusion has the same absolute configuration as rotenone (**2**), i.e. at 6a and 12a, and hence is (6aS, 12aS) as in **2**.

IX. Pachyrrhizone

Pachyrrhizone (**164**), $C_{20}H_{14}O_7$ (m.p., 272°; $(\alpha)_D^{16}+96°$ ($CHCl_3$)), occurs in Japanese yam beans (*Pachyrrhizus erosus* Urban),[127] in *Neorautenenia* (*Dolichos*) *pseudopachyrrhizus*[125] and in *N. edulis.*[125] It has been investigated by Bickel and Schmid[53] who have shown that it has a constitution distinguishing it from the other rotenoids. It is the only one containing a pyrogallol nucleus, and a securely established linear sequence of furan rings; (±)-β-malaccol has also the linear form (D/E form)[110] (cf. also dolineone). Thus within these three closely related plants a remarkable variation of the basic isoflavanoid structural theme occurs. The matter is summarized in the following table and it is almost certain that for any of the three plants the list of related compounds is incomplete due to seasonal fluctuations.[126]

Occurrence of Related Isoflavanoids and Rotenoids

Compound	Plants *N. pseudopachyrrhizus*	*N. edulis*	*P. erosus*
Pachyrrhizin	+	+	+
Neotenone	+	+	+
Dehydroneotenone			+
Rotenone			+
Dolineone	+		+
Nepseudin	+		
Neodulin		+	
Erosnin			+
Pachyrrhizone			+

The typical rotenone–like properties of pachyrrhizone are demonstrated by the color reactions, dehydrogenation by iodine or ferricyanide to dehydropachyrrhizone (**165**), further oxidation to a dehydro lactone (**166**), and the formation of an oxime and enol acetate. No free hydroxyl group could be detected, and only one methoxyl group is present.

Treatment of **164** with alkali gives an acid, $C_{10}H_8O_5$, containing one methoxyl group, one phenolic hydroxyl group, and giving a blue ferric reaction. It was proved to have structure **167** (R = COOH), because of its conversion by decarboxylation, methylation, and hydrogenation into

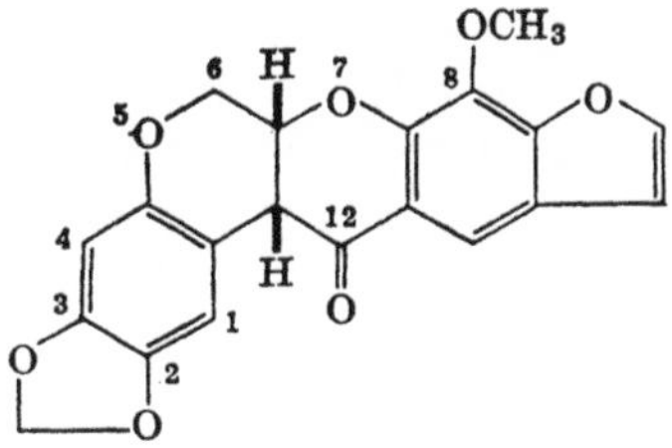

(164)

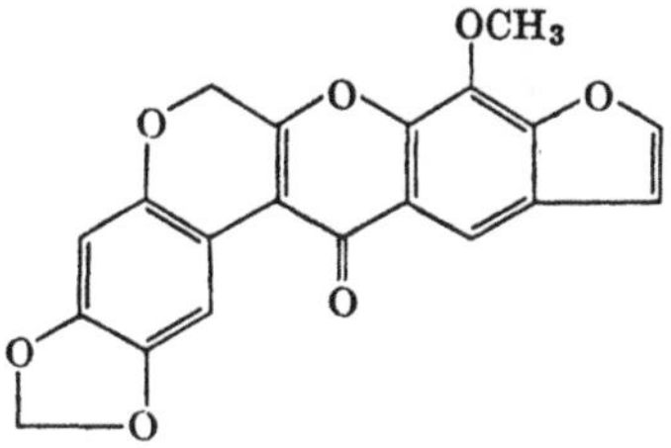

(165)

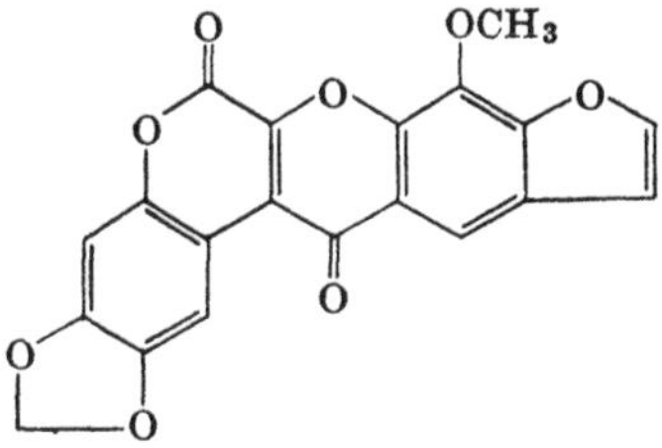

(166)

6,7-dimethoxycoumaran (**167**, R = H); thereby establishing the linear arrangement of rings B, C, D, and E.

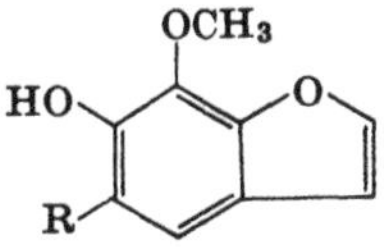

(167)

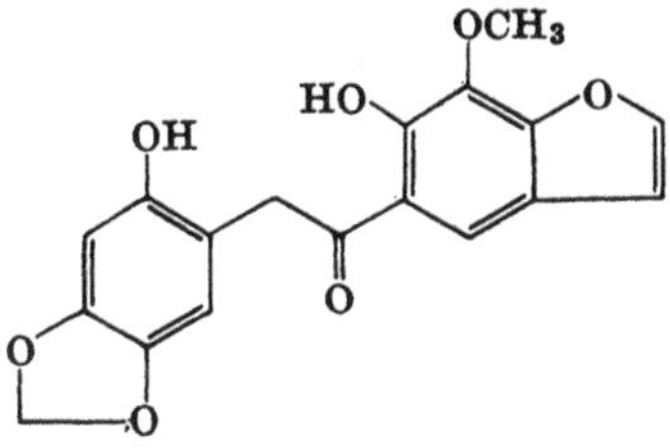

(168)

Alkaline hydrolysis of the dehydrolactone (**166**) gave oxalic acid, **167** (R = COOH), and the deoxybenzoin (**168**). This last compound gave acid (**167**, R = COOH) upon oxidation, and contained two phenolic hydroxyl groups, a carbonyl group (oxime, ferric reaction), and an active methylene group.

Structure **164** accounts for the ultraviolet spectrum which is similar to that of rotenone, but possesses an extra band at about 340 mμ (log ϵ 3.5) because of additional conjugation. Also in accordance with the proposed structure is racemization by sodium acetate in hot methyl cellosolve. The stereochemistry of pachyrrhizone is in line with that of the other rotenoids.[26]

The synthesis of the dihydrofurochromenochromone (**170**), representing the five–ring basic skeleton of pachyrrhizone (**164**), has been accomplished.[128, 129] Condensation of 6-hydroxycoumaran with 2-(carbomethoxy)methoxyphenylacetonitrile gave rise to the keto acid (**169**, R = H) and its methyl ester (**169**, R = CH_3). The keto acid suffered ring closure to yield **170** (Eq. 15).

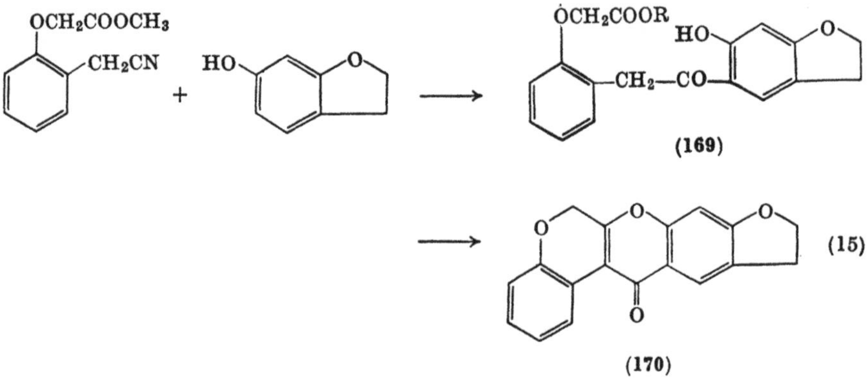

(15)

X. References

1. Oxley, *J. Indian Archipelago and East Asia*, **2**, (10) 641 (1948).
2. D. E. Buckingham, *Ind. Eng. Chem.*, **22**, 1133 (1930).
3. H. B. Haag, *J. Pharmacol.*, **43**, 193 (1931).
4. R. C. Roark, *J. Econ. Entomol.*, **34**, 684 (1941).
5. R. L. Metcalf, *Organic Insecticides*, Interscience Publishers Inc., New York, 1955, p. 23.
6. A. Butenandt, *Ann. Chem.*, **464**, 253 (1928).
7. H. L. Haller and F. B. LaForge, *J. Am. Chem. Soc.*, **52**, 4505 (1930).
8. S. Takie, S. Miyajima and M. Ono, *Chem. Ber.*, **64**, 1000 (1931).
9. S. Takie, M. Koide and S. Miyajima, *Chem. Ber.*, **63**, 1369 (1930).
10. M. Miyano and M. Matsui, *Chem. Ber.*, **91**, 2044 (1958).
11. G. Büchi, L. Crombie, P. J. Godin, J. S. Kaltenbronn, K. S. Siddalingaiah and D. A. Whiting, *J. Chem. Soc.*, 2843 (1961).
12. Geoffroy, *Ann. Inst. Colon. Marseille*, **2**, 1 (1895).
13. E. P. Clark, *Science*, **70**, 478 (1929).
14. K. Nagai, *J. Chem. Soc. Japan, Pure Chem. Sect.*, **23**, 744 (1902).
15. W. Lenz, *Arch. Pharm.*, **249**, 298 (1911).
16. T. Ishikawa, *Tokyo Igaku Zasshi*, **31**, 187 (1917).
17. S. Takie, *Chem. Ber.*, **61**, 1003 (1928).
18. A. Butenandt, *Nachr. Ges. Wiss. Göttingen, Math. Physik-Klasse*, (1) 1-8 (1928); *Chem. Abstr.*, **24**, 4784 (1930).
19. R. S. Cahn, *J. Chem. Soc.*, 1129 (1934).
20. A. Butenandt and F. Hildebrandt, *Ann. Chem.*, **477**, 245 (1930).
21. E. L. Gooden and C. M. Smith, *J. Am. Chem. Soc.*, **57**, 2616 (1935).

22. H. A. Jones and C. M. Smith, *J. Am. Chem. Soc.*, **52**, 2554 (1930).
23. H. L. Cupples and I. Hornstein, *J. Am. Chem. Soc.*, **73**, 4023 (1951).
24. H. A. Jones, *J. Am. Chem. Soc.*, **53**, 2738 (1931).
25. L. Crombie and J. W. Lown, *J. Chem. Soc.*, 775 (1962).
26. C. Djerassi, W. D. Ollis and R. C. Russel, *J. Chem. Soc.*, 1448 (1961).
27. S. H. Harper, *J. Chem. Soc.*, 309 (1940).
28. T. A. Geissman, *The Chemistry of Flavonoid Compounds*, Pergamon Press, New York, 1962, Chapter by T. R. Seshadri, pp. 27–28.
29. S. Takie, S. Miyajima and M. Ono, *Chem. Ber.*, **65**, 1041 (1932).
30. A. Butenandt and W. McCartney, *Ann. Chem.*, **494**, 17 (1932).
31. F. B. LaForge and H. L. Haller, *J. Am. Chem. Soc.*, **54**, 810 (1932).
32. A. Robertson, *J. Chem. Soc.*, 1380 (1932).
33. F. B. LaForge, H. L. Haller and L. E. Smith, *Chem. Revs.*, **12**, 182 (1932).
34. L. Feinstein and M. Jacoboon, *Fortschr. Chem. Org. Naturstoffe*, **10**, 436 (1953).
35. N. Campbell, in *Chemistry of Carbon Compounds* (Ed. E. H. Rodd), Elsevier, Amsterdam, 1959, **ivB**, p. 944.
36. H. L. Haller, L. D. Goodhue and H. A. Jones, *Chem. Revs.*, **30**, 33 (1942).
37. H. King, *Ann. Rep. Chem. Soc.*, **29**, 186 (1932).
38. L. Crombie and P. J. Godin, *J. Chem. Soc.*, 2861 (1961).
39. C. Djerassi, W. D. Ollis and R. C. Russel, *J. Chem. Soc.*, 2876 (1961).
40. M. Miyano and M. Matsui, *Chem. Ber.*, (a) **92**, 1438; (b) **92**, 2487 (1959).
41. M. Miyano and M. Matsui, *Bull. Agr. Chem. Soc. Japan*, **24**, 540 (1960).
42. M. Miyano and M. Matsui, *Bull. Chem. Soc. Japan*, **31**, 397 (1958).
43. H. Fukami, J. Oda, G. Sakata and M. Nakajima, *Bull. Agr. Chem. Soc. Japan*, **24**, 327 (1960); *Chem. Abstr.*, **55**, 1565 (1961).
44. H. Fukami, J. Oda, G. Sakata and M. Nakajima, *Agr. Biol. Chem.*, **25**, 252 (1961); *Chem. Abstr.*, **55**, 14448 (1961).
45. Y. Kawase and C. Numata, *Chem. Ind.* (*London*), 1361 (1961).
46. A. C. Mehta and T. R. Seshadri, *Proc. Indian Acad. Sci.*, **42**, 192 (1955).
47. J. R. Herbert, W. D. Ollis and R. C. Russel, *Proc. Chem. Soc.* (*London*), 177 (1960).
48. L. Crombie, L. Godin, D. A. Whiting and K. S. Siddalingaiah, *J. Chem. Soc.*, 2876 (1961).
49. F. B. LaForge and L. E. Smith, *J. Am. Chem. Soc.*, **52**, 1091 (1930).
50. L. E. Smith and F. B. LaForge, *J. Am. Chem. Soc.*, **53**, 3072 (1931).
51. L. E. Smith and F. B. LaForge, *J. Am. Chem. Soc.*, **54**, 2996 (1932).
52. H. L. Haller and F. B. LaForge, *J. Am. Chem. Soc.*, **53**, 3426 (1931).
53. H. Bickel and H. Schmid, *Helv. Chim. Acta*, **36**, 664 (1953).
54. F. B. LaForge, H. L. Haller and L. E. Smith, *J. Am. Chem. Soc.*, **53**, 4400 (1931).
55. F. B. LaForge, *J. Am. Chem. Soc.*, **54**, 3377 (1932).
56. F. B. LaForge and L. E. Smith, *J. Am. Chem. Soc.*, **52**, 2878 (1930).
57. F. B. LaForge, *J. Am. Chem. Soc.*, **53**, 3896 (1931).
58. F. B. LaForge and L. E. Smith, *J. Am. Chem. Soc.*, **52**, 3603 (1930).
59. E. P. Clark, *J. Am. Chem. Soc.*, **53**, 3431 (1931).
60. A. Robertson, *J. Chem. Soc.*, 1163 (1933).
61. H. L. Haller, *J. Am. Chem. Soc.*, **55**, 3032 (1933).
62. A. Robertson and T. S. Subramaniam, *J. Chem. Soc.*, 278 (1937).

63. T. Kariyone, Y. Kimura and K. Kondo, *J. Pharm. Soc. Japan*, **514**, 1049 (1924); *Chem. Abstr.*, **19**, 1708 (1925); T. Kariyone, and K. Kondo, *J. Pharm. Soc. Japan*, **518**, 376 (1925); *Chem. Abstr.*, **19**, 2485 (1925); *J. Pharm. Soc. Japan*, **519**, 4 (1925); *Chem. Zentr.*, **II**, 829 (1925).
64. S. Takie and M. Koide, *Chem. Ber.*, **62**, 3030 (1929).
65. H. L. Haller and F. B. LaForge, *J. Am. Chem. Soc.*, **53**, 4460 (1931).
66. S. Takie, S. Miyajima and M. Ono, *Chem. Ber.*, **65**, 279 (1932).
67. Cl. Schöpf, *Ann. Chem.*, 452 (1927); *Ann. Chem.*, **483**, 157 (1930).
68. A. Robertson and G. L. Rusby, *J. Chem. Soc.*, 1371 (1935).
69. T. Reichstein and R. Hirt, *Helv. Chim. Acta*, **16**, 121 (1933).
70. M. Miyano and M. Matsui, *Bull. Agr. Chem. Soc., Japan*, **23**, 141 (1959).
71. F. B. LaForge and L. E. Smith, *J. Am. Chem. Soc.*, **51**, 2574 (1929).
72. A. I. Scott, *Proc. Chem. Soc.*, 195 (1958).
73. A. Robertson and G. L. Rusby, *J. Chem. Soc.*, 212 (1936).
74. A. Butenandt and G. Hilgestag, *Ann. Chem.*, **495**, 172 (1932).
75. A. Butenandt and G. Hilgestag, *Ann. Chem.*, **506**, 158 (1933).
76. S. Takie, M. Koide and S. Miyajima, *Chem. Ber.*, **63**, 508 (1930).
77. H. L. Haller. *J. Am. Chem. Soc.*, **54**, 2126 (1932).
78. H. L. Haller and F. B. LaForge, *J. Am. Chem. Soc.*, **53**, 2271 (1931).
79. S. H. Harper, *J. Chem. Soc.*, 1424 (1939).
80. E. P. Clark, *J. Am. Chem. Soc.*, **53**, 2369 (1931).
81. F. B. LaForge and G. L. Keenan, *J. Am. Chem. Soc.*, **53**, 4450 (1931).
81a. H. L. Haller and F. B. LaForge, *J. Am. Chem. Soc.*, **56**, 2415 (1934).
82. H. L. Haller, *J. Am. Chem. Soc.*, **53**, 733 (1931).
83. F. B. LaForge and L. E. Smith, *J. Am. Chem. Soc.*, **52**, 1088 (1930).
84. G. W. Holton, G. Parker and A. Robertson, *J. Chem. Soc.*, 2049 (1949).
85. K. R. Hergreaves, A. McGookin and A. Robertson, *J. Chem. Soc.*, 5093 (1957).
86. F. B. LaForge, H. L. Haller and L. E Smith, *J. Am. Chem. Soc.*, **53**, 4406 (1931).
87. (a) M. Miyano and M. Matsui, *Bull. Agr. Chem. Soc. Japan*, **22**, 128 (1958); *Chem. Abstr.*, **53**, 345 (1959); (b) M. Miyano, *J. Am. Chem. Soc.*, **87**, 3958 (1965); *Chem. Abstr.*, **63**, 11525 (1965); (c) M. Miyano, *J. Am. Chem. Soc.*, **87**, 3962 (1965); *Chem. Abstr.*, **63**, 11525 (1965).
88. R. S. Cahn, R. F. Phipers and J. J. Boam, *J. Chem. Soc.*, 734 (1938).
89. M. Matsui and M. Miyano, *Proc. Jap. Acad.*, **35**, 175 (1939).
90. R. Robinson and R. C. Shah, *J. Chem. Soc.*, 1946, (1946).
91. R. C. Shah and M. C. Laiwala, *J. Chem. Soc.*, 1828 (1938).
92. L. W. Levy and R. A. Múnoz, *Inter. Am. Symposium Peaceful Appl. Nuclear Energy*, 2nd, Bueno Aires, 225 (1960); *Chem. Abstr.*, **55**, 15637 (1961).
93. M. Miyano and M. Matsui, *Chem. Ber.*, **93**, 54 (1960).
93a. A. Jennen, *Bull. Soc. Chim. Belges*, **61**, 536 (1952); *Chem. Abstr.*, **48**, 2965 (1954).
93b. G. M. Wright, *J. Am. Chem. Soc.*, **50**, 3358 (1928).
94. H. Schmid and A. Ebnöther, *Helv. Chim. Acta*, **34**, 1041 (1951).
95. E. Hardegger, H. Gempler and A. Züst, *Helv. Chim. Acta*, **40**, 1819 (1957).
96. P. A. Levene and R. E. Marker, *J. Biol. Chem.*, **101**, 413 (1933); P. A. Levene, and H. L. Haller, *J. Biol. Chem.*, **74**, 343 (1927); **65**, 49 (1925); **67**, 329 (1926).
97. K. Freudenberg, *Chem. Ber.*, **47**, 2027 (1914).
98. A. Wohl and E. Schellenberg, *Chem. Ber.*, **55**, 1404 (1922).

99. (a) R. S. Cahn, C. K. Ingold and V. Prelog, *Experienta*, **12**, 81 (1956); (b) A. Feldman, *J. Org. Chem.*, **24**, 1556 (1959).
100. R. S. Cahn and J. J. Boam, *J. Soc. Chem. Ind.*, **54**, 42 T (1935).
101. C. E. Johnson and F. A. Bovey, *J. Chem. Phys.*, **29**, 1012 (1956).
102. S. Takie, S. Miyajima and M. Ono, *Chem. Ber.*, **66**, 479 (1933).
103. F. B. LaForge and L. E. Haller, *J. Am. Chem. Soc.*, **56**, 1620 (1934).
104. L. Crombie and P. J. Godin, *Proc. Chem. Soc.*, 276 (1960).
105. W. G. Schneider, *J. Chem. Phys.*, **23**, 26 (1955).
106. E. P. Clark, *J. Am. Chem. Soc.*, 987 (1934).
107. E. P. Clark, *J. Am. Chem. Soc.*, 729 (1931).
108. A. Robertson and G. L. Rusby, *J. Chem. Soc.*, 497 (1937).
109. S. Kenny, A. Robertson and S. W. George, *J. Chem. Soc.*, 1601 (1939).
110. L. Crombie and R. Peace, *J. Chem. Soc.*, 5445 (1961).
111. S. H. Harper, *J. Chem. Soc.*, 878 (1941).
112. M. Th. Meiyer and D. R. Koolhaus, *Rec. Trav. Chim.*, **58**, 207 (1939).
113. S. H. Harper, *J. Chem. Soc.*, 593 (1942).
114. T. A. Buckley, *J. Soc. Chem. Ind.*, **55**, 285 T (1936).
115. S. H. Harper, *J. Chem. Soc.*, 1099 (1939).
116. P. S. Sarin, J. M. Sehgal and T. R. Seshadri, *Proc. Indian Acad. Sci.*, **47A**, 292 (1958).
117. K. S. Raizada, P. S. Sarin and T. R. Seshadri, *J. Sci. Ind. Res. (India)*, **19B**, 76 (1960).
118. T. R. Seshadri and S. Varadarajan, *Proc. Indian Acad. Sci.*, **37A**, 784 (1953).
119. H. Matsumoto, Y. Kawase and M. Nanbu, *Bull. Chem. Soc. Japan*, **31**, 688 (1958).
120. Y. Kawase and K. Fukui, *Bull. Chem. Soc. Japan*, **31**, 693 (1958).
121. Y. Kawase, K. Ogawa, S. Miyoshi and K. Fukui, *Bull. Chem. Soc. Japan*, **33**, 1240 (1960).
122. F. Acree, Jr., M. Jacobson and H. L. Haller, *J. Org. Chem.*, **8**, 572 (1943).
123. E. S. Kondratenko and N. K. Abubakirov, *Dokl. Akad. Nauk SSSR.*, **146**, 1340 (1962); *Chem. Abstr.*, **58**, 9037 (1963); *Dokl. Akad. Nauk Uzbek. SSR.*, 35 (1960); *Uzbek Khim. Zh.*, **6**, 60, 73 (1962); **5**, 66 (1961).
124. L. Crombie and R. Peace, *Proc. Chem. Soc.*, 246 (1963).
125. L. Crombie and D. A. Whiting, *Tetrahedron Letters*, (18) 801 (1962).
126. L. Crombie and D. A. Whiting, *J. Chem. Soc.*, 1569 (1963).
127. M.Th. Meiyer, *Rec. Trav. Chim.*, **65**, 835 (1946).
128. S. K. Pavanaram and L. R. Row, *J. Sci. Ind. Res. (India)*, **17B**, 272 (1958).
129. S. K. Pavanaram and L. R. Row, *Current Sci. (India)*, **26**, 145 (1957).
130. H. L. Haller and P. S. Schaffer, *J. Am. Chem. Soc.*, **55**, 3495 (1933).
131. H. Fukami, S. Takahashi, K. Konishi and M. Nakajima, *Bull. Agr. Chem. Soc. Japan*, **24**, 123 (1960); *Chem. Abstr.*, **54**, 16451 (1960).

CHAPTER VIII

Less Common Furopyrone Systems

I. Isogalloflavin

Natural products, containing a simple isocoumarin nucleus, are hardly known. Galloflavin (**1**), an oxidation product of gallic acid derivatives *in vitro*, is indisputably an isocoumarin[1] in which there is apparently no furan nucleus at all, until the action of potassium hydroxide under nitrogen brings one into being, namely in isogalloflavin, in an almost quantitative yield.[2,3] Isogalloflavin (**2**, R = H, R^1 = OH) is not a natural product in the rigorous sense, and may be regarded as furoisocoumarin or furoisochromen-1-one; isocoumarin is the trivial name of isochromen-1-one.

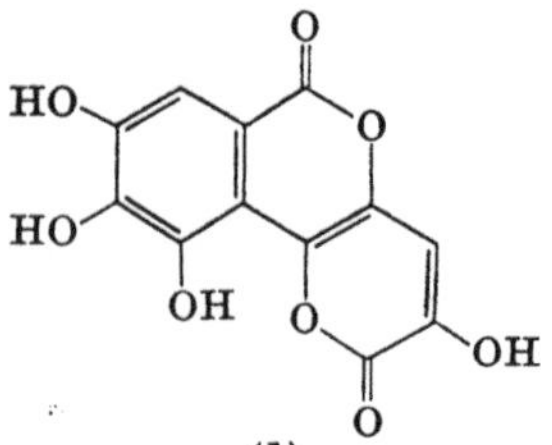

(**1**)

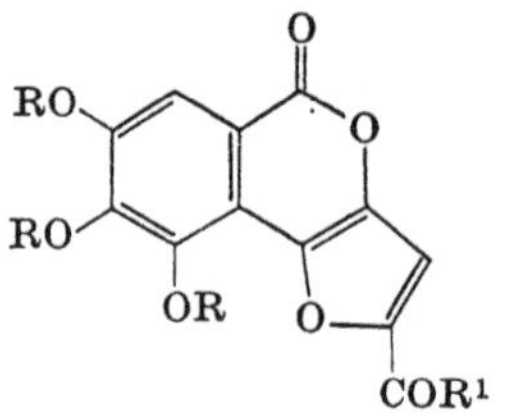

(**2**)

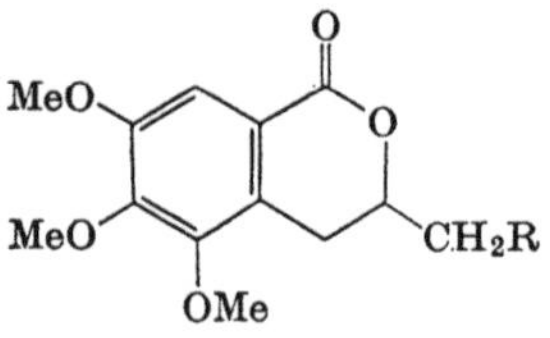

(**3**)

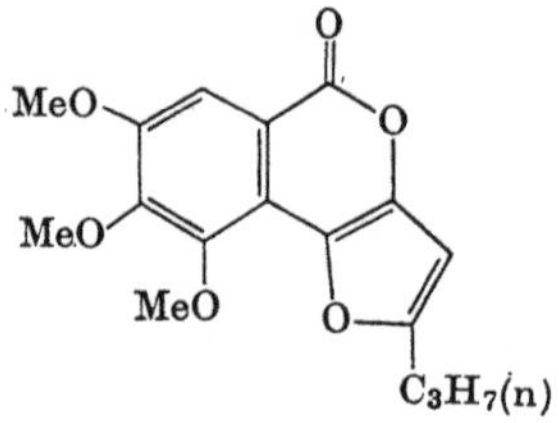

(**4**)

The assigned structure for isogalloflavin (**2**, R = H, R^1 = OH) has been confirmed by: (a) the oxidation of the hexahydro decarboxylated derivative (**3**, R = CH_2OH) to a carboxylic acid (**3**, R = COOH), identical with a synthetic specimen, and (b) the conversion of trimethyl galloflavin (**2**, $R=CH_3$, $R^1=OH$) to 5,6,7-trimethoxy-5′-n-propylfuro-(3′,2′–3,4)isocoumarin (**4**) by the sequence of reactions (Eq. 1).

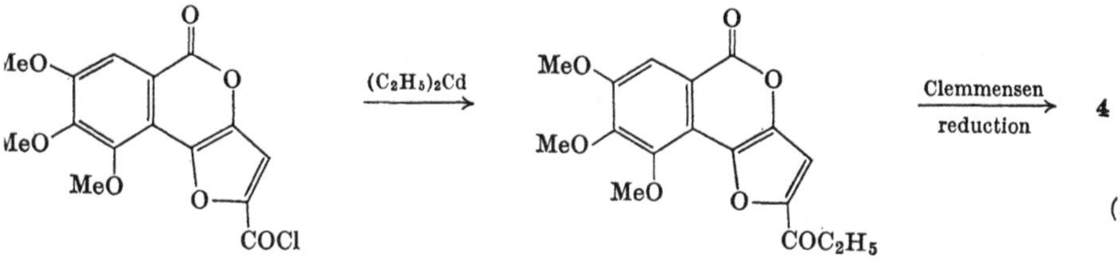

(1)

The furoisocoumarin (**4**) gave 3-acetyl-4,5,6-trimethoxyphthalide (**5**, $R=CH_3$), 4,5,6-trimethoxyphthalide-3-carboxylic acid (**5**, $R=OH$),

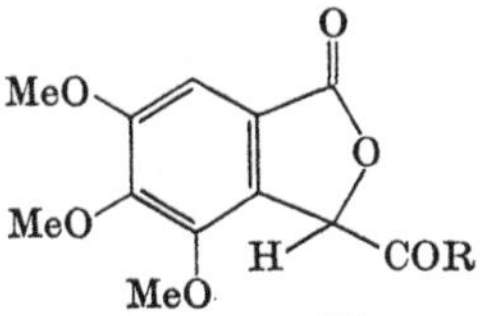

(**5**)

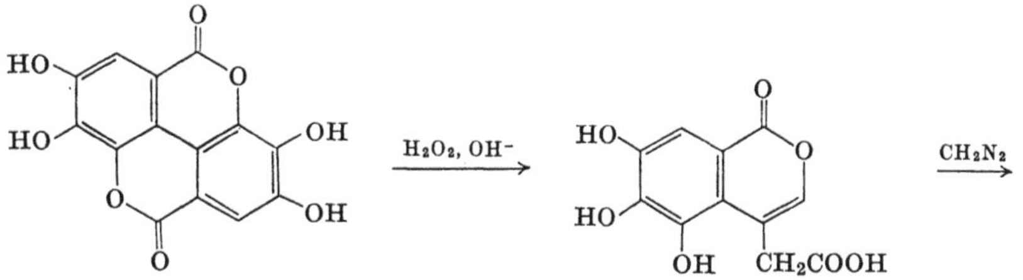

(**6**) (**7**)

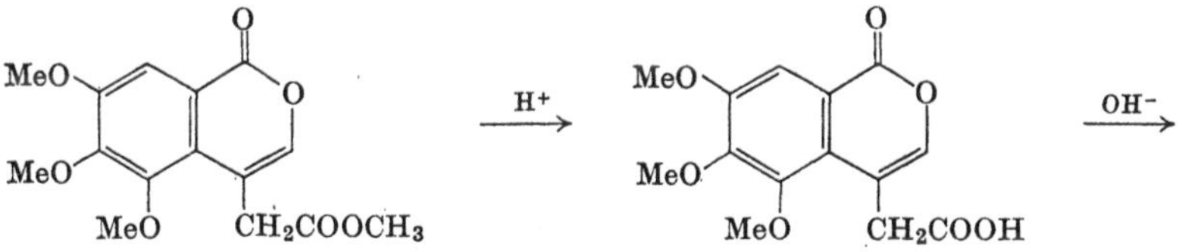

(**8**) (**9**)

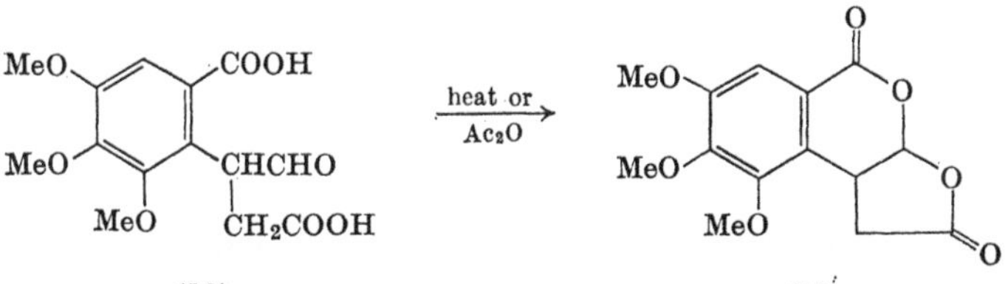

(2)

(**10**) (**11**)

methyl n-propyl ketone, and butyric acid.[1] The formation of the last two products shows that the propyl group must be attached to a carbon atom bearing an oxygen, and all four products would be expected from a structure like **4**.

An isomeric analog of isogalloflavin (**11**) was obtained by Haworth and coworkers[4] from flavellagic acid (**6**). Peroxide oxidation of the latter compound, in alkaline medium, gave **7**, which upon methylation with diazomethane to the dimorphous ether ester (**8**), followed by hydrolysis, gave the ether acid (**9**). Sublimation or treatment of the dibasic acid (**10**), obtained by rupturing the lactone ring of **9** with warm alkali, and/or treatment of **8** with methanolic potassium hydroxide solution, gave **11** (Eq. 2).

II. Enmein

The chemistry of enmein is most diverse and it has been difficult to find a heading under which a logical presentation could be made. For the purpose of classification, it may be regarded as a distant derivative of fully hydrogenated 3,4-benzocoumarin, which in accordance with the usual practice of compounds of this type has been regarded as a coumarin and not as an isocoumarin. Enmein is found in the plant and occupies a niche of its own.

Enmein[5,6] $C_{20}H_{26}O_6$, also known as isodonin,[7] is the bitter principle of *Isodon trichocarpus* KUDO (Japanese name Enmeiso). There are six oxygen atoms in enmein, two of which form a δ-lactone, one is in the cyclohexanol system (**12**) two are in the hemiacetal ring (**13**),[8,9] and one is in the α,β-unsaturated five–membered ketone system (**14**).[10] Kanatomo[11] obtained 1-ethyl-4-(3,3-dimethylcyclohexyl)benzene (**15**) by the baryta distillation of enmein, and retene by selenium dehydrogenation of the lithium aluminum hydride reduction product of α-dihydroenmein.

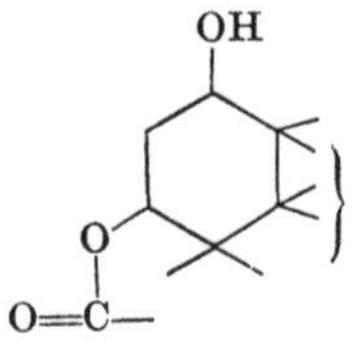

(**12**)

The structural formula **16** was then proposed for enmein in 1961. Later, however, the chemical reactions suggested that the planar

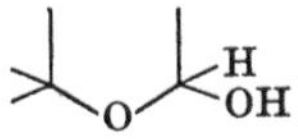

(13)

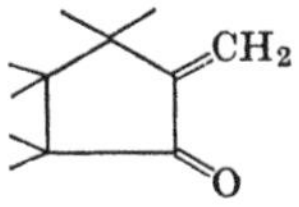

(14)

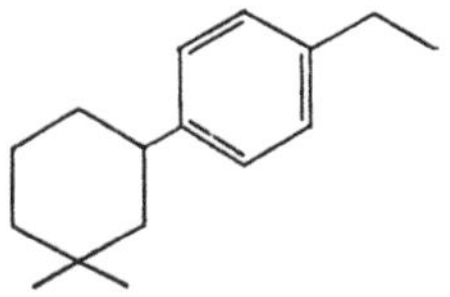

(15)

structural formula for enmein should be **12** with a five–membered acetal ring, and that its steric structure should be represented as **17** or **18**. These structures conform well with the result of X-ray analysis by Iitaka,[12] and reference to this result indicates that the steric structure of enmein should be **18**.[13]

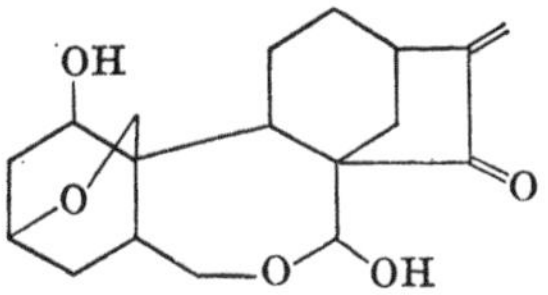

(16)

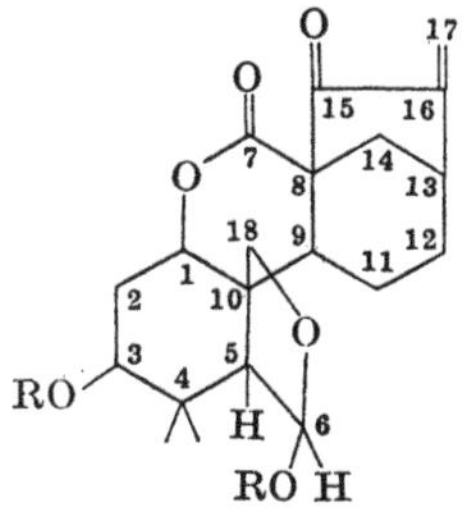

(17)

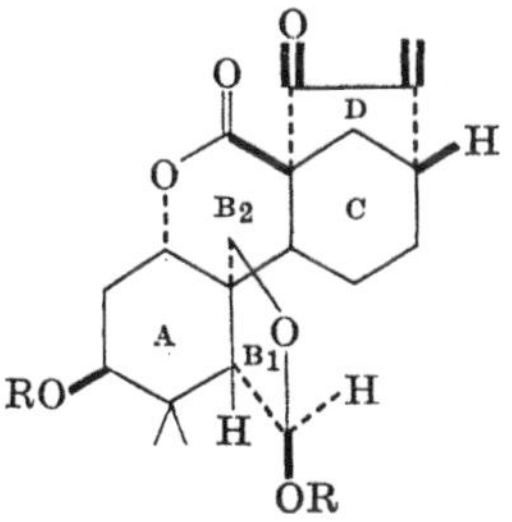

(18) boat

Oxidation of α-dihydroenmein (**19**, R = H) with chromic oxide gave bisdehydro-α-dihydroenmein (**20**),[5–7] which upon treatment with alkali gives an isomeric acid (**21**, R = H). These results can be explained well by assuming that enmein has a 1,3-dihydroxycyclohexane ring, the A ring, and that one of its hydroxyl groups takes part in the formation of a lactone[8, 9, 11] (Eq. 3).

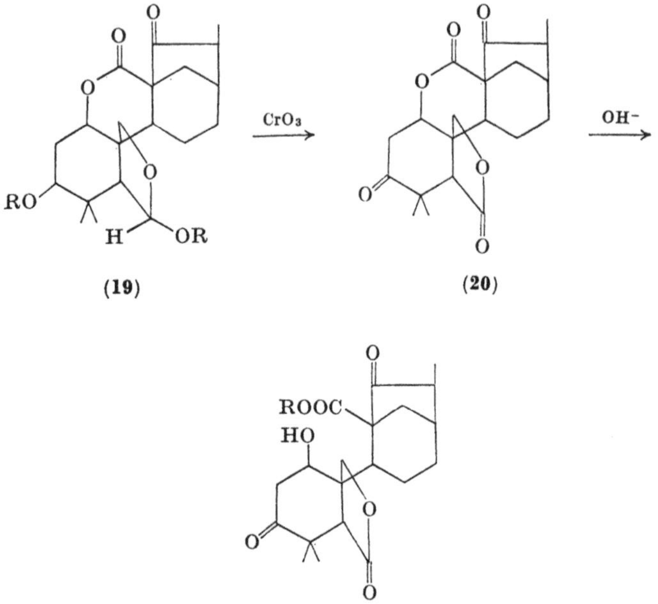

(19) (20)

(21)

(3)

The presence of an α,β-unsaturated ($=CH_2$) five-membered ketone ring,[5, 7, 10] and the assumption that the C and D rings in enmein might be phyllocladen–type because of its above mentioned selenium dehydrogenation and baryta decomposition, have been confirmed.[13] The suggestion that the hemiacetal ring (ring B_1) is seven membered (cf. **16**)[8, 9] did not satisfy the infrared absorption data of various enmein derivatives. Thus, oxalic acid hydrolysis of dihydroenmein acetate (**19**, R = Ac) or tetrahydroenmein triacetate (**24**) results in the elimination of the acetyl group in the hemiacetal alone, and the formation of the lactones (**23**) and (**26**) upon oxidation, showing a band at 1780 cm^{-1} in their infrared spectra, and hence indicating the formation of a five-membered ring. The nuclear magnetic resonance spectra confirm the hemiacetal ring B_1 (Eq. 4).

Desulfurization of the monothioketal of **19** with Raney nickel results in concurrent reduction of the acetal hydroxyl to the hydroxy monolactone (**27**). Oxidation of **27** to **28**, and its desulfurization give a monolactone (**29**). The dithioketal and monothioketal are obtained from **20** and desulfurization of these two thioketals, gives a dilactone (**30**), and a monoketodilactone (**31**), respectively (Eqs. 5 and 6). These experimental results indicate that the group adjacent to the C and D rings is not

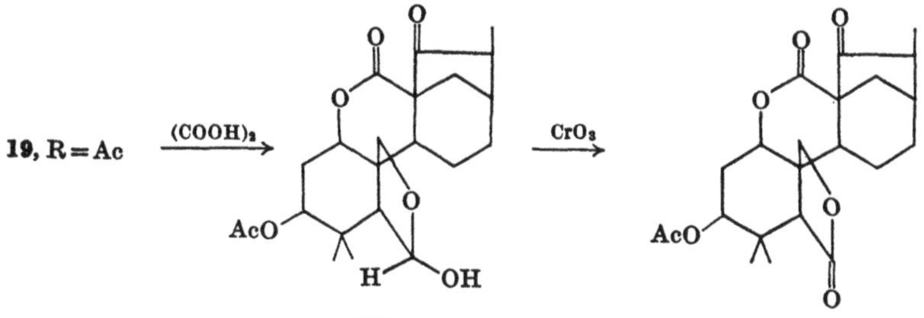

(22) (23)

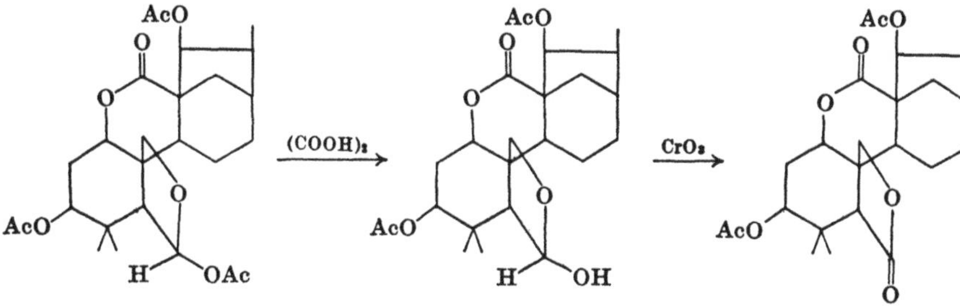

(24) (25) (26) (4)

the γ-lactone ring but, a δ-lactone ring, and it may be concluded that the planar structure of enmein is represented by **12**.

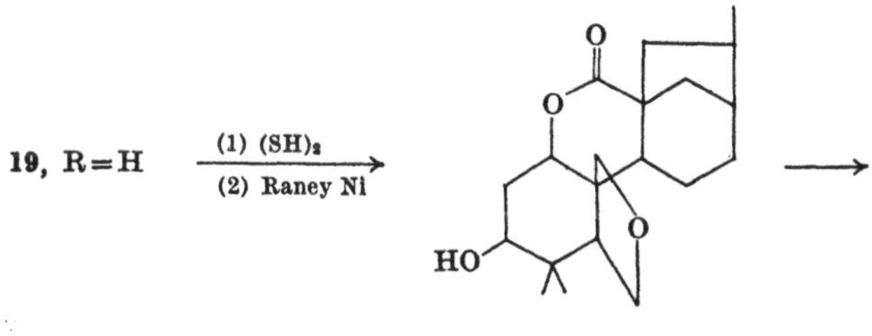

(27)

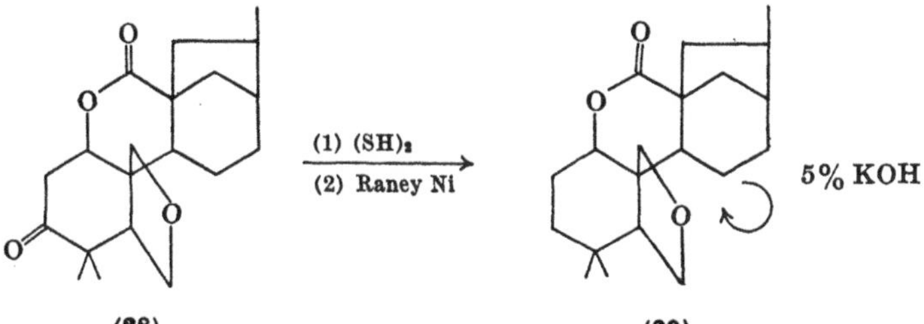

(28) (29) (5)

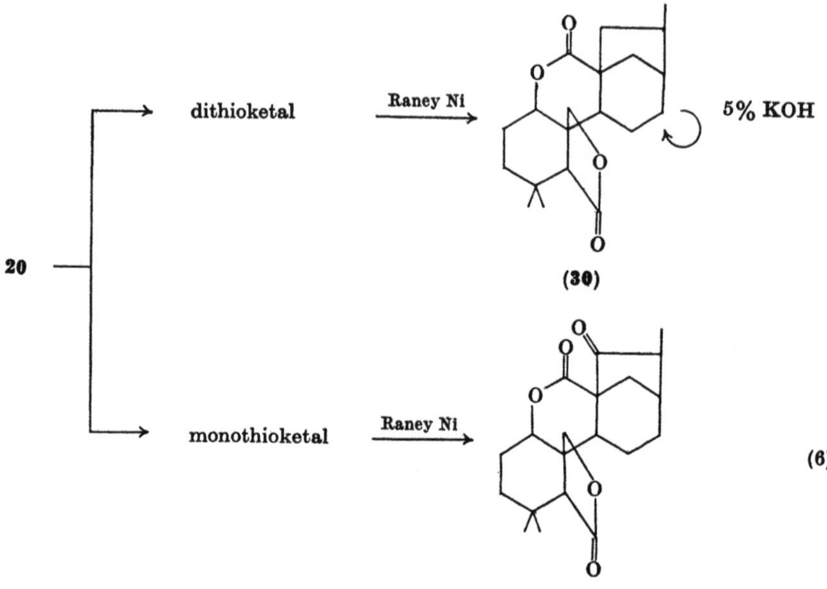

(31)

(6)

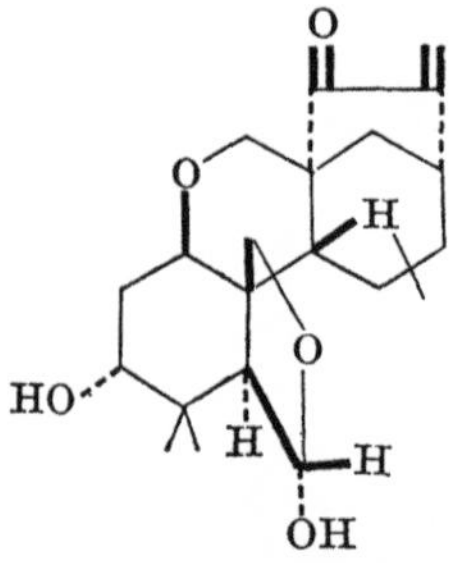

(33) chair

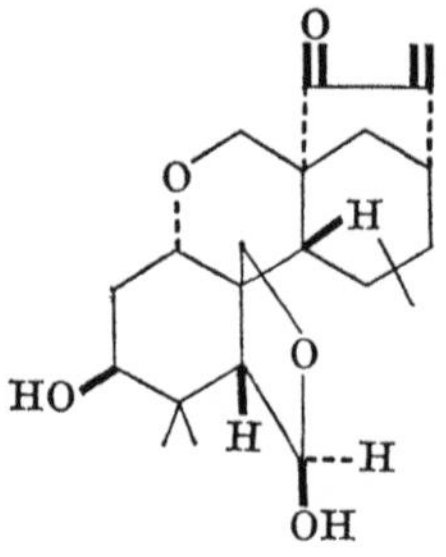

(32) chair

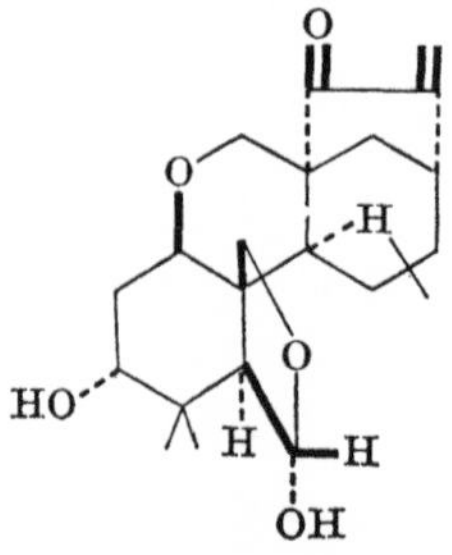

(35) boat

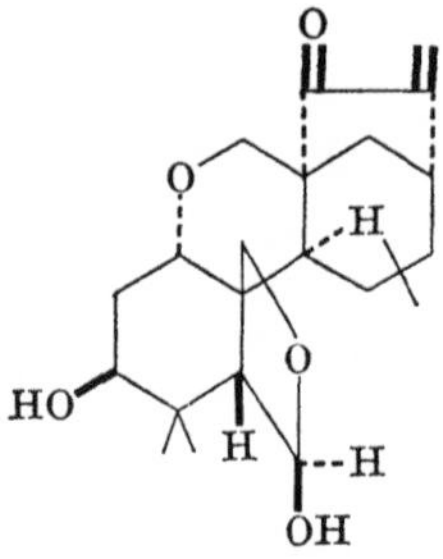

(34) boat

From the structure of C and D rings, the linkage between C_8–C_{15} and C_{13}–C_{16} take only the *cis* and the diaxial configuration. Takahashi and coworkers[10] have shown that α-dihydroenmein (**19**, R = H) has a negative Cotton effect, and with reference to the report of Kline[14] and Henderson and Hodges,[15] the C and D rings must be indicated by the absolute configuration which is the same as that of phyllocladen.[16] It follows, therefore, that the C_7–C_8 linkage is equatorial to the C ring.[13]

Furthermore, axial configuration for the C_3—OH has been suggested, and the molecular model of enmein shows that the chair and boat forms are possible for the C ring, but a more stable conformation can be obtained when the C ring is in a chair form and the C_9–C_{10} linkage is equatorial (**32–33**), and when the C ring is in a boat form and the C_9–C_{10} linkage is equatorial (**34–35**). Consequently, four formulae (**32–35**) are possible for the steric structure of enmein. The absolute configuration has been shown by X-ray analysis of enmein to be **34**.[12]

Enmein, from its structure, may also be considered as a diterpene of (–)-kaurene homolog,[17] and it is the first example of naturally occurring diterpenes formed by cleavage of the carbon linkage in the B ring.[13]

III. References

1. J. Grimshaw and R. D. Haworth, *J. Chem. Soc.*, 418 (1956).
2. J. Grimshaw, R. D. Haworth and H. K. Pindred, *J. Chem. Soc.*, 833 (1955).
3. R. D. Haworth and J. M. McLachlan, *J. Chem. Soc.*, 1583 (1952).
4. R. D. Haworth, H. K. Pindred and P. R. Jefferies, *J. Chem. Soc.*, 3617 (1954).
5. M. Ikeda and S. Kanatomo, *Yakugaku Zasshi*, **78**, 1128 (1958).
6. K. Takahashi, M. Fujita and Y. Koyama, *Yakugaku Zasshi*, **78**, 699 (1958).
7. K. Naya, *Nippon Kagaku Kaishi*, **79**, 885 (1958).
8. T. Kuboto, T. Matsuura, T. Tsutsui and K. Naya, *Bull. Chem. Soc. Japan*, **34**, 1737 (1961).
9. T. Kuboto, T. Matsuura, T. Tsutsui and K. Naya, *Nippon Kagaku Kaishi*, **84**, 353 (1963).
10. K. Takahashi, M. Fujita and Y. Koyama, *Yakugaku Zasshi*, **80**, 594, 696 (1960).
11. S. Kanatomo, *Yakugaku Zasshi*, **81**, 1049 (1961).
12. Y. Iitaka, and M. Natsume, *Tetrahedron Letters*, (20) 1257 (1964).
13. T. Kubota, T. Matsuura, T. Tsutsui, S. Uyeo, M. Takahashi, H. Irie, A. Numata, T. Fujita, T. Okomoto, M. Natsume, Y. Kawazoe, K. Sudo, T. Ikeda, M. Tomoeda, S. Kanatomo, T. Kosuge and K. Adachi, *Tetrahedron Letters*, (20) 1243 (1964).
14. W. Kline, *Tetrahedron*, **13**, 29 (1961).
15. R. Henderson and R. Hodges, *Tetrahedron*, **11**, 226 (1960).
16. P. K. Grant and R. Hodges, *Tetrahedron*, **8**, 261 (1960).
17. M. L. Briggs, B. F. Cain, R. C. Cambie, B. R. Davis, P. S. Ruttlege and J. K. Wilmhurst, *J. Chem. Soc.*, 1345 (1963).

Author Index

This author index is designed to enable the reader to locate an author's name and work with the aid of the reference numbers appearing in the text. The page numbers are printed in normal type in ascending numerical order, followed by the reference numbers in brackets. The numbers in *italics* refer to the pages on which the references are actually listed.

If reference is made to the work of the same author in different chapters, the above arrangement is repeated separately for each chapter.

Subject Index

Compounds which are referred to only in tables and which can be found by examining the tables mentioned under general headings covering the types of compounds are not mentioned individually in the index.

Lightning Source UK Ltd.
Milton Keynes UK
UKHW010614310519
343624UK00001B/49/P